Organic Synthesis Highli

Edited by Hans-Günther Schmalz

Related Titles from WILEY-VCH:

J. Mulzer / H. Waldmann (eds.)
Organic Synthesis Highlights III
1998. X. 412 pages with 302 figures
Softcover. ISBN 3-527-29500-3

K. C. Nicolaou / E. J. Sorensen
Classics in Total Synthesis
1996. XXIII. 792 pages with 444 figures
Softcover. ISBN 3-527-29231-4

H. Hopf
Classics in Hydrocarbon Chemistry
2000. XI. 547 pages with 434 figures
Hardcover. ISBN 3-527-30216-6
Softcover. ISBN 3-527-29606-9

J. Otera
Modern Carbonyl Chemistry
2000. XX. Approx 600 pages with 542 figures and 102 tables
Hardcover. ISBN 3-527-29871-1

A. Ricci
Modern Amination Methods
2000. Approx 400 pages
Hardcover. ISBN 3-527-29976-9

J. A. Gewert et al.
Problem Solving in Organic Chemistry
2000. Approx 278 pages with 284 figures
Softcover. ISBN 3-527-30187-9

Organic Synthesis Highlights IV

Edited by
Hans-Günther Schmalz

Weinheim · New York · Chichester · Brisbane · Singapore · Toronto

Prof. Dr. Hans-Günther Schmalz
Institut für Organische Chemie
der Universität zu Köln
Greinstrasse 4
D-50939 Köln
Germany

Library of Congress Card No. applied for

A catalogue record for this book is available from the British Library.

Die Deutsche Bibliothek – CIP-Catalogning-Publication-Data
A catalogue record for this publication is available from Die Deutsche Bibliothek

ISBN 3-527-29916-5

Printed on acid-free paper

Composition: Mitterweger & Partner GmbH, D-68723 Plankstadt
Printing: Strauss Offsetdruck GmbH, D-69509 Mörlenbach
Bookbinding: J. Schäffer, D-67269 Grünstadt

Printed in the Federal Republic of Germany.

Preface

During the past century, the world has changed to an unprecedented extent, and the development of the chemical sciences has greatly contributed to this change. The ability of chemists to synthesize complex organic molecules such as dyes, drugs, fragrances and crop protection agents is largely responsible for the high standard of living we enjoy today. Moreover, synthesis as a key discipline is contributing to the development of modern life sciences and materials technology. However, while the power of synthesis has led to remarkable achievements, the technology and art of organic synthesis is still far from being fully developed. Many problems remain unsolved concerning, for instance, the efficiency and atom-economy of syntheses. Organic synthesis continues to offer multifarious academic and technological challenges, and a tremendous amount of research is carried out worlwide in this field.

This fourth volume of Organic Synthesis Highlights (OSH) comprises a collection of more than 40 articles reflecting some more recent developments and achievements of organic synthesis. About half of the contributions have their origin in the review section "Synthese im Blickpunkt" in *Nachrichten aus Chemie, Technik und Laboratorium* (1994–1998), the members' journal of the GDCh; most of the others have been selected from the "Highlights" of *Angewandte Chemie* (1997–1998). The first half of the present volume concerns synthetic methodology, with emphasis on stereoselective synthesis, transition metal organometallic methods, and enantioselective catalysis. The second part focuses on applications in total synthesis of natural products and non-natural compounds and materials. In addition, a few articles reflect the recent renaissance of solid-phase synthesis and the growing importance of combinatorial chemistry.

The articles taken from "Synthese im Blickpunkt" have all been carefully updated and translated by the authors (U. Koert, O. Reiser, M. Reggelin, C. Rück-Braun). I would like to express special thanks to these colleagues and their co-workers. I am also grateful to all the other authors for their excellent and up-to-date contributions. I also have to thank the team at Wiley-VCH, especially Dr. A. Eckerle, Dr. G. Walter, Dr. A. Kessinger and P. Biel for their excellent, professional support and their patience with the editor.

I hope this new volume will find as much acceptance in the scientific community as the first three volumes of this series and will help to stimulate the interest of, in particular, young chemists in the field of synthesis.

Cologne, February 2000

Hans-Günther Schmalz

Contents

Part I. Synthetic Methods

A. New Methods in Stereoselective Synthesis

B. Transition Metal Organometallic Methods

C. Enantioselective Catalysis

Part II. Applications

A. Total Synthesis of Natural Products

B. Synthesis of Non-Natural Compounds and Materials

C. Solid Phase Synthesis and Combinatorial Chemistry

List of Contributors

Dr. P. Arya
Steacie Inst. f. Mol. Sciences
Nat. Research Council of Canada
100 Sussex Drive
CanadaK1A OR6 Ottawa

Dr. G. J. Bodwell
Department of Chemistry
Memorial University of NF
KanadaA1B 3X7 St. John's

Priv.-Doz. Dr. B. Breit
Organisch-Chemisches Institut
der Universität Heidelberg
Im Neuenheimer Feld 270
69120 Heidelberg

Dr. J. P. Clayden
Department of Chemistry
University of Manchester
Oxford Rd.
M13 9PL Manchester

Prof. A. P. Davis
Department of Chemistry
(University of Dublin
Trinity College
Ir-2 Dublin

Dr. R. Duthaler
Novartis Pharma AG
WSJ-507.109
Postfach
4002 Basel

Prof. Dr. G. Dyker
FB6-Inst. f. Synthesechemie
der Universität-GH Duisburg
Lotharstr. 1
47048 Duisburg

Dr. H. Frey
Inst. für Makromolek. Chemie
(FMF), Universität Freiburg
Stefan-Meier Str. 21/31

Prof. Dr. A. Fürstner
Max-Planck-Institut
für Kohlenforschung
Kaiser-Wilhelm-Platz 1
45470 Mülheim

Prof. Dr. N. Krause
Lehrst. für Organische Chemie
Universität Dortmund
44221 Dortmund

Prof. Dr. U. Koert
Institut für Chemie
der Humboldt-Universität
Hessische Str. 1–2
10115 Berlin

Prof. Dr. B. König
Inst. für Organische Chemie
der Universität Regensburg
Universitätsstr. 31
93053 Regensburg

Dr. T. Lindel
Pharmazeut.-chem. Institut
der Universität Heidelberg
Im Neuenheimer Feld 364
69120 Heidelberg

Dr. M. Reggelin
Institut für Organische Chemie
Universität Mainz
Düsbergweg 10–14
55099 Mainz

Prof. Dr. O. Reiser
Institut für Organische Chemie
der Universität Regensburg
Universitätsstr. 31
93053 Regensburg

Dr. K. Rück-Braun
Institut f. Organische Chemie
J. Gutenberg-Universität Mainz
Duesbergweg 10–14
55099 Mainz

Prof. Dr. H.-G. Schmalz
Inst. für Organische Chemie
Universität Köln
Greinstraße 4
50939 Köln

Dr. C. Schneider
Institut für Organische Chemie
der Universität
Tammannstr. 2
37077 Göttingen

Prof. Dr. T. Skrydstrup
Department of Chemistry
University of Aarhus
Langelandsgade 140
DK-8000 Aarhus/Denmark

Prof. Dr. R. Selke
Ifok, „Asymmetrische Katalyse"
an der Universität Rostock
Buchbinderstr. 5–6
18055 Rostock

Dr. P. Somfai
Dept. of Organic Chemistry
Stockholm University
Arrhenius Laboratory
10691 StockholmÉ
Dr. R. Stürmer BASF AG
Hauptlaborat., ZHF/D A30
67056 Ludwigshafen

Prof. Dr. D. F. Taber
Dept. of Chemistry
and Biochemistry
University of Delaware
USA19716 Newark

Prof. Dr. L. A. Wessjohann
Faculty of Chemistry (N 348)
Vrije Universiteit Amsterdam
de Boelelaan 1083
HV 1081 Amsterdam

PD Dr. T. Wirth
Institut für Organische Chemie
Universität Basel
St. Johanns-Ring 19
4056 Basel

Part I. Synthetic Methods

A. New Methods in Stereoselective Synthesis

Stereocontrolled Simmons-Smith Cyclopropanation

Julia Schuppan and Ulrich Koert

Institut für Chemie, Humboldt Universität Berlin, Germany

The cyclopropyl subunit is a frequent structural element in natural and non-natural products. In FR-900848 (**1**) [1], a natural product with fungicide bioactivity even five cyclopropane rings are found, which make its structure remarkable and its synthesis a challenging issue. Other representatives of naturally occurring cyclopropanated compounds are *allo*-coronamic acid (**2**) [2] and *cis*-chrysanthemic acid (**3**) [3]. Among the non-natural products containing cyclopropane rings the perspirocyclopropanated [3]-rotaxane (**4**) [4] and the trifunctional fullerene (**5**) [5] are worth mentioning.

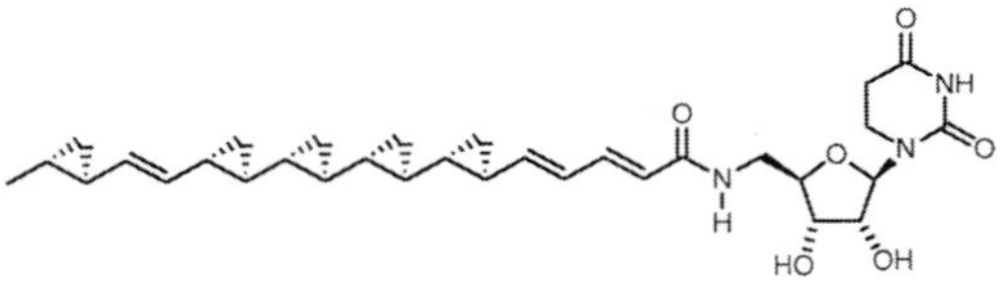

FR-900848 **1**

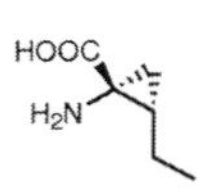

allo-coronamic acid **2**

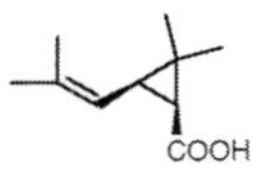

cis-chrysanthemic acid **3**

4

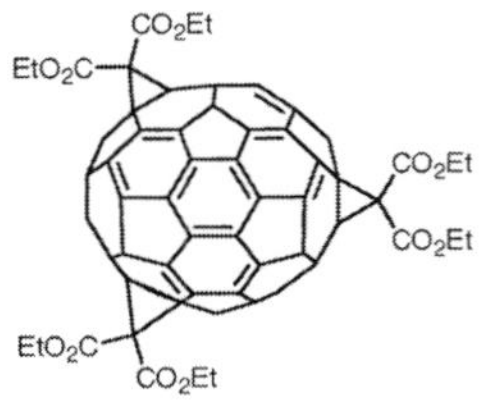

5

Figure 1

From the multitude of synthetic work in the field of cyclopropanation, we will focus on the asymmetric synthesis of cyclopropanes. Besides the known stereocontrolled addition of diazo-compounds to olefins (diazo-method) [6] the stereocontrolled Simmons-Smith cyclopropanation has received significant attention in the last ten years. The latter will be discussed further.

The reaction of activated zinc and CH_2I_2 results in the formation of a zinc carbenoid reagent "$IZnCH_2I$", which, originally introduced to literature by Simmons and Smith, converts alkenes into cyclopropanes [7]. For a successful reaction the activation of zinc metal is essential. Apart from the originally applied Cu [7] by Simmons and Smith the activation may be accomplished using Ag [8], $TiCl_4$ [9] or TMSCl/$BrCH_2$-CH_2Br (Knochel-zinc) [10]. Highly activated zinc can also be obtained by the reduction of zinc salts (Rieke-zinc [11], Fürstner-zinc [12]). Concerning commercially available zinc, the purity is of great importance. Electrolytically prepared zinc is highly pure, but pyrometallurgically made zinc, which is obtained by distillation, contains traces of lead, which can inhibit the cyclopropanation reaction [13].

Despite the various methods for activation of zinc metal, the Simmons-Smith reaction remains a heterogeneous reaction, holding all known preparative disadvantages. Hence, many efforts have been directed towards the development of homogeneous reaction conditions. Among others, two particularly successful methods will be highlighted here: the Furukawa-procedure ($Et_2Zn + CH_2I_2$) [14] and the Sawada-procedure ($EtZnI + CH_2I_2$) [15] [Eq. (1)].

Simmons-Smith
$CH_2I_2 + Zn$ 61%

Furukawa-method
$CH_2I_2 + Et_2Zn$ 79%

Sawada-method
$CH_2I_2 + EtZnI$ 92%

(1)

Another approach in the preparation of zinc carbenoids has been developed by Wittig [16]. It involves the reaction of diazomethane with a Zn(II) salt, but the delicate preparation of diazocompounds has hindered the wide spread preparative application so far (scheme 1).

DIAZO-METHOD

M L*

M = Cu, Rh

ZINC CARBENOID-METHOD

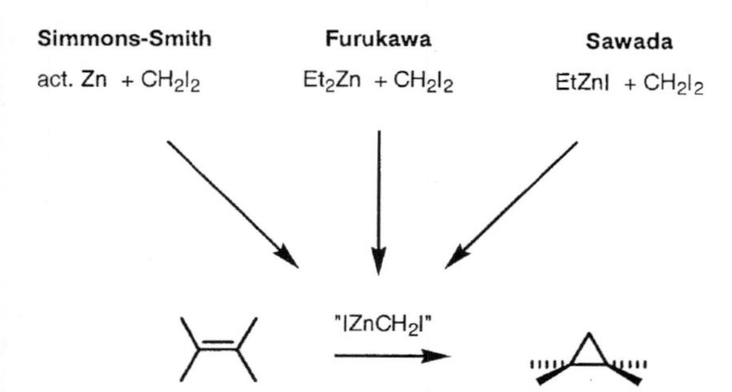

Scheme 1. Diazo method and zinc carbenoid method for cyclopropanation

The advantages of the homogeneous procedure are evident: mild conditions and low temperatures cause increased compatibility with other functional groups. The control of stoichiometry is simplified compared to the heterogeneous case by the application of an organo-zinc solution of known molarity. Furthermore, the homogeneous reaction also proceeds in non-coordinating solvents, which is of great importance especially for asymmetric synthesis. Finally, compared to a heterogeneous reaction the homogeneous procedures afford higher yields in most cases.

Although the Simmons-Smith cyclopropanation has attracted increased attention during recent years, the exact structure of the cyclopropanating reagent is still uncertain. NMR-spectroscopic investigations revealed a Schlenk equilibrium between $IZnCH_2I$ and ICH_2ZnCH_2I [Eq. (2)] [15].

$$2\ IZnCH_2I \rightleftharpoons ICH_2ZnCH_2I + ZnI_2 \quad (2)$$

Denmark et al. studied the effect of zinc iodide on the catalytic, enantioselective cyclopropanation of allylic alcohols with bis(iodomethyl)-zinc as the reagent and a bismethanesulfonamide as the catalyst [17]. They found significant rate enhancement and an increased enantiomeric excess of the product cyclopropane upon addition of 1 equivalent zinc iodide. Their studies and spectroscopic investigations showed that the Schlenk equilibrium appears to lie far on the left ($IZnCH_2I$). Charette et al. used low temperature ^{13}C-NMR spectroscopy to differentiate several zinc-carbenoid species [18]. They also found evidence that in the presence of zinc iodide, bis(iodomethyl)zinc is rapidly converted to (iodomethyl)zinc iodide. Solid-state structures of (halomethyl)zinc species have been described by Denmark for a bis(iodomethyl)zinc ether complex (**6a**) [19] and Charette for an (iodomethyl)zinc iodide as a complex with 18-crown-6 (**6b**) [20] (Fig. 2).

However, future work will show whether the cyclopropanating species actually is $IZnCH_2I$ or ICH_2ZnCH_2I.

Regarding the addition of a carbenoid **7** to an olefin **8**, resulting in the formation of cyclopropane **9**, theoretical calculations point towards a concerted mechanism involving a transition state **10** (Scheme 2) [21]. For a better understanding of the transition state of cyclopropanation with zinc carbenoids a reflection on the well-studied lithium carbenoids is profitable. Hoffmann et al.

6a **6b**

Figure 2

studied the stereochemical course of the intramolecular cyclopropanation for the carbenoids **11** and **12**, utilizing an internal stereocentre as a reference (Scheme 2) [23].

The stereochemically defined lithium carbenoid **11** forms the bicyclus **13** by intramolecular cyclopropanation even at $-110\,^{\circ}C$. In contrast, no conversion of the epimeric lithium carbenoid **14** into diastereomeric bicyclus **16** is observed under similar conditions. These results are explained on the basis of the transition state structures **12** and **15a/b**. Structure **12** allows complexation of the lithium atom by ether oxygen. This leads to activation of the carbenoid and accelerates the cyclopropanation (**11** → **12** → **13**) by a transition state choreography of type **10**. In structure **15a** the ether group is in equatorial position, which does not allow the complexation of the lithium atom. Accordingly, this carbenoid is less reactive and cyclopropanation does not proceed at $-110\,^{\circ}C$. Hoffmann et al. suggest that for this case (**14** → **15b** → **15c** → **13**) the carbolithiation competes with the concerted mechanism. In theoretical studies on the cyclopropanation of ethylene with lithium and zinc carbenoids Nakamura et al. found two competing pathways: methylene transfer and carbometallation [22]. For the lithium carbenoid, both pathways have similar activation energies and may compete in cyclopropanation, which is consistent with the results of Hoffmann's experiment [23]. However, for the zinc carbenoid, methylene transfer is found to be favored, because of a much lower activation energy compared to the carbometallation.

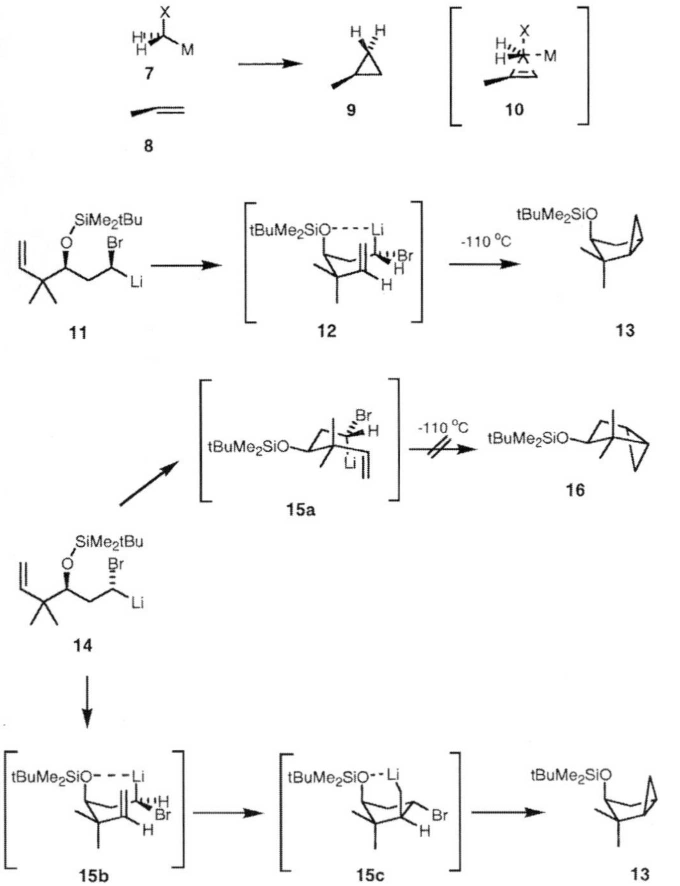

Scheme 2. Mechanistic studies on carbenoid mediated cyclopropanation

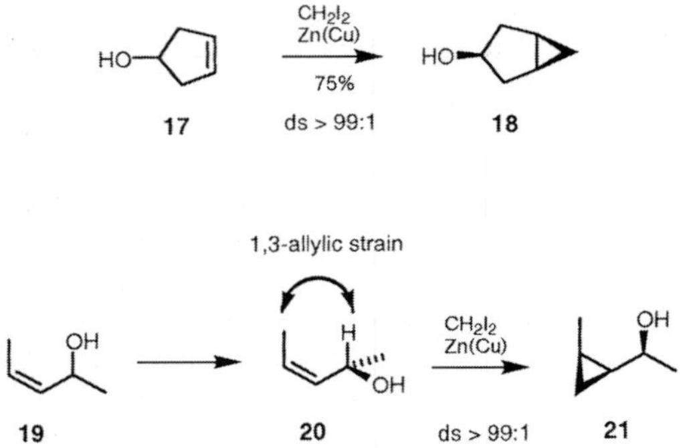

Scheme 3. The stereodirecting influence of OH-groups in cyclic and acyclic systems

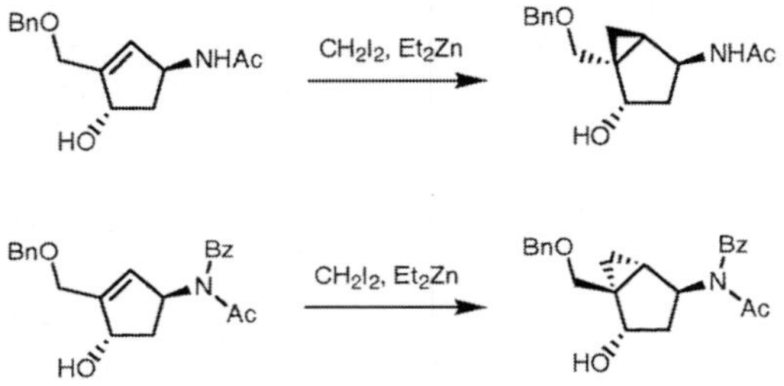

Scheme 4. The stereodirecting influence of NH-groups

As found for lithium carbenoids, an activating and directing influence of intramolecular O-donators is also known for zinc carbenoids [21]. Alcohol-groups in cyclic systems seem to have a strong syn-directing influence (**17** → **18**) (Scheme 3) [24]. High stereocontrol in acyclic systems is achieved only if conformational control restricts the rotation of single bonds. This is found for example in the cyclopropanation reaction (**19** → **21**) (Scheme 3) [25]. Herein, due to 1,3-allylic strain, the three-dimensional arrangement of the directing OH-groups in relation to the double bond is fixed.

A directing effect has also been found for intramolecular NH groups. When an OH and an NH group are in competing allylic positions, the cyclopropanation is completely directed by the NH groups. The directing influence by the OH group only comes forward after protection of the amide hydrogen (Scheme 4) [26].

For the enantioselective synthesis of cyclopropanes using zinc carbenoids, two different approaches are possible: first, by using a covalent bound chiral auxiliary or, second, by application of a chiral catalyst.

Numerous chiral auxiliaries are known today. For instance, acetals derived from tartaric acid enable the preparation of enantiomerically pure cyclopropanated aldehydes (**22** → **23** → **24** → **25**) (Scheme 5). Aldehyde **25** is a key intermediate in the synthesis of leucotriene inhibitor **26** [27].

The chiral acetonide **27** has been stereoselectively transformed into **28** by cyclopropanation. This reaction serves as the key step in the synthesis of **29** (Scheme 6) [28]. Cyclopropanated nucleosides such as **29** are interesting drug candidates for HIV therapy.

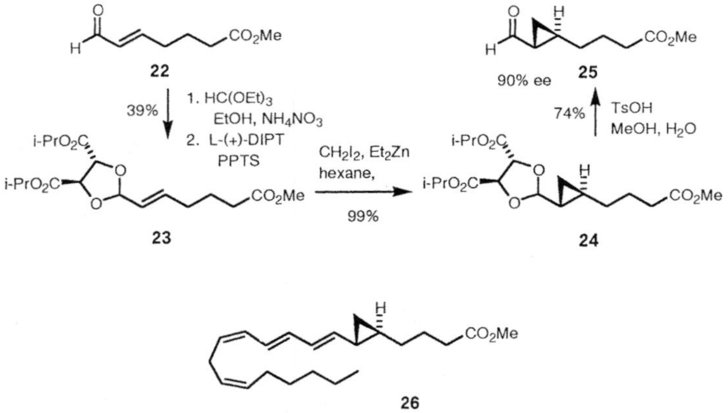

Scheme 5. Chiral acetals serving as covalent-bound auxiliaries

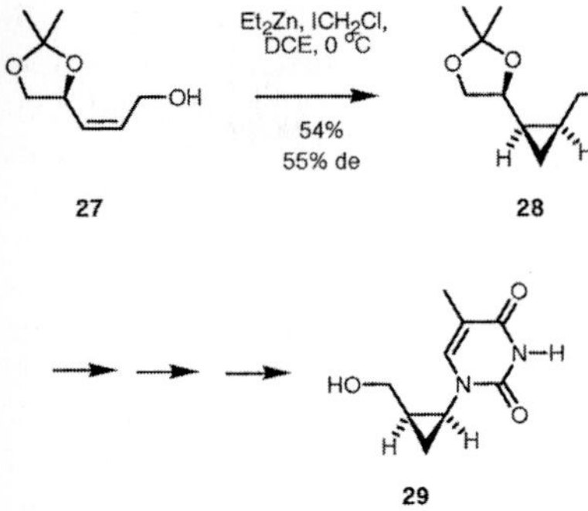

Scheme 6. Stereocontrolled cyclopropanation in the synthesis of cyclopropanated nucleosides

Carbohydrates have been used as chiral auxiliaries in cyclopropanation reactions using zinc carbenoids. The conversion of acetal **30** affords the cyclopropanated compound **31** with high diastereoselectivity [Eq. (3)] [29].

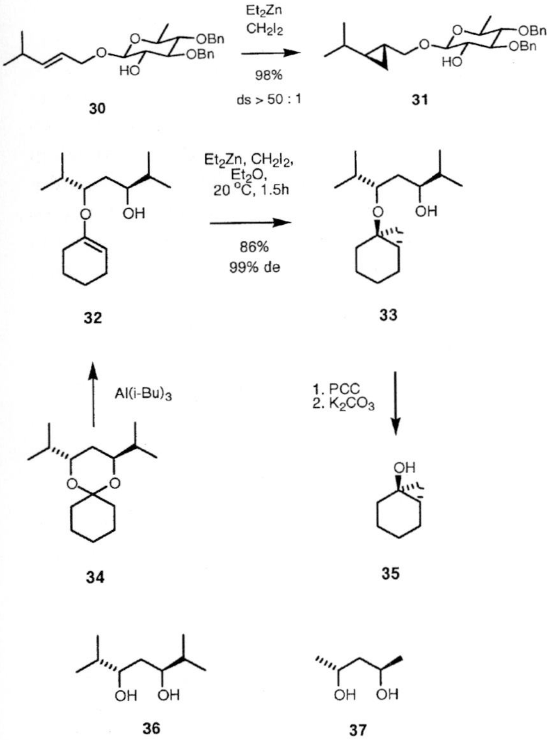

Scheme 7. Stereocontrolled cyclopropanation of enol ethers

Enol ethers may also be cyclopropanated using zinc carbenoids stereoselectively. Furukawa cyclopropanation of enol ether **32** proceeds with high stereoselection, and the obtained cyclopropyl ether **33** can be easily transformed into the enantiomerically pure cyclopropyl alcohol **35** [30]. In this case, high stereoselectivity is achieved by employing the chiral diol **36**, which is not commercially available. Using the commercially available enantiopure diol **37**, the level of stereoselectivity is significantly lower (Scheme 7).

Asymmetric Simmons-Smith cyclopropanation using no covalent-bound auxiliary but a chiral catalyst have only been successful with allylic alcohols so far. Fujisawa had shown that allylic alcohols such as **38** are converted into the corresponding alcoholate by Et_2Zn (1.1 equivalents) first [31]. Addition of diethyltartrate (1.1 equivalents) results in the formation of an intermediate **39**, which is cyclopropanated under Furukawa conditions (Et_2Zn + CH_2I_2) to give compound

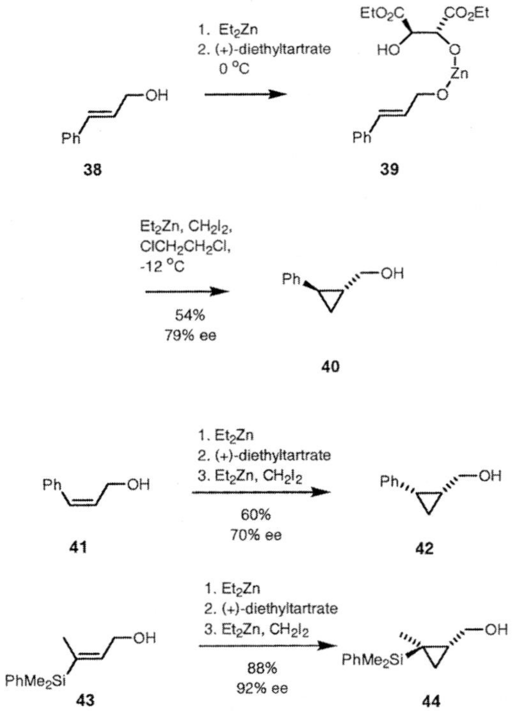

Scheme 8. Asymmetric cyclopropanation according to Fujisawa

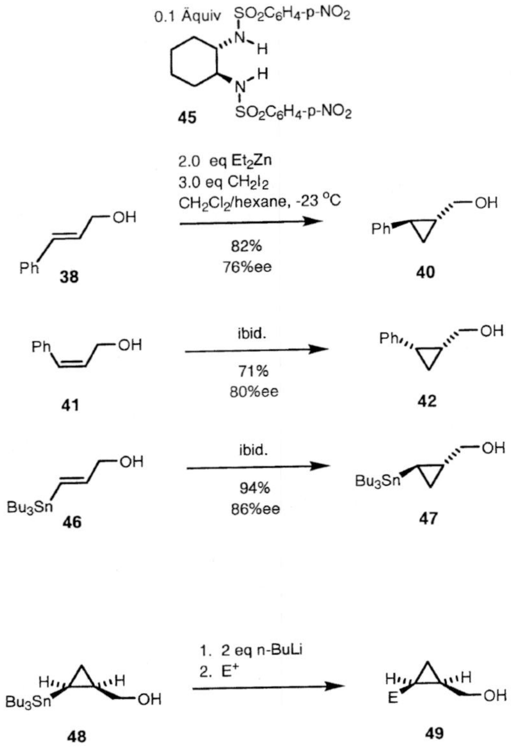

Scheme 9. Asymmetric cyclopropanation according to Kobayashi

40. No significant influence of the double bond geometry on the stereoselectivity was found. Both stereoisomers the *E*-allylic alcohol **38** and the corresponding *Z*-configured compound **41** are converted with similar enantioselectivity (ee 70–80 %). Using silyl-substituted olefins an enantiomeric excess above 90 % has been reached (Scheme 8).

Kobayashi et al. successfully performed asymmetric cyclopropanation using substoichiometric amounts of catalyst **45** (Scheme 9). [32] The levels of enantioselectivity achieved are in the 70–90 % range. Both, *E*- and *Z*-allylic alcohols are readily converted. Vinylstannanes **46** are also appropriate substrates. The resulting enantiomerically pure cyclopropanated stannanes hold great synthetic potential [33]. Thus, the cyclopropanated stannane **48** can be converted into the substituted cyclopropane **49** after successful tin-lithium exchange and electrophilic substitution.

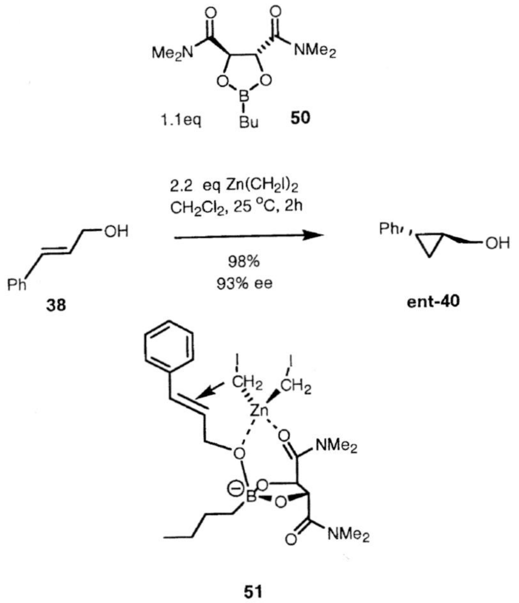

Scheme 10. Asymmetric cyclopropanation according to Charette

In an extensive study of the effect of experimental variables on the rate and selectivity of this reaction, Denmark et al. found the independent formation of ethylzinc alkoxide and bis(iodomethyl)zinc to be crucial for effective cyclopropanation [34]. They also detected an autocatalytic behavior of the reaction due to the generation of zinc iodide.

High enantioselectivity (> 90 %) and excellent yields are observed employing a method developed by Charette et al. (Scheme 10). [35] Herein, a chiral, amphoteric, bifunctional boron acid ester **50** serves as the catalyst. For example, allylic alcohol **38** can be efficiently transformed into compound **ent-40** with high enantioselectivity (ee 93 %). Unfortunately, stoichiometric amounts of **50** are necessary. With an acidic binding site at boron and a basic binding site at the carbonyl group, a transition state **51** may be reasonable. The alcoholate of the allylic alcohol and the boron acid ester form an ate-complex where the zinc carbenoid reagent now coordinates at the alcoholate-oxygen and at one of the carbonyl groups. Attack on the double bond proceeds from the direction indicated by the arrow (Scheme 10)

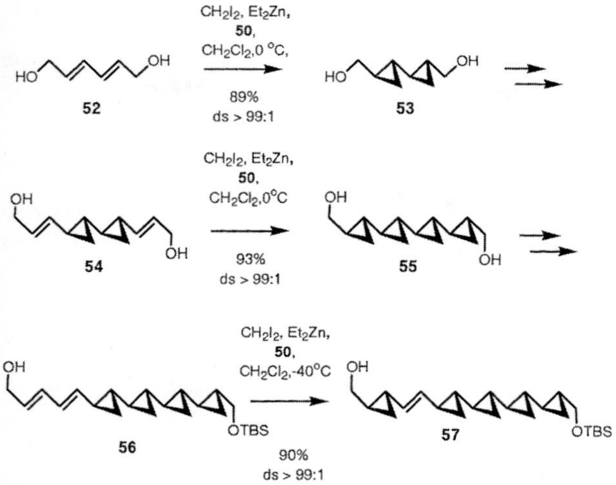

Scheme 11. Construction of the cyclopropane rings of FR-900848 (**1**) according to Barrett

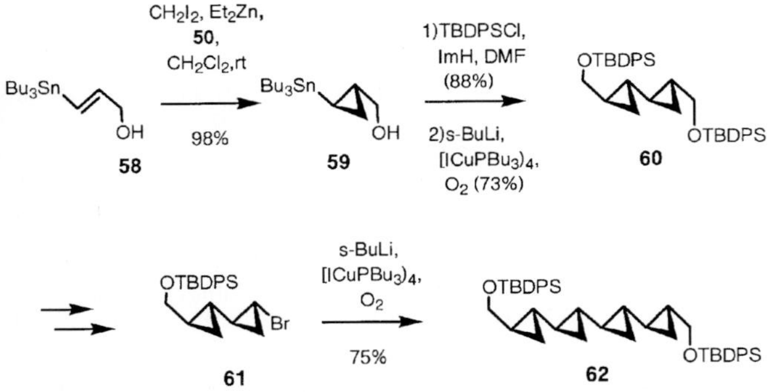

Scheme 12. Construction of the cyclopropane rings of FR-900848 (**1**) according to Falck

[35]. The formation of ate-complexes by those tailor-made ligands of type **50** was proven by X-ray crystallographic investigations [36].

With regard to the synthesis of the oligo-cyclopropane natural product FR-900848 (**1**) multiple and consecutive cyclopropanation reactions using zinc carbenoids have been applied. Thus, in the total synthesis of **1** by Barrett et al. the Furukawa-procedure was used for the conversion of **52** into the biscyclopropane **53** (Scheme 11). [37] After bidirectional elongation of the molecule, another double cyclopropanation of diene **54** using Charette's catalyst gave tetracyclopropane **55** in 93 % yield as one stereoisomer only. Finally, the olefin **56** is cyclopropanated at −40 °C to yield the desired pentacyclopropane alcohol **57**. Thus, all five cyclopropane rings of FR-900848 (**1**) are introduced with Charette's modified Furukawa method.

In the total synthesis of FR-900848 (**1**) published by Falck et al. a cyclopropane coupling strategy was successfully applied for the preparation of the tetracyclopropane backbone (Scheme 12) [38].

Summary

The examples mentioned above illustrate the progress in the field of stereocontrolled cyclopropanation. Nowadays, the asymmetric Simmons Smith cyclopropanation may well be mentioned in the line with other asymmetric reactions like epoxidation or dihydroxylation. High enantioselectivity and diastereoselectivity can be achieved applying chiral catalysts. Unfortunately, in most cases only allylic alcohols are successfully cyclopropanated using a chiral catalyst in equimolar amounts. Further work upon these problems has to be done and newly developed methods regarding the usage of substoichiometric amounts of the chiral catalyst and the application to other systems than allylic alcohols is to be expected soon. Still, great progress has been achieved so far, which is nicely represented in the recently published total syntheses of FR-900848 (**1**).

References

[1] M. Yoshida, M. Ezaki, M. Hashimoto, M. Yamashita, N. Shigematsu, M. Okuhara, M. Kohsaka, K. Horikoshi, *J. Antibiot.* **1990**, *43*, 748.

[2] K. Burgess, K.-K. Ho, D. Moye-Sherman, *Synlett* **1994**, 575.

[3] K. A. Hassal, *The Chemistry of Pesticides*, Verlag Chemie, Weinheim, **1982**, 148.

[4] S. I. Kozhushkov, T. Haumann, R. Boese, A. de Meijere, *Angew. Chem.* **1993**, *105*, 426; *Angew. Chem Int. Ed. Engl.*. **1993**, *32*, 401.

[5] A. Hirsch, I. Lamparth, H. R. Karfunkel, *Angew. Chem.* **1994**, *106*, 453; *Angew. Chem. Int. Ed. Engl.* **1994**, *33*, 437.

[6] M. P. Doyle in *Catalytic Asymmetric Synthesis* (Ed.: I. Ojima), VCH, Weinheim, **1993**, 63; A. Pfaltz, *Acc. Chem. Res.* **1993**, *26*, 339; C. Bolm, *Angew. Chem.* **1991**, *103*, 556; *Angew. Chem. Int. Ed. Engl.* **1991**, *30*, 542; G. Maa, *Top. Curr. Chem.* **1987**, *137*, 75.

[7] H. E. Simmons, R. D. Smith, *J. Am. Chem. Soc.* **1958**, *80*, 5323; H. E. Simmons, T. L. Cairns, S. A. Vladuchick, C. M. Hoiness, *Org. React.* **1973**, *20*, 1.

[8] J. M. Denis, C. Girard, J. M. Conia, *Synthesis* **1972**, 549.

[9] E. C. Friedrich, S. E. Lunetta, E. J. Lewis, *J. Org. Chem.* **1989**, *54*, 2388.

[10] P. Knochel, R. D. Singer, *Chem. Rev.* **1993**, *93*, 2117.

[11] R. D. Rieke, P. T.-J. Li, T. P. Burns, S. T. Uhm, *J. Org. Chem.* **1981**, *46*, 4323.

[12] A. Fürstner, *Angew. Chem.* **1993**, *105*, 171; *Angew. Chem. Int. Ed. Engl.* **1993**, *32*, 164.

[13] K. Takai, T. Kakiuchi, K. Utimoto, *J. Org. Chem.* **1994**, *59*, 2671.

[14] J. Furukawa, N. Kawabata, J. Nishimura, *Tetrahedron Lett.* **1966**, 3353.

[15] S. Sawada, Y. Inouye, *Bull. Chem. Japn.* **1969**, *42*, 2669.

[16] G. Wittig, K. Schwarzenbach, *Angew. Chem.* **1959**, *71*, 652.

[17] S. E. Denmark, S. P. O'Connor, *J. Org. Chem.* **1997**, *62*, 3390.

[18] A. B. Charette, J.-F. Marcoux, *J. Am. Chem. Soc.* **1996**, *118*, 4539.

[19] S. E. Denmark, J. P. Edwards, S. R. Wilson, *J. Am. Chem. Soc.* **1992**, *114*, 2592.

[20] A. B. Charette, J.-F. Marcoux, F. Bélanger-Gariépy, *J. Am. Chem. Soc.* **1996**, *118*, 6792.

[21] A. H. Hoveyda, D. A. Evans, G. C. Fu, *Chem. Rev.* **1993**, *93*, 1307; E. Nakamura, A. Hirai, M. Nakamura, *J. Am. Chem. Soc.* **1998**, *120*, 5844.

[22] A. Hirai, M. Nakamura, E. Nakamura, *Chem. Lett.* **1998**, 927.

[23] H. C. Stiasny, R. W. Hoffmann, *Chem. Euro. J.* **1995**, *1*, 619.

[24] S. Winstein, J. Sonnenberg, *J. Am. Chem. Soc.* **1961**, *83*, 3235.

[25] M. Ratier, M. Castaing, J.-Y. Godet, M. Pereyre, *J. Chem. Res. (S)* **1978**, 179.

[26] P. Russ, A. Ezzitonni, V. E. Marquez, *Tetrahedron Lett.* **1997**, *38*, 723.

[27] I. Arai, A. Mori, H. Yamamoto, *J. Am. Chem. Soc.* **1985**, *107*, 8254.

[28] Y. Zhao, T.-F. Yang, M. Lee, B. K. Chun, J. Du, R. F. Schinazi, D. Lee, M. G. Newton, C. K. Chu, *Tetrahedron Lett.* **1994**, *35*, 5405.

[29] A. B. Charette, B. Cote, J.-F. Marcoux, *J. Am. Chem. Soc.* **1991**, *113*, 8166.

[30] T. Sugimura, T. Futagawa, M. Yoshikawa, A. Tai, *Tetrahedron Lett.* **1989**, *30*, 3807.

[31] Y Ukaji, M. Nishimura, T. Fujisawa, *Chem. Lett.* **1992**, 61; Y. Ukaji, K. Sada, K. Inomata, *Chem. Lett.* **1993**, 1227.

[32] H. Takahashi, M. Yoshioka, M.Ohno, S. Kobayashi, *Tetrahedron Lett.* **1992**, *33*, 2575; N. Imai, K. Sakomato, H. Takahashi, S. Kobayashi, *Tetrahedron Lett.* **1994**, *35*, 7045; N. Imai, H. Takahashi, S. Kobayshi, *Chem. Lett.* **1994**, 177.

[33] E. J. Corey, T. M. Eckrich, *Tetrahedron Lett.* **1984**, *25*, 2419; G. Boche, H. Walborsky in *Cyclopropane Derived Reactive Intermediates* (Eds.: S. Patai, Z. Rappoport), Wiley, Chichester, **1990**.

[34] S. E. Denmark, B. L. Christenson, D. M. Coe, S. P. O'Connor, *Tetrahedron Lett.* **1995**, *36*, 2215; S. E. Denmark, B. L. Christenson, S. P. O'Connor, *Tetrahedron Lett.* **1995**, *36*, 2219.

[35] A. B. Charette, H. Juteau, *J. Am. Chem. Soc.* **1994**, *116*, 2651; A. B. Charette, H. Juteau, H. Lebel, C. Molinaro, *J. Am. Chem. Soc.* **1998**, *120*, 11943.

[36] M. T. Reetz, C. M. Niemeyer, K. Harms, *Angew. Chem.* **1991**, *103*, 1515; *Angew. Chem. Int. Ed. Engl.* **1991**, *30*, 1472.

[37] A. G. M. Barrett, K. Kasdorf, *J. Am. Chem. Soc.* **1996**, *118*, 11030.

[38] J. R. Falck, B. Mekonnen, J. Yu, J.-Y. Lai, *J. Am. Chem. Soc.* **1996**, *118*, 6096.

Oppolzer Sultams

Oliver Reiser

Institut für Organische Chemie, Universität Regensburg, Germany

On March 15, 1996, Wolfgang Oppolzer, Professor at the University of Geneva, Switzerland, died. Of his numerous important contributions to organic synthesis, the sultams, derived from campher sulfonic acid, have found widespread application especially as chiral auxiliaries.

Camphor-10-sulfonic acid (**1**) is available in large quantities in both enantiomeric forms. In only 3 steps the cyclic sulfonamide **2** (sultam) can be synthesized, which can be acylated with acid chlorides after deprotonation with sodium hydride (Scheme 1) [1, 2]. The resulting amides **3** are considerable more reactive towards nucleophiles than the corresponding carboxylic esters and the α,β-unsaturated derivatives undergo, with excellent selectivities, Diels-Alder reactions or Michael additions under mild conditions. Almost all resulting *N*-acyl derivatives are stable and can be purified by crystallization. Moreover, diastereomeric mixtures can be enriched this way. The chiral auxiliary can be cleaved under mild conditions, without erosion of the induced chirality, by saponification or reduction and subsequently reisolated in high yields and purity [3, 4].

One of the most notable properties of sultam-modified substrates is that they undergo highly selective reactions in Lewis-acid-catalyzed as well as in thermal processes. There are a number of investigations into the basic selection mechanisms of the sultam auxiliary [5], which were carried out mainly by the groups of W. Oppolzer and D. Curran. In summary, the following model has arisen, which is described here giving the

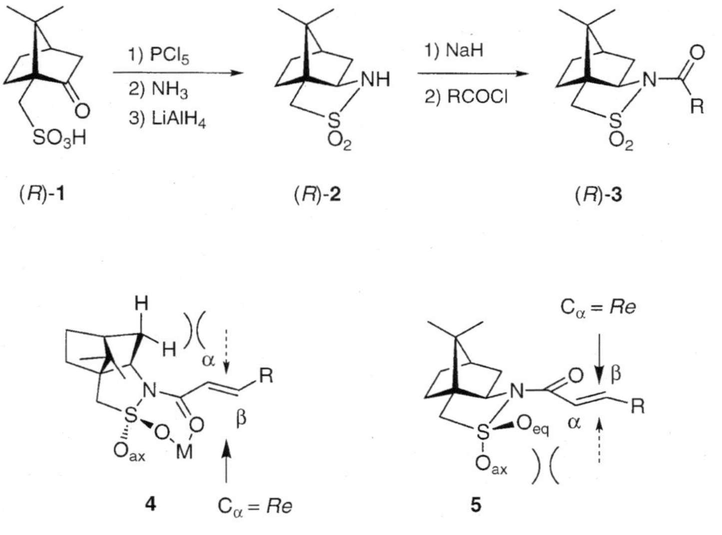

1) PCl_5
2) NH_3
3) $LiAlH_4$
SO_3H
O
(*R*)-**1**
NH
S
O_2
(*R*)-**2**
1) NaH
2) RCOCl
N
O
R
S
O_2
(*R*)-**3**

H
H
α
R
N
S
O
O
β
O_{ax}
M
4
$C_\alpha = Re$

$C_\alpha = Re$
N
O
β
R
S
O_{eq}
α
O_{ax}
5

Scheme 1

example of the acryl sultams **4** and **5**. It has to be noted that the following discussion relies on the assumption that the conformations of the ground states are similar to the conformations of the transition states. For a side-selective reaction with the C-C double bond three conditions have to be fulfilled:

- The reactive conformation of the possible rotamers (rotation around the OC–CC bond) has to be unambiguous. Of the two possible planar conformations, which allow conjugation with the π-systems, the *s-cis*-orientation of C=O/ $C_\alpha = C_\beta$ is favored based on steric reasons (O < NR_2, analogously to the well-known fact that (*Z*)-enolates of amides are more stable than (*E*)-enolates).
- The orientation of the carbonyl group has to be unambiguous, which can be parallel or antiparallel to the nitrogen-sulfur bond. Other orientations are energetically less favored because of the missing mesomeric stabilization with the amide nitrogen.
- In the most favored conformation, one side of the double bond has to be effectively blocked by the chiral auxiliary to allow an unambiguous attack of the reagent. By choosing the reaction conditions appropriately, the orientation of the carbonyl group can be influenced. Addition of a Lewis acid with *two* open coordination sites (e.g. $TiCl_4$ or $EtAlCl_2$) results in the formation of a chelate **4**. It is important to note that the two oxygen atoms connected to sulfur are *not equivalent* but that one is positioned pseudo-axial and the other pseudo-equatorial in the five-membered ring. As X-ray structure analyses show [6], the Lewis acid coordinates selectively with the equatorial oxygen atom, since this way an almost planar chelate is formed allowing one to preserve the conjugation of the π-system. In **4**, the upper side of the double bond is blocked by the camphor structure, and the attack has consequently to take place from the lower side of the molecule (chelate model). In the absence of Lewis acids or with Lewis acids having only *one* free coordination site (e.g. BF_3) no chelation is possible. Therefore, the *anti*-position of C=O and NSO_2, as shown in **5**, is favored based on steric and in particular stereoelectronic reasons (minimization of the dipole moment). In this conformation the camphor structure is too far away from the double bond to shield it effectively. However, the axially positioned oxygen atom of the SO_2 group can now take over that role, so that attack occurs mainly from the top side. Therefore, in Lewis-acid-catalyzed as well as in thermal reactions the same stereoselectivity is induced according to that model.

The direction of the induction that is shown in the following examples can be understood in almost all cases with the model described above. Only typical examples of different reaction types can be given; to comprehensively cover the vast number of applications of the sultams is beyond the scope of this article.

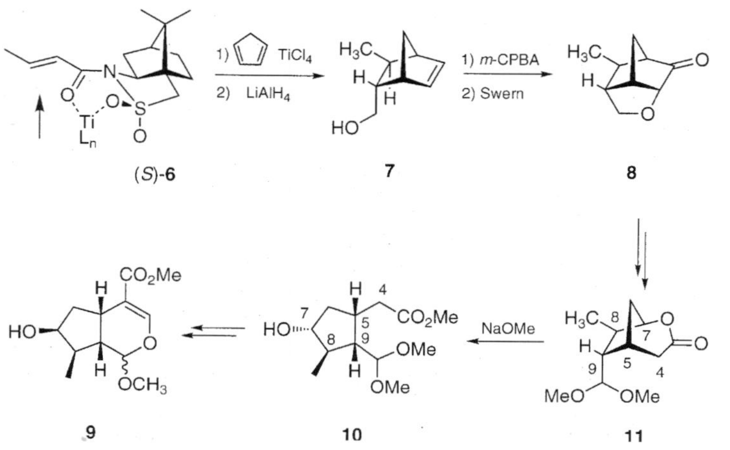

(*S*)-**6** → 1) cyclopentadiene, $TiCl_4$; 2) $LiAlH_4$ → **7** → 1) *m*-CPBA; 2) Swern → **8** ⇓ **11** → NaOMe → **10** ⇒ **9**

Scheme 2

Diels-Alder Reactions

Sultam-modified acrylates undergo [4 + 2]-cycloadditions with 1,3-dienes with excellent *endo*- and side-selectivity in the presence of $EtAlCl_2$ or $TiCl_4$ [7–9]. This feature could be used for an effective synthesis of the loganin-aglycon **9** (Scheme 2) [10]:

The Diels-Alder reaction between cyclopentadiene and the crotylsultam (*S*)-**6** and the subsequent reductive cleavage of the auxiliary gave rise to **7** in diastereo- and enantiomerically pure form, in which already three stereocenters (C5, C8 and C9) have the right configuration for the final product. Especially elegant is the subsequent regioselective opening of the norbornene structure: epoxidation with concurrent intramolecular epoxide opening of **7** followed by oxidation leads to **8**. After reductive opening of the tetrahydrofuran ring and oxidation/ketalization of the resulting CH_2-OH group at C9, the breaking point into the norbornane structure is introduced by a regioselective Baeyer-Villiger oxidation of the more highly substituted C–C bond leading to **11**. Saponification of the lactone resulted in the highly functionalized cyclopentane derivative **10**, in which its "wrong" configuration at C7 is fixed by a Mitsunobu-Inversion. The dihydropyran **9** was finally formed by formylation at C4 and Lewis-acid-catalyzed ring closure.

1,4-Additions

Acryl sultams such as (*R*)-**12** and (*R*)-**15** are also excellently suited for stereoselective 1,4-additions of various nucleophiles (Scheme 3) [11–13]. Even simple Grignard reagents can be added with excellent 1,4-regio- and good diastereoselectivity according to the chelate model [14], which seems especially useful from a preparative point of view. The resulting (*Z*)-enolates **13** and **16** can be captured with electrophiles, which proceeds also with high stereocontrol. It should be noted, that **13** reacts with *opposite* selectivity to that of **16** in this second step. The additional methyl substituent in (*R*)-**15** is sterically repelled by the camphor structure. Nevertheless, coplanarity of the acryl amide is a necessary condition for the nucleophilic addition. Therefore, despite the unfavorable interactions in the reactive conformation, chelation occurs and attack of the nucleophile takes place from the side away from the auxiliary. For the trapping of the enolate, conjugation is not necessary any longer, resulting in the formation of the more favorable enolate **16**, which is attacked from the front.

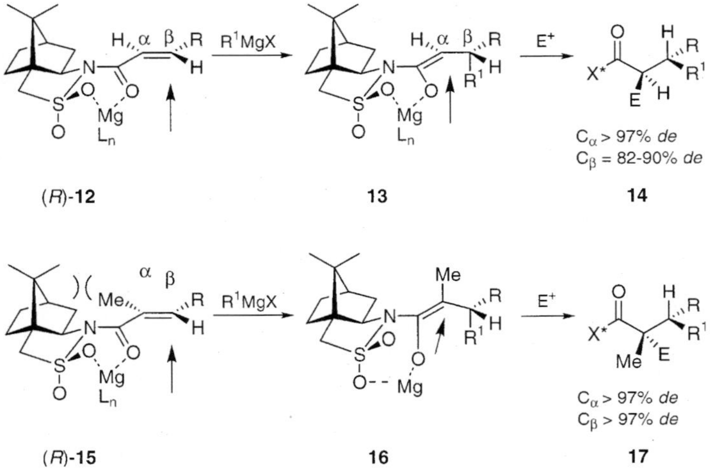

(*R*)-**12** → (R^1MgX) → **13** → (E^+) → **14**: $C_\alpha > 97\%$ *de*, $C_\beta = 82\text{-}90\%$ *de*

(*R*)-**15** → (R^1MgX) → **16** → (E^+) → **17**: $C_\alpha > 97\%$ *de*, $C_\beta > 97\%$ *de*

Scheme 3

Enolate Reactions

It became apparent from the previous examples that sultam auxiliaries could be used in stereoselective enolate reactions. Indeed, acyl derivatives (*R*)-**18** or (*R*)-**20** can be alkylated in a highly diastereoselective manner [15], which was applied e.g. for the synthesis of *α*-amino acids (Scheme 4) [16–18]. After deprotonation of the glycinate (*R*)-**18** with *n*-butyllithium, the resulting (*Z*)-enolate can be trapped with alkyl-, allyl- or benzylhalides according to the chelate model with selectivities of > 90 % *de* (> 99 % *de* after recrystallization). After acidic hydrolysis and cleavage of the auxiliary, enantiomerically pure amino acids (*S*)-**19** are obtained. A complementary method is the highly selective electrophilic amination of (*R*)-**20** with 1-chloro-1-nitrosocyclohexane, which – again according to the chelate model – leads to (*R*)-**19** [19].

Syn-, *anti-* and acetate aldol derivatives can be synthesized by choosing appropriate enolization protocols (Scheme 5) [20]. With lithium, boron and tin Lewis acids, *syn*-aldols can be obtained *via (Z)*-enolates [21]. If enolization is carried out with lithium or tin, there are enough open coordination sites available to position the aldehyde and the enolate in accordance with the chelate model for the sultam auxiliary and with the Zimmermann-Traxler model. The combination of these models predicts the formation of **22**, which is indeed experimentally obtained. If Lewis acids with only two open coordination sites are used (e.g. Et_2BOTf), chelatization is only possible between the enolate oxygen and the aldehyde. Such reactions should therefore occur through a transition state which is analogous to **5** and should also lead to **22**. However, aldol reactions catalyzed by dialkylboron triflate lead with excellent selectivities to **24**. A plausible transition state that reflects this result is depicted in **23**, in which the aldehyde reacts with the enolate from the lower face being sterically disfavored. Stereoelectronic reasons (antiperiplanar position of l_{PN} to the aldehyde) could be decisive.

If another equivalent of titanium(IV) chloride is added to the boronenolates and only subsequently the aldehyde is introduced, the reaction proceeds *via* the open transition state **25** and leads to the *anti*-aldols **26** [22]. The aldehyde attacks the enolate from the sterically favored lower face, and the group R is oriented away from the auxiliary.

The asymmetric synthesis of (−)-denticulatin A (**30**) shows an interesting application of the boron aldol chemistry (Scheme 6) [23]. In a group-selective aldol reaction between the *meso*-aldehyde **27** and (*S*)-**28**, the hydroxyaldehyde **29** was formed with > 90 % *de*, which spontaneously cyclized to the lactol **31**. The configuration at the stereocenters of C-2 and C-3 in **29** is in accordance with the induction through the sultam auxiliary as well as with preference of an *α*-chiral aldehyde to react to the *anti*-Felkin diastereomer in an aldol reaction which is controlled by the Zimmermann-Traxler model [24, 25].

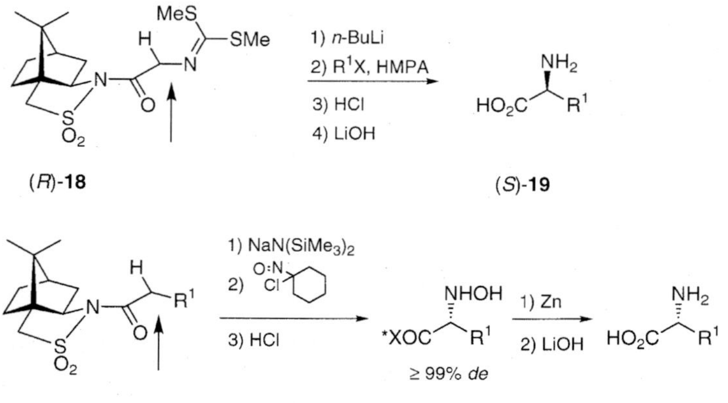

(*R*)-**18** → 1) *n*-BuLi; 2) R^1X, HMPA; 3) HCl; 4) LiOH → (*S*)-**19**

(*R*)-**20** → 1) $NaN(SiMe_3)_2$; 2) 1-chloro-1-nitrosocyclohexane; 3) HCl → **21** (≥ 99% *de*) → 1) Zn; 2) LiOH → (*R*)-**19**

Scheme 4

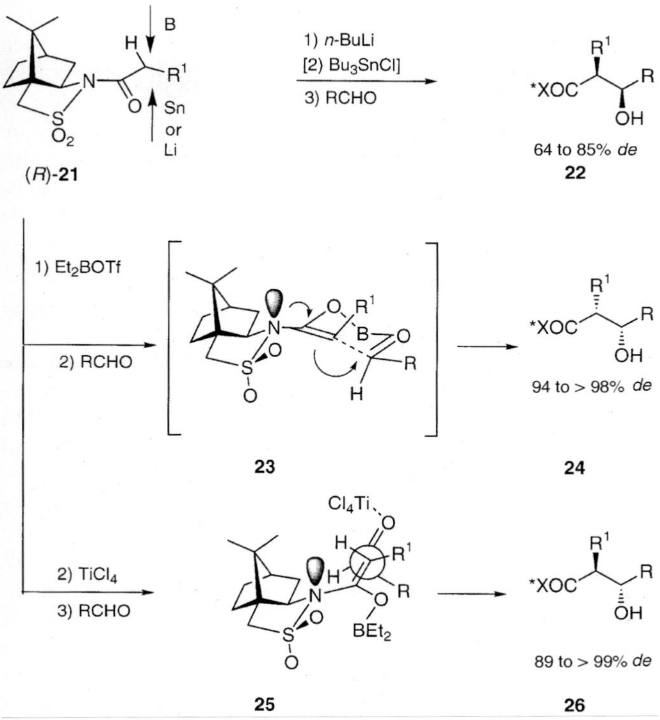

Scheme 5

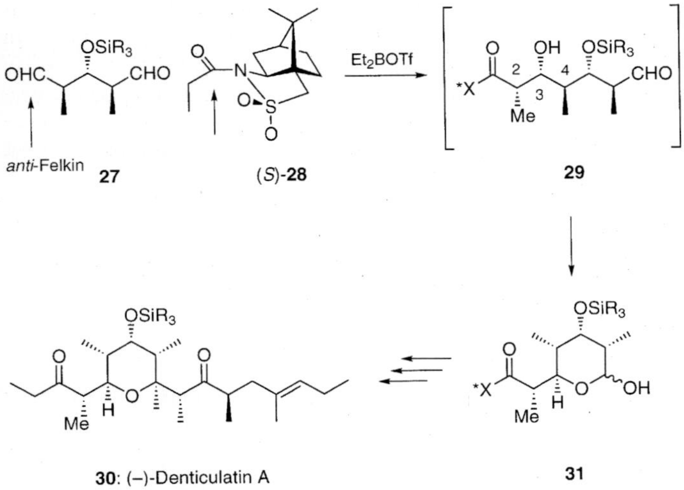

Scheme 6

The synthesis of heteroaromatic side-chain analogs masked as β-lactams of paclitaxel was efficiently accomplished by a cyclocondensation strategy between sultam-modified ester enolates and imines, demonstrating yet another strategy in sultam-enolate chemistry [26].

Radical Reactions

Also for stereoselective radical reactions such as radical additions or radical cyclizations, the camphor sultam **2** is suitable as an auxiliary (Scheme 7). The acyl radical which was gener-

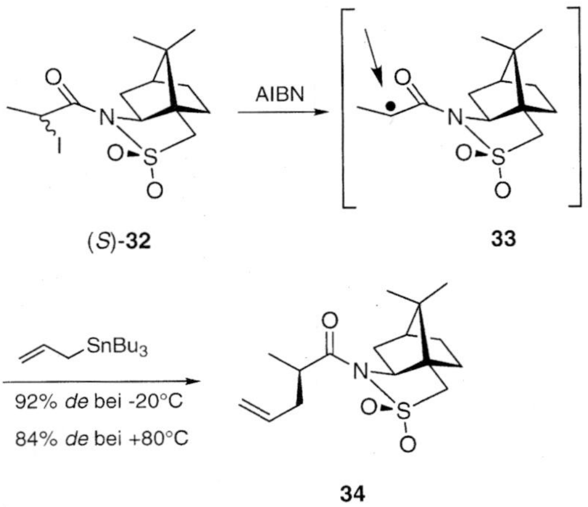

Scheme 7

ated from the iodo compound (*S*)-**32** can be allylated even at +80 °C with remarkable selectivities [27]. Alkyl radicals also add highly selectively to camphor sultam derivatives of oxime ethers to provide a convenient method for the preparation of enantiomerically pure α,β-dialkyl-β-amino acids [28].

There are many more applications of the sultams of camphor-sulfonic acid that could have been described in this article. Finally, it should be noted that recently structurally simpler sultams **35**, which are available from saccharin, have also been successfully applied as a chiral auxiliary [29].

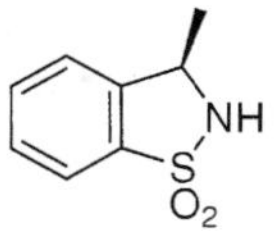

Figure 1

Without doubt, with the discovery of the sultam auxiliaries W. Oppolzer has earned himself a place in the hall of fame of chemistry. With this contribution he will never be forgotten, even by people who, like the author of this article, have never had the chance to meet him personally.

Acknowledgement: The author thanks the Fonds der Chemischen Industrie for financial support.

References

[1] W. Oppolzer, C. Chapuis, G. Bernardinelli, *Helv. Chim. Acta* **1984**, *67*, 1397–401.
[2] F. A. Davis, J. C. Towson, M. C. Weismiller, S. Lal, J. P. Caroll, *J. Am. Chem. Soc.* **1988**, *110*, 8477.
[3] W. Oppolzer, *Pure Appl. Chem.* **1990**, *62*, 1241–50.
[4] W. Oppolzer, *Tetrahedron* **1987**, *43*, 1969–2004.
[5] B. H. Kim, D. P. Curran, *Tetrahedron* **1993**, *49*, 293–318.
[6] W. Oppolzer, I. Rodriguez, J. Blagg, *Helv. Chim. Acta* **1989**, *72*, 123–30.
[7] W. Oppolzer, D. Dupuis, *Tetrahedron Lett.* **1985**, *26*, 5437–40.
[8] W. Oppolzer, M. Wills, M. J. Kelly, *Tetrahedron Lett.* **1990**, *31*, 5015–18.
[9] W. Oppolzer, B. M. Seletsky, Bernardinelli, *Tetrahedron Lett.* **1994**, *35*, 3509–12.
[10] M. Vandewalle, J. Van der Eycken, W. Oppolzer, C. Vullioud, *Tetrahedron* **1986**, *42*, 4035–43.
[11] W. Oppolzer, G. Poli, *Tetrahedron Lett.* **1986**, *27*, 4717–20.
[12] W. Oppolzer, A. J. Kingma, *Helv. Chim. Acta* **1989**, *72*, 1337–45.
[13] W. Oppolzer, A. J. Kingma, G. Poli, *Tetrahedron* **1989**, *45*, 479–88.
[14] W. Oppolzer, G. Poli, A. J. Kingma, *Helv. Chim. Acta* **1987**, *70*, 2201–14.
[15] W. Oppolzer, R. Moretti, S. Thomi, *Tetrahedron Lett.* **1989**, *30*, 5603–6.
[16] W. Oppolzer, R. Moretti, C. Zhou, *Helv. Chim. Acta* **1994**, *77*, 2363–80.
[17] W. Oppolzer, R. Moretti, S. Thomi, *Tetrahedron Lett.* **1989**, *30*, 6009–10.
[18] K. Voigt, A. Stolle, J. Saläun, A. de Meijere, *Synlett* **1995**, 226.
[19] W. Oppolzer, O. Tamura, *Tetrahedron Lett.* **1990**, *31*, 991–4.
[20] W. Oppolzer, C. Starkemann, *Tetrahedron Lett.* **1992**, *33*, 2439–42.
[21] W. Oppolzer, J. Blagg, I. Rodriguez, *J. Am. Chem. Soc.* **1990**, *112*, 2767–72.
[22] W. Oppolzer, P. Lienard, *Tetrahedron Lett.* **1993**, *34*, 4321–4.
[23] W. Oppolzer, J. De Brabander, E. Walther, B. G., *Tetrahedron Lett.* **1995**, *36*, 4413–16.
[24] W. R. Roush, *J. Org. Chem.* **1991**, *56*, 4151–4157.
[25] A. Mengel, O. Reiser, *Chem. Rev.* **1999**, *99*, 1191–1223.
[26] G. I. Georg, G. C. B. Harriman, M. Hepperle, J. S. Clowers, D. G. V. Velde, R. H. Himes, *J. Org. Chem.* **1996**, *61*, 2664–76.
[27] D. P. Curran, W. Shen, Z. Zhang, T. A. Heffner, *J. Am. Chem. Soc.* **1990**, *112*, 6738.
[28] H. Miyabe, K. Fujii, T. Naito, *Org. Lett.* **1999**, *1*, 569–572.
[29] W. Oppolzer, M. Wills, C. Starkemann, *Tetrahedron Lett.* **1990**, *31*, 4117–20.

Oxazolines: Chiral Building blocks, Auxiliaries and Ligands

Martin Glos and Oliver Reiser

Institut für Organische Chemie, Universität Regensburg, Germany

Do you need a chiral starting material which can be converted into a number of enantiopure products? Or a chiral auxiliary to perform an asymmetric transformation at a certain point of a complex molecule? Would you like to have a protecting group for an acid, which activates the *ortho*-position of an aromatic ring? Or do you need an easy-to-synthesize chiral catalyst? For all these problems oxazolines can be the solution.

Oxazolines [1] can be synthesized by several routes; two common methods are described below (Fig. 1). Readily available β-amino alcohols **1** can be coupled with an acid chloride to yield the amide **2** which is then cyclized to the oxazoline **3** in the presence of zinc(II) chloride. Alternatively a one step synthesis of **3** can be achieved by reacting **1** with nitriles. Both methods are reliable and give good yields. Twofold cyclization which leads to bis(oxazolines) **4** is also possible, giving access to a most important class of ligands for asymmetric catalysis. An obvious advantage of oxazolines is their simple synthesis; however, the synthesis of oxazolines depends on the availability of the corresponding amino alcohols, which are generally accessible from the *chiral pool* in only one enantiomeric form.

The recent development of the Sharpless aminohydroxylation [2] makes it possible to synthesize an amino alcohol in both optical antipodes (Fig. 2). For example, starting from 2-vinylnaphthalene (**5**), the amino alcohol **6** is readily synthesized by the aminohydroxylation protocol [3]. After deprotection, the free amino alcohol **8** was coupled with dimethylmalonic acid di-

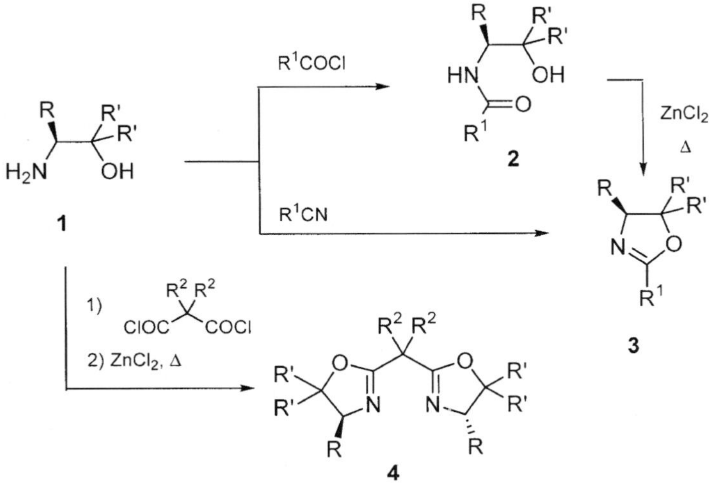

Figure 1. Synthesis of oxazolines from β-amino alcohols and carboxylic acid derivatives.

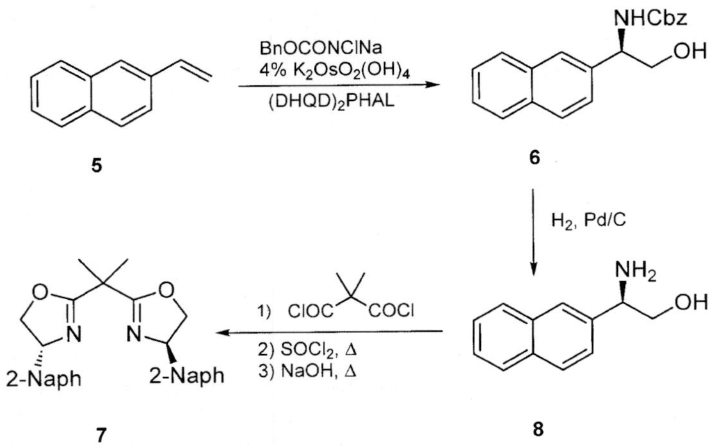

Figure 2. Use of the Sharpless aminohydroxylation to generate chiral amino alcohols.

chloride and subsequently cyclized to the corresponding bis(oxazoline) **7** which was used as chiral ligand for the Diels-Alder reaction of cyclopentadiene (**54**) with the acrylamide **55** yielding **56** in 94 % *ee* (*cf.* Fig. 9).

Oxazolines as Chiral Auxiliaries

The oxazoline **11** developed by Meyers is a versatile building block for the synthesis of chiral carboxylic acids (Fig. 3). Its synthesis is based

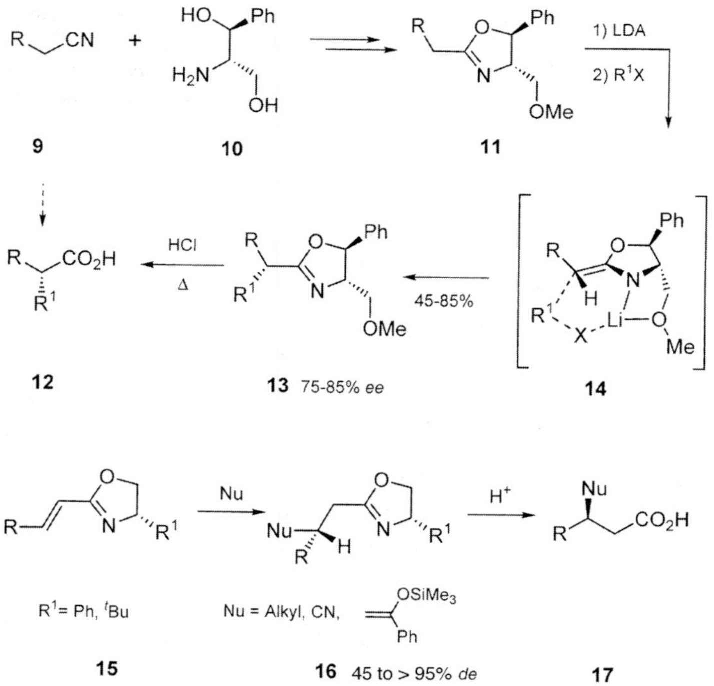

Figure 3. Asymmetric synthesis of carboxylic acids.

on the amino alcohol **10** which is accessible from serin [4]. Metallation of **11** followed by quenching with electrophiles was done in numerous variations. One example is the alkylation of **11** which leads to **13** in good optical yields. Probably the reaction proceeds through the highly ordered intermediate **14** in which a Z-enolate is formed and the 1,3-allylic strain (H/Li vs. R/Li) is minimized. Lithium is coordinated by oxygen and nitrogen and directs the electrophile R^1X to the lower side of the double bond. The phenyl group is shielding the upper side of the double bond from unwanted non-coordinated attacks of R^1X. The auxiliary is hydrolyzed to the carboxylic acid **12** by heating in dilute hydrochloric acid. When R and R^1 are introduced in the reverse order the enantiomer of **12** is also accessible.

Conversion of the auxiliary into other funcional groups is also possible. For example, **13** can be reacted with methyl triflate, reduced with sodium borohydride and then hydrolyzed with acids to yield the corresponding aldehydes [5].

1,4-Additions of nucleophiles to α,β-unsaturated oxazolines of the type **15** are generally conducted with high diastereoselectivity [6]. Good results are obtained with organolithium compounds [7] and silyl enol ethers [8a], while the addition of cyanide seems to be problematic [8b]. For these reactions it is not necessary to have a chelating moiety R^1 in the auxiliary in order to obtain good selectivities. However, there is a loss of activity, which can in some cases be compensated by activating the oxazoline with acetic anhydride [9].

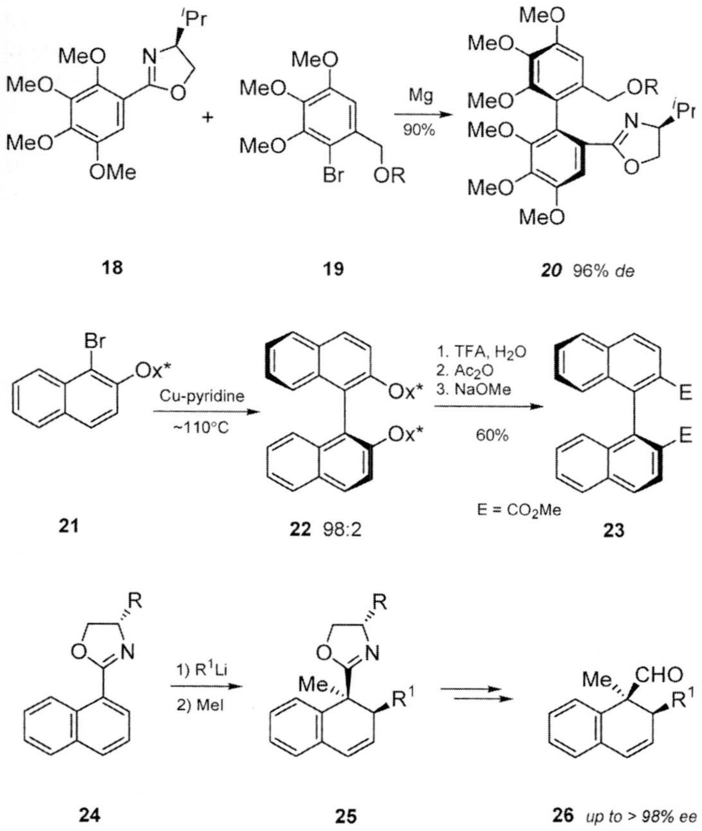

Figure 4. Reactions of aromatic systems involving chiral oxazolines.

In aromatic systems, oxazolines can have three different functions (Fig. 4). Firstly, they can be used as protecting groups for carboxylic acids. Secondly, they activate even electron-rich aromatic systems for nucleophilic substitution. Fluorine or alkoxy groups in the *ortho* position can be substituted by strong nucleophiles such as Grignard reagents. Thirdly, when biaryl compounds with axial chirality are synthesized in these reactions, oxazolines can induce the formation of only one atropisomer with excellent selectivity. These three qualities were all used in the synthesis of **20**, a precurser of the natural product isochizandrine [10].

It is also possible to conduct a diastereoselective Ullmann coupling using 1-bromo-2-oxazolylnaphthalene (**21**). Binaphthyloxazolines **22** are generated in up to 98 % *de* and can be further converted to chiral binaphthoic esters **23** [11].

Addition of alkyllithium compounds at the *ortho*-position of oxazolines is possible with heteroarenes as well as naphthalenes **24** (benzene derivatives usually tend to *ortho*-metallations) [12]. After reductive cleavage of the auxiliary, enantiopure aldehydes **26** are obtained, which have found wide application as versatile chiral precursers for complex polycyclic natural products.

Oxazolines as Ligands for Chiral Catalysts

Oxazolines are excellently suited for the complexation of metals. Based on this knowledge, a variety of metal-oxazoline complexes have been synthesized in recent years and used with outstanding results. In 1986 Pfaltz developed the semicorrin ligands **27** and thereby laid the foundation for future developments [13]. However, the more recently developed bis(oxazoline) ligands are more easily accessible and have therefore found wider application (Fig. 5).

In the beginning, the interest was focused on C_2-symmetric bis(oxazolines) such as for example **4** and **28–30** [14], and one of the first applications was the rhodium-catalyzed hydrosilylation of ketones. The generally successful concept of C_2-symmetry also proved to be advantageous here. The conversion of aromatic or aliphatic ketones **33** to the alcohols **34** was conducted with higher selectivities in the presence of ligand **30** [15] than with ligand **31** [16], although with the latter 86 % *ee* could be reached in the reduction of acetophenone (Fig. 6). Copper-bis(oxazoline)-complexes were used for cyclopropanations [17] and aziridinations [18] with great success. The latter reaction performs especially well

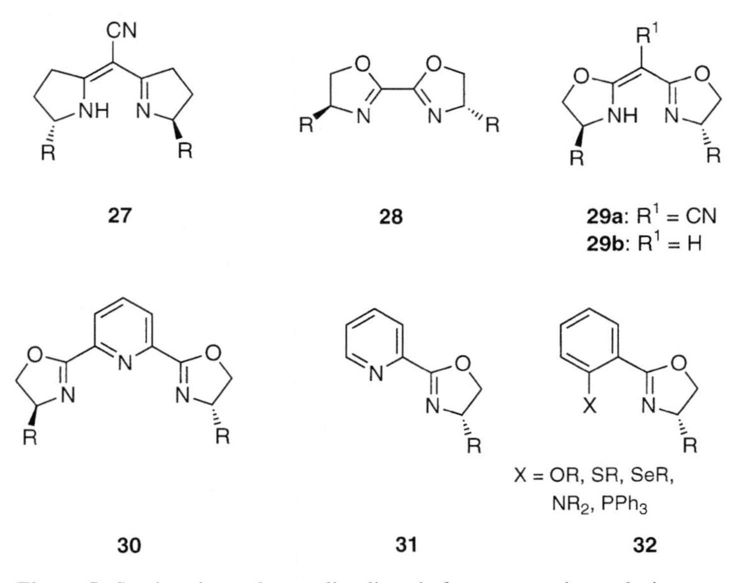

Figure 5. Semicorrin- and oxazoline-ligands for asymmetric catalysis.

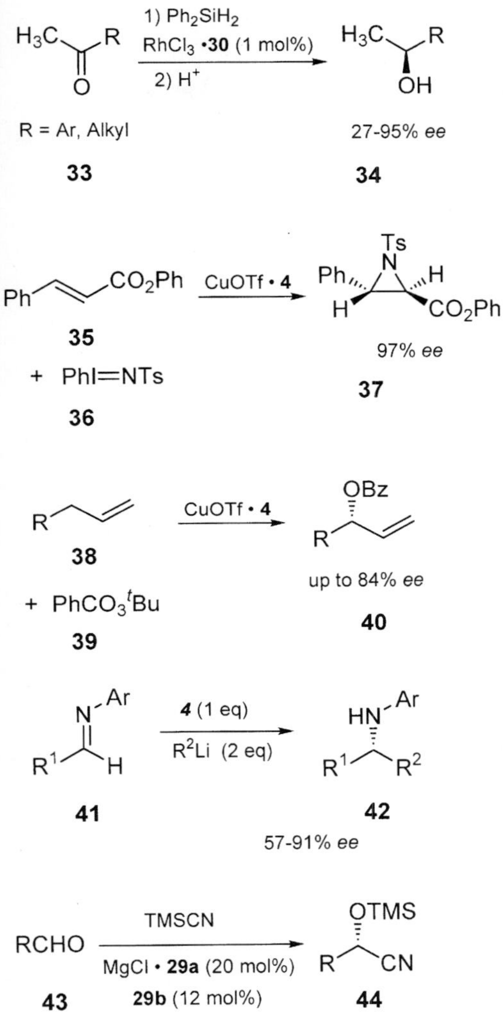

Figure 6. Application of bis(oxazolines) in asymmetric catalysis.

with cinnamic acid esters such as **35**. The aziridine **37** was obtained with excellent enantioselectivity and could be further converted to α-amino acid derivatives [19]. Oxidations are also possible with these complexes as two research groups [20] showed in recent developments. The copper(I)-catalyzed allylic oxyacylation (Kharasch reaction) of cyclic and acyclic olefines by peracid-esters was conducted in the presence of **4** with enantioselectivities up to 84 % *ee*. Another interesting application of the bis(oxazoline)-ligands was shown by Denmark. Alkyllithium compounds can be added enantioselectively to imines, but equimolar quantities of ligand **4** are necessary [21]. A catalytic system involving two bis(oxazolines) was developed by Corey and Wang [22] for the enantioselective conversion of aldehydes **43** to cyanhydrins **44**. One bis(oxazoline) (**29a**) served as ligand for magnesium which coordinates the aldehyde, while the second bis(oxazoline) (**29b**) together with TMSCN provided a source for a "chiral cyanide". The optical yields were modest for α,β-unsaturated and aromatic aldehydes (52 % *ee* for benzaldehyde) and high for aliphatic aldehydes (95 % *ee* for heptanal).

The Mukaiyama aldol reaction could be catalyzed by chiral bis(oxazoline) copper(II) complexes resulting in excellent enantioselectivities (Fig. 7) [23]. A wide range of silylketene acetals **46** and **49** were added to (benzyloxy)acetaldehyde **45** and pyruvate ester **48** in a highly stereoselective manner. The authors were also able to propose a model to predict the stereochemical outcome of these reactions.

Chiral bis(oxazolines) **51** with an oxalylic acid backbone were used for the Ru-catalyzed enantioselective epoxidation of *trans*-stilbene yielding *trans*-1,2-diphenyloxirane in up to 69 % *ee* [24]. The asymmetric addition of diethylzinc to several aldehydes has been examined with ferrocene-based oxazoline ligand **52** [25], resulting in optical yields from 78–93 % *ee*. The imide **53** derived from Kemp's triacid containing a chiral oxazoline moiety was used for the asymmetric protonation of prochiral enolates [26]. Starting from racemic cyclopentanone- and cyclohexanone derivatives, the enantioenriched isomers were obtained in 77–98 % *ee*.

Metal-bis(oxazoline) complexes were widely used as effective catalysts for enantioselective Diels-Alder reactions. Two research groups could achieve excellent diastereo- and enantioselectivity for the reaction of cyclopentadiene (**54**) and the acrylamide **55** (Fig. 9) [27]. Yet the decisive feature is only recognizable when both studies are analyzed together. In both cases the *endo* products are obtained in high selectivities using either the magnesium- or the copper-containing catalyst. However, despite the same

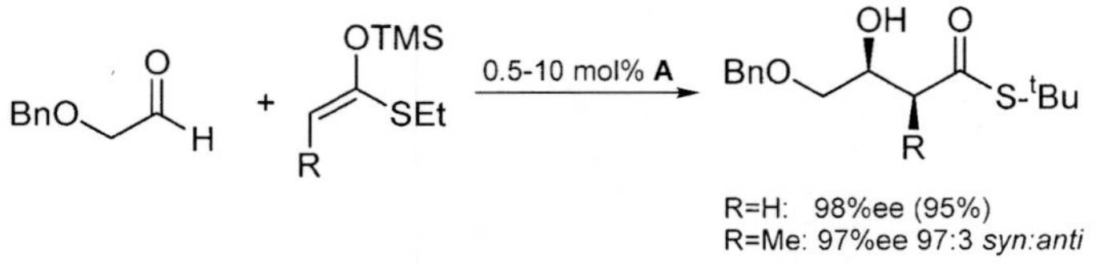

45 **46** **47**

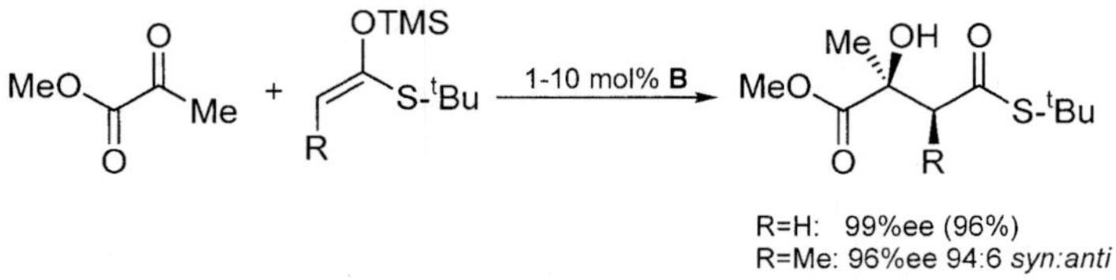

48 **49** **50**

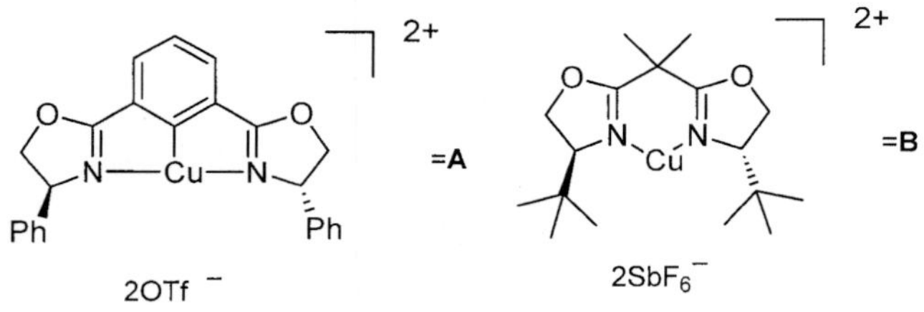

Figure 7. Enantioselective Mukaiyama aldol reaction.

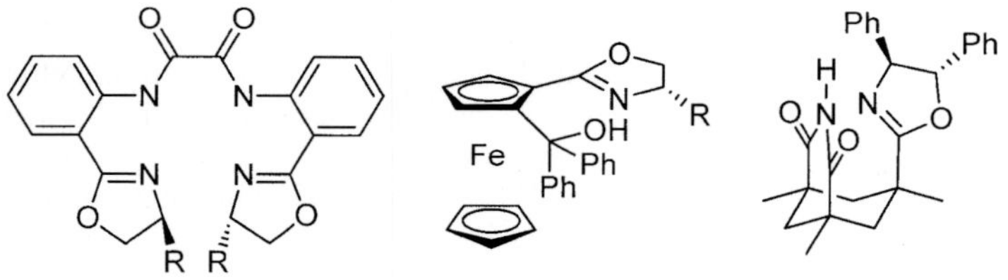

51 **52** **53**

Figure 8. Oxazoline-ligands in asymmetric catalysis.

chirality of the ligands, products of opposite absolute configuration were obtained. These results can be explained assuming the dienophile being coordinated tetrahedrally in the magnesium complex and in square planar configuration in the copper complex. In **57** the acrylate is turned by 90° compared to the coordination in **58**. The attack of the diene at the dienophile, which reacts from an *s-cis* configuration, takes place from the less hindered side opposite the bulky groups (Ph and ^{t}Bu).

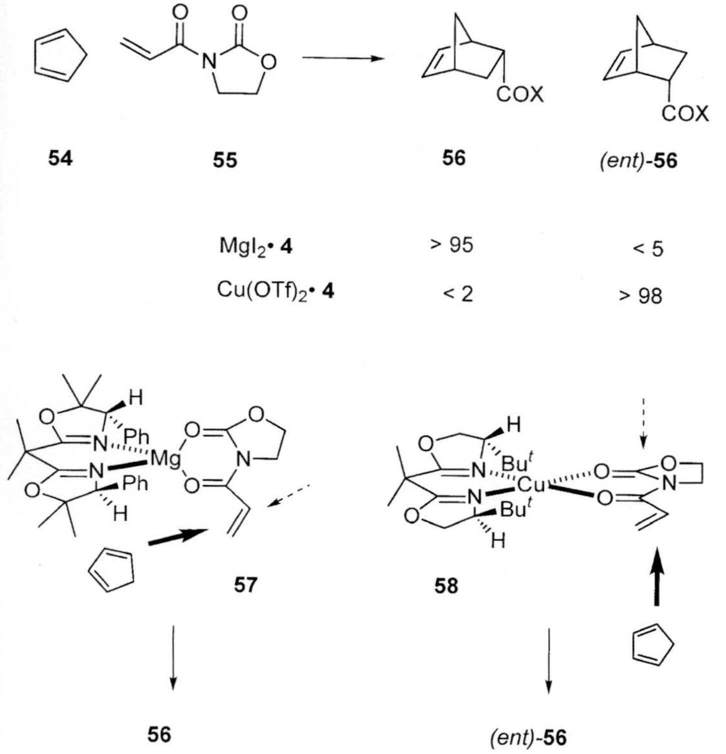

54	55	56	(ent)-56
	$MgI_2 \cdot$ **4**	> 95	< 5
	$Cu(OTf)_2 \cdot$ **4**	< 2	> 98

COX COX

57 **58**

56 *(ent)*-**56**

Figure 9. Asymmetric Diels-Alder reaction with metal-bis(oxazoline) complexes.

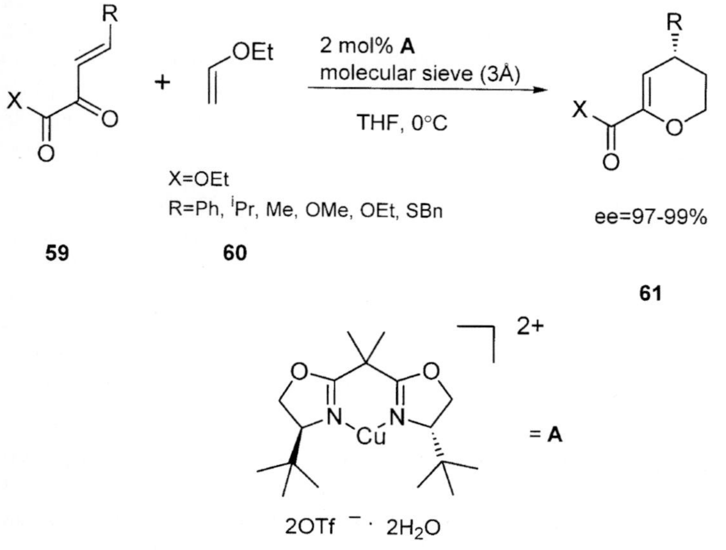

2 mol% **A**
molecular sieve (3Å)
THF, 0°C

X=OEt
R=Ph, ^{i}Pr, Me, OMe, OEt, SBn

ee=97-99%

59 **60** **61**

$2OTf^- \cdot 2H_2O$ = **A**

Figure 10. Hetero-Diels-Alder reaction catalyzed by chiral bis(oxazolines).

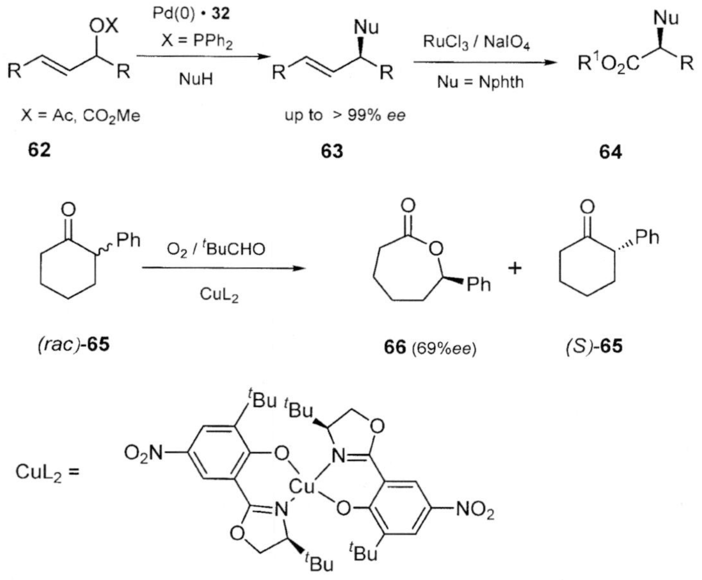

Figure 11. Non-C_2-symmetric oxazoline-ligands in asymmetric catalysis.

For the enantioselective hetero Diels-Alder reaction, very similar bis(oxazoline) complexes proved to be highly efficient [28]. The conversion of several unsaturated keto-esters **59** with ethyl-vinyl ether **60** gave cycloaddition products **61** in 97–99 % *ee*.

The enantioselective cycloaddition of thiabutadienes with the acrylate **55** yielding dihydrothiopyranes was conducted with optical yields up to 98 % *ee* using a C_2-symmetric bis(oxazoline) ligand derived from 1-amino-indan-2-ol [29].

Three research groups discovered almost at the same time that non-C_2-symmetrical oxazolines of the type **32** can be even more effective ligands for asymmetric catalysis than type **4** ligands (Fig. 11). For the palladium-catalyzed allylic substitutions on **62**, record selectivities could be reached using **32** ($X = PPh_2$) [30]. It seems that not only steric but also electronic factors, which cause different donor/acceptor qualities at the coordination centers of the ligand, seem to play a role here [31]. The reaction products can subsequently be converted to interesting molecules, for example **63** (Nu = N-phthalyl) can be oxidized to the amino acid ester **64** [32].

Optically active lactones can be obtained by a kinetic resolution catalyzed by the copper(II)-catalyst **67** in a Baeyer-Villiger-type oxidation. Molecular oxygen is used as oxidizing agent and an aldehyde has to be added in stochiometric amounts as oxygen acceptor. Again, a ligand of the type **32** proved to be better in this reaction than a bis(oxazoline)-ligand; however, the ligand had to be modified with additional groups [33].

The Diels-Alder reaction of **54** and **55** (Scheme 9) is also catalyzed in excellent enantioselectivities (92 % *ee*) [34] by a catalyst made of MeMgBr and (*ent*-**32**) (X = NTs). Interestingly the product is **56**. This suggests that the chirality transfer is – other than postulated by the authors – not analogous to the model for MgI_2*__4__.

The examples described here should give an impression of the numerous applications of oxazolines in organic synthesis. The easy accessibility combined with the excellent qualities in transferring chirality will surely lead to many more applications of oxazolines.

Acknowledgement: The authors thank the Fonds der Chemischen Industrie for financial support.

References

[1] T. G. Grant, A. I. Meyers, *Tetrahedron* **1994**, *50*, 2297.

[2] G. Li, H.-T. Chang, K. B. Sharpless, *Angew. Chem.* **1996**, *108*, 449.

[3] S. Crosignani, G. Desimoni, G. Faita, P. Righetti, *Tetrahedron*, **1998**, *54*, 15721.

[4] A. I. Meyers, G. Knaus, K. Kamata, M. E. Ford, *J. Am. Chem. Soc.* **1976**, *98*, 567.

[5] M. Reuman, A. I. Meyers, *Tetrahedron* **1985**, *41*, 837.

[6] A. I. Meyers, C. E. Whitten, *J. Am. Chem. Soc.* **1975**, *97*, 6266.

[7] A. I. Meyers, M. Shipman, *J. Org. Chem.* **1991**, *56*, 7098.

[8] (a) F. Michelon, A. Pouihes, N. v. Bac, N. Langlois, *Tetrahedron Lett.* **1992**, *33*, 1743. (b) N. Dahuron, N. Langlois, *Synlett* **1996**, 51.

[9] N. Langlois, N. Dahuron, *Tetrahedron Lett.* **1990**, *31*, 7433.

[10] A. I. Meyers, A. Meier, D. J. Rawson, *Tetrahedron Lett.* **1992**, *33*, 853.

[11] T. D. Nelson, A. I. Meyers, *J. Org. Chem.*, **1994**, *59*, 2655.

[12] D. J. Rawson, A. I. Meyers, *J. Org. Chem.* **1991**, *56*, 2292.

[13] (a) A. Pfaltz, *Synlett*, **1999**, 835. (b) H. Fritschi, U. Leutenegger, A. Pfaltz, *Angew. Chem.* **1986**, *98*, 1005.

[14] C. Bolm, *Angew. Chem.* **1991**, *103*, 556.

[15] H. Nishiyama, H. Sakaguchi, T. Nakamura, M. Horihata, M. Kondo, K. Itoh, *Organometallics* **1989**, *8*, 846.

[16] H. Brunner, U. Obermann, *Chem. Ber.* **1989**, *122*, 499.

[17] (a) R. Lowenthal, A. Abiko, S. Masamune, *Tetrahedron Lett.* **1990**, *31*, 6005. (b) D. A. Evans, K. A. Woerpel, M. M. Hinman, M. M. Faul, *J. Am. Chem. Soc.* **1991**, *113*, 726. (c) A. Pfaltz, *Acc. Chem. Res.* **1993**, *26*, 339.

[18] (a) D. A. Evans, M. M. Faul, M. T. Bilodeau, *J. Am. Chem. Soc.* **1994**, *116*, 2742. (b) R. E. Lowenthal, S. Masamune, *Tetrahedron Lett.* **1991**, *32*, 7373.

[19] D. A. Evans, M. M. Faul, M. T. Bilodeau, B. A. Anderson, D. M. Barnes, *J. Am. Chem. Soc.* **1993**, *115*, 5328.

[20] (a) A. S. Gokhale, A. B. E. Minidis, A. Pfaltz, *Tetrahedron Lett.* **1995**, *36*, 1831. (b) M. B. Andrus, A. B. Argade, X. Chen, M. G. Pamment, *ibid.* **1995**, *36*, 2945.

[21] S. E. Denmark, N. Nakajima, O. J.-C. –. Nicaise, *J. Am. Chem. Soc.* **1994**, *116*, 8797.

[22] E. J. Corey, Z. Wang, *Tetrahedron Lett.*, **1993**, *34*, 4001.

[23] a) D. A. Evans, M. C. Kozlowski, J. A. Murry, C. S. Burgey, K. R. Campos, B. T. Connell, R. J. Staples, *J. Am. Chem. Soc.*, **1999**, *121*, 669. b) D. A. Evans, C. S. Burgey, M. C. Kozlowski, S. W: Tregay, *J. Am. Chem. Soc.*, **1999**, *121*, 685.

[24] N. End, L. Macko. M. Zehnder, A. Pfaltz, *Chem. Eur. J.*, **1998**, *4*, 818.

[25] C. Bolm, K. Muniz-Fernández, A. Seger, G. Raabe, K. Günther, *J. Org. Chem.*, **1998**, *63*, 7860.

[26] A. Yanagisawa, T. Kikuchi, T. Kuribayashi, H. Yamamoto, *Tetrahedron*, **1998**, *54*, 10253.

[27] (a) E. J. Corey, K. Ishihara, *Tetrahedron Lett.* **1992**, *33*, 6807. (b) D. A. Evans, S. J. Miller, T. Lectka, *J. Am. Chem. Soc.* **1993**, *115*, 6460.

[28] (a) D. A. Evans, E. J. Olhava, J. S. Johnson, J. M. Janey, *Angew. Chem.*, **1998**, *110*, 3554. (b) M. Johannsen, K. A. Jørgensen, *Tetrahedron* **1996**, *52*, 7321.

[29] T. Saito, K. Takekawa, T. Takahashi, *Chem. Comm.*, **1999**, 1001.

[30] (a) P. v. Matt, A. Pfaltz, *Angew. Chem.* **1993**, *105*, 614. (b) J. Sprinz, G. Helmchen, *Tetrahedron Lett.* **1993**, *34*, 1769. (c) G. J. Dawson, C. G. Frost, J. M. J. Williams, *ibid.* **1993**, *34*, 3149. (d) H. Rieck, G. Helmchen, *Angew. Chem.* **1995**, *107*, 2881.

[31] (a) J. Sprinz, M. Kiefer, G. Helmchen, G. Huttner, O. Walter, L. Zsolnai, M. Reggelin, *Tetrahedron Lett.* **1994**, *35*, 1523. (b) O. Reiser, *Angew. Chem.* **1993**, *105*, 576. (c) J. M. J. Williams, *Synlett*, **1996**, 705. (d) H. Steinhagen, M. Reggelin, G. Helmchen, *Angew. Chem.*, **1997**, *109*, 2199.

[32] R. Jumnah, A. C. Williams, J. M. J. Williams, *Synlett* **1995**, 821.

[33] C. Bolm, G. Schlingloff, K. Weickhardt, *Angew. Chem.* **1994**, *106*, 1944.

New Strategies to α-Alkylated α-Amino Acids

Thomas Wirth

Universität Basel, Institut für Organische Chemie, Switzerland

Nonproteinogenic α-alkylated α-amino acids are playing an important role in natural products and for biological investigations. Because of the tetrasubstituted asymmetric carbon atom they possess high stability at the stereogenic center. They exert a remarkable influence on the conformation of peptides into which they are incorporated [1]. They can therefore be used for the investigation of enzymatic mechanisms and as enzyme inhibitors. Furthermore, they are interesting building units for the synthesis of natural products, which has already been demonstrated by several impressive examples [2].

In 1872 the simplest α,α-disubstituted amino acid (2-aminoisobutyric acid, Aib) was described [3]. In 1908 the first optically active representative of this class of compounds, (*R*)-2-ethylalanine (D-isovaline), was isolated by microbial racemic resolution [4]. Synthetic chemists have therefore been interested in the enantiopure synthesis of α-alkylated α-amino acids for some time. Their powerful methods for the construction of chiral α-amino acids can in some cases also be used for the synthesis of the α-alkylated derivatives which has been the topic of recent reviews [5].

In most of the procedures the stereogenic center is constructed by alkylation of chiral, nonracemic enolates. The established methods, which have been reviewed several times [6], are mentioned only briefly here.

One of the classics is the Schöllkopf synthesis via bislactim ethers [7]. Beside efficient methods for the preparation of bislactim ethers [8], a derivative of **1** with R = CO_2Et has been used for the synthesis of α-alkylated serines [9], and a derivative of **1** with R = H has been used in a tin-mediated aldol-type reaction to serine compounds as well [10]. The chirality of an amino acid, mostly valine, is used to create a second stereogenic center by the addition of an electrophile to the deprotonated bislactim ether **1**. Acidic cleavage of the bislactim ether makes it possible to synthesize a variety of natural and unnatural amino acids in good yields and optical purities. Similar diastereoselective alkylations of other heterocyclic enolates such as **2** can also lead to α-alkylated amino acids in high optical purities [11, 12].

The Seebach method likewise employs a chiral, cyclic enolate **3** (n = 0) (Bz = benzoyl) [13]. The required imidazolidinones are synthesized by condensation of the corresponding amides with pivalaldehyde. The *tert*-butyl group is directing the electrophile in a 1,3-induction and determines the configuration of the new stereogenic center. If six-membered heterocycles **3** (n = 1)

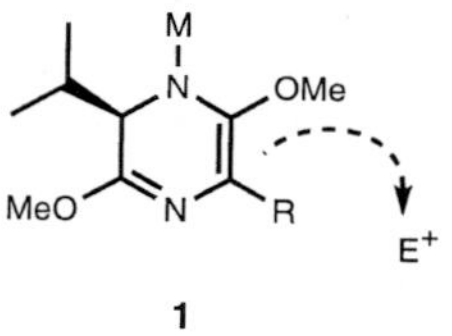

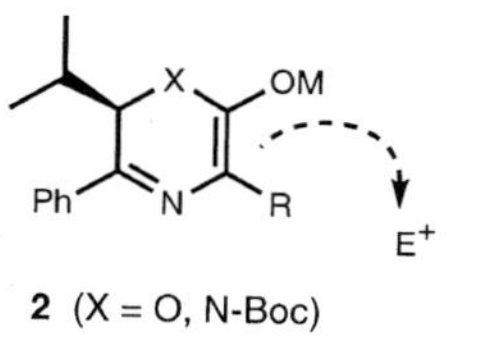

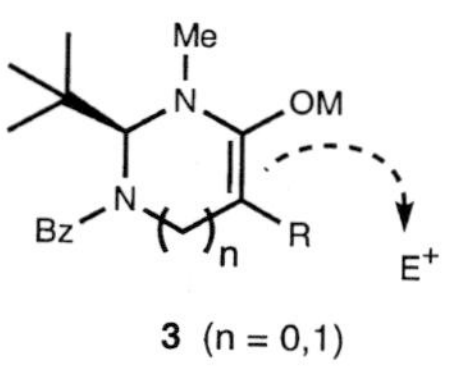

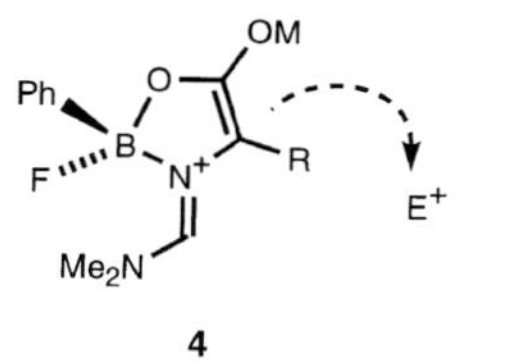

Figure 1

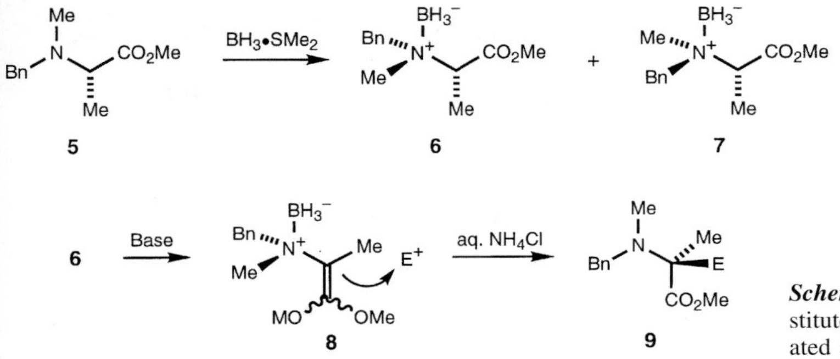

Scheme 1. Synthesis of α,α-disubstituted amino acids from α-methylated amino acids (Bn = Benzyl).

are employed in this reaction, it is possible to synthesize α,α-disubstituted β-amino acids as well [14]. If a ferrocenyl substituent is used instead of the *tert*-butyl group as a directing group, the alkylation proceeds again with excellent stereoselectivities [15]. Deprotonated oxazaborolidinones **4** can also serve as chiral equivalents of amino acid enolates [16]. It was found recently that changing the phenyl moiety to a naphthyl moiety on the boron atom in **4** resulted in a further improvement of this methodology [17]. Starting from amino acids, these compounds can be prepared in a few steps in enantiomerically pure form. As in the Seebach method, the chirality of the amino acid is used to control the stereochemistry of a second center, which then directs the attack of the electrophile at the enolate.

An interesting version of this concept was published recently [18]. After reaction of the alanine ester **5** with borane, the mixture of the diastereomers **6** and **7** could be separated (Scheme 1). The chiral nitrogen atom directs the attack of the electrophile at the enolate. The direction of the attack can be explained by the Felkin – Anh model, because the electrophile approaches *anti* to the largest (benzyl) substituent. The enolate shown in **8** should be preferred because of the repulsion of the boron and the carboxy group. The boron is removed by aqueous work-up and the α-methylated amino acid **9** is obtained in enantiomeric purities up to 82 %.

The chirality of the phenylalanine derivative **10** is used for a direct, stereoselective α-alkylation (Scheme 2) [19]. After treatment with base and reaction with an electrophile the α-alkylated amino acid **11** is obtained in up to 88 % *ee*. It is not yet clear whether the deprotonated species is an enolate with a chiral nitrogen atom (**12**) or a chiral, α-metallated compound (**13**). The protecting groups on the nitrogen seem to play an important role. It is not yet possible to alkylate other phenylalanine derivatives by means of this reaction.

A fast access to α-alkylated amino acids is also possible by the Claisen rearrangement of chelated enolates [20]. Esters of type **14** rearrange after treatment with base and chelation with a metal salt. The products **15** are obtained in good yields and with diastereoselectivities up to 99 % (Scheme 3).

Other rearrangements have been used as well to create quaternary stereocenters of α,α-disubstituted amino acids. The rearrangement of *O*-acylated azlactones **16**, described by Steglich in 1970

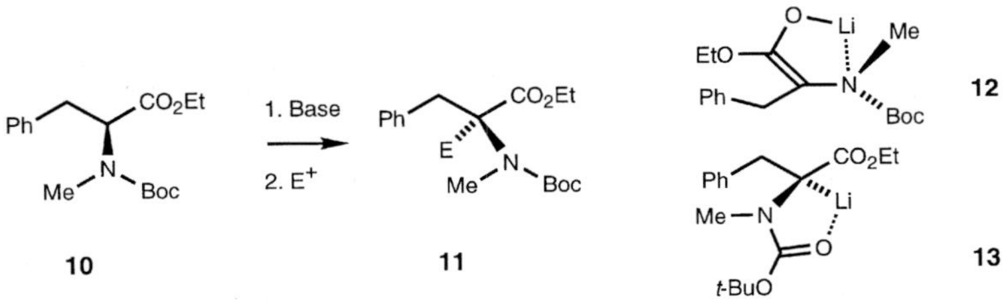

Scheme 2. Asymmetric α-alkylation of a substituted phenylalanine (Boc = *tert*-butoxycarbonyl).

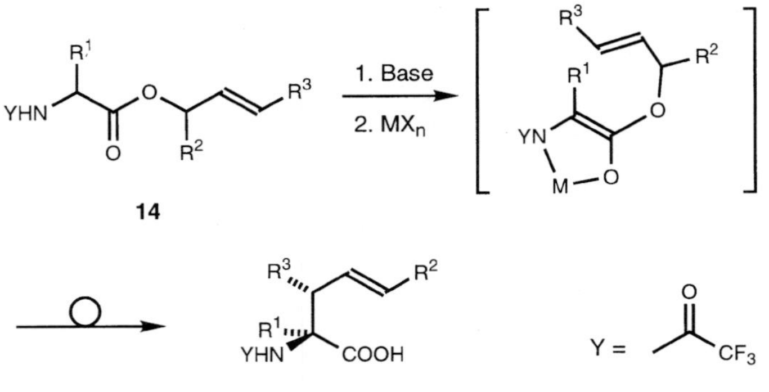

Scheme 3. α-Alkylated amino acids by Claisen rearrangement.

[21], has been developed into an efficient stereoselective method using the chiral catalyst **17**. Protected serine derivatives of type **18** are accessible in high yields and with excellent stereoselectivities as shown in Scheme 4 [22]. A Lewis acid-catalyzed rearrangement of epoxides, first reported by Jung et al. for the asymmetric synthesis of α-alkylated amino acids [23], was used recently for the synthesis of similar derivatives [24].

The amino acid synthesis from Strecker has been known since 1850 [25]. Stereoselective versions of this synthesis start with chiral amines, which are condensed with carbonyl compounds to form imines. Addition of hydrogen cyanide and subsequent hydrolysis of the amino nitriles yields the amino acids. When ketones are used for the condensation, α-alkylated amino acids are obtained in high yields and optical purities [26]. A new variant of this reaction reported recently employs valine to build a cyclic imine [27]. The cyclic imine **19** is obtained by condensation of hydroxyacetone with valine, which reacts with cyanide to yield amino nitrile **20**. With *tert*-butyl hypochlorite and triethylamine an imine is formed, which gives after acidic hydrolysis enantiopure α-methylserine **21** in an overall yield of about 55 % (Scheme 5). The stereochemical information for the asymmetric Strecker reaction can even be located outside the heterocycle, as demonstrated with the compounds **22** [28] and **23** [29] as precursors for the synthesis of serine derivatives. Because of the intramolecular protection of the hydroxy group this method allows a rapid and efficient access to hydroxy amino acids. Alternative methods of synthesizing optically active α-methyl serine derivatives have been reported recently [30].

β-Lactams are also suitable intermediates en route to α,α-disubstituted amino acids [31]. In a 1,2-induction the chiral center in **24** is created. After reductive cleavage of the C–N bond enan-

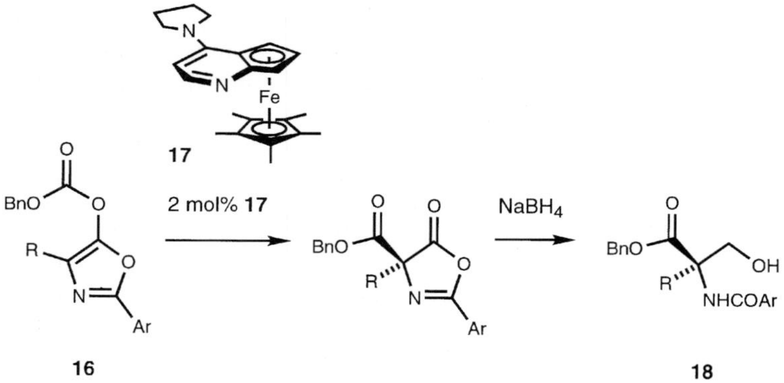

Scheme 4. Enantioselective rearrangement of *O*-acylated azlactones.

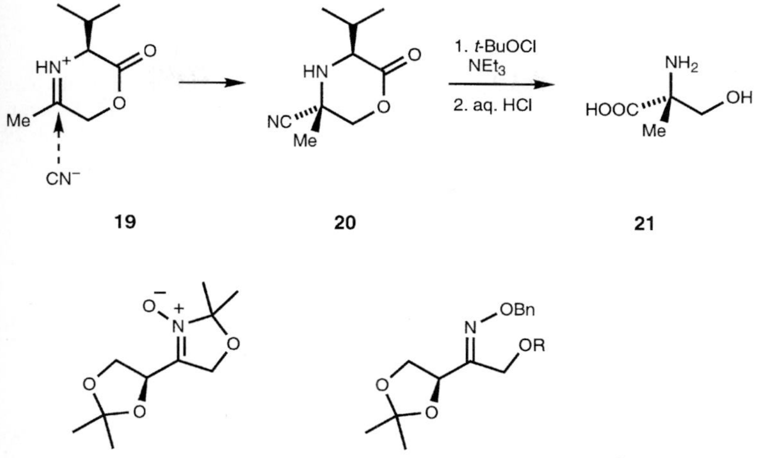

Scheme 5. Asymmetric variants of the amino acid synthesis according to Strecker (Bn = Benzyl).

tiomerically pure α-alkylated amino acids are obtained.

Two directing stereogenic centers are found in oxazinones **25** (Cbz = carbobenzyloxy), which can also be used advantageously in the synthesis of disubstituted amino acids [32]. However, during the reductive liberation of the amino acids the chiral auxiliary is destroyed.

Another conventional approach, the addition of electrophiles to enolates of chiral esters, has been applied successfully by several groups. The 8-phenylmenthol ester of *N*-benzoyl alanine can be deprotonated twice (**26**) and, after addition of various electrophiles, α-methylated derivatives are obtained. After cleavage of the chiral auxiliary they can be transformed into the enantiomerically pure α-methylated amino acids [33]. Other directing groups such as oxazolidinones **27** [34], imidazolidinones [35] or sultames [36] have been used as well. If a chiral directing group is connected to the *N*-terminus of a peptide, a stereoselective alkylation of the terminal amino acid is possible [37].

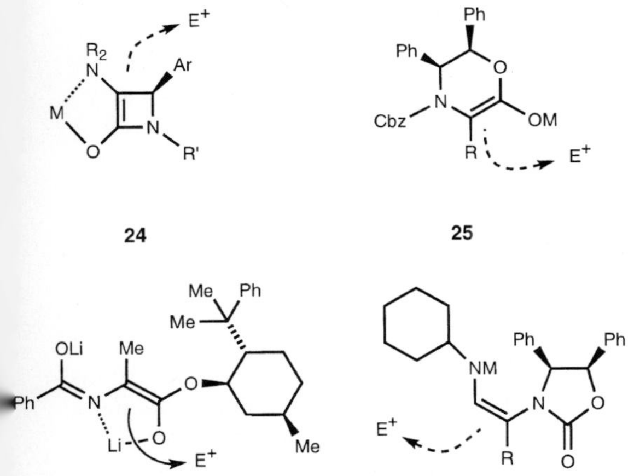

With racemic α-alkylated amino acids an enzymatic racemate resolution is possible. There are several methods to access racemic α-alkylated amino acids in high yields [38]. Different microorganisms have been applied, and the products are obtained in very high enantiomeric purities [39]. Because α-alkylated amino acids are used as building blocks for different active substances, methods for the synthesis of large quantities have been developed, especially in industry [40]. Other effective racemic resolution techniques have been described recently. Disubstituted azlactones of type **28** can react with the phenylalanine derivative **29** [41]. The diastereomers of the protected dipeptide **30** are then separated. The easy access of compounds of type **28,** together with the optimized reagent **29**, ensures

Figure 2

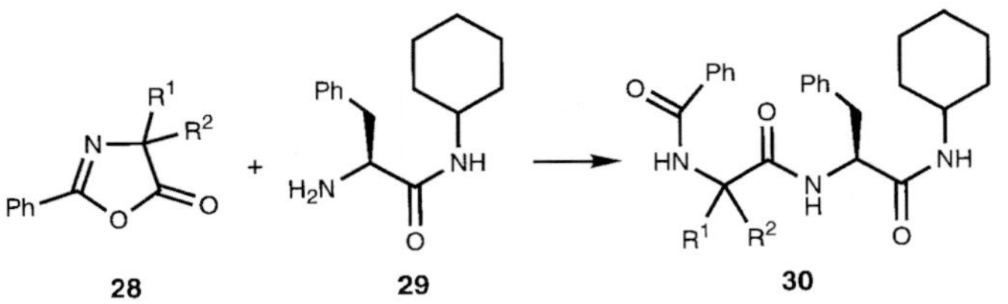

Scheme 6. Synthesis of optically pure α,α-disubstituted amino acids by treatment of racemic azlactones **28** with **29** and subsequent diastereomer separation of **30**.

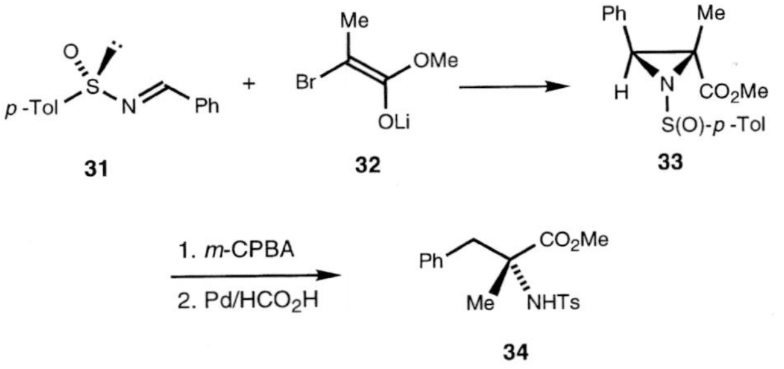

Scheme 7. Aziridines as intermediates in the synthesis of α,α-disubstituted amino acids (*p*-Tol: *p*-tolyl, Ts: *p*-toluenesulfonyl).

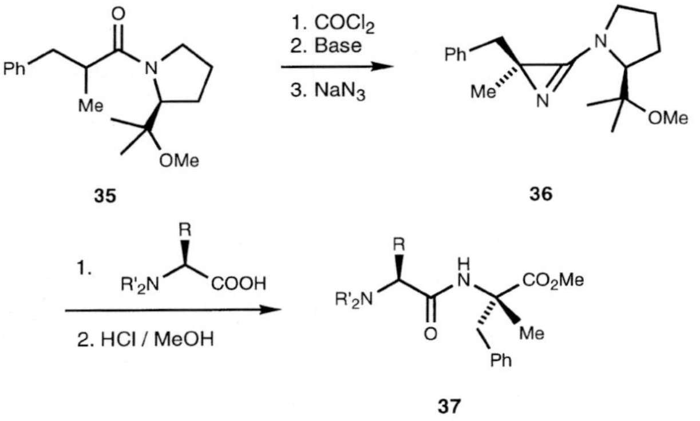

Scheme 8. Incorporation of α,α-disubstituted amino acids into peptides on treatment of diastereochemically pure azirines **30** with an amino acid and scission on the chiral residue.

that the method is powerful for the synthesis of optically pure α,α-disubstituted amino acids (Scheme 6).

Nitrogen heterocycles such as azirines and aziridines are also used effectively as building blocks for the synthesis of α,α-disubstituted amino acids. The aziridine derivative **33** is prepared in optically pure form by addition of the lithium enolate **32** to the chiral sulfinimide **31** (Scheme 7) [42]. After oxidation of the sulfoxide to the sulfone and subsequent hydration, the α-methylated phenylalanine derivative **34** is obtained in good overall yield.

The introduction of α,α-disubstituted amino acids into peptides is sometimes troublesome. A recent, easy access to enantiomerically pure 3-amino-2*H*-azirines offers an elegant way for the incorporation of α,α-disubstituted amino acids into peptides [6b, 43]. The amide **35** is treated with phosgene and a base to yield a ketene iminium salt, which then reacts with sodium azide to afford azirine **36** with loss of nitrogen (Scheme 8). Because of the chiral substituent a separation of the diastereomers is possible (only one diastereomer shown here). After reaction with the carboxylic group of an amino acid and acidic cleavage of the chiral auxiliary the dipeptide **37** is obtained. Other conformationally restricted α,α-disubstituted amino acids have been resolved and investigated also by computational methods [44].

Stoichiometric amounts of sometimes cheap chiral reagents have to be used for most of the

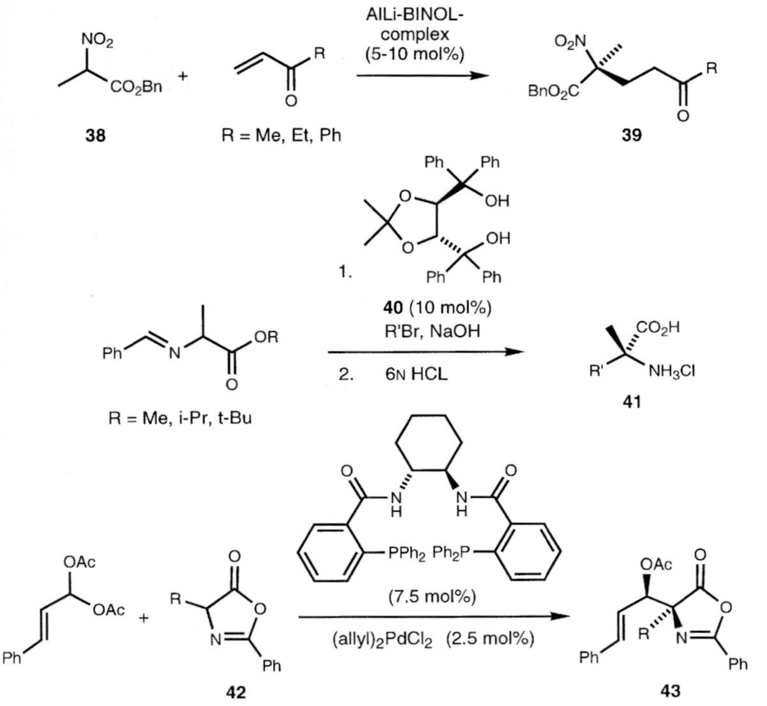

Scheme 9. Synthesis of α-alkylated amino acids with only catalytic amounts of chiral auxiliaries.

synthetic methods shown above. Very recently first reactions for the synthesis of enantiomerically enriched α-alkylated amino acids have been published in which only catalytic amounts of chiral auxiliaries are needed.

Catalytic Michael additions of α-nitroesters **38** catalyzed by a BINOL (2,2'-dihydroxy-1,1'-binaphthyl) complex were found to yield the addition products **39** as precursors for α-alkylated amino acids in good yields and with respectable enantioselectivities (8–80 %) as shown in Scheme 9 [45]. Asymmetric PTC (phase transfer catalysis) mediated by TADDOL (**40**) as a chiral catalyst has been used to synthesize enantiomerically enriched α-alkylated amino acids **41** (up to 82 % *ee*) [46]. A similar strategy has been used to access α-amino acids in a stereoselective fashion [47]. Using azlactones **42** as nucleophiles in the palladium catalyzed stereoselective allylation addition, compounds **43** were obtained in high yields and almost enantiomerically pure (Scheme 9) [48]. The azlactones **43** can then be converted into the α-alkylated amino acids as shown in Scheme 4.

References

[1] a) P. Balaram, *Curr. Opinion Struct. Biol.* **1992**, *2*, 845–851; b) A. Giannis, T. Kolter, *Angew. Chem.* **1993**, *105*, 1303–1326; *Angew. Chem. Int. Ed. Engl.* **1993**, *32*, 1244–1267; c) K. Burgess, K.-K. Ho, B. Pal, *J. Am. Chem. Soc.* **1995**, *117*, 3808–3819; d) M. P. Paradisi, I. Torrini, G. P. Zecchini, G. Lucente, E. Gavuzzo, F. Mazza, G. Pochetti, *Tetrahedron* **1995**, *51*, 2379–2386; e) A. Lewis, J. Wilkie, T. J. Rutherford, D. Gani, *J. Chem. Soc., Perkin Trans. 1* **1998**, 3777–3793.

[2] a) H. Cheng, P. Keitz, J. B. Jones, *J. Org. Chem.* **1994**, *59*, 7671–7676; b) review: U. Koert, *Nachr. Chem. Tech. Lab.* **1995**, *43*, 347–354.

[3] F. Urech, *Justus Liebigs Ann. Chem.* **1872**, *164*, 255–279.

[4] F. Ehrlich, A. Wendel, *Biochem. Z.* **1908**, *8*, 438.

[5] a) T. Wirth, *Angew. Chem.* **1997**, *109*, 235–237; *Angew. Chem. Int. Ed. Engl.* **1997**, *36*, 225–227; b) C. Cativiela, M. D. Díaz-de-Villegras, *Tetrahedron: Asymmetry* **1998**, *9*, 3517–3599; c) A. S. Franklin, *J. Chem. Soc., Perkin Trans. 1* **1998**, 2451–2465.

[6] a) R. M. Williams, Synthesis of Optically Active α-Amino Acids, Pergamon Press, Oxford, 1989; b) H. Heimgartner, *Angew. Chem.* **1991**, *103*, 271–297; *Angew. Chem. Int. Ed. Engl.* **1991**, *30*, 238–264.

[7] a) U. Schöllkopf, *Pure Appl. Chem.* **1983**, *55*, 1799–1806; b) U. Schöllkopf, U. Busse, R. Kilger, P. Lehr, *Synthesis* **1984**, 271–274; c) U. Schöllkopf, K.-O. Westphalen, J. Schröder, K. Horn, *Liebigs Ann. Chem.* **1988**, 781–786.
[8] S. D. Bull, S. G. Davies, W. O. Moss, *Tetrahedron: Asymmetry* **1998**, *9*, 321bis›327.
[9] S. Sano, M. Takebayashi, T. Miwa, T. Ishii, Y. Nagao, *Tetrahedron: Asymmetry* **1998**, *9*, 3611–3614.
[10] a) S. Sano, T. Miwa, X. Liu, T. Ishii, T. Takehisa, M. Shiro, Y. Nagao, *Tetrahedron: Asymmetry* **1998**, *9*, 3615–3618; b) S. Sano, T. Ishii, T. Miwa, Y. Nagao, *Tetrahedron Lett.* **1999**, *40*, 3013–3016.
[11] a) R. Chinchilla, L. R. Falvello, N. Galindo, C. Nájera, *Angew. Chem.* **1997**, *109*, 1036–1039; *Angew. Chem. Int. Ed.* **1997**, *36*, 995–997; b) T. Abellán, C. Nájera, J. M. Sansano, *Tetrahedron: Asymmetry* **1998**, *9*, 2211–2214; c) R. Chinchilla, N. Galindo, C. Nájera, *Tetrahedron: Asymmetry* **1998**, *9*, 2769–2772.
[12] A. Carloni, G. Porzi, S. Sandri, *Tetrahedron: Asymmetry* **1998**, *9*, 2987–2998.
[13] A. Studer, D. Seebach, *Liebigs Ann.* **1995**, 217–222.
[14] a) E. Jurasti, M. Balderas, Y. Ramírez-Quirós, *Tetrahedron: Asymmetry* **1998**, *9*, 3881–3888; b) E. Jurasti, H. López-Ruiz, D. Madrigal, Y. Ramírez-Quirós, J. Escalante, *J. Org. Chem.* **1998**, *63*, 4706–4710.
[15] F. Alonso, S. G. Davies, A. S. Elend, J. L. Haggitt, *J. Chem. Soc., Perkin Trans. 1* **1998**, 257–264.
[16] a) E. Vedejs, S. C. Fields, M. R. Schrimpf, *J. Am. Chem. Soc.* **1993**, *115*, 11612–11613; b) E. Vedejs, R. W. Chapman, S. C. Fields, S. Lin, M. R. Schrimpf, *J. Org. Chem.* **1995**, *60*, 3020–3027; c) E. Vedejs, S. C. Fields, S. Lin, M. R. Schrimpf, *J. Org. Chem.* **1995**, *60*, 3028–3034.
[17] E. Vedejs, S. C. Fields, R. Hayashi, S. R. Hitchcock, D. R. Powell, M. R. Schrimpf, *J. Am. Chem. Soc.* **1999**, *121*, 2460–2470.
[18] V. Ferey, L. Toupet, T. Le Gall, C. Mioskowski, *Angew. Chem.* **1996**, *108*, 475–477; *Angew. Chem. Int. Ed. Engl.* **1996**, *35*, 430–432.
[19] a) T. Kawabata, T. Wirth, K. Yahiro, H. Suzuki, K. Fuji, *J. Am. Chem. Soc.* **1994**, *116*, 10809–10810; b) T. Kawabata, T. Wirth, K. Yahiro, H. Suzuki, K. Fuji, *ICR Ann. Rep.* **1996**, *3*, 36–37; c) K. Fuji, T. Kawabata, *Chem. Eur. J.* **1998**, *4*, 373–376.
[20] a) U. Kazmaier, *Angew. Chem.* **1994**, *106*, 1046–1047; *Angew. Chem. Int. Ed. Engl.* **1994**, *33*, 998–999; b) U. Kazmaier, S. Meier, *Tetrahedron* **1996**, 52, 941–954; c) U. Kazmaier, *J. Org. Chem.* **1996**, 61, 3694–3699; d) U. Kazmaier, C. Schneider, *Synthesis* **1998**, 1321–1326.
[21] W. Steglich, G. Höfle, *Tetrahedron Lett.* **1970**, 4727–4730.
[22] J. C. Ruble, G. C. Fu, *J. Am. Chem. Soc.* **1998**, *120*, 11532–11533.
[23] M. E. Jung, D. C. D'Amico, *J. Am. Chem. Soc.* **1995**, *117*, 7379–7388.
[24] M. Matsushita, H. Maeda, M. Kodama, *Tetrahedron Lett.* **1998**, *39*, 3749–3752.
[25] A. Strecker, *Justus Liebigs Ann. Chem.* **1850**, *75*, 27.
[26] a) K. Weinges, H. Blackholm, *Chem. Ber.* **1980**, *113*, 3098–3102; b) P. K. Subramanian, R. W. Woodard, *Synth. Commun.* **1986**, *16*, 337–342; c) D. Ma, H. Tian, G. Zou, *J. Org. Chem.* **1999**, *64*, 120–125.
[27] a) S.-H. Moon, Y. Ohfune, *J. Am. Chem. Soc.* **1994**, *116*, 7405–7406; b) Y. Ohfune, S.-H. Moon, M. Horikawa, *Pure Appl. Chem.* **1996**, *68*, 645–648; c) M. Horikawa, T. Nakajima, Y. Ohfune, *Synlett* **1997**, 253–254; d) Y. Ohfune, M. Horikawa, *J. Synth. Org. Chem. Jpn.* **1997**, *55*, 982–993.
[28] J. A. Marco, M. Carda, J. Murga, R. Portolés, E. Falomir, J. Lex, *Tetrahedron Lett.* **1998**, *39*, 3237–3240.
[29] a) J. A. Marco, M. Carda, J. Murga, S. Rodríguez, E. Falomir, M. Oliva, *Tetrahedron: Asymmetry* **1998**, *9*, 1679–1701; b) M. Carda, J. Murga, S. Rodríguez, F. González, E. Castillo, J. A. Marco, *Tetrahedron: Asymmetry* **1998**, *9*, 1703–1712.
[30] a) P. Wipf, S. Venkatraman, C. P. Miller, *Tetrahedron Lett.* **1995**, *36*, 3639–3642; b) D. Obrecht, M. Altorfer, C. Lehmann, P. Schönholzer, K. Müller, *J. Org. Chem.* **1996**, *61*, 4080–4086; c) R. Grandel, U. Kazmaier, *Eur. J. Org. Chem.* **1998**, 409–417.
[31] I. Ojima, *Acc. Chem. Res.* **1995**, *28*, 383–389.
[32] a) R. M. Williams, M.-N. Im, *J. Am. Chem. Soc.* **1991**, *113*, 9276–9286; b) J. E. Baldwin, V. Lee, C. J. Schofield, *Synlett* **1992**, 249–251.
[33] D. B. Berkowitz, M. K. Smith, *J. Org. Chem.* **1995**, *60*, 1233–1238.
[34] S. Wenglowsky, L. S. Hegedus, *J. Am. Chem. Soc.* **1998**, *120*, 12468–12473.
[35] G. Guillena, C. Nájera, *Tetrahedron: Asymmetry* **1998**, *9*, 3935–3938.
[36] A. López, R. Pleixats, *Tetrahedron: Asymmetry* **1998**, *9*, 1967–1977.
[37] K. Miyashita, H. Iwaki, K. Tai, H. Murafuji, T. Imanishi, *Chem. Commun.* **1998**, 1987–1988.
[38] a) D. L. Griffith, M. J. O'Donnell, R. S. Pottorf, W. L. Scott, J. A. Porco, *Tetrahedron Lett.* **1997**, *38*, 8821–8824; b) P. Dauban, R. H. Dodd, *Tetrahedron Lett.* **1998**, *39*, 5739–5742; c) S. W. Kim, Y. S. Shin, S. Ro, *Bioorg. Med. Chem. Lett.* **1998**, *8*, 1665–1668.
[39] S. Sano, K. Hayashi, T. Miwa, T. Ishii, M. Fujii, H. Mima, Y. Nagao, *Tetrahedron Lett.* **1998**, *39*, 5571–5574.

[40] a) W. H. Kruizinga, J. Bolster, R. M. Kellogg, J. Kamphuis, W. H. J. Boesten, E. M. Meijer, H. E. Schoemaker, *J. Org. Chem.* **1988**, *53*, 1826–1827; b) J. J. Lalonde, D. E. Bergbreiter, C.-H. Wong, *J. Org. Chem.* **1988**, *53*, 2323–2327; c) J. Kamphuis, W. H. J. Boesten, B. Kaptein, H. F. M. Hermes, T. Sonke, Q. B. Broxterman, W. J. J. van den Tweel, H. E. Schoemaker, in Chirality in Industry, Eds. A. N. Collins, G. N. Sheldrake, J. Crosby, Wiley, 1992, 187–208; d) W. Liu, P. Ray, S. A. Benezra, *J. Chem. Soc., Perkin Trans. 1*, **1995**, 553–559; e) B. Westermann, I. Gedrath, *Synlett* **1996**, 665–666.

[41] a) D. Obrecht, U. Bohdal, C. Broger, D. Bur, C. Lehmann, R. Ruffieux, P. Schönholzer, C. Spiegler, K. Müller, *Helv. Chim. Acta* **1995**, *78*, 563–580; b) D. Obrecht, C. Abrecht, M. Altorfer, U. Bohdal, A. Grieder, M. Kleber, P. Pfyffer, K. Müller, *Helv. Chim. Acta* **1996**, *79*, 1315–1337.

[42] F. A. Davis, H. Liu, G. V. Reddy, *Tetrahedron Lett.* **1996**, *37*, 5473–5476.

[43] C. B. Bucher, A. Linden, H. Heimgartner, *Helv. Chim. Acta* **1995**, *78*, 935–946.

[44] a) A. Avenoza, J. H. Busto, C. Cativiela, J. M. Peregrina, F. Rodríguez, *Tetrahedron* **1998**, *54*, 11659–11674; b) A. Avenoza, P. J. Campos, C. Cativiela, J. M. Peregrina, M. A. Rodríguez, *Tetrahedron* **1999**, *55*, 1399–1406.

[45] E. Keller, N. Veldman, A. L. Spek, B. L. Feringa, *Tetrahedron: Asymmetry* **1997**, *8*, 3403–3413.

[46] Y. N. Belokon, K. A. Kochetkov, T. D. Churkina, N. S. Ikonnikov, A. A. Chesnokov, O. V. Larionov, V. S. Parmár, R. Kumar, H. B. Kagan, *Tetrahedron: Asymmetry* **1998**, *9*, 851–857.

[47] a) E. J. Corey, F. Xu, M. C. Noe, *J. Am. Chem. Soc.* **1997**, *119*, 12414–12415; b) E. J. Corey, M. C. Noe, F. Xu, *Tetrahedron Lett.* **1998**, *39*, 5347–5350.

[48] B. M. Trost, X. Ariza, *Angew. Chem.* **1997**, *109*, 2749–2751; *Angew. Chem. Int. Ed.* **1997**, *36*, 2635–2637.

New Sequential Reactions with Single Electron Transferring Agents

Troels Skrydstrup

Department of Chemistry, University of Aarhus, Denmark

The construction of complex organic compounds via sequential transformations is by far the most impressive and efficient synthetic strategy to be reported, clearly demonstrating the imaginative and creative abilities of the synthetic chemist. The advantages in the use of such tandem reactions are many and have been thoroughly discussed by Tietze and Beifuss in a review on the subject in *Angew. Chem.* from 1993 [1]. Now through the work of the groups of Molander and Murphy three newly discovered sequential reactions may be added to this growing list of transformations making use of single-electron-transferring agents such as samarium diiodide and tetrathiofulvalene. These new sequential reactions provide an effective and elegant route to a series of polycyclic compounds.

Of the plethora of contemporary organic synthetic reagents now at our disposal, perhaps the most remarkable to stand out in this last decade is the divalent lanthanide reagent, samarium diiodide [2]. This unique and polyvalent reducing agent has been applied to a multitude of important individual synthetic transformations which are generally associated with high levels of stereochemical control. The secret to the success of this one-electron-donating agent lies in its intermediate reducing potential, such that one of either a radical or carbanionic reaction may be efficiently performed through the judicious choice of both reactants and conditions. As has been demonstrated by several groups, these two reaction types may furthermore be combined in a tandem process, known as a radical/polar sequence [3]. That is, an initial SmI_2-induced radical cyclization may be succeeded by a second electron transfer from an additional equivalent of SmI_2 affording a corresponding organosamarium with concomitant reactions such as β-elimination and intermolecular nucleophilic addition or substitution [3]. Radical cyclizations are thereby terminated by the addition of functionalization rather than its loss, as is characteristic for normal tin hydride-ring forming chemistry.

In an extension of this chemistry, Molander and Harris have recently demonstrated that SmI_2 may additionally be employed in either one of an anionic/anionic or an anionic/radical sequential reaction for the stereocontrolled construction of complex bicyclic and tricyclic ring systems, which enhances even further the versatility and possibilities of this reagent [4, 5]. The Molander team, having greatly contributed to the popularity of this reagent over the last de-

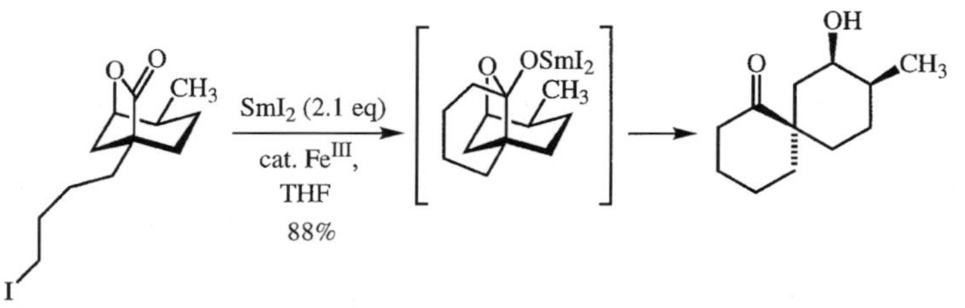

SmI_2 (2.1 eq), cat. Fe^{III}, THF, 88%

Scheme 1. SmI_2-induced nucleophilic acyl substitution.

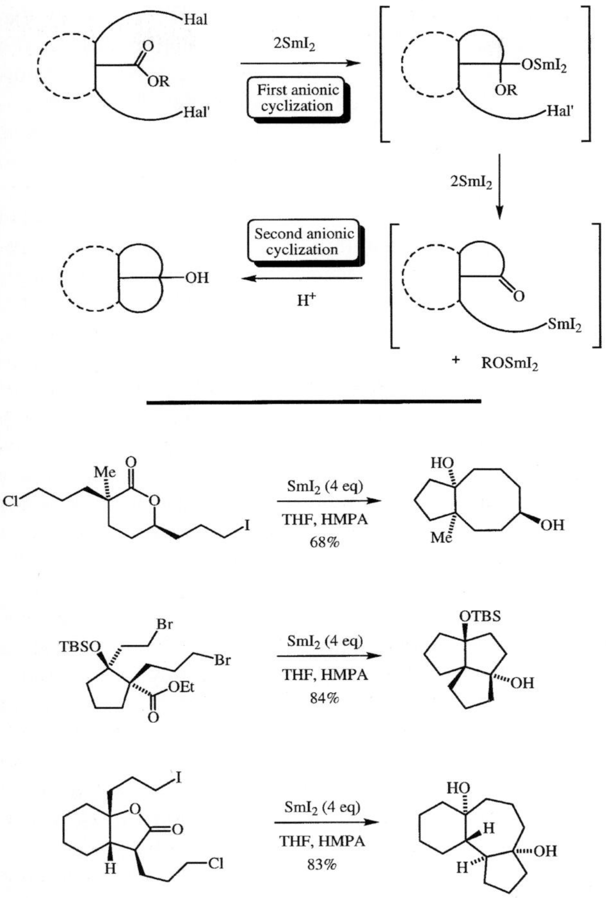

Scheme 2. Several examples of sequential anionic/anionic reactions promoted by SmI_2.

cade, had previously observed that nucleophilic acyl substitution reactions proceed exceptionally well to afford cyclic or acyclic ketones, as exemplified in Scheme 1 [6]. Bearing in mind the ability of SmI_2 to promote Barbier-type cyclization reactions with alkyl halides, it seemed possible that these two transformations might be combined in a sequential manner. This was exactly the case when simple-to-prepare substrates as illustrated in Scheme 2 were subjected to four equivalents of SmI_2 rather than the usual two, leading to a diverse array of cyclic products depending on the substitution pattern chosen [4]. The mechanism proposed for these reactions involves a tetrahedral intermediate which is formed upon nucleophilic addition of an organosamarium generated upon SmI_2 addition. Liberation of the ketone is then followed by an additional intramolecular attack upon reduction of the second alkyl halide side chain. Crucial to the success of some of these reactions is the sequenced formation of the organosamarium intermediates, controlled by the different reduction rates of alkyl halides displayed by SmI_2 ($k_{red(R\text{-}I)} > k_{red(R\text{-}Cl)}$). In the other cases employing the same halide in the two side chains, it is apparently their length which determines the sequence of attack to the carbonyl functionality. A wide variety of ring systems are made available by this methodology including seven- and eight-membered rings. Highly impressive are the efficient transformations of the last two examples in Scheme 2 to the corresponding tricyclic systems, related to certain naturally occurring sesqui- and sesterterpenes, respectively.

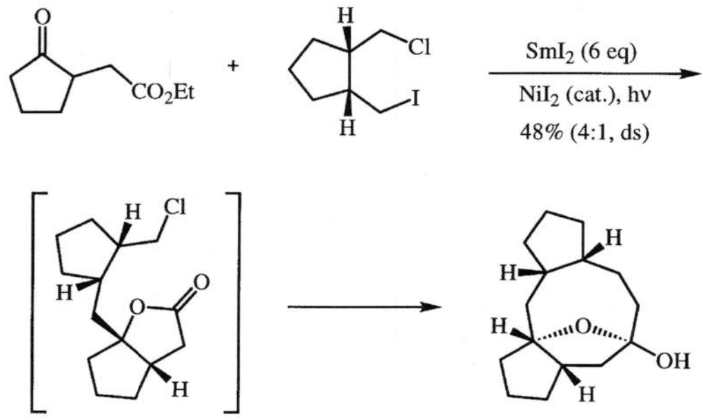

Scheme 3. An example of an intermolecular sequential anionic/anionic reaction promoted by SmI_2.

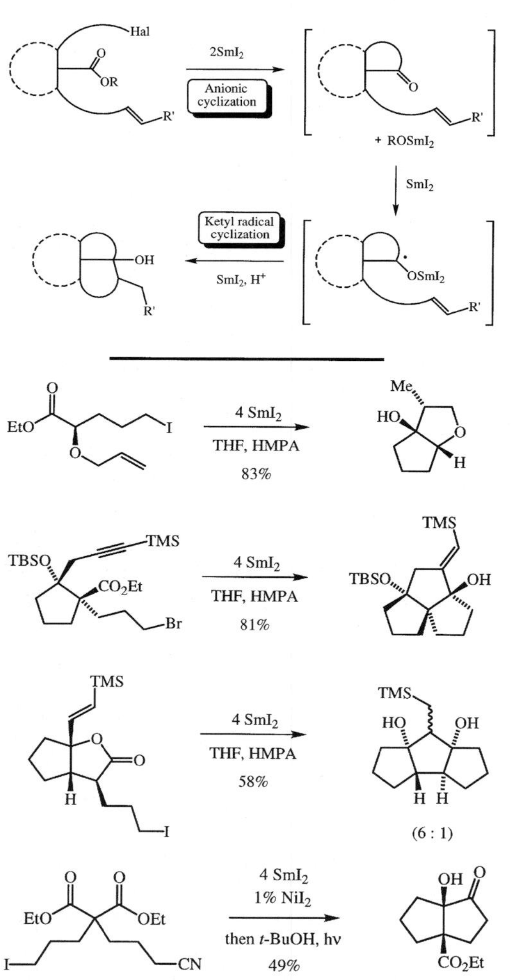

Scheme 4. Several examples of sequential anionic/radical reactions promoted by SmI_2.

An interesting intermolecular version of this reaction has likewise been put forward for the preparation of seven-, eight-, and nine-membered carbocycle, as illustrated with a sole example in Scheme 3 [7]. In contrast to the above, these reactions begin with a carbonyl addition reaction of chloroiodoalkanes to cyclic or acyclic keto esters leading to the formation of an intermediate lactone. An intramolecular nucleophilic acyl substitution then terminates the sequence. The example in Scheme 3 represents a simple method for the construction of the 5: 8:5 tricyclic ring system.

In another series of papers, the Molander team demonstrate the efficiency of the SmI_2-promoted anionic/radical sequence as a viable approach to similar ring systems [5]. Basically the same starting materials are exploited again, with the minor but important modification that one of the alkyl halide side chains has been replaced with that of an alkenyl. The logic behind these examples is that in the absence of a second alkyl halide reduction step, the intermediate ketone formed after acyl substitution is reduced to its corresponding ketyl radical with subsequent cyclization onto the unsaturation. As shown by the substrates in Scheme 4, this approach also provides an efficient and facile access to numerous substituted carbocycles, including heterocycles, and in general with high diastereoselective display. However, the best yields of this bicyclization process were noted with substrates possessing activating groups on the alkene. A slight variation of this class of tandem reactions was reported employing a cyanide group as the ketyl radical trap rather than an unsaturated alkane, as shown in the last example in Scheme 4 [8]. In this way, access to bicyclic hydroxy ketones is allowed via one step.

A nice extension of this chemistry to sequential anionic/radical/anionic sequence was also provided [5, 9]. Normally after acyl addition and radical cyclization onto a C = C bond, the newly formed carbon radical is reduced to an organosamarium intermediate which is subsequently protonated. However, as depicted in Scheme 5, this organosamarium may be trapped in the presence of a ketone substrate, thus terminating this three-step process. In another demonstration of how such anions may be further exploited, substrates possessing vinyl ethers as the radical acceptor were found to under-

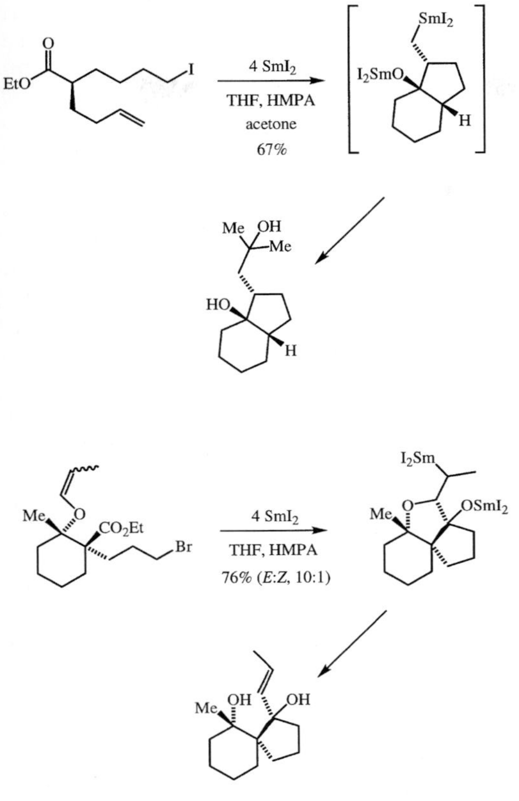

Scheme 5. Examples of sequential anionic/radical/anionic reactions promoted by SmI_2.

go a final β-elimination step resulting in the transfer of an alkenyl side chain. These examples certainly suggest the possibility of extending these SmI_2-induced sequential reactions even further.

Another impressive array of radical/polar sequential reactions have likewise been in the developmental stage employing an alternative one-electron-transferring agent, which previously has not been exploited in organic synthesis, namely tetrathiafulvalene (TTF). In contrast to the traditional radical/polar reactions induced by SmI_2 [3], Murphy and collaborators have demonstrated that radical cyclizations promoted by TTF may be terminated by S_N1–type nucleophilic substitution at the new exocyclic center [10]. The principles of this sequence are illustrated in Scheme 6. After single-electron reduction of a suitable substrate **A** by TTF and subsequent radical cyclization as with the SmI_2-promoted reactions, the newly formed carbon radical center is formally oxidized by combination with the $TTF^{+\cdot}$ radical cation and formation of the corresponding sulfonium ion, rather than being reduced by a second TTF, as is the case when employing divalent samarium. Substitution at the sulfonium ion-bearing carbon center with external nucleophiles (solvents) such as H_2O, MeOH or CH_3CN were found to follow S_N1-type kinetics. These reactions have so far only been applied to aryl diazonium salts and are hence restricted to aryl-type radical cyclizations. Nevertheless an intriguing facet of this chemistry is the ability of this radical/polar sequence to be carried out in the presence of a catalytic amount of TTF, as the one-electron transferring agent is regenerated after nucleophilic attack. This represents a clear distinction from the samarium(II)-induced chemistry, where two equivalents are necessary for the radical/anionic process. Further

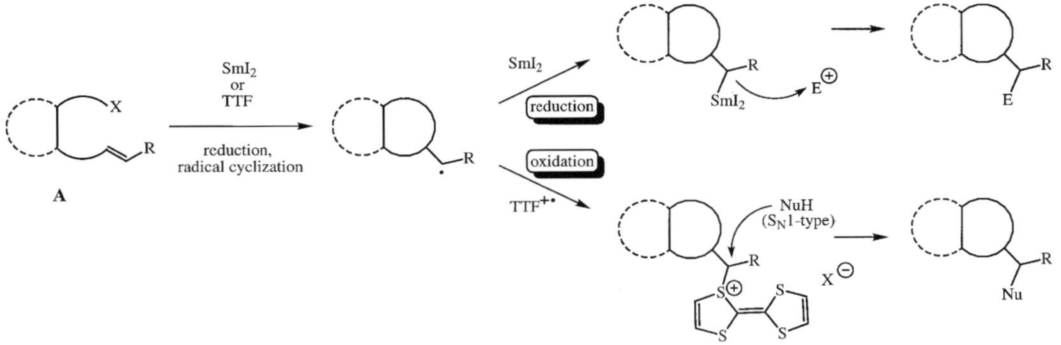

Scheme 6. Comparison of SmI_2- and TTF-induced radical/polar sequential reactions.

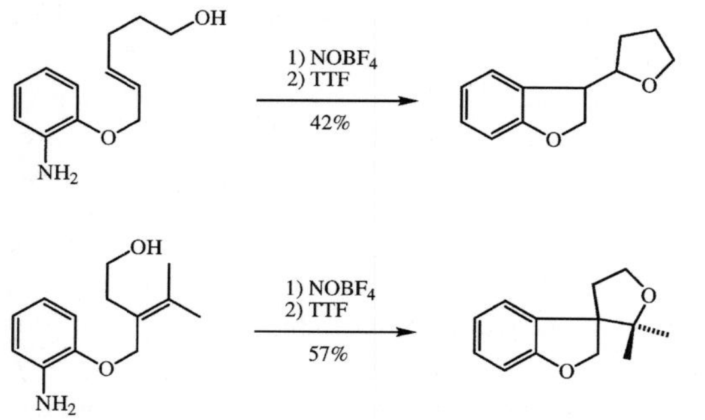

Scheme 7. Examples of sequential radical/cationic reactions promoted by TTF with internal nucleophiles.

studies along this line are necessary, as the best yields, as well as appreciable reaction times, were obtained when one equivalent of TTF was employed.

A more interesting application of this chemistry involves substrates containing internal nucleophiles allowing the construction of more complex ring systems such as the first two examples shown in Scheme 7 [11]. In these cases, radical cyclization is terminated by a substitution reaction with the appending primary hydroxyl group.

A cunning demonstration of this approach was provided by the Murphy group in their model studies for the preparation of the ABCE tetracyclic substructure of the *Aspidosperma* alkaloids, such as aspidospermidine and strychnine [12]. Diazotization of precursors **B** and **C** and their subsequent reaction with TTF were performed *in situ* and furnished the desired tetracyclic ring systems in good yields (Scheme 8). Most importantly, complete stereochemical control at the three contiguous stereocenters was observed affording the all *cis* product and providing further evidence for the S_N1 substitution reaction with the sulfonium ion intermediate. This radical/polar sequential reaction therefore nicely complements the tandem radical cyclization approach reported by the same group [13].

However, in order to prepare the E ring of aspidospermidine containing the correct heteroatom, it was found that the use of internal nitrogen nucleophiles was not efficient for such TTF-mediated cyclizations. To overcome this problem, these reactions were performed in moist acetone in order to introduce a hydroxyl group at the C ring (Scheme 9) [14]. Manipulations of this compound then allowed for the introduction of both the E and D rings, thus completing the total synthesis of this natural product. This example represents the first application of this new and powerful ring-forming methodology for the efficient synthesis of a complex natural product. Other examples will no doubt follow.

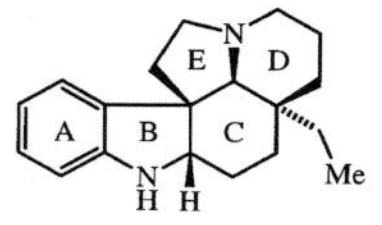

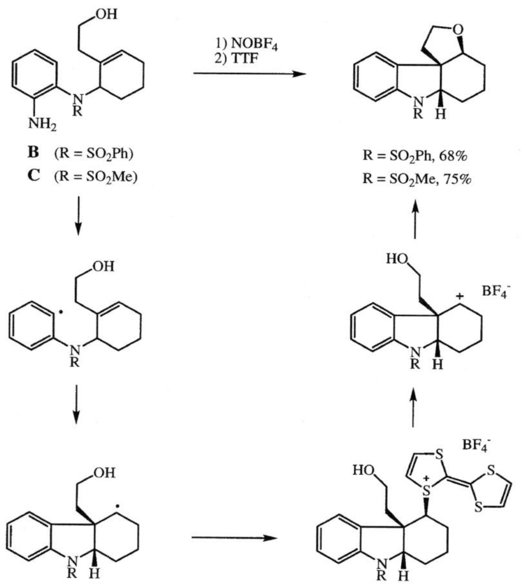

Scheme 8. Synthesis of the ABCE tetracyclic structure of aspidospermidine.

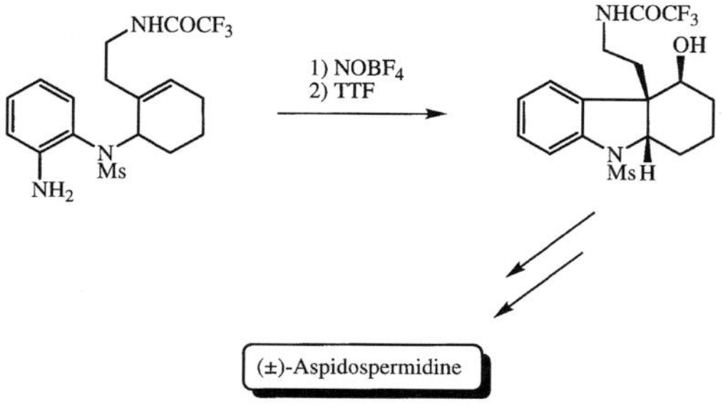

Scheme 9. Application of the radical/polar crossover reaction to the total synthesis of aspidospermidine.

References

[1] L.F. Tietze, U. Beifuss, *Angew. Chem.* **1993**, *105*, 137; *Angew. Chem., Int. Ed. Engl.* **1993**, *32*, 131.

[2] For several reviews, see: a) A. Krief, A.-M. Laval, *Chem. Rev.* **1999**, *99*, 745; b) G.A. Molander, C.R. Harris, *Tetrahedron* **1998**, *54*, 3321; c) G.A. Molander, C.R. Harris, *Chem. Rev.* **1996**, *96*, 307; d) G.A. Molander in *Organic Reactions, Vol. 46*, (Ed.: L.A. Paquette), Wiley, New York, **1994**, p. 211; e) G.A. Molander, *Chem. Rev.* **1992**, *92*, 29; f) D.P. Curran, T.L. Fevig, C.P. Jaspersen, M.L. Totleben, *Synlett* **1992**, 943; g) H.B. Kagan, J.L. Namy, *Tetrahedron* **1986**, *42*, 6573.

[3] For some applications of this sequential reactions, see: a) G.A. Molander, L.S. Harring, *J. Org. Chem.* **1990**, *55*, 6171; b) D.P. Curran, T.L. Fevig, M.L. Totleben, *Synlett* **1990**, 773; c) D.P. Curran, M.L. Totleben, *J. Am. Chem. Soc.* **1992**, *114*, 6050. d) M.L. Totleben, D.P. Curran, P. Wipf, *J. Org. Chem.* **1992**, *57*, 1740; e) G.A. Molander, J.A. McKie, *ibid.* **1992**, *57*, 3132; f) G.A. Molander, J.A. McKie, *ibid.* **1994**, *59*, 3186; g) G.A. Molander, J.A. McKie, *ibid.* **1995**, *60*, 872; h) G.A. Molander, C. Kenny, *ibid.* **1991**, *56*, 1439; i) G.A. Molander, S.R. Shakya, *ibid.* **1996**, *61*, 5885; j) G.A. Molander, J.C. McWilliams, *J. Am. Chem. Soc.* **1997**, *119*, 1265; k) E.J. Enholm, A. Trivellas, *Tetrahedron. Lett.* **1994**, *35*, 1627; l) Z. Zhou, S.M. Bennett, *Tetrahedron. Lett.* **1997**, *38*, 1153; m) D.P. Curran, B. Yoo, *Tetrahedron. Lett.* **1992**, *33*, 6931; n) G.A. Molander, C.R. Harris, *J. Org. Chem.* **1998**, *63*, 812; o) M. Sasaki, J. Collin, H.B. Kagan, *Tetrahedron. Lett.* **1988**, *29*, 6105.

[4] G.A. Molander, C.R. Harris, *J. Am. Chem. Soc.* **1995**, *117*, 3705.

[5] a) G.A. Molander, C.R. Harris, *J. Am. Chem. Soc.* **1996**, *118*, 4059; b) G.A. Molander, C.R. Harris, *J. Org. Chem.* **1997**, *62*, 2944.

[6] a) G.A. Molander, J.A. McKie, *J. Org. Chem.* **1993**, *58*, 7216; b) G.A. Molander, S.R. Shakya, *ibid.* **1994**, *59*, 3445.

[7] G.A. Molander, C. Alonso-Alija, *J. Org. Chem.* **1998**, *63*, 4366.

[8] G.A. Molander, C.N. Wolfe, *J. Org. Chem.* **1998**, *63*, 9031.

[9] G.A. Molander, C.R. Harris, *J. Org. Chem.* **1998**, *63*, 4374.

[10] a) C. Lampard, J.A. Murphy, N. Lewis, *J. Chem. Soc., Chem. Commun.* **1993**, 295; b) R.J. Fletcher, C. Lampard, J.A. Murphy, N. Lewis, *J. Chem. Soc. Perkin Trans. 1* **1995**, 623; c) J.A. Murphy, M. Kizil, C. Lampard, *Tetrahedron. Lett.* **1996**, *37*, 2511; d) T. Koizumi, N. Bashir, J.A. Murphy, *Tetrahedron. Lett.* **1997**, *38*, 7635; e) N. Bashir, O. Callaghan, J.A. Murphy, T. Ravishanker, S.J. Roome, *Tetrahedron. Lett.* **1997**, *38*, 6255; f) O. Callaghan, X. Franck, J.A. Murphy, *Chem. Commun.* **1997**, 1923; g) J.A. Murphy, F. Rasheed, S, Gastaldi, T. Ravishanker, N. Lewis, *J. Chem. Soc. Perkin Trans. 1* **1997**, 1549.

[11] a) J.A. Murphy, F. Rasheed, S.J. Roome, N. Lewis, *J. Chem. Soc., Chem. Commun.* **1996**, 737; b) J.A. Murphy, F. Rasheed, S.J. Roome, K.A. Scott, N. Lewis, *J. Chem. Soc. Perkin Trans. 1* **1998**, 2331.

[12] a) R.J. Fletcher, D.E. Hibbs, M. Hursthouse, C. Lampard, J.A. Murphy, S.J. Roome, *J. Chem. Soc., Chem. Commun.* **1996**, 739; b) R.J. Fletcher, M. Kizil, C. Lampard, J.A. Murphy, S.J. Roome, *J. Chem. Soc. Perkin Trans. 1* **1998**, 2341.

[13] M. Kizil, J.A. Murphy, *J. Chem. Soc., Chem. Commun.* **1995**, 1409; This work has been reviewed in a recent Highlight: U. Koert, *Angew. Chem.* **1996**, *108*, 441; *Angew. Chem., Int. Ed. Engl.* **1996**, *35*, 405.

[14] O. Callaghan, C. Lampard, A.R. Kennedy, J.A. Murphy, *Tetrahedron. Lett.* **1999**, *40*, 161.

Deracemisation by Enantiodifferentiating Inversion in 1,3- and 1,2-Diols

Anthony P. Davis

Department of Chemistry, Trinity College, Dublin, Ireland

The conversion of achiral molecules into enantiomerically pure chiral compounds [1] is generally accomplished using one of two types of method. On the one hand, the application of a chiral reagent, catalyst or auxiliary may be used to effect an enantioselective synthesis, in which asymmetric induction desymmetrises the starting material to give molecules which are chiral in predominantly one sense. On the other, the use of simple achiral reagents gives a racemic mixture of chiral molecules, which may then be resolved by interaction with a chiral agent. While the first of these methods may give, in principle, 100 % conversion of achiral educt into enantiomerically pure product, the second is limited to 50 % conversion in most cases. Only certain circumstances allow "deracemisation", whereby a racemate may be converted to a single enantiomer in up to 100 % yield. One possibility is "dynamic resolution", wherein equilibration of the starting enantiomers permits the enantiodifferentiating reagent, catalyst or complexing agent to deliver a quantitative yield of enantiomerically pure product (e.g. Scheme 1) [2]. An alternative involves destruction of the asymmetric centre in the racemate, followed by enantioselective regeneration [3]; this is, however, only trivially different from conventional enantioselective synthesis. More distinctive, conceptually, are methods for deracemisation in which enantiomers are differentiated in an initial step and then carried along separate paths to the same enantiopure final product. Such methods may be unwieldy if separations are required, but elegant and practical if the material can be kept together throughout the procedure.

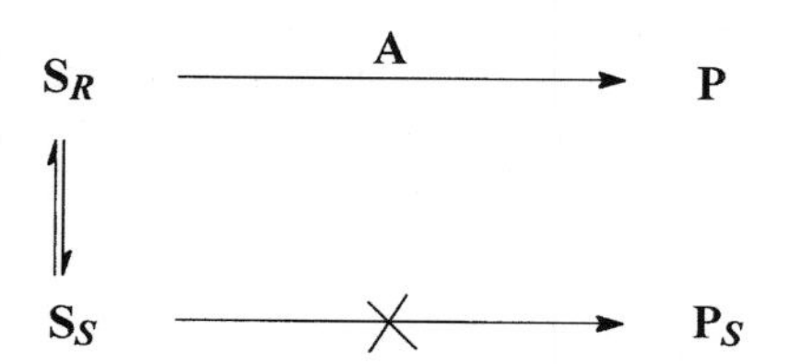

Scheme 1. Dynamic resolution of an equilibrating mixture of enantiomers. $\mathbf{S}_R$ and $\mathbf{S}_S$ represent substrate enantiomers, **A** represents the resolving agent (reagent, catalyst or complexer) and $\mathbf{P}_R$ the enantiomerically pure product.

1a **1b** **2**

While most of these "divergent-convergent" deracemisations rely on biocatalysis [2a, b], there is no reason why chemical methodology should not be used, provided selectivities are sufficiently high. An impressive example was provided by Harada, Oku and co-workers [4], in research leading up to a landmark paper in 1995 [4c]. The methodology developed by the Japanese group was based on the chemistry of menthone-derived spiroketals **1** [4a]. The [6, 6] bicyclic framework of this system sets up interactions and dispositions which rigorously control the stereochemistry of many of its reactions. The major source of stereochemical bias is the isopropyl group, which determines the conformation of the menthone-derived ring (see **1a**) and also fixes that of the 1,3-dioxane ring (see **1b**) by prohibiting **2**. Important consequences are that (a) substituents at positions 2 and 4 of the dioxane must take just one of the two possible

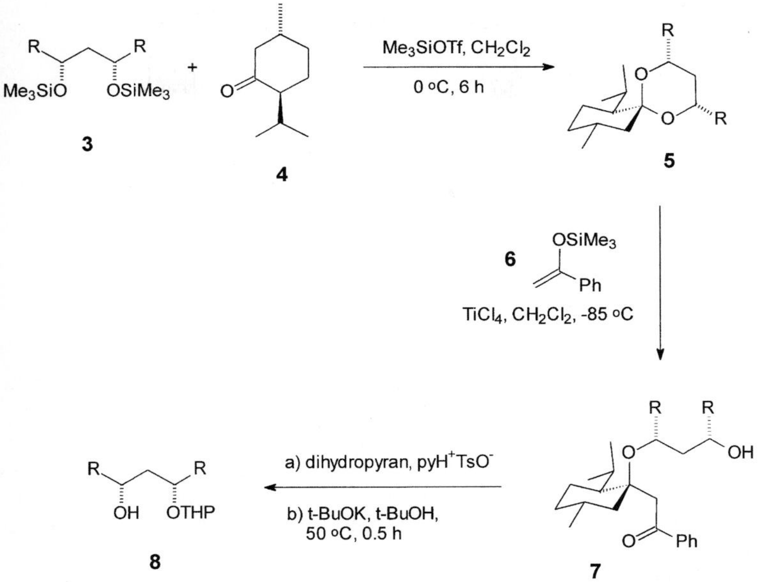

Scheme 2

orientations (marked eq. in **1b**) to avoid 1,3-diaxial interactions with a menthone ring carbon, and (b) the dioxane oxygens occupy very different steric environments, the equatorial oxygen (see **1a**) being far more accessible to an approaching electrophile.

The well-defined preferences of **1** offer various opportunities for the development of stereoselective methodology, not least the desymmetrisation of *meso*-1,3-diols which proceeds as shown in Scheme 2 [5]. Ketalisation of TMS ethers **3** with L-menthone **4** under thermodynamic control leads exclusively to dioxanes **5**, in which both groups R are equatorial. When **5** are treated with silyl enol ether **6** and $TiCl_4$ at low temperature, the less-hindered equatorial ketal oxygen is attacked preferentially by the Lewis acid. The resulting oxonium ion is quenched selectively from the equatorial direction [6] to give "monoprotected diol" **7** as a single stereoisomer. Derivatisation of the free hydroxyl group followed by base-induced removal of the chiral auxiliary can give the alternative protection pattern as in **8**. Notably, this sequence can be carried out on *meso*-1,3-diols even when contaminated by the D,L-diastereomer. The latter can only form "eq.-ax." acetals such as **9** in the initial step, and therefore prefer to remain unreacted.

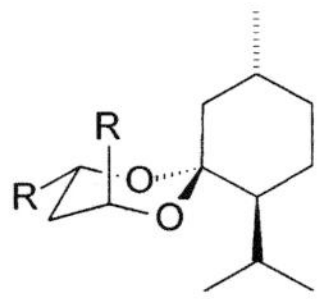

This procedure is elegant and useful, but conventional in that, like most enantioselective syntheses, it involves an achiral starting material desymmetrised by a chiral reagent. However, as shown in Scheme 3, the methodology was extended in a remarkable and unusual direction through the application of similar chemistry to *racemic* starting materials [7]. To allow a generalised discussion of this and subsequent procedures, it is helpful to introduce special stereochemical descriptors for use with 1,3-diols and their de-

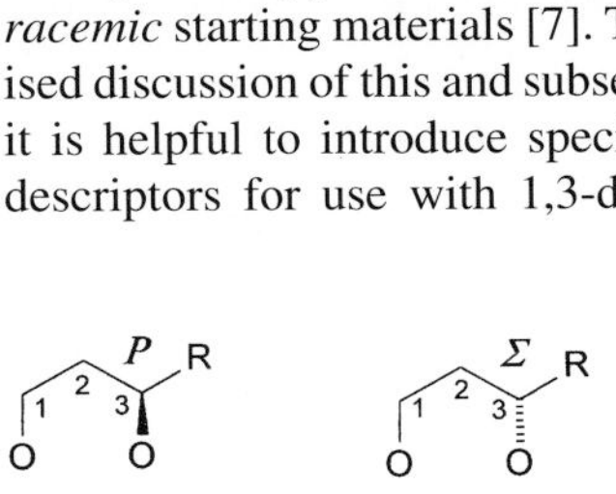

Figure. 1. Σ and P centres in 1,3-dioxygenated molecules, as defined for the purposes of this article. External substituent R takes lower priority than central carbon C(2).

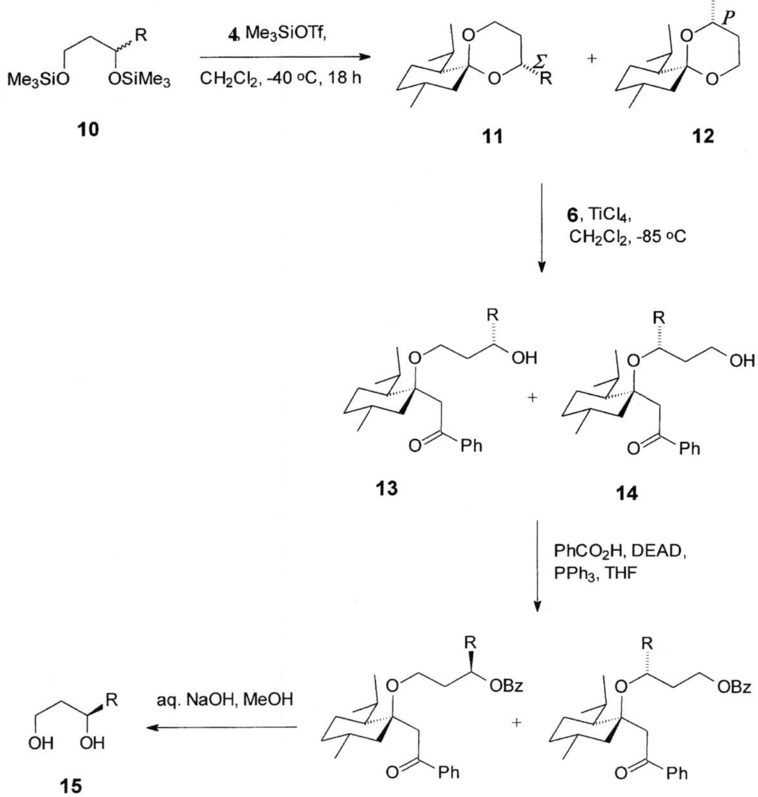

Scheme 3

rivatives. P and Σ asymmetric centres are defined as equivalent to R and S with the proviso that the central carbon of the 1,3-dioxygenated system is assigned higher priority than all external substituents (see Fig. 1) [8]. Turning to Scheme 3, when TMS ethers **10** are derivatised with *l*-menthone diaxial interactions can be avoided only if Σ-**10** reacts to give **11**, while P-**10** reacts to give **12**.

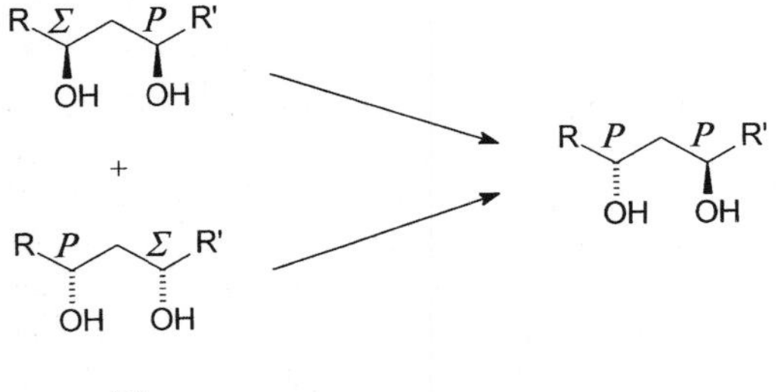

Scheme 4

Accordingly, **11** and **12** are the only stereoisomers detectable [7]. Now if this mixture is treated with enolsilane **6**/$TiCl_4$, the course of reaction is again dictated by the conformation of the bicyclic system. Only the equatorial C–O bonds are cleaved, so that **11** leads to **13** and **12** leads to **14** [4b]. The diol enantiomers have been differentiated, such that the secondary hydroxyl groups in Σ and P isomers are respectively exposed and protected. A Mitsunobu reaction inverts the former, while performing a "stereoneutral" displacement in **14**. On base-induced deprotection, the products (now regioisomers) converge on P-diol **15**, which is formed in $\geq$ 95 % enantiomeric excess.

The essential basis of this method is that the equatorial oxygen in an L-menthone derivative must be attached to a Σ centre while the axial oxygen must be connected to a P centre. As the equatorial C–O is cleaved in the second

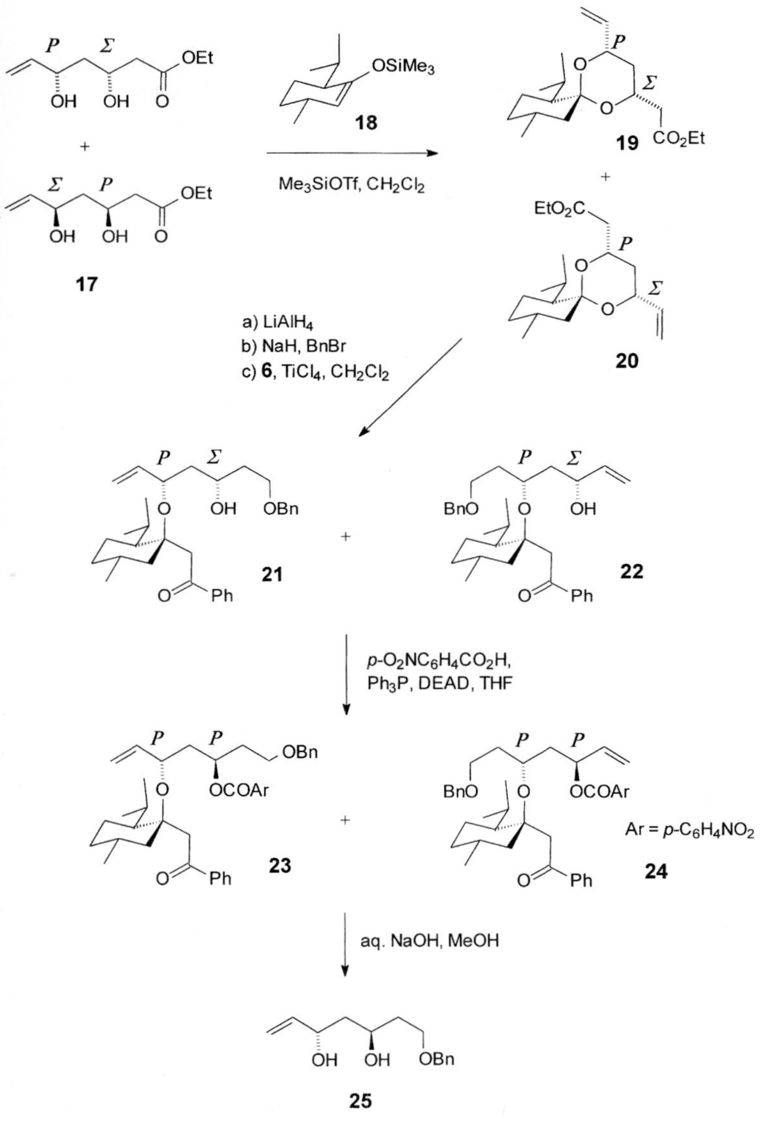

Scheme 5

step, the Σ centre is selectively exposed and converted to P, resulting in deracemisation. This analysis encourages consideration of racemates **16**, in which both Σ and P centres are present. If the Σ centres in both enantiomers could be converted to P, deracemisation would again be possible (Scheme 4).

A remarkable *tour de force* by the Japanese group showed that this process could indeed be effected [4c]. In an initial demonstration (Scheme 5), diol enantiomers **17** were converted directly to menthonides **19** and **20** by treatment with silyl enol ether **18** and Me_3SiOTf. After conversion of the esters to benzyloxy groups, the Σ centres were revealed using **6**/$TiCl_4$ to give **21** and **22**. Mitsunobu inversion of this mixture gave **23** + **24**, which on deprotection gave enantiopure (> 95 % ee) diol **25**.

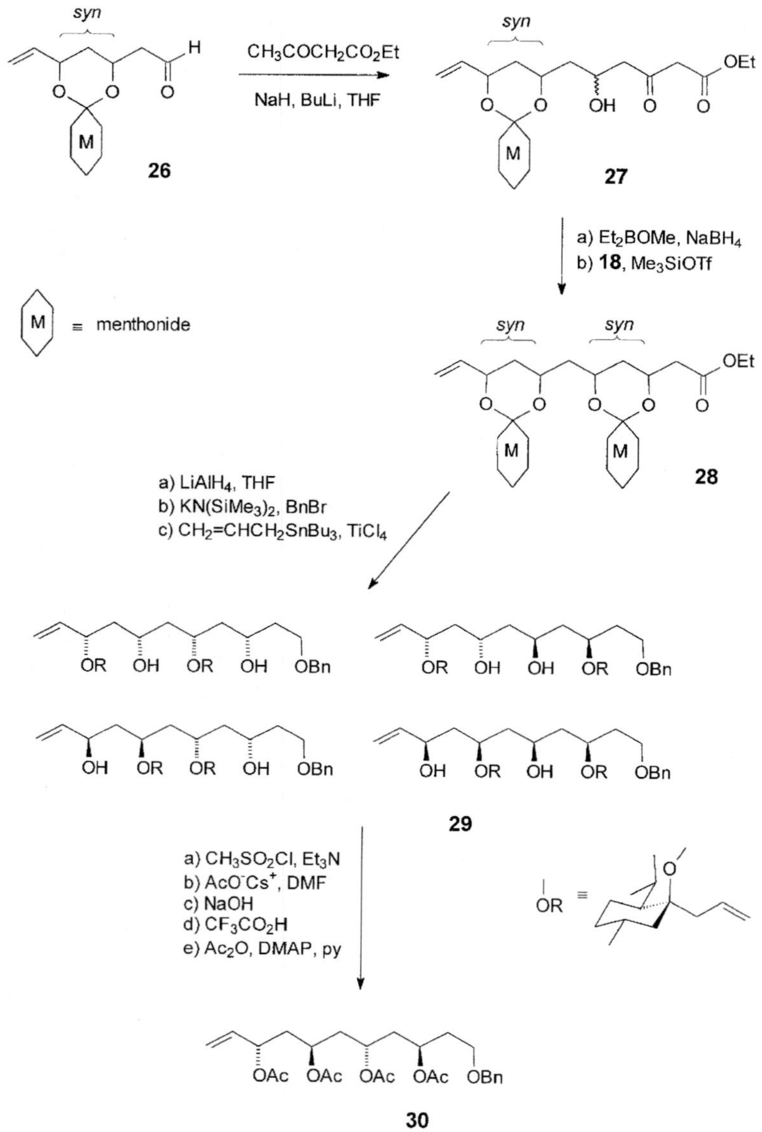

Scheme 6

The scope of the strategy was then extended in spectacular fashion (Scheme 6). The mixture of **19** and **20** was transformed into the corresponding aldehydes **26** through reduction/oxidation of the ester groups. Treatment with the dianion derived from ethyl acetoacetate then led to hydroxyketones **27**. Chelation-controlled, *syn*-selective reduction [9] of this mixture followed by menthonide formation gave **28** as four diastereomers, each containing two Σ and two P centres. After ester reduction and *O*-benzylation, the stage was set for Σ-selective unmasking, inversion and convergence. The usual conditions employing **6** were unsuccessful, but the combination of $TiCl_4$ with allyltributylstannane produced the desired effect. Equatorial-selective ring-opening yielded the mixture of compounds **29**, in which all the hydroxyl groups were derived

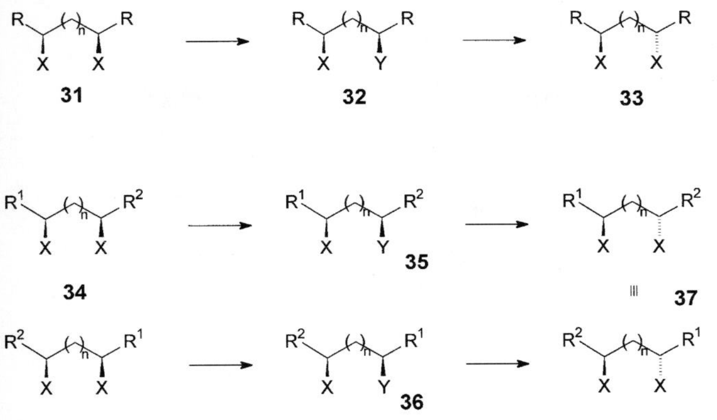

31 → 32 → 33

34 → 35 → 37

36

Scheme 7

from Σ centres in **28**. Hydroxyl inversion, deprotection (including acidic cleavage of the menthyl-derived tertiary ethers) and reacetylation yielded **30** as a single diastereomer. The enantiomeric purity of **30** was not confirmed, but may be inferred with confidence from the diastereoselectivity and the previous record of the methodology.

Intellectually this methodology is elegant and distinctive, especially in its uninhibited use of complex mixtures as intermediates. Judgement on its practicality is best delayed, but it undoubtedly contributed to an area of real importance in organic synthesis. The 1,3-diol substructures generated are relevant to important classes of natural products [10] most notably the polyene macrolide antibiotics [11]. Indeed, **30** contains a stereochemical pattern present in several such targets, including rotaxicin and mycoticin A and B. From another perspective, the work further demonstrated the value of organo silicon- and orga-

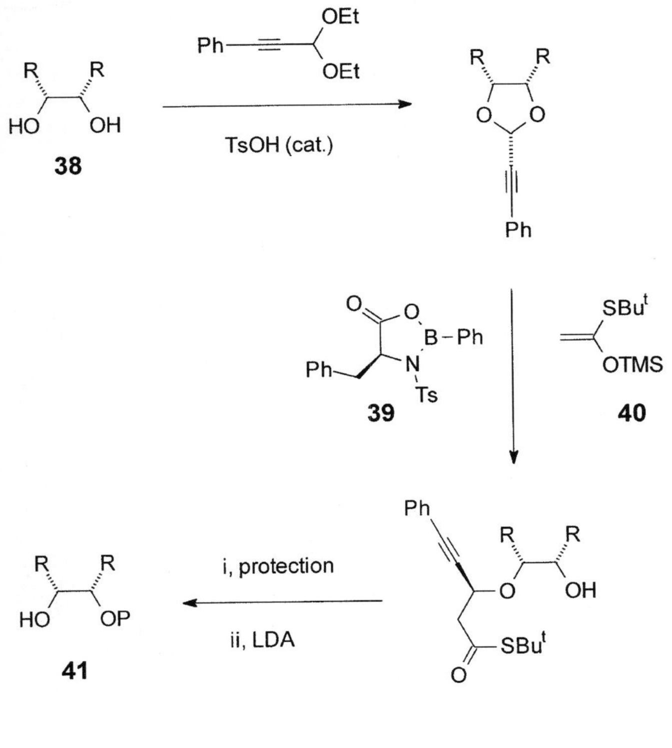

38 → TsOH (cat.)

39 40

i, protection
ii, LDA

41

Scheme 8

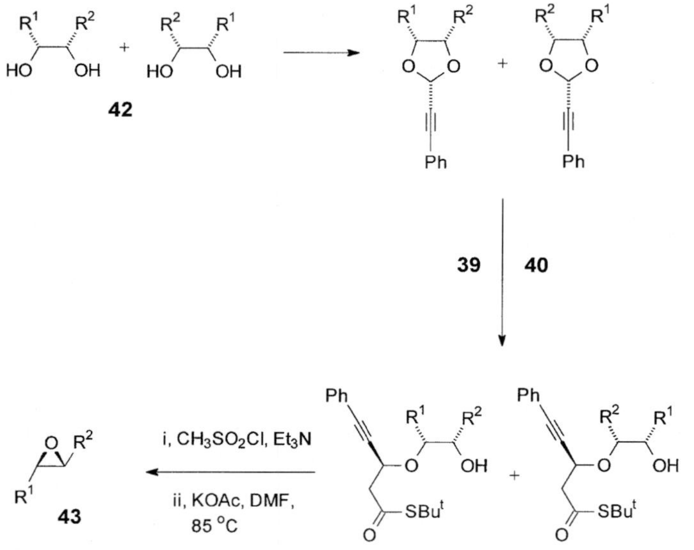

Scheme 9

no- tin-based nucleophiles in organic synthesis [12]. Their comparatively low reactivity often generates the need for catalysis, providing a dimension of control which may be exploited to various stereochemical ends [13]. The application in this case was delightfully oblique (C–C bond formation is not usually employed to elaborate a protecting group in a synthetic intermediate), and nicely illustrated the versatility of these widely-used reagents.

The progression from desymmetrisation (Scheme 2) to deracemisation (Schemes 3–6) leads to a useful general insight. The desymmetrisation of σ-symmetric difunctional compounds (**31** $\rightarrow$ **32**, Scheme 7) is a common approach to asymmetric synthesis [14], and there may be various circumstances where the regeneration of functional symmetry, but with stereoinversion, is also possible (**32** $\rightarrow$ **33**). If **31** is now allowed to "mutate" into the asymmetrical **34**, present as a racemate, it is likely that the desymmetrising reaction will yield **35** + **36**, converging on **37** after the final step. Any desymmetrisation of σ-symmetric (e.g. *meso*) diols may thus be extended, potentially, to deracemisation, and other substrates may lend themselves to analogous sequences.

Harada, Oku et al. have themselves illustrated this point by developing a second deracemisation, this time of 1,2-diols. The desymmetrisation method shown in Scheme 8 relies on chiral reagent **39**, rather than a covalently-bonded chiral auxiliary, and proceeds to give a range of asymmetrically protected products **41** in $>$ 70 % yields and $\geq$ 95 % *ee* [15]. Having been applied to meso-diols **38**, the methodology was then extended to racemates **42** (Scheme 9). The necessary inversion was achieved *via* an intramolecular displacement, to give epoxides **43** in, for most cases, $>$ 90 % *ee* [16].

Reference

[1] General reference: G. Helmchen, R. W. Hoffmann, J. Mulzer, E. Schaumann (eds.), *Houben-Weyl Methods of Organic Chemistry, Vol. E 21, Stereoselective Synthesis*, Thieme, Stuttgart, **1995**. See also: R. Noyori, *Asymmetric Catalysis in Organic Synthesis*, John Wiley, Chichester, **1994**; M. Nógrádi, *Stereoselective Synthesis*, VCH, Weinheim, **1995**; R. S. Atkinson, *Stereoselective Synthesis*, John Wiley, Chichester, **1995**; G. Procter, *Asymmetric Synthesis*, OUP, Oxford, **1996**.

[2] For leading references, see: a) U. T. Strauss, U. Felfer, K. Faber, *Tetrahedron-Asym.* **1999**, *10*, 107; b) H. Stecher, K. Faber, *Synthesis* **1997**, 1; c) R. S. Ward, *Tetrahedron-Asym.* **1995**, *6*, 1475; d) M. Kitamura, M. Tokunaga, R. Noyori, *J. Am. Chem. Soc.* **1993**, *115*, 144.

[3] For examples involving deprotonation-protonation, see: C. Fehr, *Angew. Chem.* **1996**, *108*, 2726; *Angew. Chem. Int. Ed. Engl.* **1996**, *35*, 2567.

[4] a) T. Harada, A. Oku, *Synlett* **1994**, 95; b) T. Harada, H. Kurokawa, A. Oku, *Tetrahedron Lett.* **1987**, *28*, 4847; c) T. Harada, T. Shintani, A. Oku, *J. Am. Chem. Soc.* **1995**, *117*, 12346.

[5] T. Harada, K. Sakamoto, Y. Ikemura, A. Oku, *Tetrahedron Lett.* **1988**, *29*, 3097.

[6] T. Harada, T. Hayashiya, I. Wada, N. Iwa-aki, A. Oku, *J. Am. Chem. Soc.* **1987**, *109*, 527.

[7] T. Harada, H. Kurokawa, A. Oku, *Tetrahedron Lett.* **1987**, *28*, 4843.

[8] Note that in polyoxygenated systems a given centre may be both Σ and P, depending on which 1,3-dioxygenated fragment is under consideration.

[9] K.-M. Chen, K. G. Gunderson, G. E. Hardtmann, K. Prasad, O. Repic, M. J. Shapiro, *Chem. Lett.* **1987**, 1923.

[10] T. Oishi, T. Nakata, *Synthesis* **1990**, 635.

[11] S. D. Rychnovsky, *Chem. Rev.* **1995**, *95*, 2021.

[12] E. W. Colvin, *Silicon Reagents in Organic Synthesis.*, Academic Press, New York, 1988; M. Pereyre, J.-P. Quintard, A. Rahm, *Tin in Organic Synthesis.*, Butterworths, London, 1987.

[13] Leading references: D. A. Evans, M. C. Kozlowski, J. A. Murry, C. S. Burgey, K. R. Campos, B. T. Connell, R. J. Staples, *J. Am. Chem. Soc.* **1999**, *121*, 669; H. Groger, E. M. Vogl, M. Shibasaki, *Chem. Eur. J.* **1998**, *4*, 1137; A. P. Davis, S. J. Plunkett, J. E. Muir, *Chem. Commun.* **1998**, 1797; K. Iseki, S. Mizuno, Y. Kuroki, Y. Kobayashi, *Tetrahedron Lett.* **1998**, *39*, 2767; I. Fleming, A. Barbero, D. Walter, *Chem. Rev.* **1997**, *97*, 2063; R. O. Duthaler, A. Hafner, *Angew. Chem.* **1997**, *109*, 43; *Angew. Chem. Int. Ed. Engl.* **1997**, *36*, 43; G. Stork, J. J. LaClair, *J. Am. Chem. Soc.* **1996**, *118*, 247; M. P. Sibi, J. G. Ji, *Angew. Chem.* **1996**, *108*, 198; *Angew. Chem., Int. Ed. Engl.* **1996**, *35*, 190; L. F. Tietze, C. Schünke, *Angew. Chem.* **1995**, *107*, 1901; *Angew. Chem., Int. Ed. Engl.* **1995**, *34*, 1731; L. F. Tietze, K. Schiemann, C. Wegner, *J. Am. Chem. Soc.* **1995**, *117*, 5851; T. Bach, *Angew. Chem.* **1994**, *106*, 433; *Angew. Chem., Int. Ed. Engl.* **1994**, *33*, 417.

[14] For a selection of methods applicable to diols, see H. Fujioka, Y. Nagatomi, N. Kotoku, H. Kitagawa, Y. Kita, *Tetrahedron Lett.* **1998**, *39*, 7309, and references therein.

[15] M. Kinugasa, T. Harada, A. Oku, *J. Am. Chem. Soc.* **1997**, *119*, 9067; M. Kinugasa, T. Harada, A. Oku, *Tetrahedron Lett.* **1998**, *39*, 4529.

[16] T. Harada, T. Nakamura, M. Kinugasa, A. Oku, *Tetrahedron Lett.* **1999**, *40*, 503.

Non-Biaryl Atropisomers: New Classes of Chiral Reagents, Auxiliaries and Ligands?

Jonathan Clayden

Department of Chemistry, University of Manchester, U.K.

While many chemists are familiar with the problems posed by abnormally large barriers to rotation about single bonds when it comes to interpreting NMR spectra, few have sought to make use of these barriers as tools for stereoselective synthesis. A renewal of interest in this prospect followed Fuji's remarkable observation, published in 1991, of stereospecific alkylation of ketone **1**. The stereochemistry of the starting material was retained in the product **2** despite the intermediacy of an apparently achiral enolate (Scheme 1) [1].

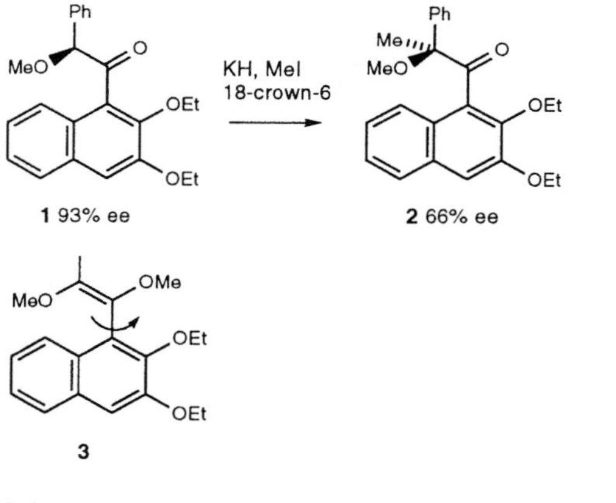

KH, MeI, 18-crown-6

1 93% ee → **2** 66% ee

3

Scheme 1

Et_3N, MeOH, Δ

4 → **5** 65%

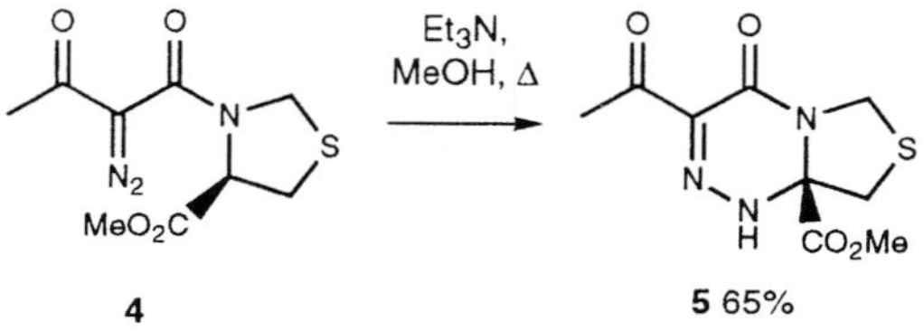

Scheme 2

The conveyor of "chiral memory" in the reaction turned out to be restricted rotation in the enolate intermediate, and Fuji was able to show that the enol ether **3**, formed as a by-product in the reaction, was chiral because of restriction to rotation about the arrowed bond, and could be recovered enantiomerically enriched. The enantiomeric excess of **3** decayed with a half-life of about 1 h at ambient temperature, and the enantiomers of **3** are atropisomers [2].

A similar phenomenon was observed by Stoodley and co-workers in the stereospecific cyclisation of **4** to a single isomer of **5** [3]. Again, the most likely explanation was an intermediate with a barrier to racemisation sufficiently high to prevent it interconverting with its enantiomer on the timescale of the reaction (Scheme 2).

Reactions like these, in which stereoselectivity is the consequence of steric hindrance to bond rotation, are most well known among the biaryls, and derivatives of binaphthyl have provided chemists with a valuable range of chiral ligands [4–6]. But the biaryls are only a small subset of axially chiral compounds containing two trigonal centres linked by a rotationally restricted single bond. Many others are known, some with much greater barriers to rotation than Fuji's enol ether [7]. Yet until quite recently there were no reports of reactions in which non-biaryl atropisomers were the source, conveyor, or product of asymmetric induction.

An early attempt to exploit the atropisomerism inherent in aromatic amides [8] drew inspiration from the suggestion that the orientation of nicotinamide's $C=O$ bond was fundamental to the stereoselectivity of hydride transfer to or from NADH [9]. Ohno and his co-workers managed to separate a quinolinamide from its atropisomeric diastereoisomer and showed that the face-selectivity of hydride transfer to its methiodide salt **6** was stereoselective (Scheme 3) [10]. When the diastereoisomer of **6** with the opposite configuration about the C–CO bond was used as the starting material, the diastereoselectivity was inverted, so the configuration of **7** must be controlled by the axial chirality of **6**. Similar experiments on enantiomeric atropisomers of dihydropyridines have confirmed this [11].

The resulting dihydroquinoline **7** turned out to be an enantioselective reducing agent, since it transformed methyl benzoylformate to methyl-(*S*)-mandelate with 99 % *ee*. This reaction is not controlled by axial chirality because with an sp^3 hybridised C4 atom there is free rotation about the C–C = O bond of **7**. Yet the re-oxidised quinolinium salt **6** returned by these reactions recovers its axial chirality with almost complete stereospecificity, making this reaction the first example of an asymmetric synthesis of a non-biaryl atropisomer. The compound is also unique in being what we could term a "chiral shuttle", in which chirality is transferred from the C–CO bond to the chiral centre at C4 and vice versa, with the source of asymmetric induction being destroyed each time.

There are as yet no reports of successful translation of non-biaryl axial chirality directly into

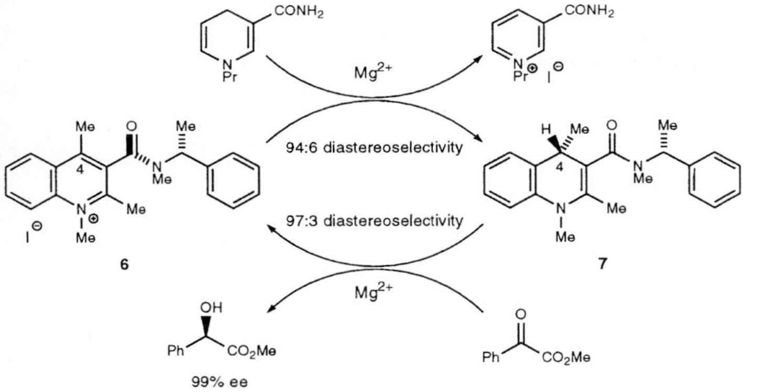

Scheme 3

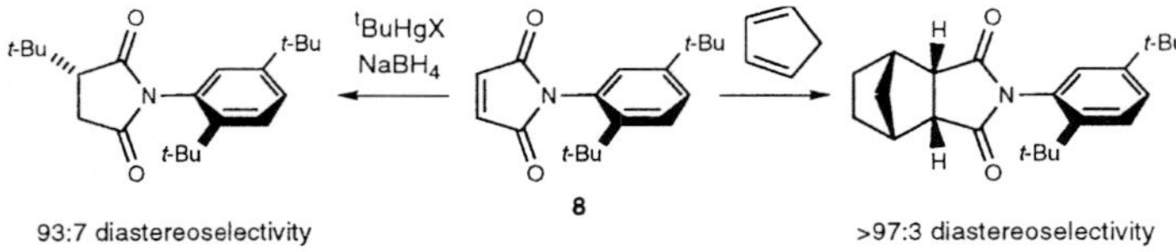

Scheme 4

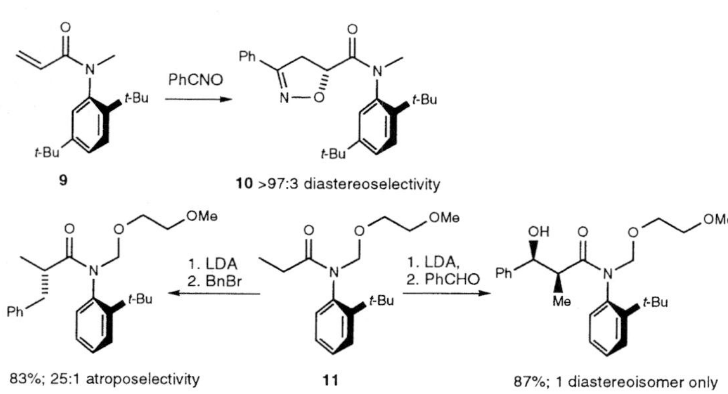

Scheme 5

chiral centres in other molecules. However, some recent publications indicate the potential of non-biaryl atropisomers, and in particular atropisomeric amides, for intramolecular asymmetric induction.

Curran [12] showed in 1994 that maleimides bearing ortho-substituted aryl groups could react diastereoselectively, favouring a single atropisomer of the product (Curran termed these reactions "atroposelective"). Racemic maleimide **8** underwent radical reactions and cycloadditions from the face unshielded by the tert-butyl group, as shown in Scheme 4.

Acyclic compounds will also react atroposelectively: amide **9** gave the isoxazoline **10** with >97 : 3 diastereoselectivity [12], and only one diastereoisomer was obtained from the alkylations or aldol reactions of **11** [13] (Scheme 5).

These high stereoselectivities bode well for the use of axially chiral anilides as chiral auxiliaries, provided they can be made in enantiomerically pure form. One approach [13] employs a kinetic resolution: enolate formation is faster from one enantiomer of the starting anilide **12** than from the other (Scheme 6).

An alternative employs classical resolution using lactic acid as the source of asymmetry. Amide formation from (*S*)-*O*-acetyllactic acid and **13** gave a separable mixture of **14** [14, 15]. The lactanilides could be eliminated or reduced [16] to remove the stereogenic centre to give optically active analogues of **9** and **11**. Nonetheless, a serious problem with the effective use of anilides as auxiliaries is their recovery in enantiomerically pure form.

In our own group, we [17] have shown that rotationally restricted aromatic amides ($ArCONR_2$) can be functionalised atroposelectively, both by nucleophiles and electrophiles [18]. Atroposelective additions to carbonyl groups are possible:

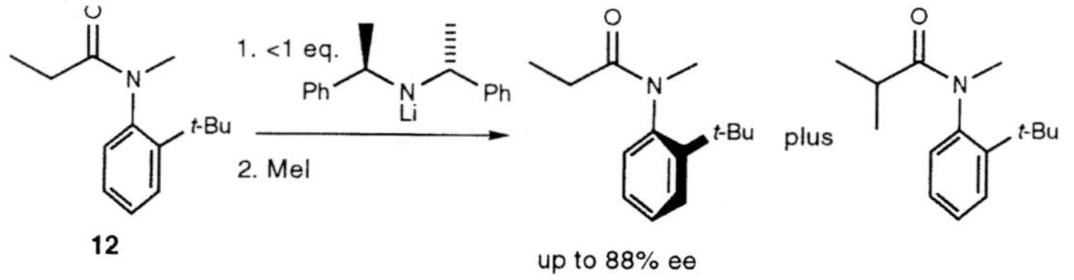

Scheme 6

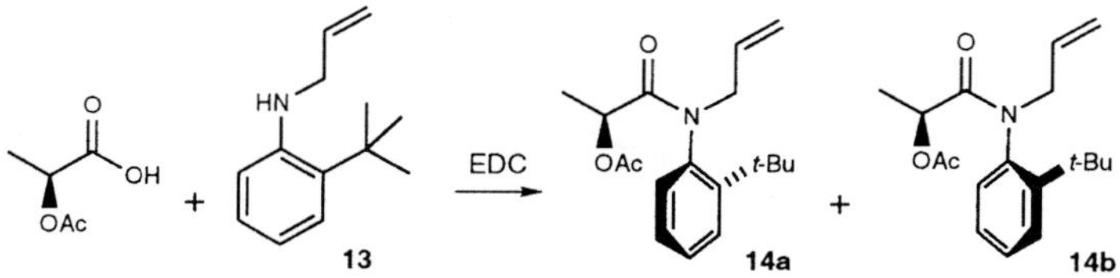

Scheme 7

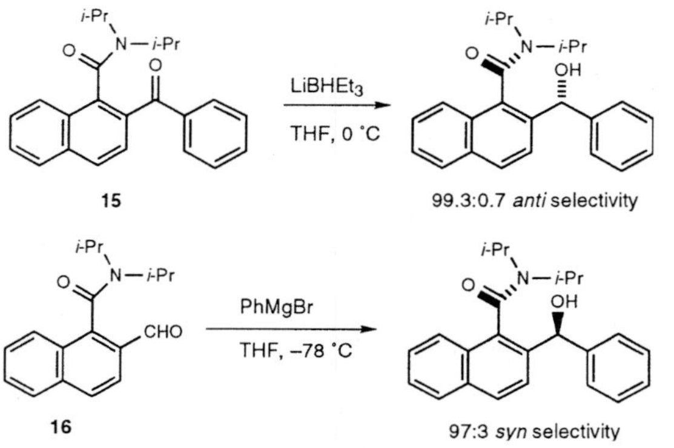

Scheme 8

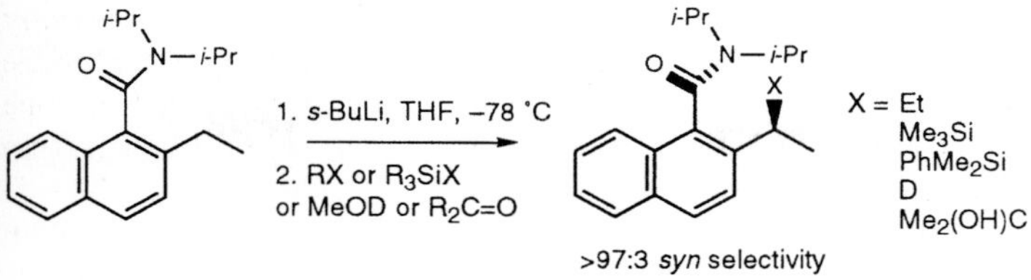

1. *s*-BuLi, THF, –78 °C
2. RX or R_3SiX or MeOD or $R_2C{=}O$

>97:3 *syn* selectivity

X = Et, Me_3Si, $PhMe_2Si$, D, $Me_2(OH)C$

Scheme 9

reduction of keto-naphthamides such as **15** with bulky reducing agents gives very high selectivities for attack *anti* to the bulky *N,N*-dialkyl group (Scheme 8). Interestingly, reaction of the aldehyde **16** with phenylmagnesium bromide generates the other diastereoisomer atroposelectively.

Lateral benzylic lithiation of 2-alkylnaphthamides gives a single atropisomer of the configurationally stable benzylic organolithium, and can be used to introduce electrophiles atroposelectively (Scheme 9).

This class of amides has been made enantioselectively – albeit in low yield and low *ee* – by an ortholithiation reaction (Scheme 10). More successful is Uemura's demonstration [19] that the arene-chromium tricarbonyl complex chemistry that works with atroposelective biaryl couplings is also successful with atroposelective amide-forming reactions (Scheme 11). These reactions, and other successful resolutions (both classical [20, 21] and on chiral stationary phase [8, 22, 23]) of atropisomeric aromatic amides means that this class of non-biaryl atropisomers are now available enantiomerically enriched.

Atropisomers should be suited to enantioselective synthesis using thermodynamic control, and Curran has proposed using anilides as "prochiral auxiliaries", responding to stereochemistry within

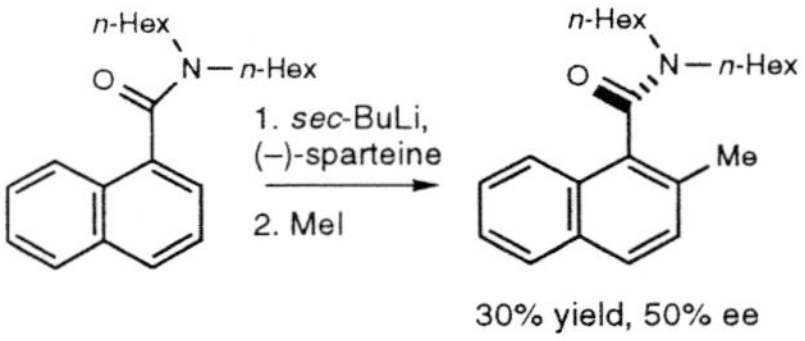

1. *sec*-BuLi, (–)-sparteine
2. MeI

30% yield, 50% ee

Scheme 10

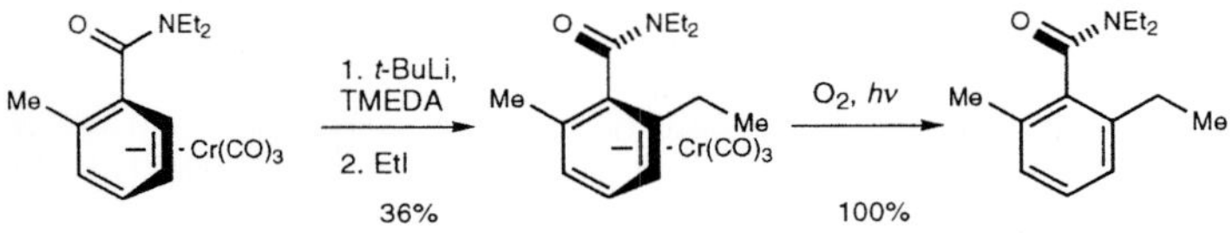

1. *t*-BuLi, TMEDA
2. EtI
36%

O_2, *hν*
100%

Scheme 11

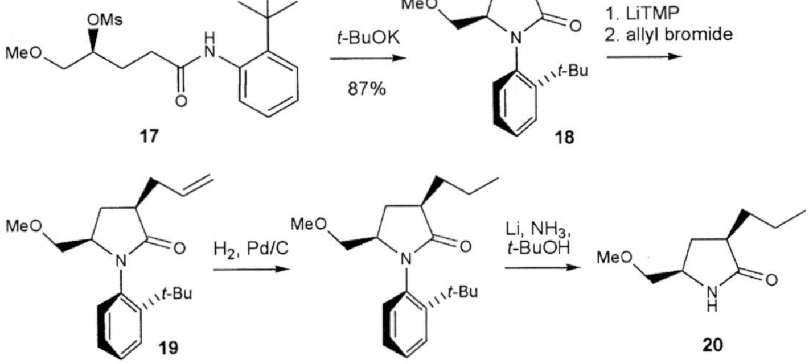

t-BuOK
87%

1. LiTMP
2. allyl bromide

H_2, Pd/C

Li, NH_3, *t*-BuOH

Scheme 12

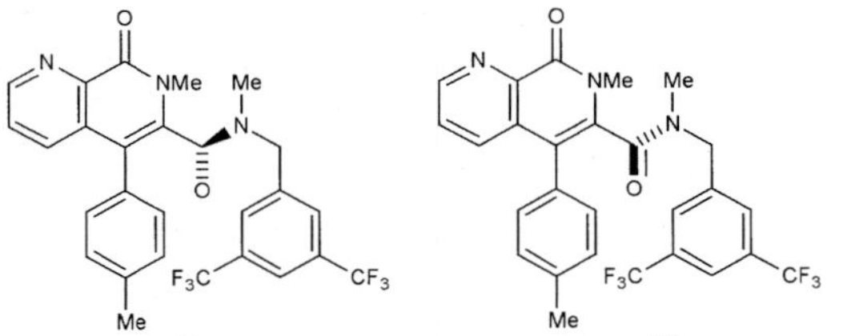

Scheme 13

a molecule and relaying stereochemical information from one centre to another [24]. Taguchi [25] has demonstrated this idea (Scheme 12), with the stereogenic centre of **17** exerting control over the axis of **18** during a cyclisation reaction. The axis can now control the enolate chemistry of the product: **19** is formed stereoselectively as the *syn* stereoisomer. It was possible to remove the anilide portion reductively (with loss of axial chirality) to give the pyrrolidone **20**.

The search for a general synthesis of enantiomerically pure non-biaryl atropisomers has been given added impetus by the discovery [26, 27] that the absolute configuration of some atropisomeric amides affects their biological activity. The enantiomeric atropisomers **21** and **22** (Scheme 13) differed in activity at the tachykinin NK_1 receptor by a factor of 6–13, with **21** being the more active. Atroposelectivity must now be an important new consideration in drug synthesis.

The contrasted electronic properties of the carbonyl oxygen atom and steric bulk of the dialkylamino group suggest that these amides would make good candidates as asymmetric ligands for metals, and it should not be long before we see novel classes of ligands based on these structures.

References

[1] T. Kawabata, K. Yahiro, K. Fuji, *J. Am. Chem. Soc.* **1991**, *113*, 9694.
[2] M. Oki, *Topics in Stereochem.* **1983**, *14*, 1.
[3] B. Beagley, M. J. Betts, R. G. Pritchard, A. Schofield, R. J. Stoodley, S. Vohra, *J. Chem. Soc., Perkin Trans. 1* **1993**, 1761.
[4] K. Tomioka, *Synthesis* **1990**, 541.
[5] C. Rosini, L. Franzini, A. Raffaelli, P. Salvadori, *Synthesis* **1992**, 503.
[6] R. Noyori, *Asymmetric Catalysis in Organic Synthesis*, Wiley, New York **1994**.
[7] E. L. Eliel, S. H. Wilen, *Stereochemistry of Organic Compounds*, Wiley, New York **1994**.
[8] M. A. Cuyegkeng, A. Mannschreck, *Chem. Ber.* **1987**, *120*, 803.
[9] M. C. A. Donkersloot, H. M. Buck, *J. Am. Chem. Soc.* **1981**, *103*, 6549, 6554.
[10] A. Ohno, M. Kashiwagi, Y. Ishihara, S. Ushida, S. Oka, *Tetrahedron* **1986**, *42*, 961.
[11] P. M. T. de Kok, L. A. M. Bastiaansen, P. M. van Lier, J. A. J. M. Vekemans, H. M. Buck, *J. Org. Chem.* **1989**, *54*, 1313.
[12] D. P. Curran, H. Qi, S. J. Geib, N. C. DeMello, *J. Am. Chem. Soc.* **1994**, *116*, 3131.
[13] A. D. Hughes, D. A. Price, O. Shishkin, N. S. Simpkins, *Tetrahedron Lett.* **1996**, *37*, 7607.
[14] O. Kitagawa, H. Izawa, T. Taguchi, M. Shiro, *Tetrahedron Lett.* **1997**, *38*, 4447.
[15] O. Kitagawa, H. Izawa, K. Sato, A. Dobashi, T. Tagichi, *J. Org. Chem.* **1998**, *63*, 2634.
[16] A. D. Hughes, N. S. Simpkins, *Synlett* **1998**, 967.
[17] J. Clayden, N. Westlund, F. X. Wilson, *Tetrahedron Lett.* **1996**, *37*, 5577.
[18] J. Clayden, *Synlett* **1998**, 810.
[19] H. Koide, M. Uemura, *J. Chem. Soc., Chem. Commun.* **1998**, 2483.
[20] J. H. Ackerman, G. M. Laidlaw, G. A. Snyder, *Tetrahedron Lett.* **1969**, 3879.
[21] P. M. T. de Kok, M. C. A. Donkersloot, P. M. van Lier, G. H. W. M. Meulendijks, L. A. M. Bastiaansen, H. J. G. van Hooff, J. A. Kanters, H. M. Buck, *Tetrahedron* **1986**, *42*, 941.
[22] C. Kiefl, H. Zinner, T. Burgemeister, A. Mannschreck, *Rec. Trav. Chim. Pays-Bas* **1996**, *115*, 125.
[23] W. H. Pirkle, C. J. Welch, A. J. Zych, *J. Chromatography* **1993**, *648*, 101.
[24] D. P. Curran, G. R. Hale, S. J. Geib, A. Balog, Q. B. Cass, A. L. G. Degani, M. Z. Hernandes, L. C. G. Freitas, *Tetrahedron Asymmetry* **1997**, *8*, 3955.
[25] M. Fujita, O. Kitagawa, H. Izawa, A. Dobashi, H. Fukaya, T. Taguchi, *Tetrahedron Lett.* **1999**, *40*, 1949.
[26] Y. Ikeura, Y. Ishichi, T. Tanaka, A. Fujishima, M. Murabayashi, M. Kawada, T. Ishimaru, I. Kamo, T. Doi, H. Natsugari, *J. Med. Chem.* **1998**, *41*, 4232.
[27] Y. Ikeura, T. Ishimaru, T. Doi, M. Kawada, A. Fujishima, H. Natsugari, *J. Chem. Soc., Chem. Commun.* **1998**, 2141.

Amino Acid Derivatives by Multicomponent Reactions

Gerald Dyker

Institut für Synthesechemie, Universität Duisburg, Germany

Amino acids constitute one of the most important classes of naturally occurring substances and possess a variety of biological functions. The areas of application of the amino acid derivatives range from sweeteners through pharmaceuticals to crop and plant protection. Widely applicable methods for the synthesis of α-amino acids [1, 2] are of great interest, especially for the construction of compound libraries by combinatorial chemistry. Multicomponent reactions, as a special case of a domino process [3], are particularly fascinating since they facilitate rapid syntheses from simple building blocks. A classic example of such a reaction is the Strecker synthesis, which has been known for almost 150 years but is still as topical as ever [4]: an aldehyde **1** (or a ketone) is condensed, using an acid catalyst, with an amine **2** and an alkali metal cyanide such as **3** to give an α-aminonitrile **4**, which can be hydrolyzed to the amino acid **5**. In recent investigations on the Strecker synthesis attempts have been made to optimize the reaction conditions [4b] and to achieve stereoselective syntheses [4c–h]. Usually chiral amines such as **7** have been employed, which can, for example, on reaction with **6** be converted to the thiophene-substituted amino acid **8** [4c].

In Ugi's four-component condensation, imine formation from an aldehyde **1** and an amine **2** is likewise the initiating step [5, 6]; a carboxylic acid **9** and an isonitrile **10** are the other reaction components, which finally yield the bisamide **11**. Both for this reaction and the Strecker synthesis, the galactosylamine **12** is particularly suitable for carrying out a stereoselective reaction (synthesis of **13**) [4d–e, 5f]. With an aminoglucopyranose as a chiral auxiliary, the stereoselectivity of the reaction can be further increased [5b]. Amino acids as condensation components yield particularly impressive results. For instance, the imino-

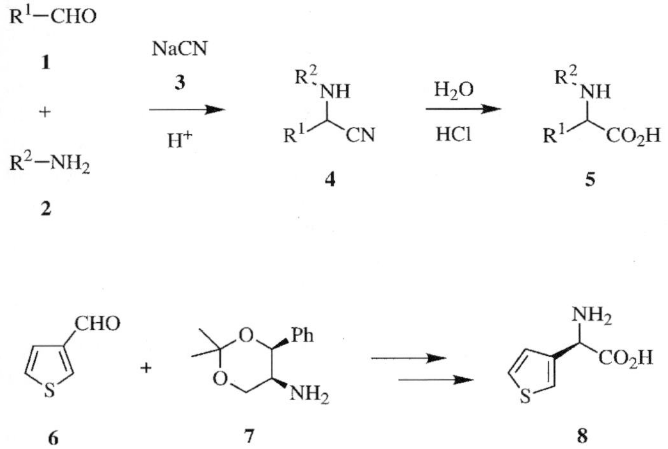

R^1–CHO **1** + R^2–NH_2 **2** → (NaCN **3**, H^+) → **4** → (H_2O, HCl) → **5**

NH, CN, CO_2H, CHO, Ph, O, S, NH_2

6 + **7** → **8**

Scheme 1

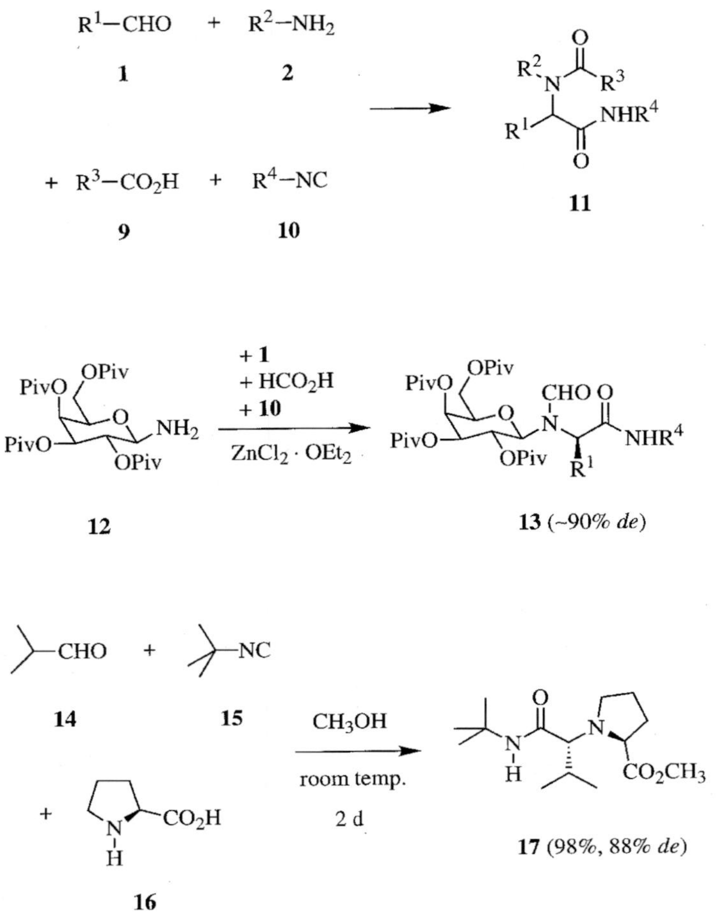

Scheme 2

dicarboxylic acid **17** can be synthesized from components **14–16** at room temperature and obtained in excellent yield and with remarkable diastereoselectivity [5c]. Readily available reagents were used exclusively in both the Strecker synthesis and the Ugi reaction.

A convincing reaction from organometallic chemistry was a long time coming. Petasis and Zavialov now report [7a–b] that vinylboronic acids **18** [7c–d], which are readily accessible and easy to handle, can participate in a three-component reaction with primary and secondary amines **19** and with α-keto acids **20**. The process, which proceeds at room temperature, leads to β,γ-unsaturated α-amino acids **21** in good yields. The reaction of vinylboronic acid **22** with the chiral amino alcohol **23** and glyoxylic acid hydrate **24** illustrates that the reaction is not sensitive to air or humidity, can be carried out under mild conditions, and has impressive diastereoselectivity.

Conceptually this new three-component reaction resembles the methods applied by Steglich, Enders et al. [8a] and Yamamoto et al. [8b], who treated the preformed iminoesters **26** and **29** at low temperature with the electron-rich olefins **27** and **30**, respectively [9]. However, the use of boronic acid derivatives appears to be a versatile method with a great deal of potential. As Petasis and Zavialov note at the end of their paper, unprotected amino acids and peptides, aryl boronic acids, and chiral boronates can also participate in their three component condensation [7a]. Transition metal catalysis also

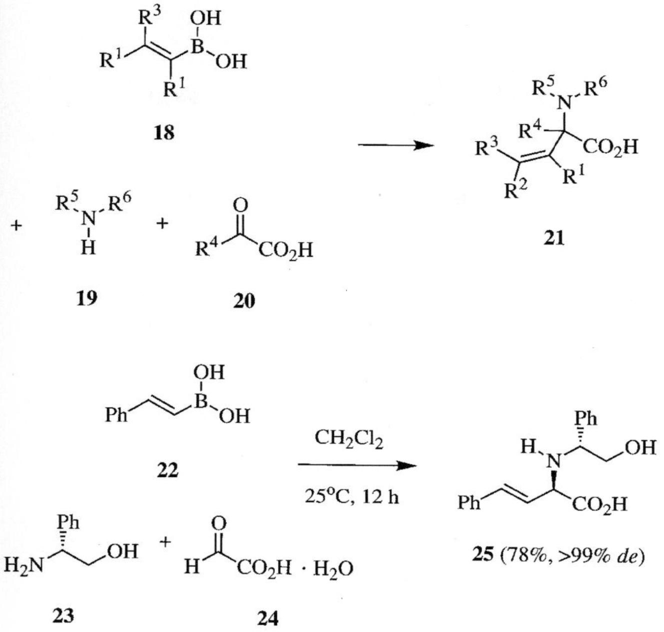

Scheme 3

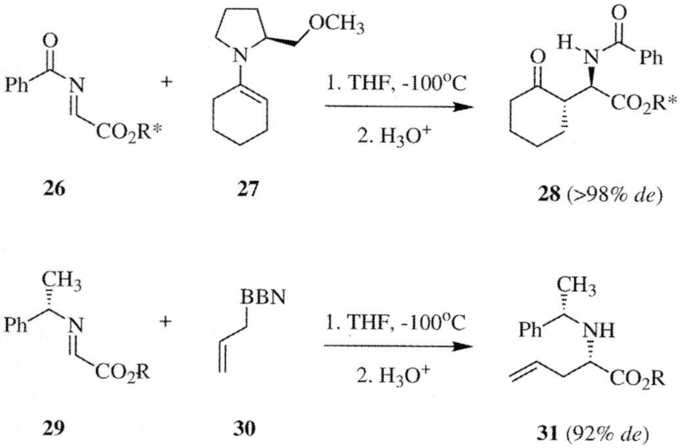

Scheme 4

provides surprising solutions for the synthesis of amino acids by multicomponent reactions [10].

More than 25 years ago Wakamatsu et al. [11a] found that aldehyde **1** could be coupled with carboxamides **32** and carbon monoxide under pressure to give *N*-acylamino acids **33**, a process catalyzed by octacarbonyldicobalt and known as amidocarbonylation [11b]. In the presence of hydrogen, the reaction conditions are suitable for hydroformylation; thus, alkenes **34** could be used in place of the aldehyde **1** [11b]. Although amidocarbonylation appears to be very favorable from an economic point of view, it has not yet been realized in industry. The main problem con-

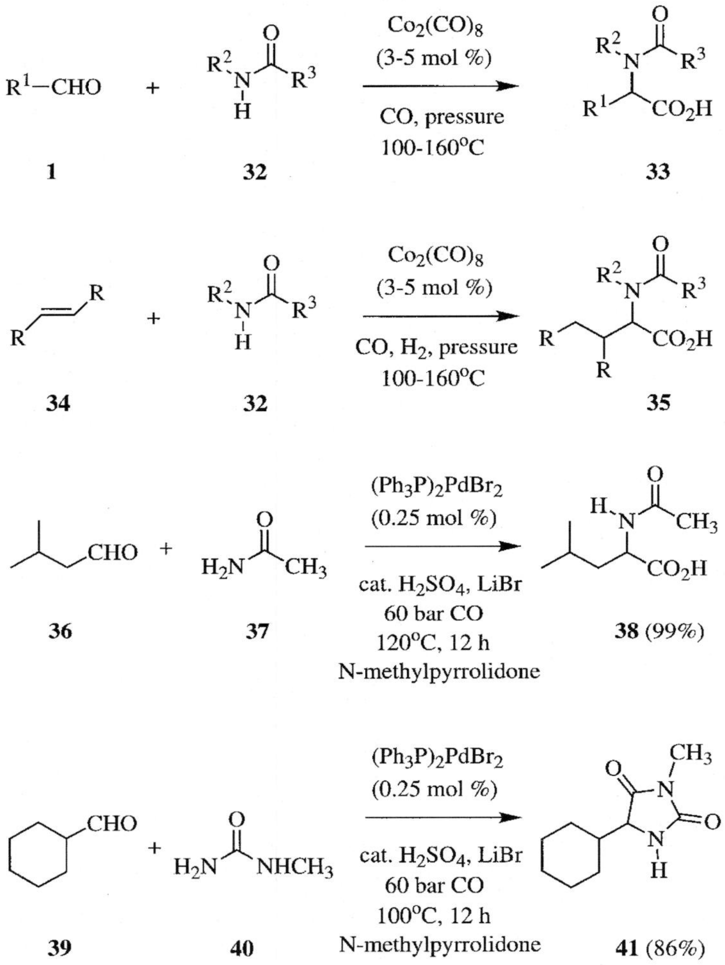

Scheme 5

cerns low catalyst efficiency, but in addition it does not appear to be possible to run the enantio/diastereoselective process with the previously applied cobalt catalysts. Beller et al. have now found [12] that amidocarbonylation of aldehydes can also be achieved by using palladium catalysts with sulfuric acid and halide ions as cocatalysts, as shown by the reaction of isovaleraldehyde **36** with acetamide **37**.

This palladium-catalyzed synthesis is superior to that catalyzed by cobalt in many respects. Both the catalytic efficiency and the attainable yields are considerably higher. Several components, such as aromatic aldehydes and monoalkyl amides [11c], which the cobalt-catalyzed method failed to convert, can now be used without any problems, and even ureas such as **40** are suitable coupling components, opening up an efficient access to hydantoins (for instance **41**) [12e]. In addition, the reaction conditions required are milder: very good yields can be obtained at 80 °C and a CO pressure of only 10 bar. Most importantly, perhaps, the change of catalyst allows the possibility of controlling the process through choice of ligand and of running an enantio/diastereoselective process.

References

[1] Current examples of the synthesis of α-amino acids: a) J. Mulzer, H.-J. Altenbach, M. Braun, K. Krohn, H.-U. Reissig, *Organic Synthesis Highlights,* VCH, Weinheim, **1990**, pp. 300–305; b) R. O. Duthaler, *Tetrahedron* **1994**, *50,* 1539–1650; c) U. Kazmaier, *Liebigs Ann./Recueil* **1997**, 285–295; d) M. Braun, K. Opdenbusch, *ibid.* **1997**,141–154; e) L. S. Hegedus, *Acc. Chem. Res.* **1995**, *28*, 299–305; f) D. Enders, R. Funk, M. Klatt, G. Raabe, E. R. Hovestreydt, *Angew. Chem.* **1993**, *105*, 418–420; *Angew. Chem. Int. Ed. Engl.* **1993**, *32*, 418–420.

[2] Current examples of the synthesis of β-amino acids: a) D. Enders, J. Wiedemann, *Liebigs Ann./ Recueil* **1997**, 699–706; b) J. Voigt, M. Noltemeyer, O. Reiser, *Synlett* **1997**, 202–204; c) H. Kunz, A. Burgard, D. Schanzenbach, *Angew. Chem.* **1997**, *109*, 394–396; *Angew. Chem. Int. Ed. Engl.* **1997**, *36*, 386–387; d) G. Cardillo, S. Casolari, L. Gentilucci, C. Tomasini, *ibid.* **1996**, ***108***, 1939–1941 and **1996**, *35*, 1848–1849; e) J. Podlech, D. Seebach, *ibid.* **1995**, *107*, 507–509 and **1995**, *34*, 471–472; f) J. Podlech, D.Seebach, *Liebigs Ann.* **1995**, 1217–1228; g) P. Gmeiner, E. Hummel, C. Haubmann, *ibid.* **1995**, 1987–1992; h) J. Escalante, E. Juaristi, *Tetrahedron Lett.* **1995**, **36**, 4397–4400; i) D. C. Cole, *Tetrahedron* **1994**, *50*, 9517–9582.

[3] L. F. Tietze, U. Beifuss, *Angew. Chem.* **1993**, *105*, 137–170; *Angew. Chem. Int. Ed. Engl.* **1993**, *32*, 131–163.

[4] a) A. Strecker, *Justus Liebigs Ann. Chem.* **1850**, *75*, 27: b) M. P. Georgiadis, S. A. Haroutounian, *Synthesis* **1989**, 616–618; c) K. Weinges, H. Brachmann, P. Stahnecker, H. Rodewald, M. Nixdorf, H. Irngartinger, *Liebigs Ann.* **1985**, 566–578; d) H. Kunz, W. Sager, D. Schanzenbach, M. Decker, *ibid.* **1991**, 649–654; e) H. Kunz, K. Rück, *Angew. Chem.* **1993**, *105,* 355–377; *Angew. Chem. Int. Ed. Engl.* **1993**, *32*, 336–358; f) F. A. Davis, P. S. Portonovo, R. E. Reddy, Y. Chiu, *J. Org. Chem.* **1996**, *61*, 440–441; g) T. K. Chakraborty, K. A. Hussain, G. V. Reddy, *Tetrahedron* **1995**, *51*, 9179–9190; h) M. S. Iyer, K. M. Gigstad, N. D. Namdev, M. Lipton, *J. Am. Chem. Soc.* **1996**, *118*, 4910–4911.

[5] a) I. Ugi, *Angew. Chem.* **1982**, *94*, 826–835; *Angew. Chem. Int. Ed. Engl.* **1982**, *21*, 810–819; b) S. Lehnhoff, M. Goebel, R. M. Karl, R. Klösel, 1. Ugi, *ibid.* **1995**, *107,* 1208–1211 and **1995**, *34*, 1104–1107; c) A. Demharter, W. Hörl, E. Herdtweck, Ugi, *ibid.* **1996**, *108*, 185–187 and **1996**, *35*, 173–175; d) A. Dömling, M. Starnecker, I. Ugi, *ibid.* **1995**, *107*, 2465–2467 and **1995**, *34*, 2238–2239; e) T. Yamada, T. Yanagi, Y. Omote, T. Miyazawa, S. Kuwata, M. Sugiura, K. Matsumoto, *J. Chem. Soc. Chem. Commun.* **1990**, 1640–1641; f) H. Kunz, W. Pfrengle, *J. Am. Chem. Soc.* **1988**, *110*, 651–652; g) P. A. Tempest, S. D. Brown, R. W. Armstrong, *Angew. Chem.* **1996**, *108*, 689–691; *Angew. Chem. Int. Ed. Engl.* **1996**, *35*, 640–642; h) T. A. Keating, R. W. Armstrong, *J. Am. Chem. Soc.* **1996**, *118*, 2574–2583; i) H. Quast, S. Aldenkortt, *Chem. Eur. J.* **1996**, *2*, 462–469.

[6] Synthesis of an amino acid derivative by using a seven-component reaction: A. Dömling, I. Ugi, *Angew. Chem.* **1993**, *105*, 634–635; *Angew. Chem. Int. Ed. Engl.* **1993**, *32*, 563–564.

[7] a) N. A. Petasis, I. A. Zavialov, *J. Am. Chem. Soc.* **1997**, *119*, 445–446; b) N. A. Petasis, A. Goodman, I. A. Zavialov, *Tetrahedron* **1997**, *53*, 16463–16470; c) *Tetrahedron Lett.* **1996**, *37*, 567–570; d) N. A. Petasis, I. Akritopoulou, *ibid.* **1993**, *34*, 583–586; e) for a related synthesis of chiral β-aminoalcohols, see: N. A. Petasis, I. A. Zavialov, *J. Am. Chem. Soc.* **1998**, *120*, 11798–11799.

[8] a) R. Kober, K. Papadopoulos, W. Miltz, D. Enders, W. Steglich, H. Reuter, H. Puff, *Tetrahedron* **1985**, *41*, 1693–1701; b) Y. Yamamoto, W. Ito, K. Maruyama, *J. Chem. Soc. Chem. Commun.* **1985**, 1131–1132.

[9] Further examples of amino acid syntheses starting with glyoxylic acid derivatives: a) P. Münster, W. Steglich, *Synthesis* **1987**, 223–225; b) M. J. O Donnell, W. D. Bennett, *Tetrahedron* **1988**, *44*, 5389–5401; c) E. C. Roos, M. C. Lopez, M. A. Brook, H. Hiemstra, W. N. Speckamp, B. Kaptein, J. Kamphuis, H. E. Schoemaker, *J. Org. Chem.* **1993**, *58*, 3259–3268.

[10] M. Beller, B. Cornils, C. D. Frohning, C. W. Kohlpaintner, *J. Mol. Cat.* **1995**, *44*, 237–273.

[11] a) H. Wakamatsu, J. Uda, N. Yamakami, *J. Chem. Soc. Chem. Commun.* **1971**, 1540; b) "Amidocarbonylation": J. F. Knifton in *Applied Homogeneous Catalysis with Metal Complexes* (Eds.: B. Cornils, W. A. Herrmann), VCH, Weinheim, **1996**, pp. 159–168; c) P. Magnus, M. Slater, *Tetrahedron Lett.* **1987**, *28*, 2829–2832.

[12] a) M. Beller, M. Eckert, F. Vollmüller, S. Bogdanovic, H. Geissler, *Angew. Chem.* **1997**, *109*, 1534–1536; *Angew. Chem. Int. Ed. Engl.* **1997**, *36*, 1494–1496; b) M. Beller, M. Eckert, F. Vollmüller, *J. Mol. Cat. A* **1998**, *135*, 23–33; c) M. Beller, M. Eckert, E. W. Holla, *J. Org. Chem.* **1998**, *63*, 5658–5661; d) M. Beller, M. Eckert, W. A. Moradi, *Synlett* **1999**, 108–110; e) M. Beller, M. Eckert, W. A. Moradi, H. Neumann, *Angew. Chem.* **1999**, *111*, 1562–1565; *Angew. Chem. Int. Ed. Engl.* **1999**, *38*, 1460–1463.

New Polyol Syntheses

Christoph Schneider

Institut für Organische Chemie, Universität Göttingen, Germany

Among the natural products of polyketide origin the polyene macrolide antibiotics have only recently attracted attention from synthetic chemists [1]. More than 200 members of this class of natural products have been identified, but the constitution and configuration of many have only been partially elucidated. Best known are probably amphotericin B and nystatin, which are used clinically for the treatment of systemic fungal infections. Their biological activity rests on their ability to damage cytoplasmic membranes of eucaryotic cells by various mechanisms with consequent loss of ions, amino acids, and carbohydrates. Structurally they consist of a polyene moiety with up to seven, mostly conjugated double bonds and a polyol moiety with nine secondary, mainly 1,3-orientated hydroxy groups. New strategies for the stereoselective synthesis of the polyol structures are of great interest to synthetic chemists.

Rychnovsky and his group have recently developed new synthetic methods that lead to the total syntheses of the polyene macrolides roxaticin [2], roflamycoin [3], and filipin III [4]. The polyol chains of all three natural products were constructed by iterative, stereoselective alkylation of lithiated cyanohydrin acetonides and subsequent reductive decyanation, illustrated here by the synthesis of the polyol framework of filipin III (**1**) (Scheme 1). The bifunctional cyanohydrin acetonide **2**, prepared by ruthenium/BINAP catalyzed enantioselective hydrogenation of the corresponding *β*-keto ester (BINAP = [1,1'-binaphthyl]-2,2'-diylbis(diphenylphosphane)), is deprotonated with $LiNEt_2$ and alkylated with 2-benzyloxy-1-iodoethane. The alkylation product **3** is converted by a Finkelstein reaction into the iodide **4**, which is used to alkylate a second molecule of **2**. After a second Finkelstein reaction, the protected tetraol **6** is coupled with the lithiated cyanohydrin acetonide **7** in a third alkylation to form the complete polyol segment **8** of filipin III.

Each alkylation step proceeds in good yield (70–80 %), and the 1,3-dioxanes **9**, in which the slim cyano group assumes the axial position, are formed with high stereoselectivity. Once they have fulfilled their role for the alkylations, the three cyano groups are removed with Li/NH_3 with retention of configuration to afford the 1,3-*syn*-diol acetonides **11** (Scheme 2). Control experiments revealed, however, that the reductive decyanation always gives 1,3-*syn* configured diol acetonides irrespective of the cyanohydrin acetonide configuration. The reason is that the intermediate radical **10** prefers a configuration in which the unpaired electron assumes the axial position at the anomeric center. The carbanion formed by the next electron transfer is protonated at this position so the H atom is also in an axial position.

Protected 1,3-*anti*-diols **14** are accessible by the highly stereoselective Lewis acid promoted addition of dialkylzinc compounds to 4-acetoxy-1,3-dioxanes **13** (Scheme 3) [5]. The two *cis*-orientated alkyl substituents at C2 and C6 fix the carboxonium ion **15** in the half-chair conformation, which undergoes preferential axial attack by the dialkylzinc under stereoelectronic control. The 4-acetoxy-1,3-dioxanes **13** may be synthesized from the Seebach 1,3-dioxan-4-ones **12** by reduction with diisobutylaluminum hydride (DIBAH) and acetylation. Since dialkylzinc compounds are now readily available and are compatible with many functional groups, this

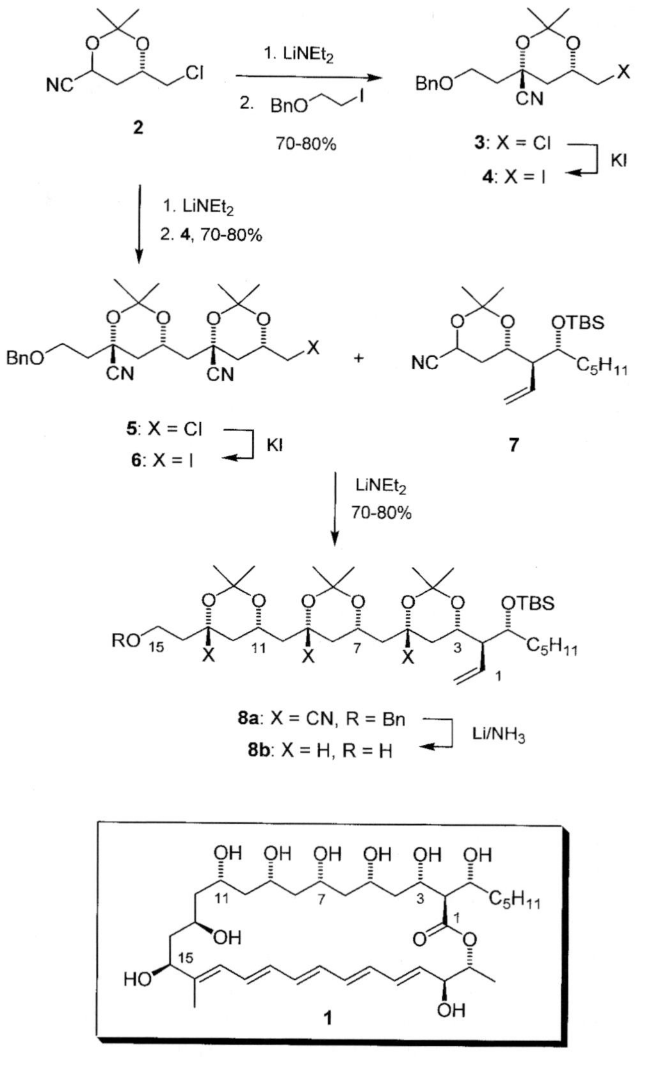

Scheme 1. Synthesis of the C(1)–C(15) polyol fragment **8b** of filipin III (**1**) according to Rychnovsky et al. Bn = benzyl, TBS = *tert*-butyldimethylsilyl.

method should be widely applicable to the preparation of 1,3-*anti*-diols.

The most direct method for the preparation of polyol frameworks is without doubt the aldol reaction. The diastereofacial selectivity of the reaction can be controlled by β-alkoxy groups in both the methylketone enolate and the aldehyde. As investigations by Evans [6] and Paterson [7] and their groups have demonstrated, the correct selection of enolization conditions and the protective group for the β-hydroxy group are important for the stereocontrol of the reaction.

The boron-aldol reaction of the *p*-methoxybenzyl(PMB)-protected methylketone **16** proceeds with excellent 1,5-*anti*-selectivity (Scheme 4). In cases where the asymmetric induction is lower it may be improved by a double stereodifferential aldol reaction with chiral boron ligands [7]. The reason for this high stereoselectivity is currently unknown. Ab initio calculations suggest the involvement of twisted boat structures rather than chair transition structures [6].

If the chiral information is contained within the aldehyde, a Mukaiyama aldol reaction is the

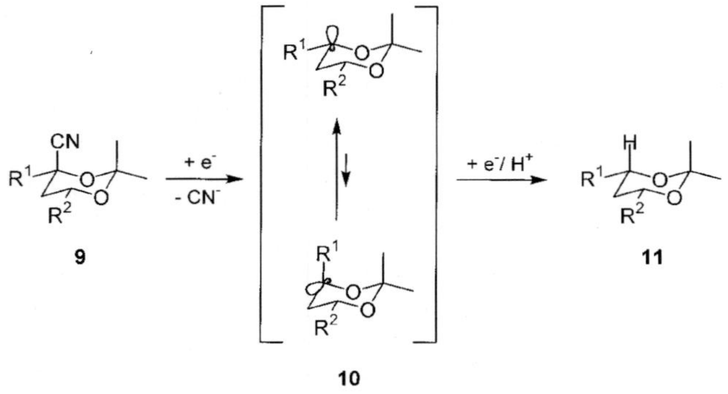

Scheme 2. Reductive decyanation of the cyanohydrin acetonide **9** according to Rychnovsky et al.

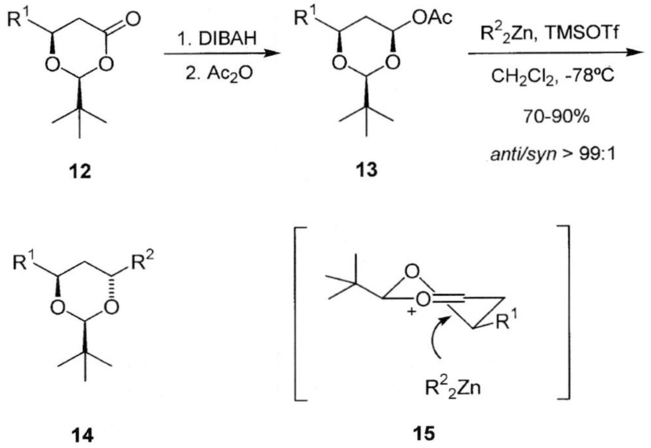

Scheme 3. Preparation of the 1,3-*anti*-diol acetal **14** by dialkylzinc addition to 4-acetoxy-1,3-dioxanes **13** according to Rychnovsky et al. TMSOTf = trimethylsilyltrifluoromethane sulfonate.

method of choice [8]. The BF_3-catalyzed addition of the silyl enol ether **18** to the chiral aldehyde **19** affords the 1,3-*anti* product **20** in high yield and stereoselectivity. Evans has developed a modified Felkin-Anh model as a working hypothesis in which repulsive interactions between the carbonyl group and the β-alkoxy group and steric interactions between the carbonyl group and the β-alkyl group are minimized. The complementary reaction conditions permit selective product formation in a double stereodifferential aldol reaction of a chiral ketone and a chiral aldehyde, even in the mismatched situation [6]. Thus the boron enolate **21** adds to the aldehyde **22** with excellent stereoselectivity and forms almost exclusively the 1,5-*anti* product **23** (enolate control). In the Mukaiyama reaction of the corresponding silyl enol ether **24**, however, the aldehyde **22** exercises stereochemical control, and the 1,5-*syn* product **25** is the main product. The ketone carbonyl function of the aldol products can be reduced with high diastereoselectivity to the 1,3-*syn*-diols with $Et_2BOMe/NaBH_4$ [9] and DIBAH [10], and to the 1,3-*anti*-diols with $Me_4NBH(OAc)_3$ [11] and SmI_2/RCHO[12].

A catalytic, enantioselective approach towards the synthesis of polyol chains has recently been reported by Carreira et al. and has been applied in a synthesis of the polyol subunit of amphotericin B (Scheme 5) [13]. Aldol addition of the silyl dienolate **26** to furfural **(27)** catalyzed by the Tol-BINAP-CuF_2-complex (2 mol-%) gives rise to the addition product **28** in 95 % yield and >99 % *ee* after one recrystallization. Spectroscopic evidence indicates that a copper dienolate is formed in situ from the silyl dienolate **26** and is actually the active nucleophile [14]. Standard transformations including a *syn*-selective reduction of the β-hydroxy ketone by the method of Prasad and the oxidative conversion of the furan ring to the carboxyl group with subsequent reduction furnish the aldehyde **29** as one half of the target molecule. The other subunit **30** is assembled from the enantiomeric aldol product by the same set of transformations and conversion to the alkyne. Both fragments are then joined through the addition of the lithium salt of **30** to the aldehyde **29** which gives a 78 : 22 mixture of diastereomers with the wrong stereoisomer predominating. Subsequent hydrogenation of the triple bond and installation of the correct stereochemistry at the central hydroxyl-bearing carbon atom through an oxidation-reduction sequence produce the complete polyol fragment **32** of amphotericin B.

Brückner and et al. use the oxidative degradation of butyrolactones by a Criegee rearrangement as the key step in the synthesis of different polyol systems (Scheme 6) [15]. In their convergent route to the pentamethyl ether **33** from *Tolypothrix conglutinata* they synthesize the homoallylic alcohols **34** and **36** by catalytic, enantioselective allylstannane additions and convert them into the two central building blocks **35** and **37** by ozonolysis and *syn*-selective ketone reduction. Coupling of these two fragments to form the complete carbon framework **38** of the polyol chain is achieved by Peterson olefination and stereoselective hydrogenation of the conjugated double

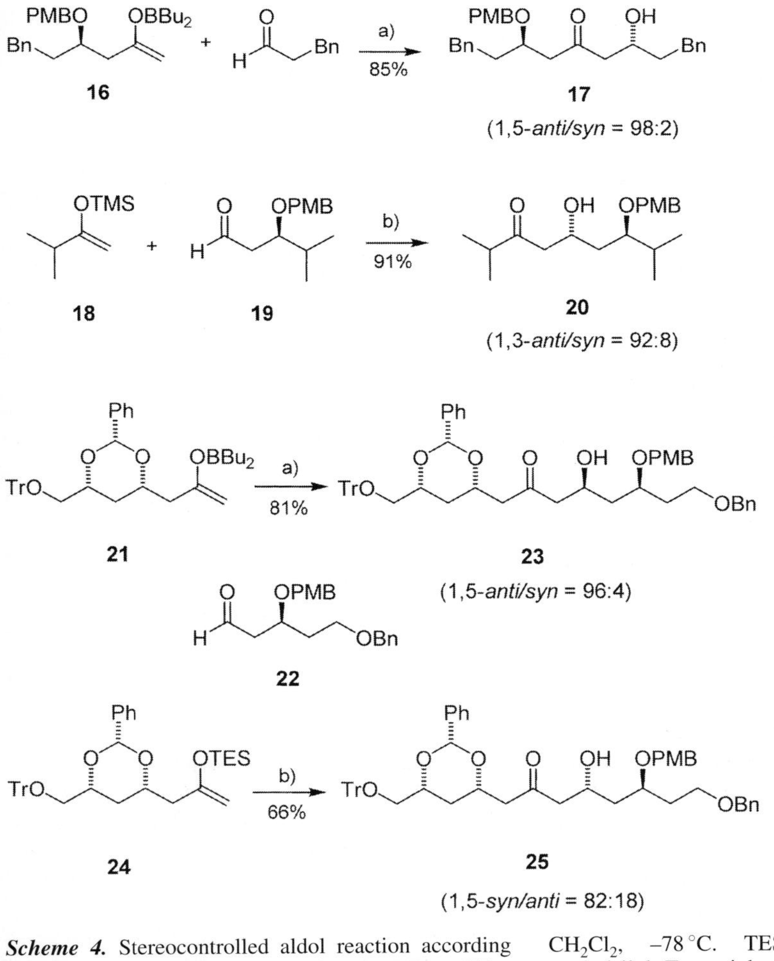

PMBO OBBu2 Bn 16 + O H Bn a) 85% PMBO O OH Bn Bn 17 (1,5-*anti/syn* = 98:2)

OTMS 18 + O OPMB H 19 b) 91% O OH OPMB 20 (1,3-*anti/syn* = 92:8)

Ph O O OBBu2 TrO 21 a) 81% Ph O O O OH OPMB TrO OBn 23 (1,5-*anti/syn* = 96:4)

O OPMB H OBn 22

Ph O O OTES TrO 24 b) 66% Ph O O O OH OPMB TrO OBn 25 (1,5-*syn/anti* = 82:18)

Scheme 4. Stereocontrolled aldol reaction according to Evans et al. a) Et_2O, −115 °C; b) $BF_3 \cdot OEt_2$, CH_2Cl_2, −78 °C. TES = triethylsilyl, TMS = trimethylsilyl, Tr = triphenylmethyl.

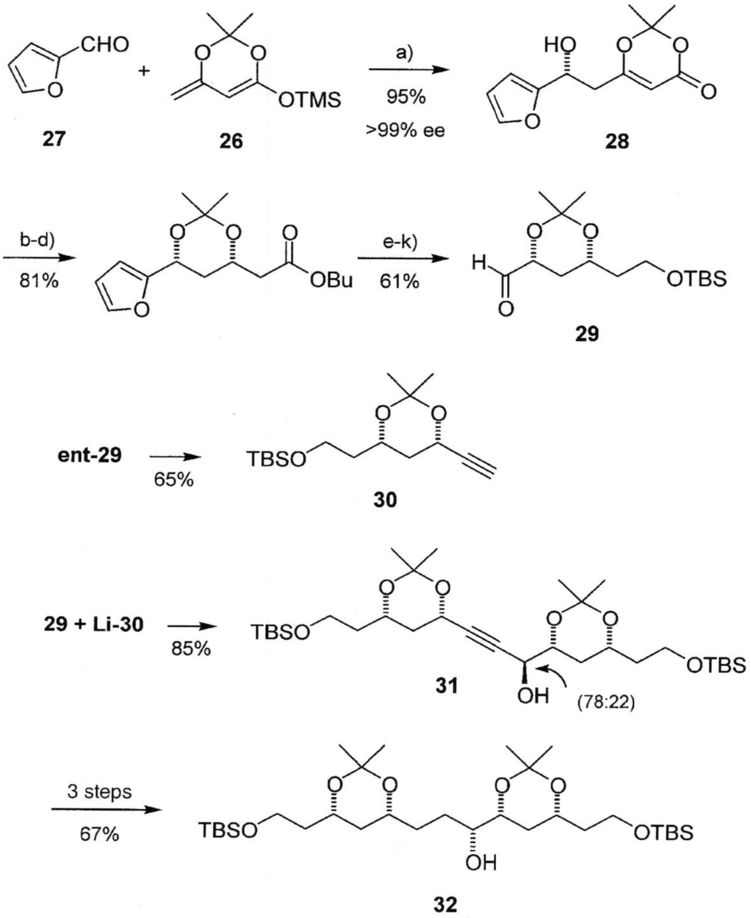

Scheme 5. Catalytic, enantioselective synthesis of the polyol fragment of amphotericin B according to Carreira et al. a) 2 mol % TolBINAP-CuF_2, −78 °C, then CF_3COOH; b) nBuOH, 110 °C; c) Et_2BOMe, $NaBH_4$, THF, −78 °C; d) $Me_2C(OMe)_2$, PPTS; e) $LiAlH_4$, THF; f) TBSCl, imidazole, DMF; g) O_3, CH_2Cl_2, −78 °C; h) Me_3SiCHN_2; i) $LiAlH_4$, THF; k) Dess-Martin periodinane.

bond. The butyrolactone **38** is now converted into the peroxosulfonate **39** by addition of MeLi, oxidation with H_2O_2, and reaction with *p*-nitrobenzenesulfonyl chloride. This sulfonate undergoes a Criegee rearrangement at room temperature to form the resonance-stabilized, cyclic carboxonium ion **40** by cleavage of the unstable peroxo compound and stereospecific migration of the neighboring carbon atom. Alkaline hydrolysis yields the *syn*-diol **41**, and standard methods then give the natural product in a few steps.

Smith et al. have developed a very elegant route to complex polyol structures by sequential dithiane-epoxide coupling reactions (Scheme 7) [16]. Following the work of Tietze [17], 2-silyl-1,3-dithianes **42** are deprotonated with *t*BuLi in ether and converted into the stable lithium alkoxides **43** with enantiomerically pure epoxides. A fast 1,4-Brook rearrangement occurs only after the addition of 0.3 equivalents of hexamethylphosphoramide (HMPA) or 1,3-dimethylhexahydro-2-pyrimidone (DMPU) to the reaction mixture. A new lithiated dithiane **44** that can undergo

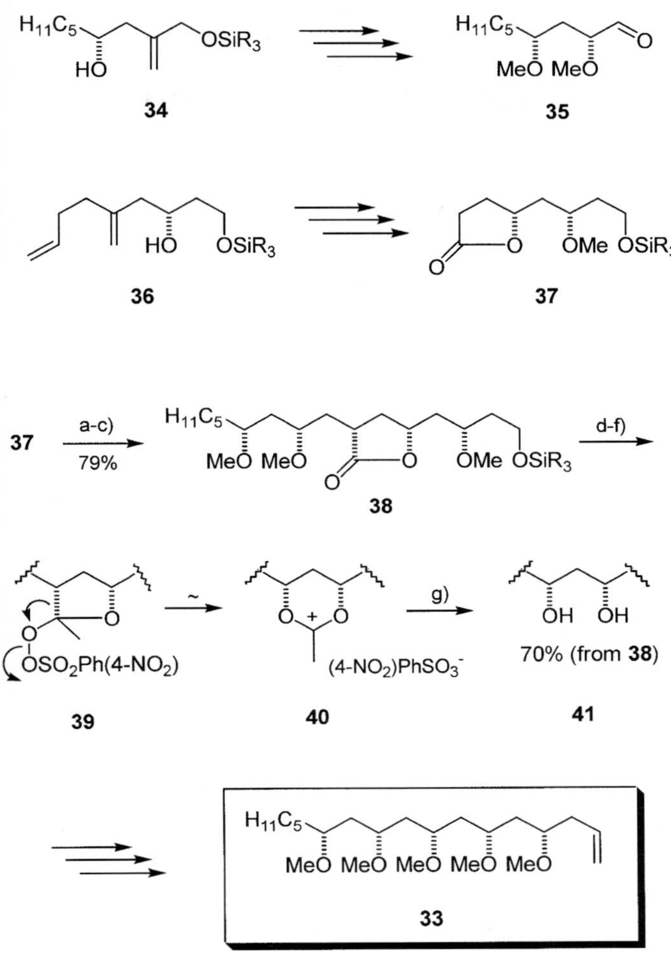

Scheme 6. Synthesis of the Tolypothrix pentamethyl ether **33** according to Brückner et al. a) 2 equiv LDA, 1 equiv $MePh_2SiCl$, −78 °C–rt; b) 35, −78 °C–rt; c) $Rh/C/H_2$; d) MeLi, −78 °C; e) H_2O_2, PPTS; f) $(4-NO_2)C_6H_4SO_2Cl$, NEt_3; g) H_2O, K_2CO_3, MeOH. LDA = lithium diisopropylamide, PPTS = pyridinium toluene-4-sulfonate.

nucleophilic addition to a second epoxide is formed. Thus, careful tuning of the reaction conditions allows unsymmetrical bisalkylation of dithianes with two different epoxides. The suitable selection of stoichiometric ratios and the insertion of additional leaving groups into the second epoxide allows a third dithiane-epoxide coupling to be achieved. The new epoxide **46**, which is formed from the doubly alkylated product **45** by intramolecular S_N reaction, is opened by unreacted Brook rearrangement product. Thus a C_{11}-polyol chain (**47**) with five free or protected secondary alcohol functions is formed from five components in a one pot reaction. However, this remarkable efficiency is achieved only if the complete stereochemical information of the polyol chain has already been introduced by the chiral epoxides.

A general, asymmetric polyol synthesis has been developed by Schneider et al. [18] which takes advantage of the benefits of the silyloxy-Cope rearrangement of the chiral aldol product

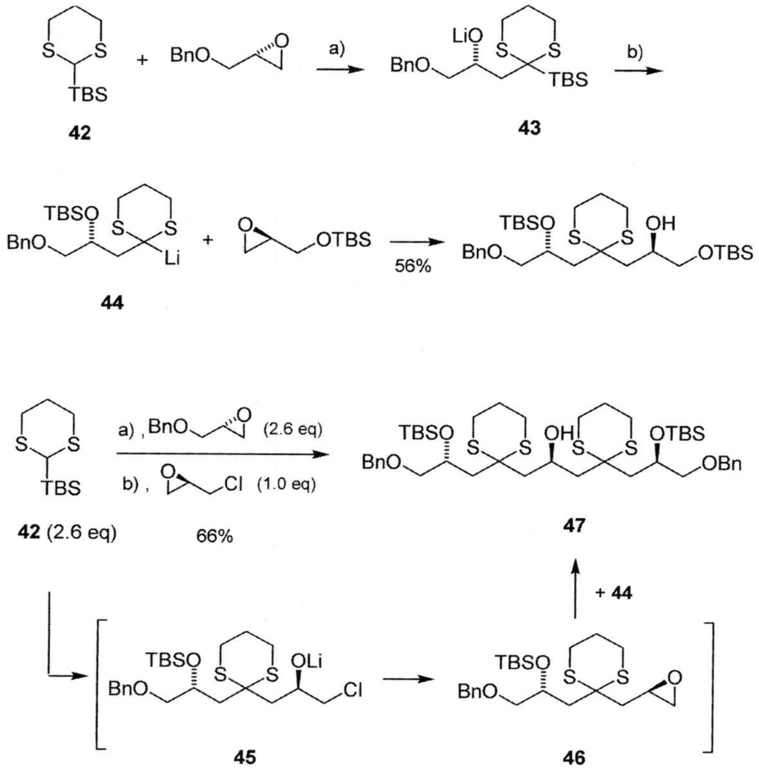

Scheme 7. Sequential dithiane-epoxide coupling according to Smith et al. a) *t*BuLi, Et_2O, $-78\,°C \rightarrow -45\,°C$; b) HMPA, Et_2O, $-78\,°C \rightarrow -25\,°C$.

48 (Scheme 8). The enantiopure 7-oxo-5-phenyldimethylsilyl-2-enimide **49** used as the chiral key intermediate is produced in a stereospecific [3.3]-sigmatropic rearrangement in good yield and carries three masked hydroxy groups in the required 1,3,5 relationship. The first hydroxy group can be liberated through reagent-controlled allylboration of the aldehyde moiety of **49**, which leads to either stereoisomer depending on the borane reagent used. The second hydroxy group is introduced into the molecule via the conjugate addition of a silyl cuprate reagent to the conjugated double bond. This Michael-type reaction proceeds with high stereocontrol to deliver the addition products **51** with a 3,5-*syn*-stereochemistry in the case of the imide **50** (auxiliary control) and **54** with a 3,5-*anti*-stereochemistry when using the ester **52** (substrate control). Finally, oxidative desilylation of the phenyldimethylsilyl groups furnishes the second and third hydroxy groups which are protected as their acetonides, in one step. The terminal double bond of the protected triols may be selectively oxidized to a methyl ketone or an aldehyde – two subunits which can easily be joined in aldol reaction to get access to larger polyol chains.

Although the absolute configuration has only been established for a few polyene macrolide antibiotics, the search for new, efficient, and selective strategies for the synthesis of their polyol structures is in full swing. A number of the synthetic procedures presented here will certainly be used in future syntheses of this class of natural products [19].

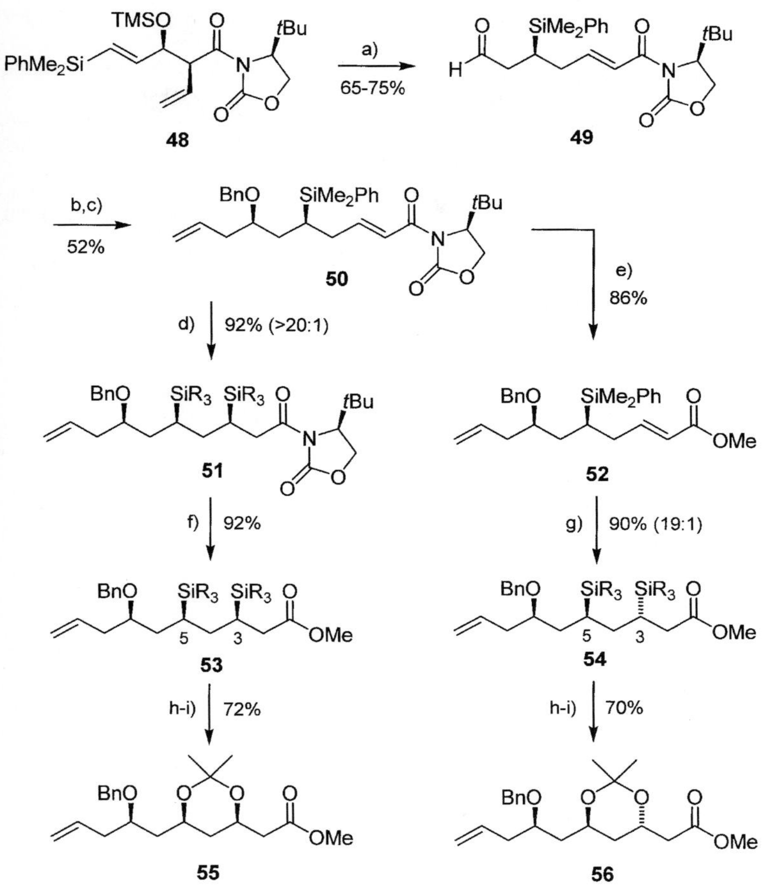

Scheme 8. General, asymmetric synthesis of 1,3,5-triols according to Schneider et al. a) toluene, 170 °C; b) allyl $(2\text{-}i\text{Car})_2$borane, Et_2O, −78 °C; c) $Cl_3CC(=NH)OBn$, TfOH, CH_2Cl_2, rt; d) $PhMe_2SiLi$, CuI, Me_2AlCl, THF, −78 °C; e) MgClOMe, CH_2Cl_2, 0 °C; f) MgClOMe, CH_2Cl_2, 40 °C; g) $PhMe_2SiLi$, CuCN, $BF_3{\cdot}OEt_2$, THF, −78 °C; h) $BF_3{\cdot}(AcOH)_2$, CH_2Cl_2, rt, then *m*CPBA, KF, DMF, 0 °C; i) $Me_2C(OMe)_2$, PPTS, rt.

Reference

[1] S. Omura, *Macrolide Antibiotics: Chemistry, Biology, Practice*, Academic Press, New York, **1984**; S. D. Rychnovsky, *Chem. Rev.* **1995**, *95*, 2021–2040.

[2] S. D. Rychnovsky, R. C. Hoye, *J. Am. Chem. Soc.* **1994**, *116*, 1753–1765.

[3] S. D. Rychnovsky, U. R. Khire, G. Yang, *J. Am. Chem. Soc.* **1997**, *119*, 2058–2059.

[4] T. I. Richardson, S. D. Rychnovsky, *J. Am. Chem. Soc.* **1997**, *119*, 12360–12361.

[5] S. D. Rychnovsky, N. A. Powell, *J. Org. Chem.* **1997**, *62*, 6460–6461.

[6] D. A. Evans, P. J. Coleman, B. Cote, *J. Org. Chem.* **1997**, *62*, 788–789.

[7] I. Paterson, K. R. Gibson, R. M. Oballa, *Tetrahedron Lett.* **1996**, *37*, 8585–8588.

[8] D. A. Evans, J. L. Duffy, M. J. Dart, *Tetrahedron Lett.* **1994**, *35*, 8537–8540.

[9] K. Narasaka, F.-C. Pai, *Tetrahedron* **1984**, *40*, 2233–2238; K.-M. Chen, G. E. Hardtmann, K. Prasad, O. Repic, M. J. Shapiro, *Tetrahedron Lett.* **1987**, *28*, 155–158.

[10] S. Kiyooka, H. Kuroda, Y. Shimasaki, *Tetrahedron Lett.* **1986**, *27*, 3009–3012.

[11] D. A. Evans, K. T. Chapman, E. M. Carreira, *J. Am. Chem. Soc.* **1988**, *110*, 3560–3578.

[12] D. A. Evans, A. H. Hoveyda, *J. Am. Chem. Soc.* **1990**, *112*, 6447–6449.

[13] J. Krüger, E. M. Carreira, *Tetrahedron Lett.* **1998**, *39*, 7013–7016.

[14] B. Pagenkopf, J. Krüger, A. Stojanovic, E. M. Carreira, *Angew. Chem.* **1998**, *110*, 3312–3314; *Angew. Chem. Int. Ed.* **1998**, *37*, 3122–3124.

[15] S. Weigand, R. Brückner, *Liebigs Ann./Recueil* **1997**, 1657–1666.

[16] A. B. Smith, A. M. Boldi, *J. Am. Chem. Soc.* **1997**, *119*, 6925–6926.

[17] L. F. Tietze, H. Geissler, J. A. Gewert, U. Jakobi, *Synlett* **1994**, 511–512.

[18] C. Schneider, M. Rehfeuter, *Tetrahedron Lett.* **1998**, *39*, 9–12; C. Schneider, M. Rehfeuter, *Chem. Eur. J.* **1999**, *5*, 2850–2858.

[19] For a comprehensive review of polyol synthesis up to 1990 see: T. Oishi, T. Nakata, *Synthesis* **1990**, 635–645.

Stereoselection at the Steady State: The Design of New Asymmetric Reactions

Thomas Wirth,

Institut für Organische Chemie, Universität Basel, Switzerland

A reaction leading to an optically pure product in quantitative yield is the dream of every chemist working in the field of asymmetric synthesis. If only catalytic amounts of a chiral compound are needed to achieve this goal, then it is like paradise. But only very few asymmetric reactions have been developed thus far.

Compounds with asymmetric centers can be obtained from prochiral starting molecules by either face-selective reactions [1] (stereoheterotopic facial addition) or group-selective reactions (stereoheterotopic ligand substitution). The transition states of these selective stereodivergent reactions must be diastereomeric, and the kinetics are the same as those of parallel reactions with different products (enantiomers or diastereomers). The selectivity in the stereoselective event leading to the different transition states can never be exceeded by the final yield of the major stereoisomer.

In two recent publications Curran et al. described the theoretical as well as the mathematical background of stereoconvergent reactions [2]. They give further evidence for their analysis by providing some examples from the field of radical chemistry to demonstrate this strategy called complex stereoselection. The process of stereoconvergent synthesis was proposed by Fischli et al. in 1975 [3]. The first step in this process is the nonselective monoprotection of a reactive group (X in **1**) by a chiral agent (G*, Scheme 1). After separation of the diastereomers **2a** and **2b**, the still reactive groups are converted into "c" and "d". These derivatives generate **3** with a theoretical enantiomeric excess of 100 % and without destroying any material. The ratio **2a**:**2b** has no influence on the outcome of the reaction.

Although the subsequent discussion describes the stereoselection at the steady state through the example of radical reactions, the analysis and principles are general for any reaction profile that fits into the scheme of complex stereoselective reactions. In the process proposed and analyzed by Curran et al., the activation of compounds of type **1** is done, for example, by radical formation. The group selectivity in this first step has again no effect on the stereomeric nature of the product. To obtain a stereoconvergent process it is crucial, however, that the reaction is operating at the steady state. This means that the concentrations of the radial intermediates (compounds in brackets in Scheme 2) is low and stationary, while their absolute concentrations are determined by the different rates of reaction.

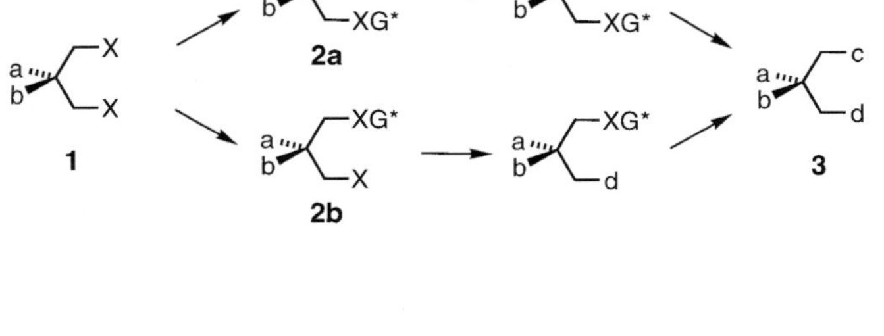

Scheme 1

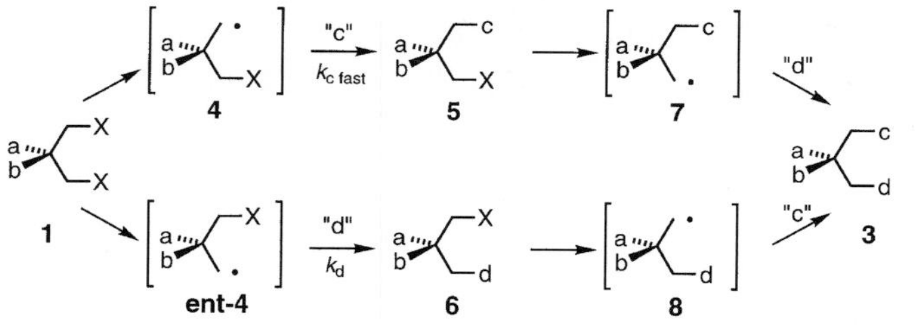

Scheme 2

The radicals **4** and *ent*-**4** formed in the first reaction must have competing reaction rates for c and d: the second process must have a reaction rate in competition with the two rates of the first process ($k_{c\ fast} > k_d > k_{c\ slow}$). This means that radical **4** reacts much faster with c than with d yielding product **5**. Radical *ent*-**4**, however, will react faster with d than with c, and **6** will be produced. Compounds **5** and **6** are then activated again by conversion into **7** and **8**. Radical **7** is mismatched for the reaction with c ($k_{c\ slow} < k_d$), and reaction with d provides product **3**. Radical **8** is converted also into **3** as shown in Scheme 2. The achiral products (disubstitution of X with either c or d) result from a leakage out of one convergence and are only minor products because of the reaction rates. Furthermore, it is necessary for the stereoconvergence at the steady state that at least one of these processes (reaction with either c or d) is stereoselective.

The mathematical description of this and related reaction schemes are complex and will not be discussed here. To verify the analysis of the stereoselection at the steady state experimentally, diastereoselective radical cyclization reactions were selected. Although face-selective radical cyclizations are much more common than the group-selective counterparts [4], several model compounds were suggested by Curran et al [2]. Molecules which have two radical precursors and one radical acceptor group such as **9** were selected to provide experimental verification

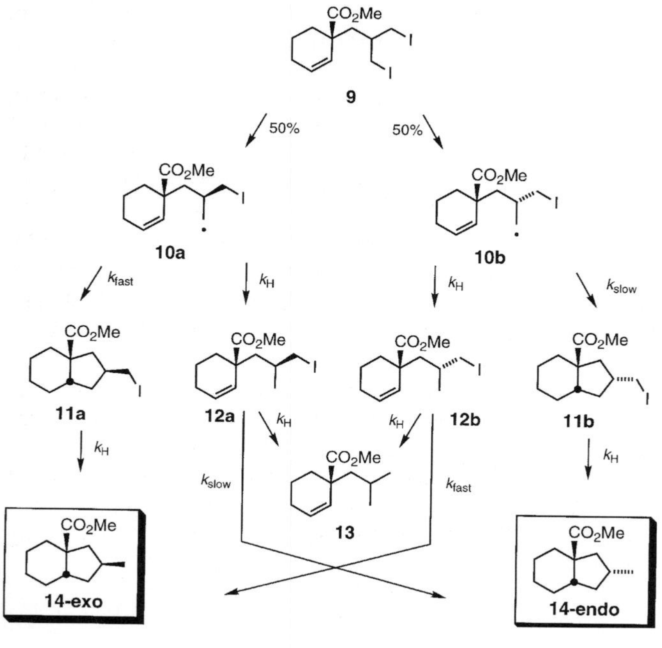

Scheme 3

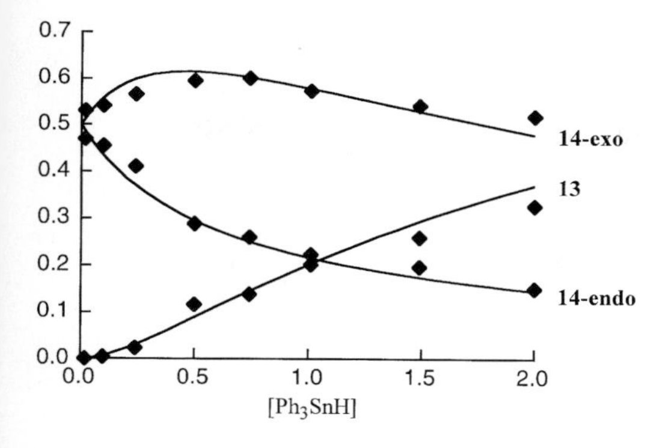

Figure 1. The effect of the tin hydride concentration on the formation of **13**, **14**-*exo*, and **14**-*endo* (selectivity approximately five). The experimental results (◆) are compared with the calculated ones (*lines*) [6].

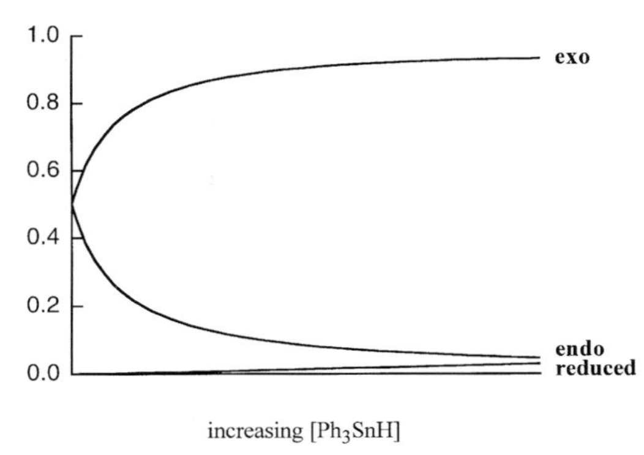

Figure 2. Calculated curves showing the effect of the tin hydride concentration on the formation of **13**, **14**-*exo*, and **14**-*endo* when a selectivity of 500 is assumed.

(Scheme 3). This choice also seems wise from a different point of view: because the intermediates are radicals, the two competing reactions are an intramolecular cyclization and an intermolecular hydrogen transfer reaction. The latter is dependent upon the tin hydride concentration. In other words, by varying the tin hydride concentration it is possible to influence the sterochemical outcome of the reaction.

At low tin hydride concentrations the product ratio of **14**-*exo*:**14**-*endo* is approximately 1 : 1; this ratio confirms that there is no selectivity in the first abstraction of iodine by the tin radical. The radicals **10a** and **10b** are formed in equal amounts. Product **13** is not detected. The rate for the subsequent cyclization reaction is, however, different for **10a** and **10b**. Cyclization of radical **10a** to the *exo*-derivative **11a** is faster than the generation of the *endo*-derivative **11b** from **10b**. With increasing tin hydride concentration the competing hydrogen transfer reaction is opening new paths on the reaction topography. If the faster cyclizing radical **10a** is reduced to **12a** before cyclization occurs, most of the product **12a** will end up at the doubly reduced compound **13**. When the slower cyclizing radical **10b** is reduced, however, most of the resulting product **12b** follows the pathway to the major product **14**-*exo*. The yield of **14**-*exo* is, therefore, both eroded (**12a** → **13**) and supplemented (**12b** → **14**-*exo*). However, because the concentration of the slower cyclizing radical **10b** always exceeds the concentration of the faster cyclizing radical **10a**, the yield of **14**-*exo* is supplemented faster than it is eroded (Scheme 3) [5]. This situation is additionally visualized in Fig. 1.

The data fits nicely into a k_{fast}:k_{slow} ratio of about five. With this selectivity between the major and the minor stereoconvergence, the maximum yield of **14**-*exo* is about 60 %. The importance of this concept, however, becomes appearent when a selectivity of 500 is assumed. Figure 2 shows the calculated curves for this scenario, whereby an increase in the tin hydride concentration leads to a large *exo*:*endo* ratio.

The reaction of **9** to products **14** is one simple example for a complex stereoselective reaction. Other reactions based on radical chemistry are suggested by Curran et al., but these reactions with a stereoselection at the steady state are by no means restricted to radical chemistry. Organometallic reactions can also involve transient intermediates which may fulfill the crucial need for competing reactions with different rates, with the rate of the nonselective reaction lying between the rates of the two selective ones. In this context catalytic enantioselective reactions are especially challenging and could have a high preparative value. The stage is now set for the discovery of new stereoconvergent reactions orchestrated by stereoselection at the steady state.

References

[1] The dependence of reaction conditions on the facial diastereoselectivity of nucleophilic addition was reported recently: G. Cainelli, D. Giacomini, P. Galletti, *Chem. Commun.* **1999**, 567–572.

[2] a) N. C. DeMello, D. P. Curran, *J. Am. Chem. Soc.* **1998**, *120*, 329–341. b) D. P. Curran, C.-H. Lin, N. DeMello, J. Junggebauer, *J. Am. Chem. Soc.* **1998**, *120*, 342–351.

[3] A. Fischli, M. Klaus, H. Meyer, P. Schönholzer, R. Rüegg, *Helv. Chim. Acta* **1975**, *58*, 564–584.

[4] a) D. P. Curran, S. J. Geib, C. H. Lin, *Tetrahedron: Asymmetry* **1994**, *5*, 199–202. b) D. P. Curran, W. Shen, J. Zhang, S. J. Gieb, C.-H. Lin, *Heterocycles* **1994**, *37*, 1773–1788.

[5] The pairs of fast and slow rate constants are assumed to be equal to simplify the theoretical analysis.

[6] The following rate constants have been used for the calculation: $k_{fast} = 4.6 \times 10^7$ s^{-1}; $k_{slow} = 1.0 \times 10^7$ s^{-1}; $k_H = 2.0 \times 10^7$ s^{-1}.

B. Transition Metal Organometallic Methods

Photolysis of Fischer Carbene Complexes

Oliver Kiehl and Hans-Günther Schmalz

Institut für Organische Chemie, Universität Köln, Germany

Introduction

Heteroatom-stabilized carbene complexes of type **1**, first discovered by E.O. Fischer in 1964 [1], nowadays belong to the best investigated classes of transition metal compounds. Such complexes are coordinatively saturated, intensely colored solids ($\lambda_{max} = 350-400$ nm), which exhibit a sufficient stability for normal preparative use. Especially chromium carbene complexes (**2**) enjoy increasing importance in organic synthesis, and it must be added that *thermal* reactions such as benzannulations (i.e. the Dötz reaction), cyclopropanations and additions to α,β-unsaturated complexes clearly predominate [2].

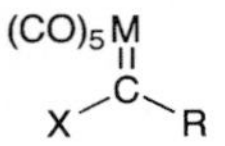

M = Cr, Mo, W
X = OR, NR_2

The synthetic utility of these compounds is based on their easy accessibility from commercially available $[Cr(CO)_6]$ by treatment with an organolithium compound and subsequently with Meerweins salt [1b]. Alternatively, the anionic intermediates **3** can be generated by reaction of acid chlorides with $Na_2[Cr(CO)_5]$ (Scheme 1) [3].

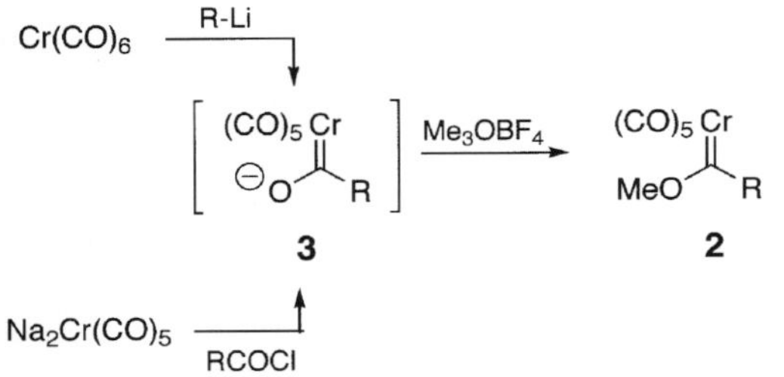

Scheme 1

Besides the thermal reactions of Fischer carbenes, photochemical transformations of such complexes also deserve attention. Since the discovery of McGuire and Hegedus [4] in the early 1980s that irradiation of chromium carbene complexes (**4**) with visible light in the presence of N-substituted benzaldimines (**5**) leads to (racemic) β-lactams of type **6**, photolytic reactions of Fischer carbene complexes have been intensively investigated by L.S. Hegedus and his group in Fort Collins, Colorado [5].

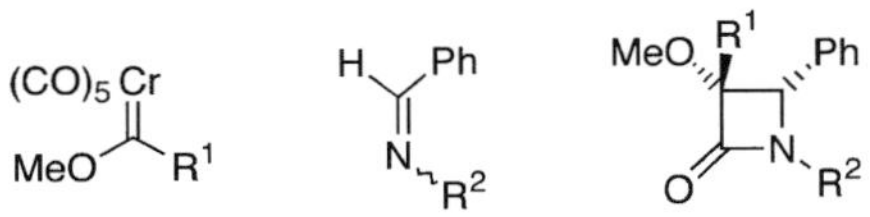

Nowadays, it is an accepted mechanistic model [5, 6] that the photolysis step (which proceeds under thermo-reversible CO insertion) leads to species best described as chromium ketene complexes of type **7** (Scheme 2). Indeed, these intermediates exhibit a ketene-like reactivity: they undergo [2 + 2] cycloaddition reactions with olefins, imines and enol ethers, whereas reaction with nucleophiles leads to carboxylic acid derivatives.

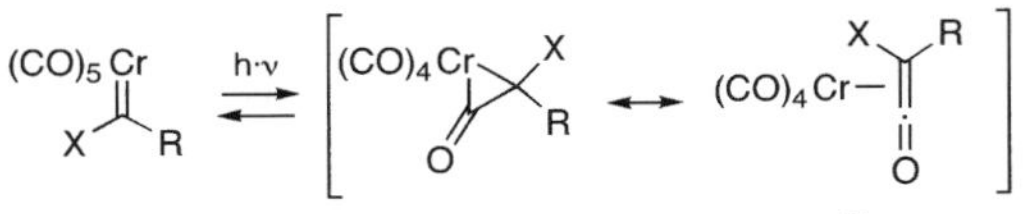

Scheme 2

Thus, the photolysis of Fischer carbenes opens up, under extremely mild (neutral) conditions, a synthetic access to electron-rich alkoxy- or amino ketenes which are difficult or impossible to synthesize through other means. These ketene intermediates are generated as metal-bound species in low stationary concentrations (less side reactions!), but can be further reacted in a variety of synthetically useful ways. In recent years, several synthetic methods exploiting this chemistry have been developed to a remarkable level of maturity. This article wants to briefly highlight this chemistry by discussing a few selected applications.

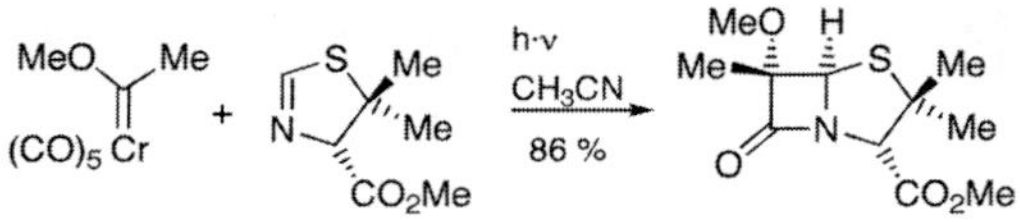

Scheme 3

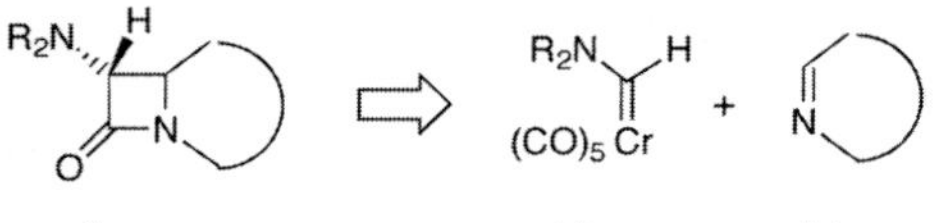

Scheme 4

Synthesis of *β*-Lactams

As mentioned above, the irradiation of Fischer carbenes in the presence of imines offers access to *β*-lactams [4–8]. The preparation of compound **8** is a prominent example [6b] which also demonstrates that it is possible to achieve high diastereoselectivities if chiral imines are employed as starting materials (Scheme 3). It should be mentioned that $[Cr(CO)_6]$ can be recovered if the photolysis is carried out under CO pressure.

The most biologically active *β*-lactams are *α*-amino substituted derivatives (**9**), which according to Hegedus retrosynthetically derive from amino carbene complexes of type **10** and cyclic imines (**11**) (Scheme 4).

For complexes of type **10** (with a hydrogen at the carbene carbon) a synthesis was worked out in which a formamide is first reacted with $K_2[Cr(CO)_5]$ followed by reaction with TMSCl [7]. This way, the non-racemic formamide **12** leads to the chirally modified amino carbene complex **13**, which serves as starting material for the diastereoselective synthesis of various optically active *β*-lactams [8]. An example is the (formal) total synthesis of 1-carbacephalothin **16**, a carbon analog of the cephalosporins (Scheme 5) [8b]. In this case, the complex **13** is irradiated in the presence of in situ prepared imine **14** to afford the *β*-lactam with high diastereoselectivity but only in modest yield. The product (**15**) could (in principle) be converted in to the target compound **16**.

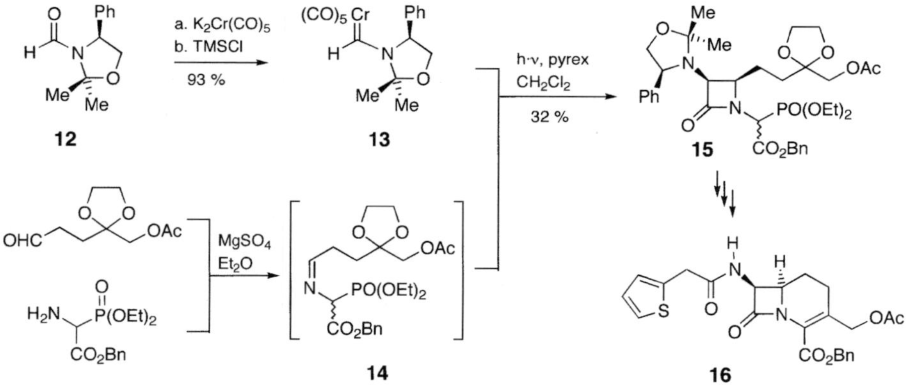

Scheme 5. Formal total synthesis of (+)-1-carbacephalothin according to L. S. Hegedus.

Synthesis of Cyclobutanones

Similar to the thermal reaction of ketenes with alkenes, the photolysis of alkoxycarbenes **18** in the presence of (electron-rich) olefins **19** leads to cyclobutanones (Scheme 6) [9]. In these reactions the sterically more strained [2 + 2] cycloadducts of type *rac*-**17** are generally formed with good regio- and diastereoselectivity. Starting from complexes of type **21**, the intramolecular version of this reaction affords bicyclic products of type *rac*-**20** (Scheme 7) [9].

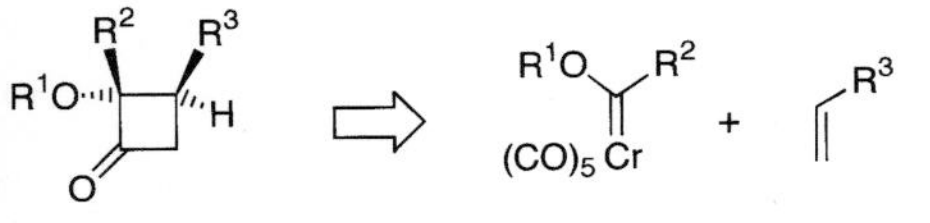

Scheme 6

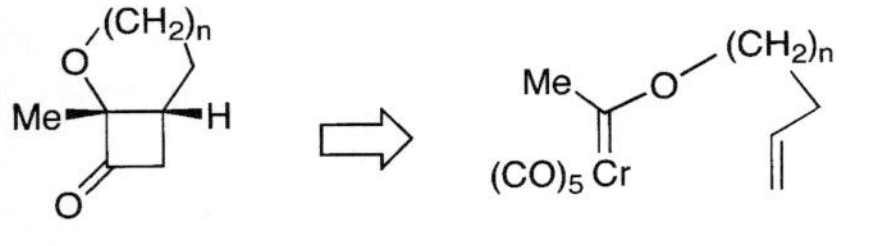

Scheme 7

If suitable chiral alkenes are employed it is additionally possible to control the absolute configuration of the newly formed chiral centers [10]. The efficiency of this method was demonstrated in a total synthesis of the antibiotic (+)-tetrahydrocerulenin **28** (Scheme 8) and (+)-cerulenin [11]. Irradiation of complex **22** in the presence of the chiral *N*-vinyl-oxazolidinone **24**, which is easily prepared from the amino carbene complex **23** [12], leads to the cyclobutanone **25** with high diastereoselectivity. Regioselective Baeyer-Villiger oxidation followed by base-induced elimination of the chiral carbamate yields the butenolide **26** in high enantiomeric purity. This is finally converted, using Nozoe's protocol [13], to the target molecule **28** by diastereoselective epoxidation (→ **27**) and subsequent aminolysis.

In an analogous fashion (using substrates with modified alkyl chains), the total synthesis of the structurally related butenolides **29** and **30** was achieved [11a]. These compounds are metabolites of the marine sponge *Plakortis lita*. In these cases, the chiral olefin *ent*-**24** was used in the photolytic step. Again, the target molecules were obtained in high enantiomeric purity (>95 % *ee*) with impressively high overall yields of 48 % and 38 %, respectively, based on chromium hexacarbonyl.

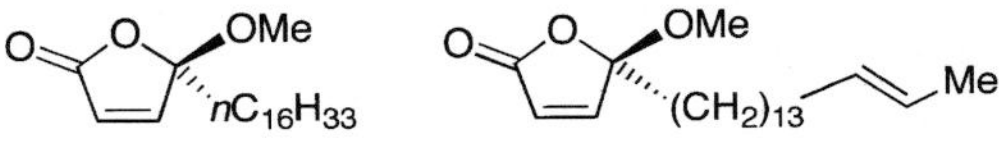

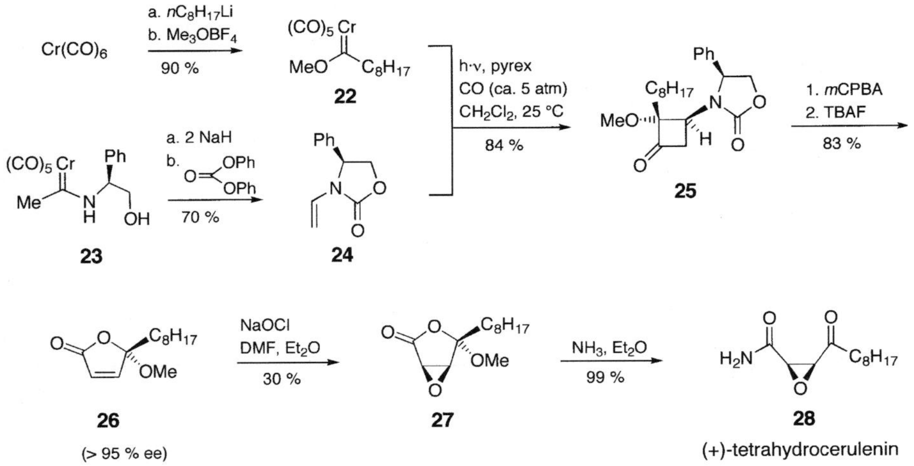

Scheme 8. Total synthesis of (+)-tetrahydrocerulenin according to L. S. Hegedus.

Synthesis of Amino Acid Derivatives

The photolysis of chiral amino carbene complexes of type **31** [14] in the presence of alcohols leads diastereoselectively to products of type **32** and thus opens an efficient access to a wide range of natural and non-natural amino acid derivatives in optically active form [15]. These reactions proceed via the intermediates **33** and **34**, the products being formed by highly diastereoselective protonation of **34** (Scheme 9).

As an example, the reaction of complex **13** with MeOD gives compound **35** in excellent yield which can easily be converted to optically active [2-$^2H_1$] glycine (Scheme 10) [16].

A special highlight of this chemistry is the possibility to construct compounds with two adjacent ^{13}C-labeled carbon atoms [17], which are of great interest for the investigation of biological systems by means of ^1H NMR spectroscopic methods. For this purpose, the starting complexes (i.e. **36**) are prepared from ^{13}C-labelled chromium hexacarbonyl (Scheme 11).

Photolysis of **36** in the presence of the carbonyl-protected alanine derivative **37** allows the diastereoselective assembly of the dipeptide **38** in a one pot synthesis [13]. This process is remarkable for several reasons: on the one hand the peptide bond and one new stereogenic center are established in a single preparative step; on the other hand visible light serves as the coupling reagent. This opens up a possibility for the direct incorporation of various (even very unusual) amino acids into peptides without the requirement to synthesize the amino acid separately.

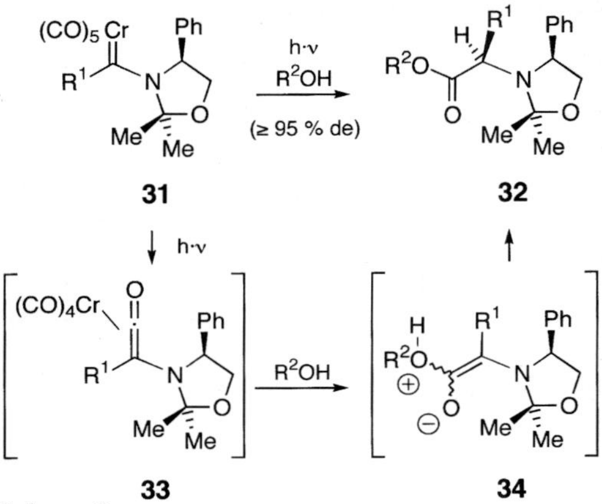

Scheme 9

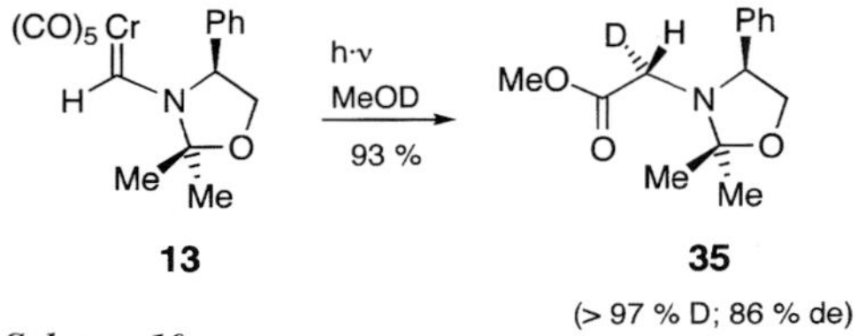

Scheme 10

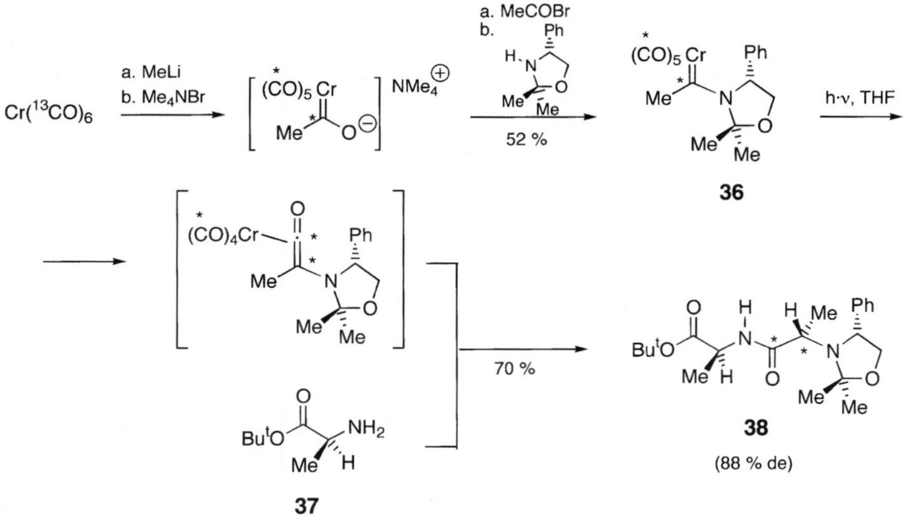

Scheme 11. Preparation of a double ^{13}C labelled dipeptide according to L. S. Hegedus.

Synthesis of *ortho*-Substituted Phenol Derivatives

In recent years the group of C. A. Merlic has reported photochemically induced cyclizations of dienyl carbene complexes of type **39** to produce phenol derivatives **40** [19]. In these very intelligently designed reactions, which are related to the Dötz reaction, the primary, photochemically generated intermediates of type **41** undergo a (formal) electrocyclic ring-closure to form linear, conjugated cyclohexadienones **42**, which then tautomerize to the phenols (Scheme 12).

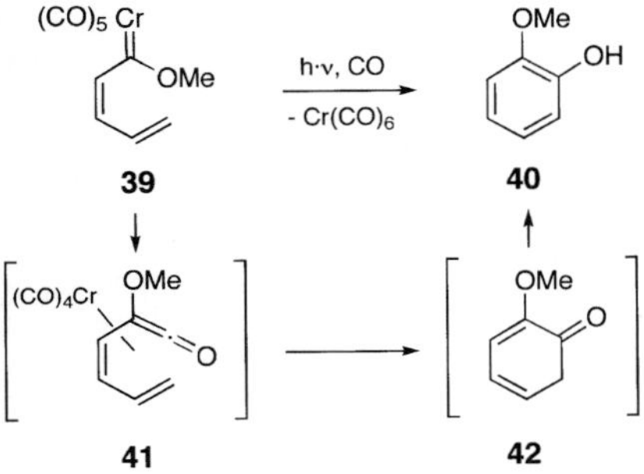

Scheme 12

Using this principle, especially benzannulations can be accomplished in high yields, for instance the reactions of **43** to **44** and of **45** to **46**, respectively (Scheme 13). As demonstrated by the second example, it is possible to get high regioselectivities (>25 : 1) in cases of substrates with unsymmetrically substituted aromatic rings [19b].

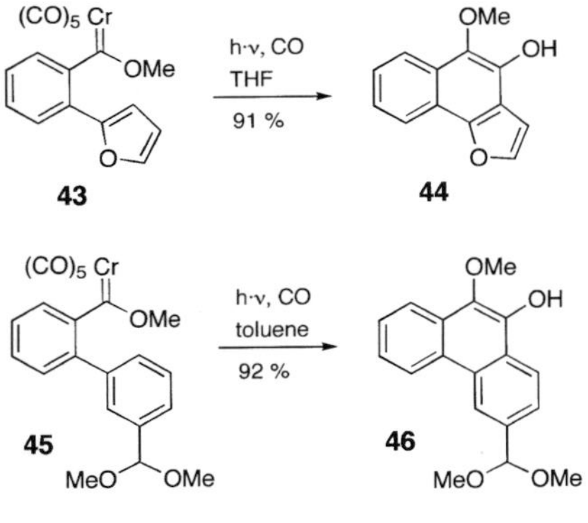

Scheme 13

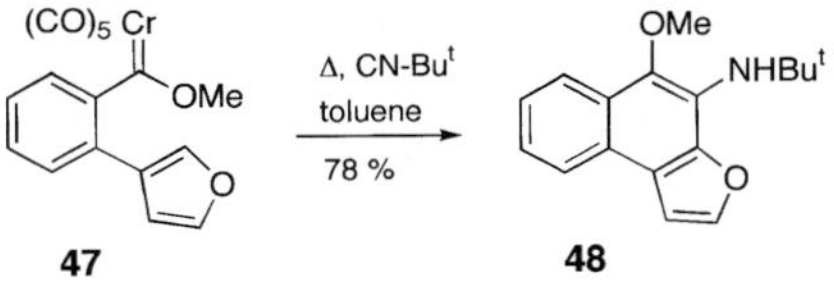

Scheme 14

Finally, it should be mentioned that the reaction of complexes of type **39** with isonitriles leads under mild thermal conditions to aromatic amines (i.e. **47** → **48**) [20]. The comparably easy formation of the intermediate ketimines (due to the nucleophilicity of the isonitrile) even permits reactions without photochemical activation (Scheme 14).

Outlook

This brief article, which because of its limited size does not refer to all the work which has been done in the field, is intended to convince the reader that the photolytic generation of electron-rich ketene equivalents from Fischer carbene complexes represents quite a general and highly valuable synthetic methodology. One should expect that there will be a lot of interesting and useful applications arising from this chemistry in the future.

References

[1] a) E. O. Fischer, A. Maasböl, *Angew. Chem.*, **1964**, *76*, 645; *Angew. Chem. Int. Ed. Engl.* **1964**, *3*, 580; b) R. Aumann, E. O. Fischer, *Chem. Ber.*, **1968**, *101*, 954.

[2] For selected reviews, see: a) K. H. Dötz, *Angew. Chem.*, **1984**, *96*, 573; *Angew. Chem. Int. Ed. Engl.* **1984**, *23*, 587; b) K. H. Dötz in *Organometallics in Organic Synthesis* (Eds: A. de Meijere, H. tom Dieck), Springer, Berlin, **1988**, p. 85; c) W. D. Wulff, *Advances in Metal-Organic Chemistry*, **1989**, *1*, 209; d) W. D. Wulff in *Comprehensive Organic Synthesis, Vol. 5* (Eds.: B. M. Trost, I. Fleming), Pergamon, Oxford, **1991**, p. 1065; W. D. Wulff in *Comprehensive Organometallic Chemistry II, Vol. 12* (Eds.: E.W. Abel, F.G.A. Stone, G. Wilkinson), Pergamon, Oxford, **1995**, p. 470.

[3] M. F. Semmelhack, G. R. Lee, *Organometallics* **1987**, *6*, 1839.

[4] a) M. A. McGuire, L. S. Hegedus, *J. Am. Chem. Soc.* **1982**, *104*, 5538; b) L. S. Hegedus, *Pure Appl. Chem.* **1983**, *55*, 1745; c) L. S. Hegedus,

M. A. McGuire, L. M. Schultze, C. Yijun, O. P. Anderson, *J. Am. Chem. Soc.* **1984**, *106*, 2680; c) L. S. Hegedus, M. A. McGuire, L. M. Schultze, *Org. Synthesis* **1987**, *65*, 140.

[5] For an excellent review, see: a) L. S. Hegedus, *Tetrahedron* **1997**, *53*, 4105.

[6] a) L. S. Hegedus, G. deWeck, S. DAndrea, *J. Am. Chem. Soc.* **1988**, *110*, 2122; b) L. S. Hegedus, *Phil. Trans. R. Soc. Lond. A.* **1988**, *326*, 505.

[7] a) R. Imwinkelried, L. S. Hegedus, *Organometallics* **1988**, *7*, 702; b) M. A. Schwindt, T. Lejon, L. S. Hegedus, *Organometallics* **1990**, *9*, 2814.

[8] a) L. S. Hegedus, R. Imwinkelried, M. Alarid-Sargent, D. Dvorak, Y. Satoh, *J. Am. Chem. Soc.* **1990**, *112*, 1109; b) Y. Narukawa, K. N. Juneau, D. Snustad, D. B. Miller, L. S. Hegedus, *J. Org. Chem.* **1992**, *57*, 5453; c) B. Roman, L. S. Hegedus, *Tetrahedron* **1993**, *49*, 5549; d) P.-C. Colson, L. S. Hegedus, *J. Org.Chem.* **1993**, *58*, 5918.

[9] a) B. C. Söderberg, L. S. Hegedus, M. Sierra, *J. Am. Chem. Soc.* **1990**, *112*, 4364; b) A. G. Riches, L. A. Wernersbach, L. S. Hegedus, *J. Org. Chem.* **1998**, *63*, 4691.

[10] L. S. Hegedus, R. W. Bates, B. C. Söderberg, *J. Am. Chem. Soc.* **1991**, *113*, 923;

[11] a) M. Miller, L. S. Hegedus, *J. Org. Chem.* **1993**, *58*, 6779; b) T. E. Kedar, M. W. Miller, L. S. Hegedus, *J. Org. Chem.* **1996**, *61*, 6121.

[12] J. Montgomery, G. M. Wieber, L. S. Hegedus, *J. Am. Chem. Soc.* **1990**, *112*, 6255.

[13] T. Ohta, H. Tsuchiyama, S. Nozoe, *Heterocycles* **1986**, *24*, 1137.

[14] For a brief (limited) review on the use of aminocarbene complexes, see: M. A. Schwindt, J. R. Miller, L. S. Hegedus, *J.Organomet. Chem.* **1991**, *413*, 143.

[15] a) L. S. Hegedus, M. A. Schwindt, S. DeLombaert, R. Imwinkelried, *J. Am. Chem. Soc.* **1990**, *112*, 2264; b) J. M. Vernier, L. S. Hegedus, D. B. Miller, *J. Org.Chem.* **1992**, *57*, 6914.

[16] L. S. Hegedus, E. Lastra, Y. Narukawa, D. C. Snustad, *J. Am. Chem. Soc.* **1992**, *114*, 2991.

[17] E. Lastra, L. S. Hegedus, *J. Am. Chem. Soc.* **1993**, *115*, 87.

[18] J. R. Miller, S. R. Pulley, L. S. Hegedus, S. DeLombaert, *J. Am. Chem. Soc.* **1992**, *114*, 5602; b) S. R. Pulley, L. S. Hegedus, *J. Am. Chem. Soc.* **1993**, *115*, 9037.

[19] a) C. A. Merlic, D. Xu, *J. Am. Chem. Soc.* **1991**, *113*, 7418; b) C. A. Merlic, W. M. Roberts, *Tetrahedron Lett.* **1993**, *34*, 7379; c) C. A. Merlic, D. Xu, B. G. Gladstone, *J. Org. Chem.* **1993**, *58*, 538.

[20] a) C. A. Merlic, E. E. Burns, D. Xu, S. Y. Chen, *J. Am. Chem. Soc.* **1992**, *114*, 8722; b) C. A. Merlic, E. E. Burns, *Tetrahedron Lett.* **1993**, *34*, 5401.

Zr-Catalyzed Carbomagnesation of Alkenes

Florian Blume and Hans-Günther Schmalz

Institut für Organische Chemie, Universität Köln, Germany

Carbometallations, i.e. reactions in which a new metallo-organic species **3** is formed by (formal) 1,2-addition of a metallo-organic species **1** along a multiple C-C bond **2**, exhibit a great synthetic potential [1]. After all, not only is a new C-C bond formed during the course of the reaction, but also a carbon-metal bond that is directly available for further synthetic exploitation.

R–M + C=C ⟶ R–C–C–M

1 **2** **3**

While organo-magnesium species (Grignard reagents) can be counted amongst the most utilized organometallic reagents due to their facile availability [2], carbomagnesations of non-activated alkenes have not received very much attention in the past. The reason can be attributed to the fact that such reactions usually only proceed, if at all, under rather drastic conditions [3]. Exceptions are allylmagnesium halides **4**, which react with olefins under comparatively mild conditions via an ene-type mechanism to yield (intermediate) products of type **5**. Intramolecular versions of such "magnesium-ene reactions" often proceed with high regio- and diastereoselectivity. They have been employed by Oppolzer and co-workers with tremendous success as key steps in natural product syntheses [4].

MgX "ene" ⟶ MgX

4 **5**

A few years ago, Dzhemilev and co-workers reported that the ethyl-magnesation of terminal (non-activated) alkenes (**6**) takes place under relatively mild conditions if catalytic amounts of zirconocene dichloride (Cp_2ZrCl_2) are present [5, 6]. These reactions lead with high regioselectivity to (racemic) products of type **7**. Best results are achieved with diethylmagnesium: other ethylmagnesium derivates (e.g. EtMgX) react more slowly, and longer-chain dialkylmagnesium compounds give rise to undesired dimerization and elimination products. The high tolerance of the Zr-catalyzed reactions towards other functionalities, such as trimethylsilyl, dialkylamino, ketal (e.g. **8** → **9**), alkoxy, and even hydroxy groups in the substrates is remarkable.

Et_2Mg, 1 mol % Cp_2ZrCl_2, Et_2O, 4 h: R (**6**) ⟶ R, Et, MgEt (**7**)

Et_2Mg, 1 mol % Cp_2ZrCl_2, Et_2O, 4 h, 95 %: O, O, Me (**8**) ⟶ O, O, Me, Et, MgEt (**9**)

Zr-catalyzed carbomagnesation has gained much importance since 1991 as a consequence of the work of Hoveyda and co-workers [7, 8]. These researchers discovered that reactions of terminal alkenes (**10**) with an excess of the simple Grignard reagent EtMgCl proceeded with practically complete regioselectivity (>99 : 1).

Also, they demonstrated that the Grignard intermediates **11** could easily be further converted to a series of valuable functionalized products.

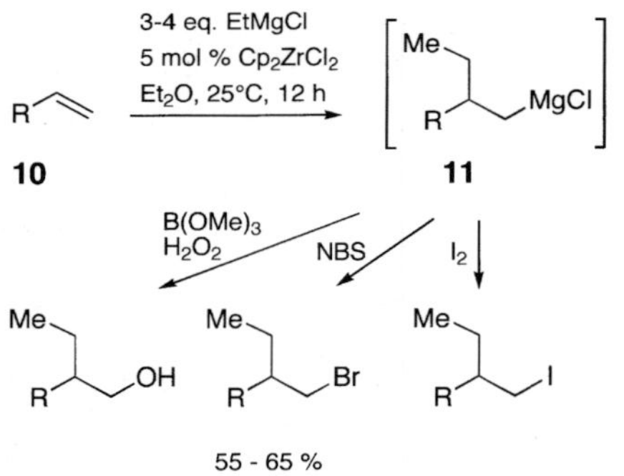

3-4 eq. EtMgCl
5 mol % Cp_2ZrCl_2
Et_2O, 25°C, 12 h

10 **11**

$B(OMe)_3$
H_2O_2
NBS
I_2

55 - 65 %

In a simplified picture, the mechanism of the Zr-catalyzed ethylmagnesation can be rationalized as shown in Scheme 1 [8]. At first, the zirconocene-ethene complex **12** is generated from the catalyst precursor Cp_2ZrCl_2. Complex **12** can also be regarded as a metallacyclopropane **16**. After coordination and insertion of the alkene **10**, a metalla-cyclopentane **13** is formed, which subsequently reacts with the Grignard reagent regioselectively to the open-chain intermediate **14** (transmetallation). After β-H elimination to give **15**, the product **11** is released by reductive elimination while the active catalyst **12** is regenerated.

A mechanistically related reaction that is worth mentioning is the cyclomagnesation of dienes [9]. In this case, treatment of α,ω-dienes of type **17** with 2 equivalents of butylmagnesium chloride in the presence of catalytic amounts of Cp_2ZrCl_2 gives cyclized products (1,4-bis-Grignard-reagents) of type **18**.

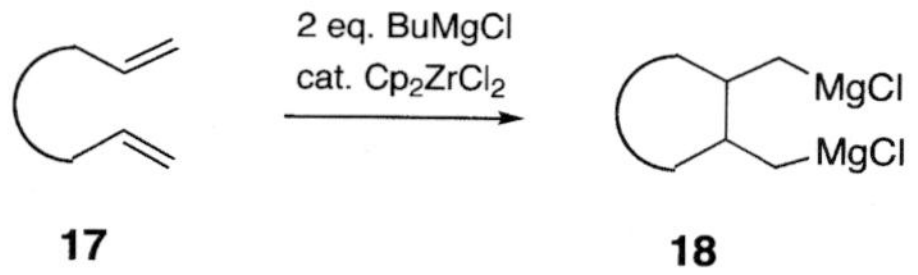

2 eq. BuMgCl
cat. Cp_2ZrCl_2

17 **18**

From the mechanism sketched in Scheme 1 one can conclude that carbomagnesation is unsuitable for a methyl group transfer. Homologous Grignard reagents, as seen in the transformation **19** to **20**, however, can be used successfully in some cases. This example directs attention to a further aspect of this chemistry, i.e. the high stereoselectivity of many Zr-catalyzed carbomagnesations.

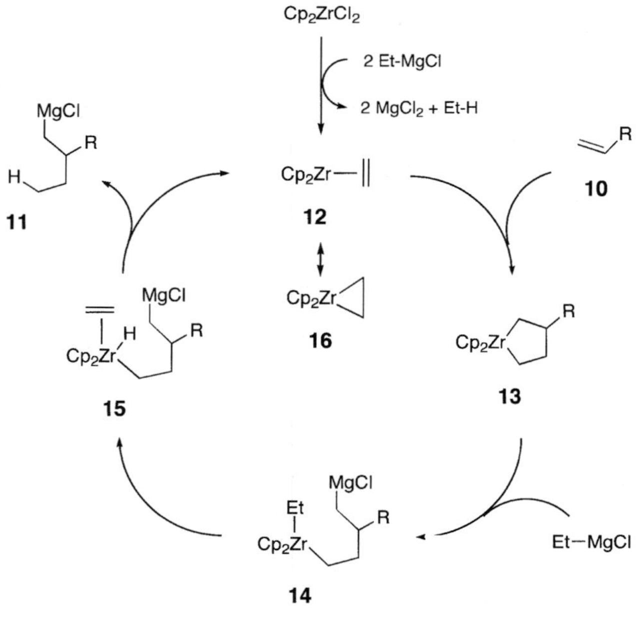

Cp_2ZrCl_2
2 Et-MgCl
2 $MgCl_2$ + Et-H
Et−MgCl

Scheme 1. Mechanism of the Zr-catalyzed ethylmagnesation of alkenes (simplified).

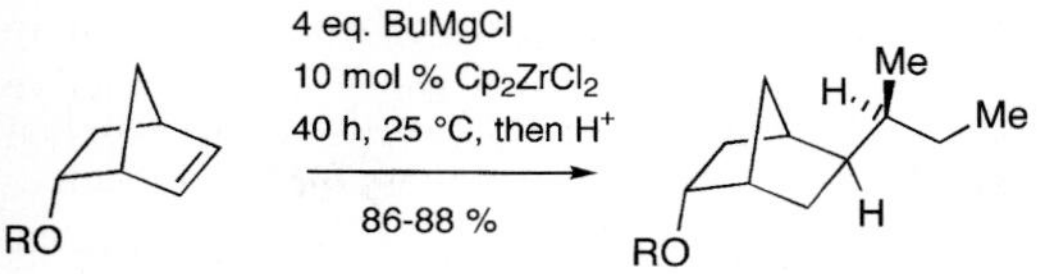

4 eq. BuMgCl
10 mol % Cp_2ZrCl_2
40 h, 25 °C, then H^+
86-88 %

19 (R = H, MEM) → **20** (> 99 % *de*)

Diastereoselective Reactions

An important contribution from Hoveyda is the realization that carbomagnesations only occur smoothly and, more importantly, diastereoselectively, when the substrate provides a Lewis base functionality in the proximity of the double bond. Especially allylic and homoallylic alcohols and the corresponding ethers can be transformed with high diastereoselectivity [7]. Remarkably, the Zr-catalyzed addition of ethyl magnesium chloride to secondary allylic alcohols of type **21** gives mainly *syn*-configured products, while methyl ethers (**23**) lead predominantly to the *anti*-configured products. The formed Grignard derivatives can be subsequently oxidized, for instance, to 1,3-diol derivatives of type **22** or **24** (Scheme 2).

The stereochemistry of the products can be controlled by optional protection of the OH moiety. The highly selective transformation of **25** to **26** demonstrates that other ethers, such as methoxyethoxymethyl (MEM), are also suitable.

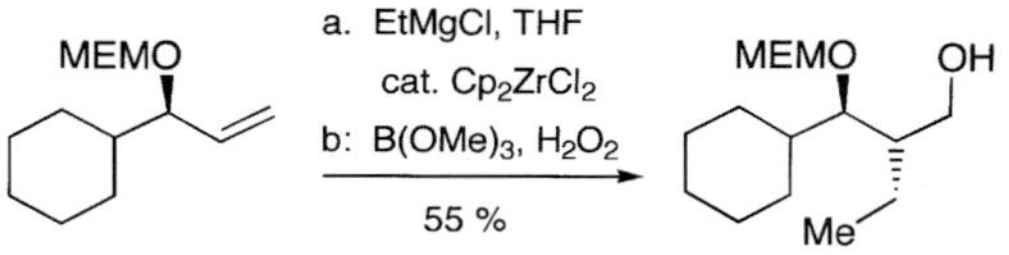

a. EtMgCl, THF
cat. Cp_2ZrCl_2
b: $B(OMe)_3$, H_2O_2
55 %

25 → **26** (> 99 % *de*)

The relative reactivity of allyl ethers depends on steric factors. For instance, compound **27** reacts with virtually complete regioselectivity at the less hindered double bond, as indicated by the formation of **28**.

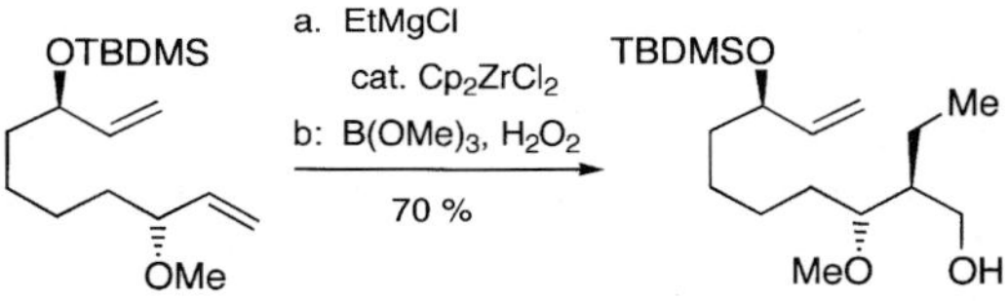

a. EtMgCl
cat. Cp_2ZrCl_2
b: $B(OMe)_3$, H_2O_2
70 %

27 → **28** (80 % *de*)

Homoallylic alcohols also react with high diastereoselectivity (e.g. **29**→**30**), and their ethers give products with the same configuration (albeit with slightly lower selectivity).

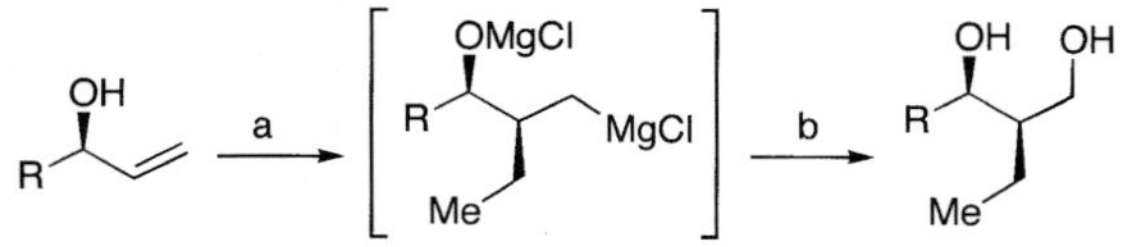

21 —a→ [*syn*] —b→ **22**

R	yield (%)	*syn* / *anti*
n-C_9H_{19}	70	95 : 5
c-hex-CH_2	72	95 : 5
benzyl	53	85 : 15

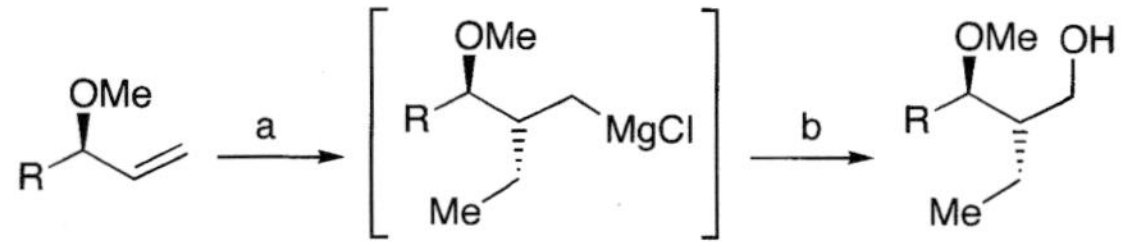

23 —a→ [*anti*] —b→ **24**

R	yield (%)	*syn* / *anti*
n-C_9H_{19}	80	11 : 89
c-hexyl	90	4 : 96
benzyl	90	12 : 88

a: 3 eq. EtMgCl, Et_2O, 5-10 mol % Cp_2ZrCl_2, 25°C, 8 - 12 h
b: $B(OMe)_3$, H_2O_2, -78 °C

Scheme 2. Diastereoselective ethylmagnesation according to A. H. Hoveyda.

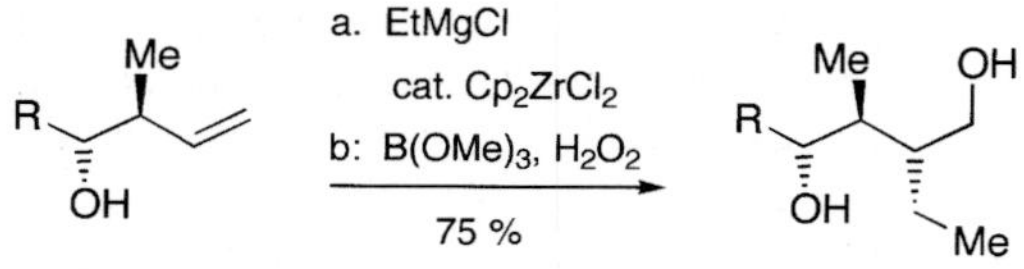

a. EtMgCl cat. Cp_2ZrCl_2
b: $B(OMe)_3$, H_2O_2
75 %

29 → 30 (> 90 % *de*)

Enantioselective Reactions

Because of the availability of the chiral catalyst **31**, it was obvious to investigate the enantioselective carbomagnesation of prochiral substrates. Hoveyda and co-workers demonstrated that asymmetric carbomagnesations of cyclic allylic ethers (or allylic amines) can be achieved with high enantioselectivities (Scheme 4) [12]. In this manner several synthetically valuable compounds, such as those of type **34** have been synthesized from precursors of type **35**.

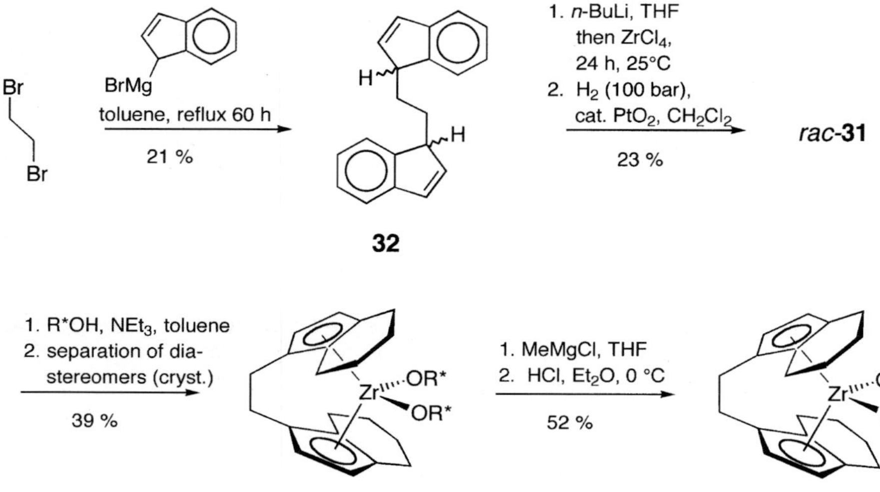

BrMg (indenyl), toluene, reflux 60 h, 21 % → **32**

1. *n*-BuLi, THF then $ZrCl_4$, 24 h, 25°C
2. H_2 (100 bar), cat. PtO_2, CH_2Cl_2
23 % → *rac*-**31**

1. R*OH, NEt_3, toluene
2. separation of diastereomers (cryst.)
39 % → **33**

1. MeMgCl, THF
2. HCl, Et_2O, 0 °C
52 % → **31**

Scheme 3. Synthesis of the chiral zirconocene catalyst **31** according to H.-H. Brintzinger.

The diastereoselectivities that can be achieved in the transformations of chiral, non-racemic substrates can be improved in certain cases by use of a chiral catalyst [10]. Because of the effect of double diastereoselection the correct absolute configuration of the catalyst is important. Hoveyda et al. chose the chiral *ansa*-zirconocene derivative **31**, which can be synthesized according to a method described by Brintzinger (Scheme 3) [11].

At first, reaction of 1,2-dibromethane with an excess of indenyl magnesium bromide leads to the ethano-bridged bis-tetrahydroindenyl ligand **32** (mixture of isomers), from which the racemic complex (*rac*-**31**) is obtained. The separation of the enantiomers is finally achieved by fractionated crystallization of the derivative **33** (R*OH = (*R*)-O-acetylmandelic acid).

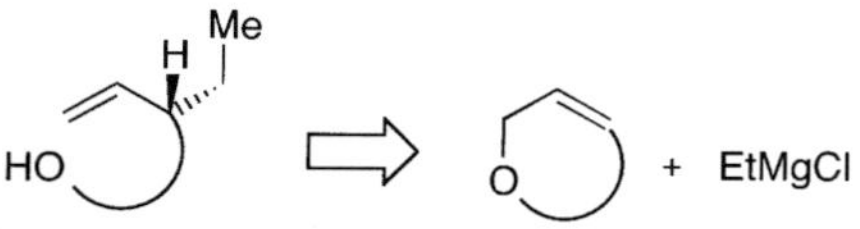

34 ⇨ **35** + EtMgCl

The fact that these reactions proceed under ring opening can easily be understood. For instance, the dihydrofuran **36** initially gives rise to **38** (via the intermediate **37**), which easily fragments. Concomitant with the regeneration of the catalyst, the magnesium salt of the isolated product (**39**) is formed.

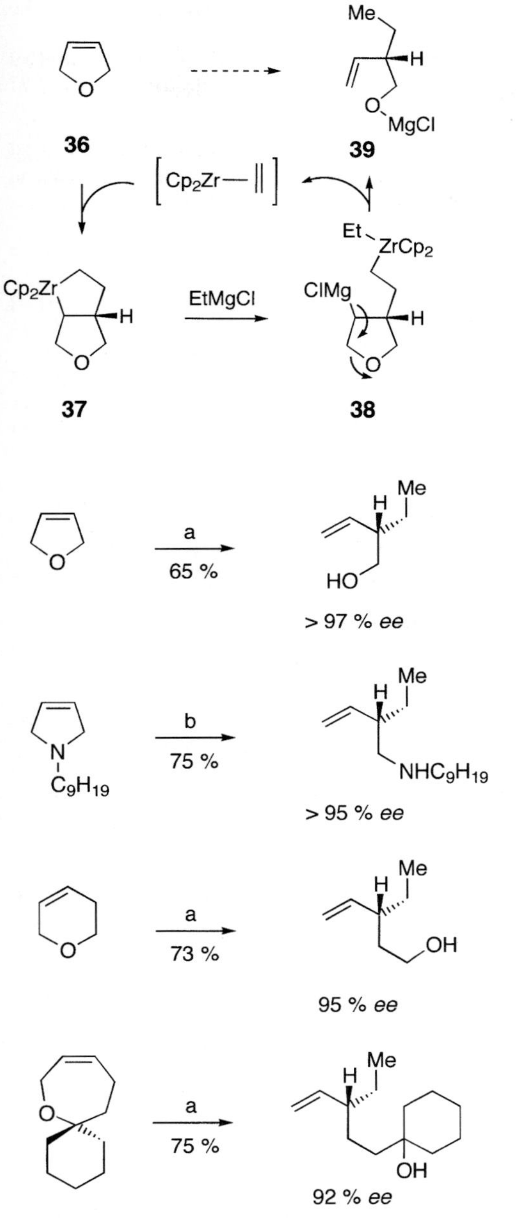

a:	10 mol % **31**, 5 eq. EtMgCl, THF, 25 °C, 6-12 h
b:	10 mol % **31**, 5 eq. EtMgCl, THF, 4 °C, 12 h

Scheme 4. Enantioselective ethylmagnesation according to A. H. Hoveyda.

The examples listed in scheme 4 illustrate the reliability of this remarkable new methodology for catalytic enantioselective C-C bond formation. It should be mentioned that isopropylated products are generated when *n*-propyl magnesium chloride is employed – with good selectivity, as the transformation **40** to **41** indicated.

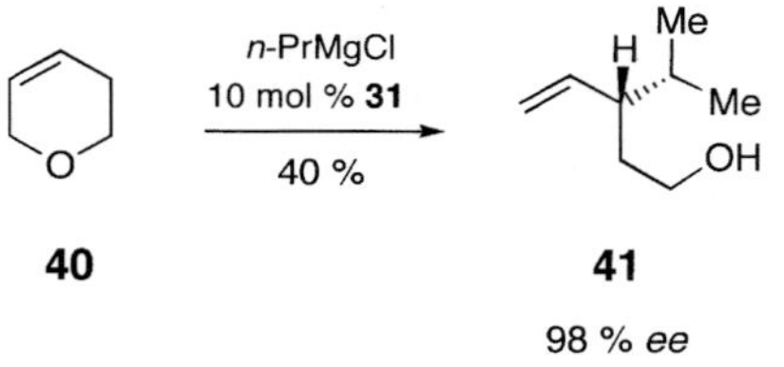

n-PrMgCl, 10 mol % **31**, 40 %; **40** → **41**, 98 % *ee*

In more recent work, Hoveyda et al. go a step further and utilize the enantioselective carbomagnesation (catalyzed by **31**) for the kinetic resolution of racemic dihydropyrans of type *rac*-**42**, *rac*-**43**, or *rac*-**44**.

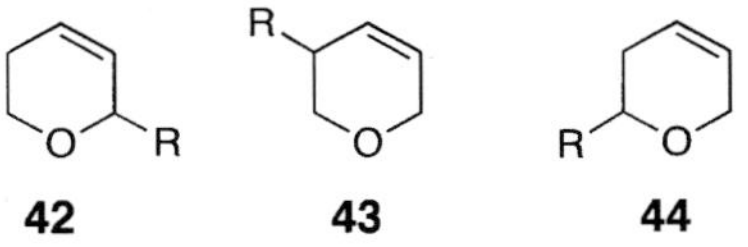

When these reactions are stopped at about 60 % conversion, it is possible in most cases to isolate the unreacted dihydropyrans (e.g. **45**) with enantiomeric purities of >99 % *ee*. Interestingly, these selectivities can only be achieved at elevated temperatures (70 °C).

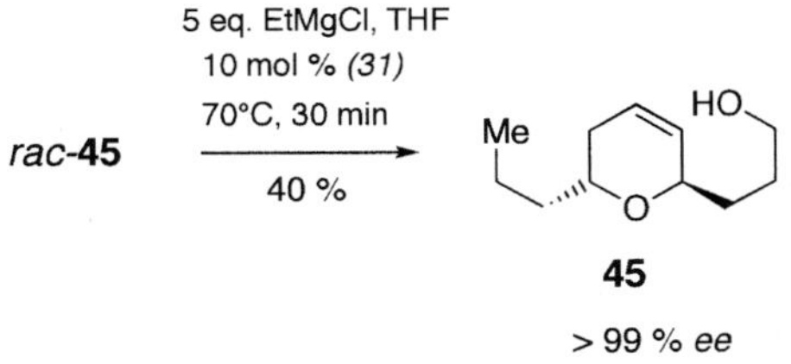

rac-**45**: 5 eq. EtMgCl, THF, 10 mol % *(31)*, 70°C, 30 min, 40 % → **45**, > 99 % *ee*

The chiral product (**39**) obtained from the asymmetric ethylmagnesation of dihydrofuran was used as chiral building block in a total syntheses of the Sch 38516 [14].

Outlook

The work discussed in this brief article demonstrates that the Zr-catalyzed carbomagnesation of alkenes has developed over the last few years into a powerful synthetic methodology, allowing the synthesis of valuable products with high regio-, diastereo- and enantioselectivities. So far, carbomagnesations have been studied predominantly. Therefore, it will be an interesting task in the future to expand the method for higher alkyl Grignard substrates also. Moreover, the combination of the kinetic resolution methodology with subsequent ring-closing metathesic reactions opens up new powerful strategies for the construction of complex carbocycles [13c].

References

[1] Review: P. Knochel in *Comprehensive Organic Synthesis, Vol. 4* (Eds.: B. M. Trost, I. Fleming), Pergamon, Oxford, **1991**, p. 865.
[2] Grignard compounds are produced on an industrial scale and marketed e.g. by Chemetall GmbH, Frankfurt am Main.
[3] H. Lehmkuhl, *Bull. Soc. Chim. Fr., Part 2*, **1981**, 87.
[4] see, for instance: W. Oppolzer in *Selectivity – A Goal for Synthetic Efficiency* (Eds.: W. Bartmann, B. M. Trost), Verlag Chemie, Weinheim, **1984**, p. 137.
[5] a) U. M. Dzhemilev, O. S. Vostrikova, *J. Organomet. Chem.* **1985**, *285*, 43; b) U. M. Dzhemilev, O. S. Vostrikova, G. A. Tolstikov, *J. Organomet. Chem.* **1986**, *304*, 17, and refs. cited therein.
[6] For a review on newer developments in the field of Zr-mediated C-C coupling, see: M. E. Maier, *Nachr. Chem. Tech. Lab.* **1993**, *41*, 811.
[7] a) A. H. Hoveyda, Z. Xu, *J. Am. Chem. Soc.* **1991**, *113*, 5079; b)A. H. Hoveyda, Z. Xu, J. P. Morken, A. F. Houri, *J. Am. Chem. Soc.* **1991**, *113*, 8950; c) A. F. Houri, M. T. Didiuk, Z. Xu, N. R. Horan, A. H. Hoveyda, *J. Am. Chem. Soc.* **1993**, *115*, 6614.
[8] A. H. Hoveyda, J. P. Morken, A. F. Houri, Z. Xu, *J. Am. Chem. Soc.* **1992**, *114*, 6692.
[9] a) T. Takahashi, T. Seki, Y. Nitto, M. Saburi, C. J. Rousset, E.-i. Negishi, *J. Am. Chem. Soc.* **1991**, *113*, 6266; b) K. S. Knight, R. M. Waymouth, *J. Am. Chem. Soc.* **1991**, *113*, 6268; c) U. Wischmeyer, K. S. Knight, R. M. Waymouth, *Tetrahedron Lett.* **1992**, *33*, 7735; d) K. S. Knight, D. Wang, R. M. Waymouth, J. Ziller, *J. Am. Chem. Soc.* **1994**, *116*, 1845.
[10] A. H. Hoveyda, J. P. Morken, *J. Org. Chem.* **1993**, *58*, 4237.
[11] a) F. R. W. P. Wild, L. Zsolnai, G. Huttner, H.-H. Brintzinger, *J. Organomet. Chem.* **1982**, *232*, 233; b) F. R. W. P. Wild, M. Wasiucionek, G. Huttner, H.-H. Brintzinger, *J. Organomet. Chem.* **1985**, *288*, 63; c) A. Schäfer, E. Karl, L. Zsolnai, G. Huttner, H.-H. Brintzinger, *J. Organomet. Chem.* **1987**, *328*, 87; for an overview on applications of **31** in organic synthesis, see: d) A. H. Hoveyda, J. P. Morken, *Angew. Chem.* **1996**, *108*, 1378; *Angew. Chem. Int. Ed. Engl.* **1996**, *35*, 1262.
[12] a) J. P. Morken, M. T. Didiuk, A. H. Hoveyda, *J. Am. Chem. Soc.* **1993**, *115*, 6997; b) M. T. Didiuk, C. W. Johannes, J. P. Morken, A. H. Hoveyda, *J. Am. Chem. Soc.* **1995**, *117*, 7097; c) M. S. Visser, N. M. Heron, M. T. Didiuk, J. F. Sagal, A. H. Hoveyda, *J. Am. Chem. Soc.* **1996**, *117*, 4291; for a brief review, see: d) A. H. Hoveyda in: *Transition Metals for Organic Synthesis, Vol. 1*, (Eds.: M. Beller, C. Bolm), Wiley-VCH, **1998**, p. 195.
[13] a) J. P. Morken, M. T. Didiuk, M. S. Visser, A. H. Hoveyda, *J. Am. Chem. Soc.* **1994**, *116*, 3123; b) M. S. Visser, A. H. Hoveyda, *Tetrahedron* **1995**, 4383; c) N. M. Heron, J. A. Adams, A. H. Hoveyda, *J. Am. Chem. Soc.* **1997**, *119*, 6205; d) J. A. Adams, J. G. Ford, P. J. Stamatos, A. H. Hoveyda, *J. Org. Chem.* **1999**, *64*, 9691.
[14] a) A. F. Houri, Z. Xu, D. A. Cogan, A. H. Hoveyda, *J. Am. Chem. Soc.* **1995**, *117*, 10926; b) Z. Xu, C. W. Johannes, A. F. Houri, D. S. La, D. A. Cogan, G. E. Hofilena, A. H. Hoveyda, *J. Am. Chem. Soc.* **1997**, *117*, 10302.

Intramolecular Alkoxypalladation

Oliver Geis and Hans-Günther Schmalz

Institut für Organische Chemie, Universität Köln, Germany

Ever since the initial discovery of the Wacker process [1], i.e. the Pd/Cu-catalyzed oxidation of ethylene to acetaldehyde (**1**) in water, methods for the palladium (II) – mediated oxidative functionalization of alkenes have found widespread application in the synthesis of complex molecules [2].

$H_2C{=}CH_2$ + 1/2 O_2 → (cat. $PdCl_2$, cat. $CuCl_2$, H_2O) MeCHO **1**

The basic principle of this chemistry is that η^2-alkene-Pd(II) complexes, usually generated in situ, are easily attacked by nucleophiles to form σ-alkyl-Pd species, which in turn are able to react further in a variety of ways. In general, several kinds of nucleophiles (e.g. alcohols, amines and enolethers) are able to attack the alkene complex intermediates in an intra- or intermolecular fashion. This article, however, focusses exclusively on intramolecular alkoxypalladations, i.e. transformations of the type **2** → **3**, which are of particular synthetic relevance.

2 → (- HX) **3**

The first example of such a reaction was reported by Hosokawa et al. in 1973 [3]. Since then, many further examples have appeared in the literature [2]. In recent years, this chemistry has received increasing recognition after reliable experimental procedures had been developed and several synthetic applications had demonstrated the value of this methodology.

Alkoxypalladation/Carbonylation

After James and Stille had shown the general possibility to link (intermolecular) alkoxypalladation and carbonylation processes [4], Semmelhack et al. were the first to achieve Pd-catalyzed intramolecular alkoxycarbonylation reactions and to apply them in the synthesis of specific target molecules [5]. In a typical reaction, the substrate is stirred at room temperature in methanol under an atmosphere of CO (1.1 atm) in the presence of 5–10 mol% of $PdCl_2$ and an excess (3 eq.) of $CuCl_2$. As the conversion of the simple model compound **4** to the products **5** and **6** implies, such reactions often proceed in good yields but not necessariliy with high diastereoselectivies (**5** : **6** = 3 : 1).

4 → (5 mol % $PdCl_2$, 2 eq. $CuCl_2$, 1.1 atm CO, ROH, 2 h; 82 - 87 %) **5** + **6**

The presumed reaction mechanism is shown in Scheme 1. First, the in situ generated η^2-alkene-Pd(II) complex **2** is intramolecularly attacked by the nucleophile to form an alkoxypalladated species **3**. In the presence of carbon monoxide this intermediate rapidly undergoes migratory inser-

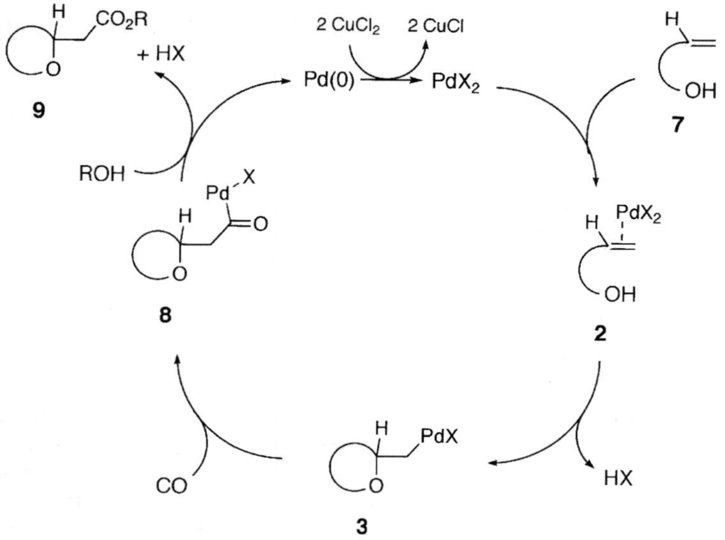

Scheme 1. Mechanism of the Pd-catalyzed alkoxycarbonylation of alkenes.

tion, usually faster than the competing β-hydride elimination. The resulting acylpalladium derivative **8** is finally trapped by a second nucleophile (usually the alcoholic solvent) to give the product **9**. To close the catalytic cycle, the released Pd(0) must be converted back to Pd(II). This is usually achieved by the addition of stoichiometric amounts of $CuCl_2$. In principle, oxygen can be used as stoichiometric oxidant as in the Wacker oxidation; however, this seems to be efficient only in certain cases.

First applications of this chemistry were reported by Semmelhack et al., who synthesized the (racemic) naphthoquinone antibiotics nanaomycin A (*rac*-**10**, R = Me) and desoxyfrenolicin (*rac*-**10**, R = propyl). These target molecules are retrosynthetically traced back to a precursor of type **11** [5]. In these cases, the Pd-catalyzed cyclizations proceeded only with low diastereoselectivity; the desired *trans*-stereochemistry, however, can be set up by subsequent equilibration.

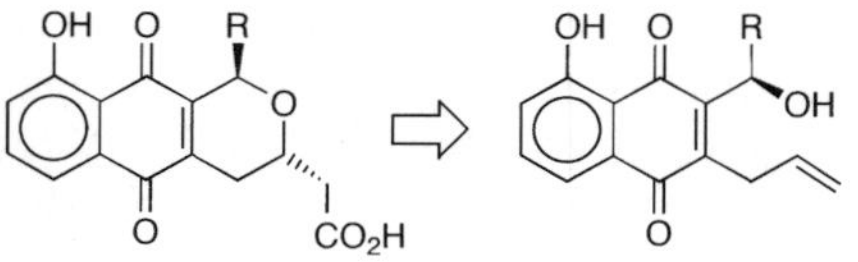

If other hydroxy functionalities are appropriately placed in the substrate, the Pd-acyl intermediates are intramolecularly trapped to form lactones [5c]. In a remarkably short enantioselective total synthesis of frenolicin B (**17**) (Scheme 2) Kraus et al. successfully exploited Semmelhack's tandem methodology for the construction of rings C and D [6a].

Starting from the (prochiral) ketone **12**, the alcohol **13** is generated by enantioselective reduction with (+)-diisopinocampheylchloroborane [(+)-IPC_2BCl)]. This reaction represents the chirogenic step of the synthesis. The alcohol **13** is then converted with *n*-butyllithium into a dianion, which in turn reacts with acrolein to give a 1 : 1.5-mixture of two diastereomeric diols. The desired major product **14** can be purified by flash chromatography. The Pd-catalyzed cyclization of **14** leads diastereoselectively to lactone **15**, which is first oxidized to the corresponding quinone before the missing ring is attached by a regioselective (!) Diels-Alder reaction. In the final step, the Diels-Alder product **16** is converted to the natural product **17** by Jones oxidation. It should be mentioned that Kraus et al. succeeded in controlling the regioselectivity of a Diels-Alder reaction (with the help of strategically placed substituents) also in a synthesis of hingconine [6b].

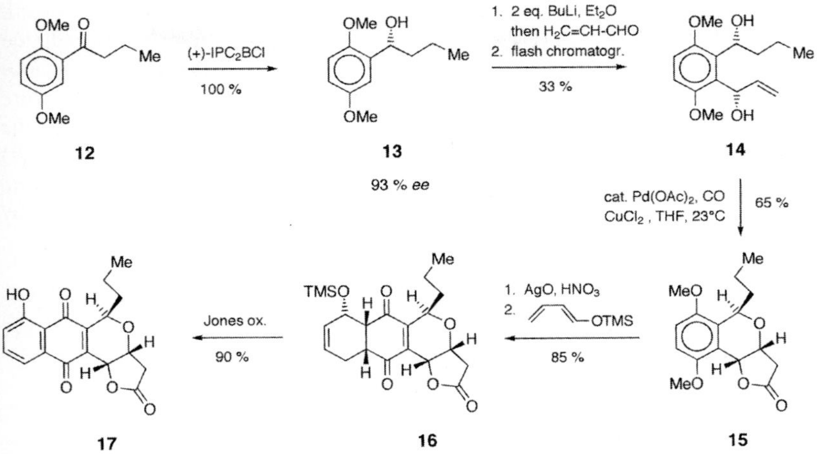

Scheme 2. Synthesis of frenolicin B according to G. A. Kraus.

Based on the results of mechanistic studies on the stereochemical course of the Wacker reaction and related processes [7], one would expect the alkoxypalladation of alkenes to proceed stereospecifically as a *trans*-addition. Indeed, this was confirmed in the highly diastereoselective formation of compound *rac*-**18** from **20** and of *rac*-**19** from **21**.

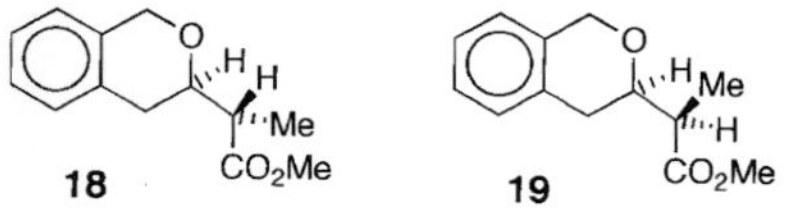

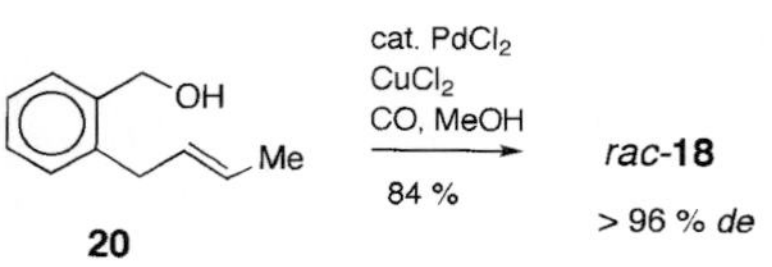

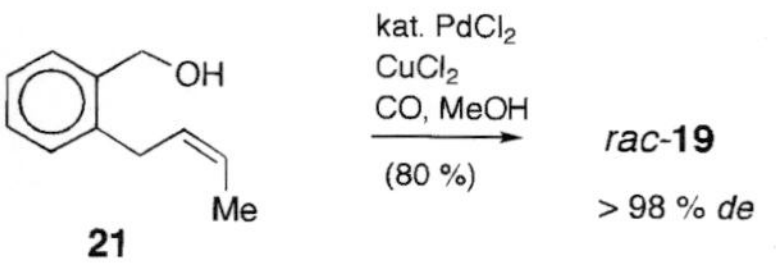

Purely aliphatic systems also react with good diastereoselectivity, but only if remote substituents cause an energetic differentiation of the competing transition states [8, 9]. For instance, *Z*-configurated substrates of type **22** are suitable starting materials for the construction of 2,6-disubstituted tetrahydropyrans of type **23** [8a].

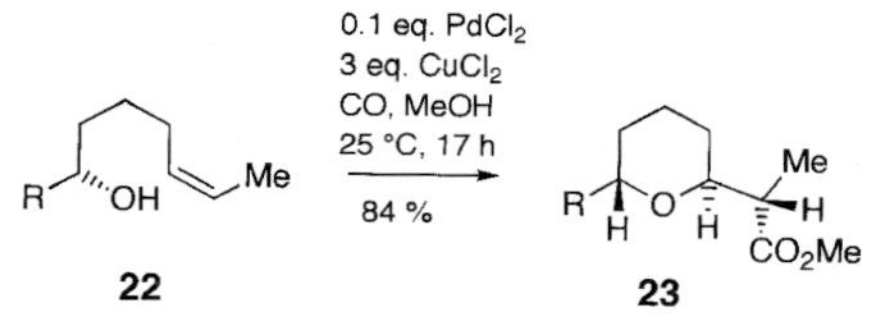

Even the preparation of 2,3,5-trisubstituted tetrahydrofurans is possible in good yield as the transformations **24** to **25** and **26** to **27** exemplify [8b, 9].

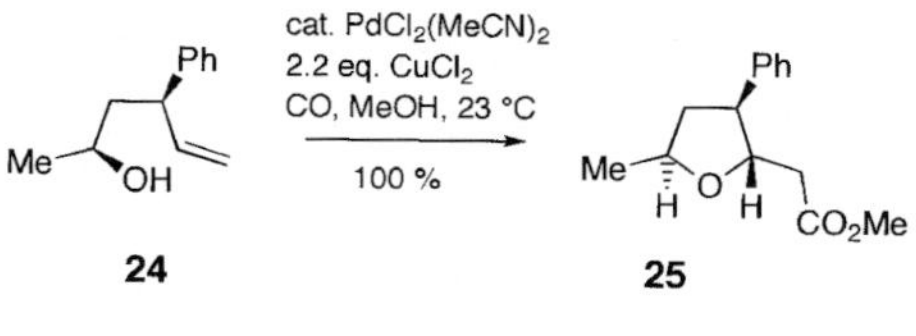

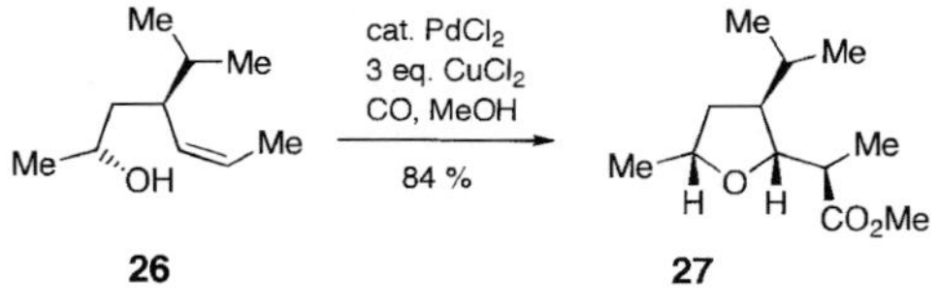

As Jaeger and co-workers demonstrated, the Pd-catalyzed alkoxycarbonylation of unsaturated polyols is also useful for the ω-homologization of aldoses. In general, *cis*-fused bicyclic lactones are obtained with high selectivities [10]. As an example, the D-*gluco*-configurated product **29** is generated in high yield starting from the D-lyxose derivative **28**.

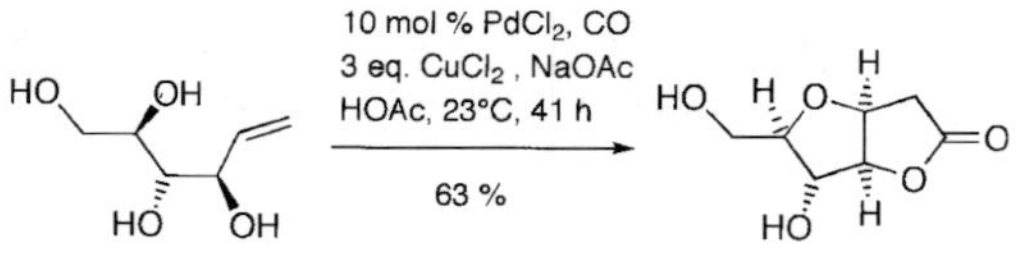

10 mol % $PdCl_2$, CO
3 eq. $CuCl_2$, NaOAc
HOAc, 23°C, 41 h
63 %

28 **29**

This chemistry was applied in a synthesis of (−)-goniofufuron (**35**), the unnatural enantiomer of a cytotoxic compound isolated from the bark of the Thai tree *goniothalamus giganteus* [10b]. Starting from the D-glucose derivative **30**, the synthesis (scheme 3) leads, via the hexenose derivatives **31** and **32**, to the epimeric mixture **33**, which is converted in excellent yield to the bicyclic lactones **34** (mixture of diastereomers) by means of Pd-catalyzed alkoxycarbonylation. Finally, the desired epimer **35** is separated by flash chromatography.

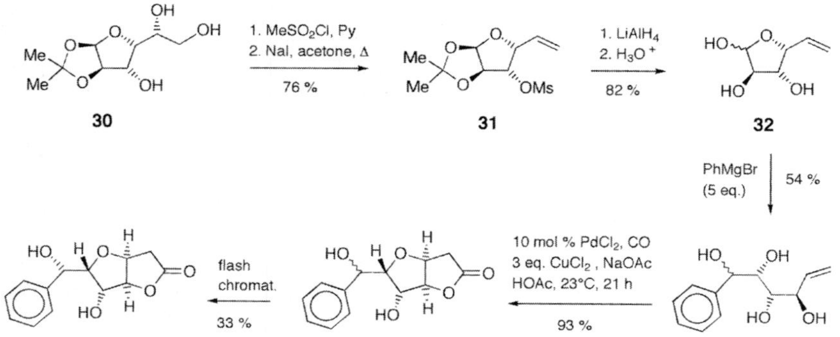

1. $MeSO_2Cl$, Py
2. NaI, acetone, Δ
76 %

1. $LiAlH_4$
2. H_3O^+
82 %

PhMgBr (5 eq.) 54 %

10 mol % $PdCl_2$, CO
3 eq. $CuCl_2$, NaOAc
HOAc, 23°C, 21 h
93 %

flash chromat.
33 %

30 **31** **32** **33** **34** **35**

Scheme 3. Synthesis of (−)-goniofufurone according to V. Jäger.

Alkoxypalladation/β-H-Elimination

If the alkoxypalladations are conducted in the absence of carbon monoxide, the primarily generated σ-alkyl species often react to olefins by β-hydride elimination [2b]. Based on this fact, a novel method for the synthesis of 2-alkenyl-tetrahydropyrans **36** starting from hydroxyalkenes of type **37** was developed [11].

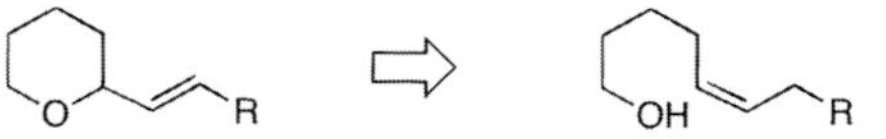

36 **37**

In these reactions, a remarkable solvent effect became apparent: while in DMF, THF and acetonitrile only low yields and a high proportion of isomerized products are observed, in DMSO, however, the reactions proceed very smoothly. Obviously, in this solvent the β-hydride elimination proceeds in a very controlled manner to selectively give the *E*-configurated alkenes. Usually, *Z*-configurated substrates of type **38** give the best results, leading to products of type **39**. By comparing the two competing chair-like transition states **41** and **43** (both having two equatorial substituents), it seems that **41** is preferred because of an unfavorable 1,5-interaction in **43**. Indeed, the configuration of the major product (**39**) corresponds to a transition state of type **41** and a resulting intermediate of type **42**.

A certain disadvantage of this method is that stoichiometric amounts of the palladium "catalyst" have to be employed. While it is possible to reoxidize the in situ generated Pd(0) in the above-mentioned alkoxycarbonylations, the development of efficient catalytic procedures for the alkoxypalladation/β-hydride elimination reactions still represents an unsolved problem. However, first successes [12] in special systems suggest that this problem should not be insoluble.

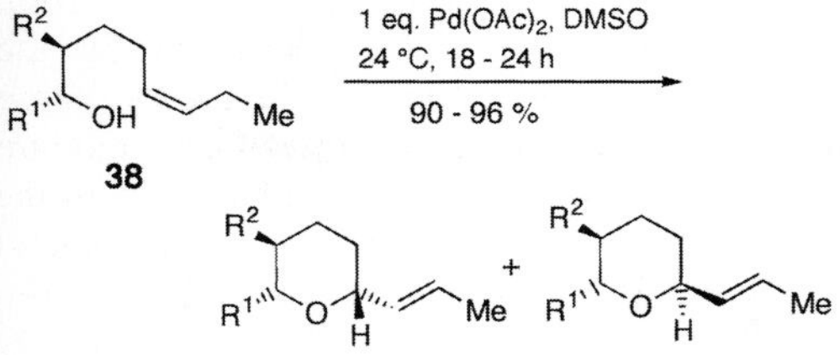

1 eq. $Pd(OAc)_2$, DMSO, 24 °C, 18 - 24 h, 90 - 96 %

38 → **39** + **40**

	39		40
R^1 = Me, R^2 = H :	> 90	:	< 5
R^1 = H, R^2 = Me :	98	:	2
R^1 = R^2 = Me :	97	:	3

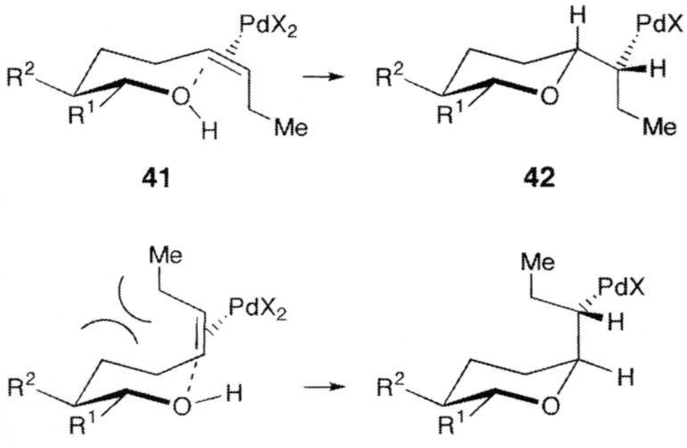

41 → **42**

43 → **44**

Synthesis of Tetronomycin

Ionophores (polyether antibiotics) [13] such as tetronomycin (**45**) are challenging target molecules for organic synthesis and ideal test compounds for the trial of new methods for the stereoselective construction of oxygen heterocycles.

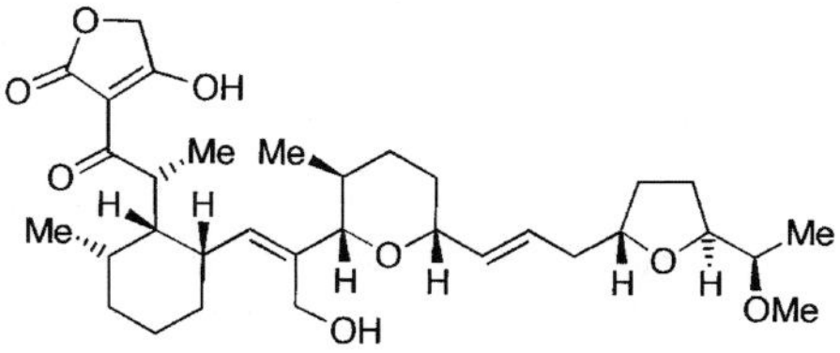

45

In a synthesis of tetronomycin (**45**) published in 1994 [14], Semmelhack et al. probe the scope of intramolecular alkoxypalladations. The retrosynthetic analysis (Scheme 4) shows that the chosen strategy exploits such Pd-catalyzed transformations even twice. The pre-target structure **46** is formally derived from **47** by Pd-mediated cyclization. Compound **47** can be traced back via **48** to the tetrahydrofuran derivative **49**, which in turn should be available by alkoxycarbonylation from a precursor of type **50**.

In the following, the realization of this concept will be briefly described. The synthesis of the tetronomycin building block **55** (Scheme 5) starts from the chiral aldehyde **51,** which is easily accessed from D-arabinose. Reaction with vinyllithium leads to a 1 : 1 mixture of the epimeric alcohols **52a** and **52b**, which are separated by HPLC. Interestingly, both diastereomers can be converted (on separate routes) into the building block **55**. Both the Pd-catalyzed key steps (**53** → **54** and **56** → **57**) proceed with high diastereoselectivity.

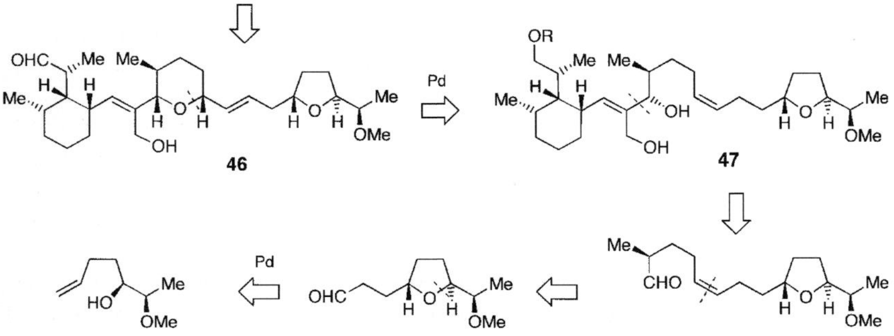

Scheme 4. Retrosynthesis of tetronomycin according to M. F. Semmelhack.

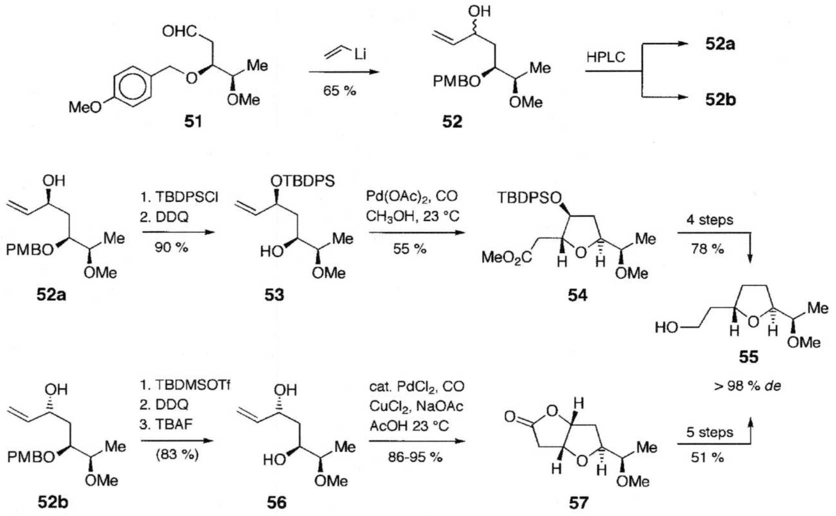

Scheme 5. Synthesis of the tetronomycin building block **55** according to M. F. Semmelhack.

For the assembly of the tetronomycin skeleton (Scheme 6), the dilithiated species prepared from **58** with butyllithium is reacted with the aldehyde **48** to give a mixture of diastereomers (1 : 2), from which the desired isomer **59** can be separated by chromatography. The undesired epimer is converted into **59** through protection of the primary OH function, Swern oxidation and diastereoselective reduction. The crucial Pd(II)-mediated cyclization of **59** in DMSO (in the presence of 8–10 equivalents of acetic acid) then occurs with controlled β-H elimination to diastereoselectively afford compound **60**. After protection of the primary hydroxy function, cleavage of the silyl group and Swern oxidation give the advanced tetronomycin precursor **61**. Although the final preparation of the target molecule was not reported, the synthesis of **61** represents a convincing demonstration of the value of intramolecular alkoxypalladation techniques.

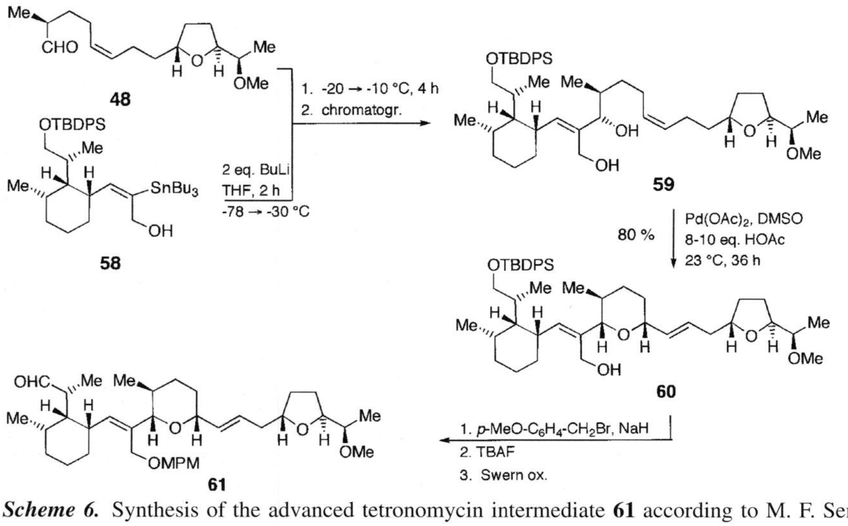

Scheme 6. Synthesis of the advanced tetronomycin intermediate **61** according to M. F. Semmelhack.

Enantioselective Reactions

Although diastereoselective intramolecular alkoxypalladations have been investigated intensively and have found application in synthesis (see above), there are few examples of enantioselective alkoxypalladations [2b]. For instance, Hosokawa et al. were able to cyclize unsaturated phenol derivatives of type **62** in the presence of chiral (η^3-allyl-PdOAc complexes, i.e. **63**), but only with modest enantioselectivities. Under the same conditions the conversion of the phenol **65** to chroman **66** (a compound related to vitamin E) proceeded in acceptable yields, but with only low asymmetric induction. Newer results by Uozumi et al., for instance the Pd-catalyzed cyclization of **67** to **68** in the presence of a chiral bis-oxazolin ligand [15], show that much higher enantioselectivities can be achieved, at least for certain substrates.

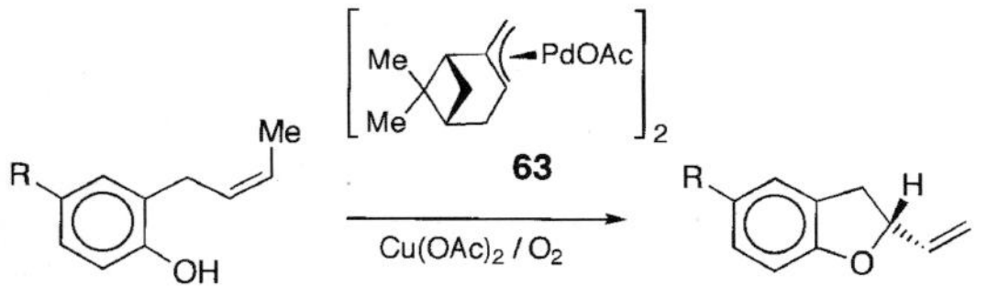

Conclusion

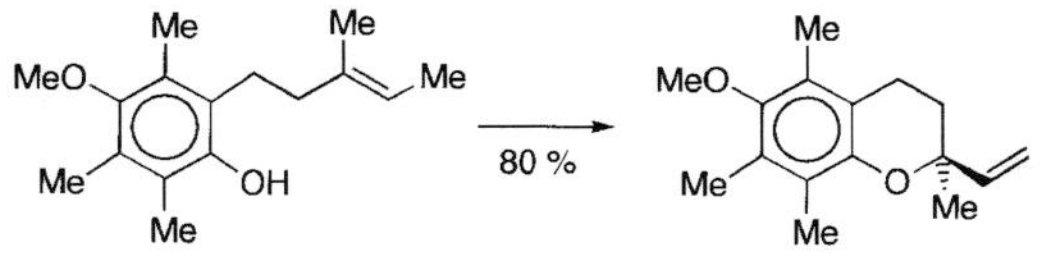

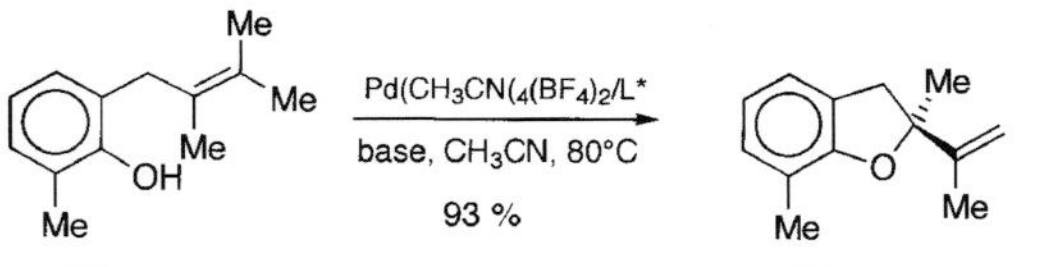

The intramolecular alkoxypalladation has been developed into a remarkable (catalytic) methodology which exhibits a great synthetic potential, especially in combination with carbonylation. This is reflected by convincing applications in natural product syntheses. This chemistry opens reliable and highly diastereoselective approaches to several hydropyran and hydrofuran systems. The development of efficient enantioselective protocols for the various chirogenic transformations is still a challenging goal for the future.

References

[1] J. Smidt, W. Hafner, R. Jira, R. Sieber, J. Sedlmeier, A. Sabel, *Angew. Chem.* **1959**, *71*, 176; *Angew. Chem.* **1962**, *74*, 93.

[2] Reviews: a) L. S. Hegedus, *Tetrahedron* **1984**, *40*, 2415; b) T. Hosokawa, S.-I. Murahashi, *Acc. Chem. Res.* **1990**, *23*, 49; c) L. S. Hegedus, Transition Metals in the Synthesis of Complex Organic Molecules (2nd ed.), University Science Books, Sausalito, CA, **1999**, p. 192; d) B. L. Feringa in: *Transition Metals for Organic Synthesis*, *Vol.2* (M. Beller, C. Bolm; eds.), Wiley-VCH, Weinheim **1998**, chapter 2.8.

[3] T. Hosokawa, K. Maeda, K. Koga, I. Moritani, *Tetrahedron Lett.* **1973**, 739.

[4] D. E. James, J. K. Stille, *J. Am. Chem. Soc.* **1976**, *98*, 1810.

[5] a) M. F. Semmelhack, J. J. Bozell, T. Sato, W. Wulff, E. Spiess, A. Zask, *J. Am. Chem. Soc.* **1982**, *104*, 5850; b) M. F. Semmelhack, A. Zask, *J. Am. Chem. Soc.* **1983**, *105*, 2034; c) M. F. Semmelhack, C. Bodurow, M. Baum, *Tetrahedron Lett.* **1984**, *25*, 3171.

[6] a) G. A. Kraus, J. Li, M. S. Gordon, J. H. Jensen, *J. Am. Chem. Soc.* **1993**, *115*, 5859;
b) G. A. Kraus, J. Li, M. S. Gordon, J. H. Jensen, *J. Org. Chem.* **1994**, *59*, 2219.

[7] a) J. K. Stille, D. E. James, L. F. Hines, *J. Am. Chem. Soc.* **1973**, *95*, 5062; b) J. K. Stille, R. Divakaruni, *J. Am. Chem. Soc.* **1978**, *100*, 1303; c) J. E. Bäckvall, B. Åkermark, S. O. Ljunggren, *J. Am. Chem. Soc.* **1979**, *101*, 2411.

[8] a) M. F. Semmelhack, C. Bodurow, *J. Am. Chem. Soc.* **1984**, *106*, 1496; b) M. F. Semmelhack, N. Zhang, *J. Org. Chem.* **1989**, *54*, 4483.

[9] a) M. McCormick, R. Monahan III, J. Soria, D. Goldsmith, D. Liotta, *J. Org. Chem.* **1989**, *54*, 4485; b) C. P. Holmes, P. A. Bartlett, *J. Org. Chem.* **1989**, *54*, 98.

[10] a) T. Gracza, T. Hasenöhrl, U. Stahl, V. Jäger, *Synthesis*, **1991**, 1108; b) T. Gracza, V. Jäger, *Synlett* **1992**, 191.
[11] M. F. Semmelhack, C.R. Kim, W. Dobler, M. Meier, *Tetrahedron Lett.* **1989**, *30*, 4925.
[12] S. Saito, T. Hara, N. Takahashi, M. Hirai, T. Moriwake, *Synlett* **1992**, 237.
[13] J. W. Westley, Polyether Antibiotics: Naturally Occuring Acid Ionophores, Marcel Dekker, New York, **1991**.
[14] M. F. Semmelhack, W. R. Epa, A. W.-H. Cheung, Y. Gu, C. Kim, N. Zhang, W. Lew, *J. Am. Chem. Soc.* **1994**, *116*, 7455.
[15] Y. Uozumi, K. Kato, T. Hayashi, *J. Am. Chem. Soc.* **1997**, *119*, 5063; *J. Org. Chem.* **1998**, *63*, 5071.

Ring-closing Olefin Metathesis

Michael Karle and Ulrich Koert

Institut für Chemie, Humboldt-Universität, Berlin, Germany

Olefin metathesis is the transformation of two olefins **1** and **2** into two new olefins **3** and **4**. Formally, the reaction represents a mutual exchange of alkylidene groups [Eq. (1)]. The reaction is catalyzed by various transition metal complexes.

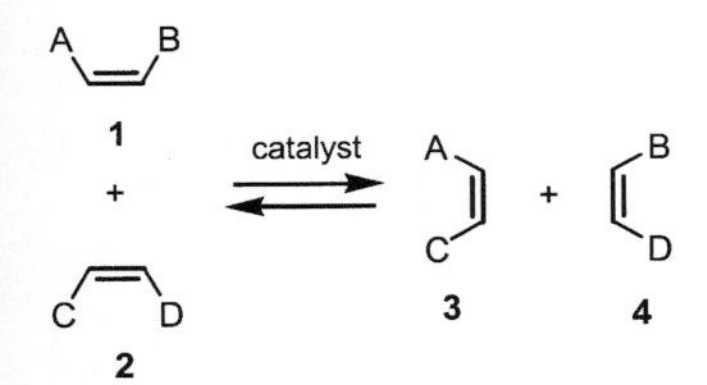

The reaction is applied in industrial processes (Phillips triolefin process, Shell higher olefin process) and has importance in ring opening-metathesis polymerization (ROMP) in polymer chemistry [1]. In the past, olefin metathesis was not commonly applied in organic synthesis [2] because of the reversibility of the reaction, leading to olefin mixtures. In contrast, industrial processes often handle product mixtures easily. In ROMP, highly strained cyclic olefins allow the equilibrium of the reaction to be shifted towards the product side.

Attention paid to olefin metathesis in organic synthesis has increased over recent years because of the progress in ring-closing metathesis [3]. This metathesis variation consists of the cyclization of an acyclic (α, ω)-diene **5** and formation of

X catalyst X +

5 6 7

X = C, N, O

cyclic olefin **6** and ethylene **7** under metathesis conditions [Eq. (2)].

In this cyclization, the equilibrium of the reaction is shifted towards the product side for entropic reasons, and removal of the volatile ethylene moves the equilibrium further in the same direction.

Several catalysts for ring-closing metathesis are now known. Prominent examples are shown in Scheme 1: the molybdenum carbene complex **8**, introduced by Schrock et al. [4], methyltrioxorhenium **9**, discovered by Herrmann et al. [5] and the ruthenium carbene complex **10**, developed by Grubbs et al. [6]

Recently, chiral molybdenum catalysts such as **11** have been synthesized, giving access to asymmetric ring-closing metathesis [7]. Water-soluble variations of catalyst **10**, like catalyst **12**, perform metathesis in water or methanol [8], extending the scope of the reaction to substrates which are poorly soluble in organic solvents.

Currently the most promising catalyst is **10**. Its synthesis was first described by Grubbs et al. [6] Recently, Fürstner et al. reported a simple method for in situ preparation of a catalyst for ring-closing metathesis reactions by heating a solution of the diene with catalytic amounts of commercially available $[(p\text{-cymene})RuCl_2]_2$ and $P(cy)_3$ in CH_2Cl_2 under neon light [9]. The same group also reported ring-closing metathesis at ambient temperatures promoted by conveniently accessible coordinatively unsaturated allenylidene Ru-complexes [9].

Scheme 2 shows a simplified description of the ring-closing metathesis mechanism [3]. In the first step, one double bond of starting diene **13** forms with the catalyst **20** via the π-complex

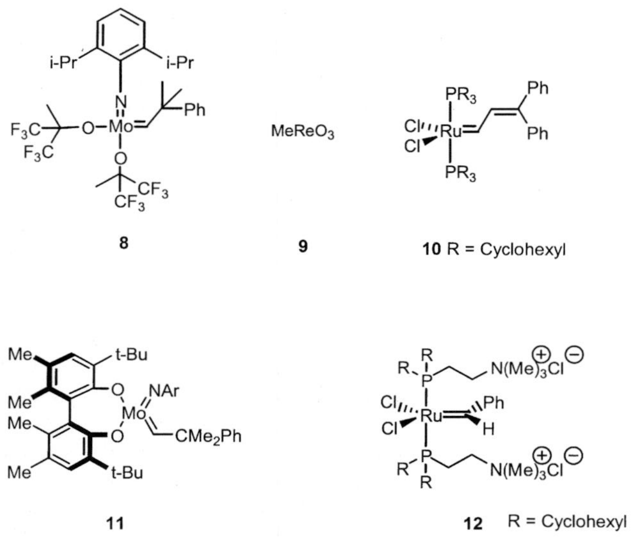

Scheme 1.
Catalysts for ring-closing metathesis.

14 the metallacyclobutane **15**. (2 + 2)-Cycloreversion of **15** leads to carbene complex **17.** Here the by product ethylene **7** is formed. The π-complex **16**, first postulated as an intermediate by Grubbs, was meanwhile confirmed by X-ray structural analysis [10]. Intramolecular (2+2)-cycloaddition of **17** leads to the intermediate **18**. Upon the release of the product **19**, the catalytic cycle is closed. Scheme 2 shows only the catalytic cycles pathway leading to product **19**. In regard to the reversibility of the single steps, there is another catalytic cycle, which is not shown, describing the reactions pathway in the other direction.

Catalyst **10** allows the production of cyclopentene (**21** $\rightarrow$ **22**) and cyclohexene derivatives (**23** $\rightarrow$ **24**) in high yields [11] (Scheme 3). The tolerance of many different functional groups is remarkable. For example, dienes of the type **23**, possessing an unprotected carboxylic acid, aldehyde, or alcohol function, can be used in ring-closing metathesis, employing catalyst **10**.

The application of the reaction is not limited to carbocycles. *O*- and *N*-heterocycles are acces-

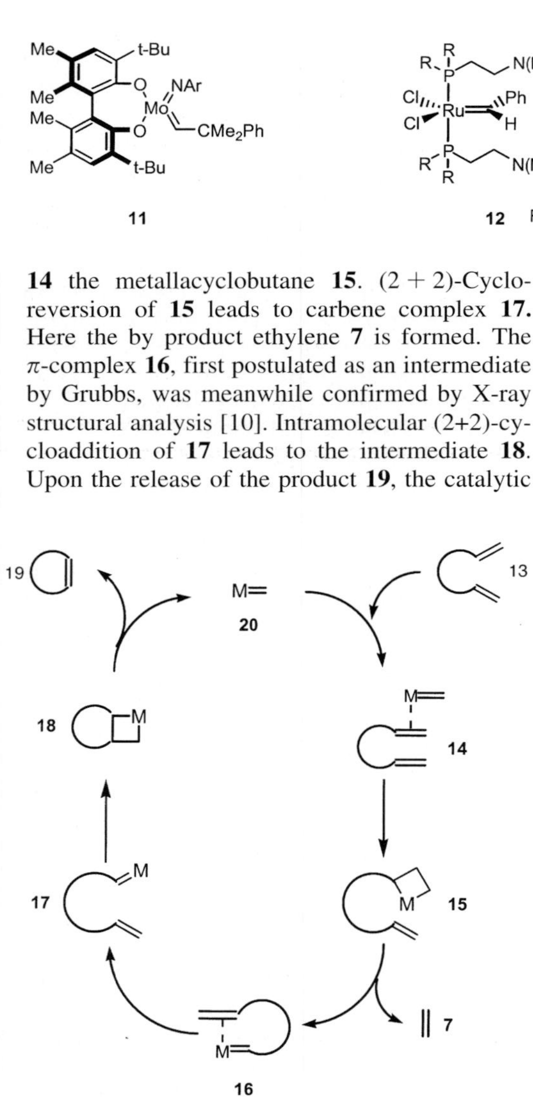

Scheme 2. Mechanism of ring-closing metathesis.

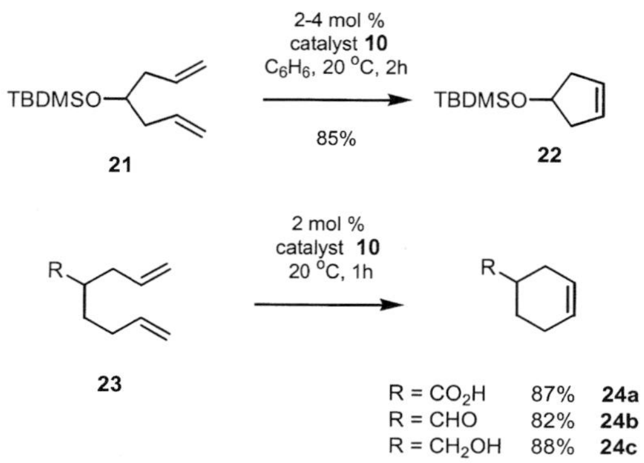

Scheme 3. Formation of cycloalkenes.

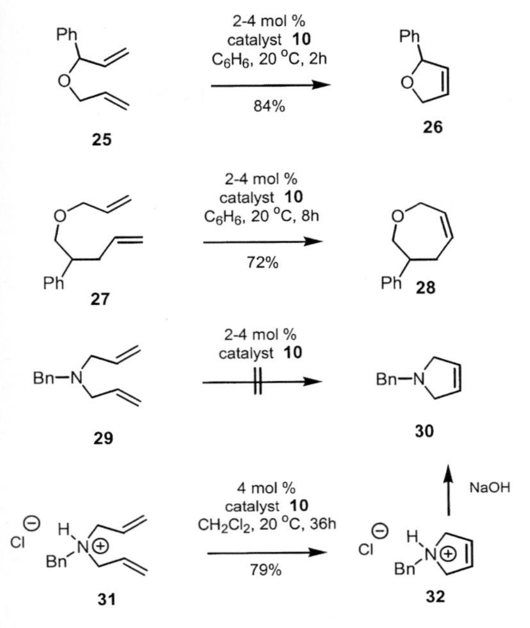

Scheme 4. Formation of *O*- and *N*-heterocycles.

sible in a facile way too [11] (Scheme 4). Thus, the cyclization of the bisallyl ether **25** to the dihydrofuran derivative **26** succeeds with a yield of 84 %. Even the synthesis of oxepines (**27** → **28**) is possible.

There are limitations with *N*-heterocycles. Amines are not compatible with the catalyst. Thus, the cyclization of bisallylamine **29** to dihydropyrrole **30** is not possible.

Since the amine nitrogen is too basic, it blocks the coordination sites at the transition-metal of the catalyst. This can be avoided by protonation. For this, the amine **29** is converted into the corresponding hydrochloride **31**, which can be cyclized to the protonated dihydropyrrole **32**. In this case, a longer reaction time is required.

In the previous examples, disubstituted double bonds have been synthesized via ring-closing metathesis. As shown in Scheme 5, it is possible to extend the method to tri- or tetrasubstituted double bonds. Thus, bisallylic ether **33** leads to the trisubstituted olefin **34** and compound **35** gives the tetrasubstituted olefin **36** [3]. Here, the additional substituents are simple methyl groups, but silyl groups for example, can be introduced too. The volatile byproducts in these cases are not ethylene but butene or propene.

Because of the excellent performance of the new catalysts, many research groups use ring-closing metathesis as the key step in natural product synthesis [12]–[18]. Scheme 6 shows some examples. Via ring-closing metathesis of the olefin **37** to the hydroazulene **38**, Blechert et al. [12] succeeded in synthesizing a cyclic system which is part of many sesquiterpenes. Cyclooctane derivatives, whose synthesis is the main problem in taxol synthesis, can be obtained in good yields (**39** → **40**), as demonstrated by Grubbs et al. [13].

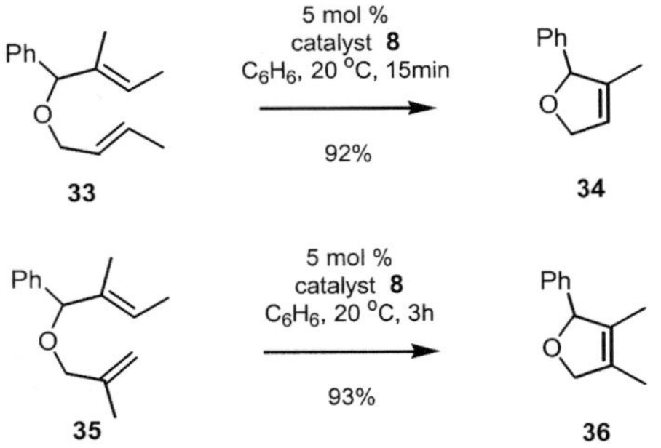

Scheme 5. Formation of tri- and tetrasubstituted olefins.

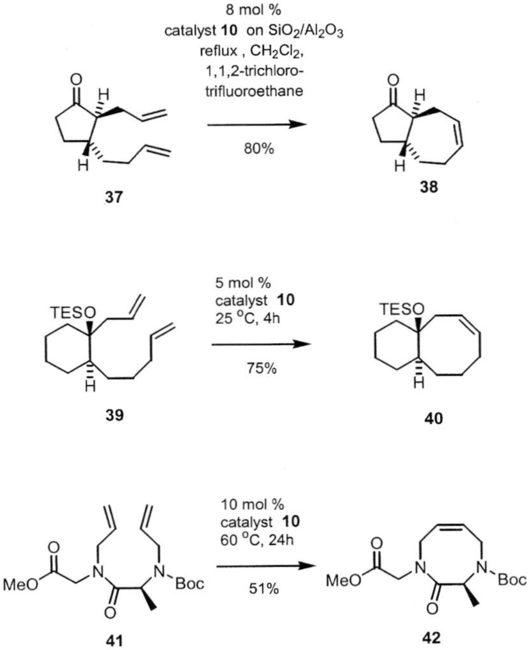

Scheme 6. Examples of synthetic applications.

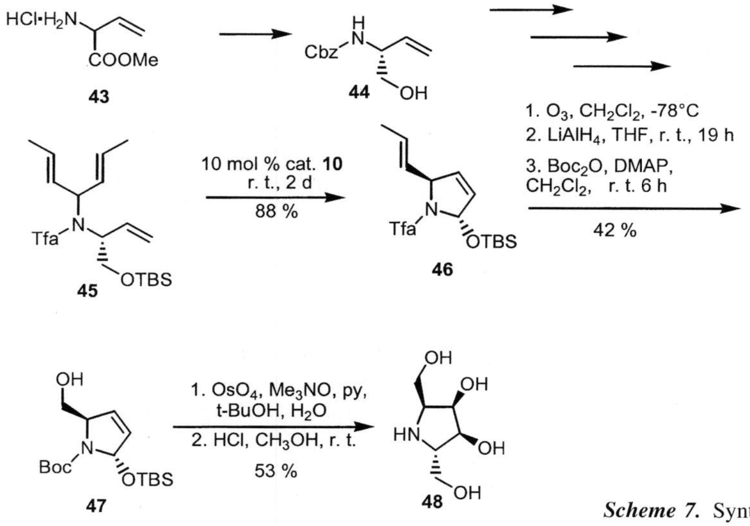

Scheme 7. Synthesis of azasugars.

With catalyst **10**, the bisallyl dipeptide **41** can be converted into the conformationally constrained but not fully rigid dipeptide **42** [13]. Artificial dipeptides of type **42** are discussed as enzyme inhibitors.

The double bonds formed in ring-closing metathesis reactions can be the basis for various further transformations (e.g. epoxidation, bis-hydroxylation). One example of further functionalization of the metathesis product is the synthesis of azasugars of type **48** by Blechert et al. [14] (Scheme 7). The starting point for the synthesis of **48** is vinylglycine methyl ester hydrochloride **43**, which is converted into the Cbz-protected amino alcohol **44**. Within three steps, employing standard reactions, the Tfa-protected bisallylamine **45** is synthesized. Ring-closing metathesis with catalyst **10** leads to compound **46**. Regioselective ozonolysis of **46**, followed by reductive work up with lithium aluminium hydride and introduction of a Boc group gave the protected diol **47**. Catalytic dihydroxylation of **47** with osmium tetroxide and subsequent removal of the protecting groups with HCl resulted in the desired azasugar **48**.

Ring-closing metathesis is now applied as the key step in many natural product syntheses. Prominent examples are the syntheses of manzamine A [15,17] and epothilone A [18] (Scheme 8).

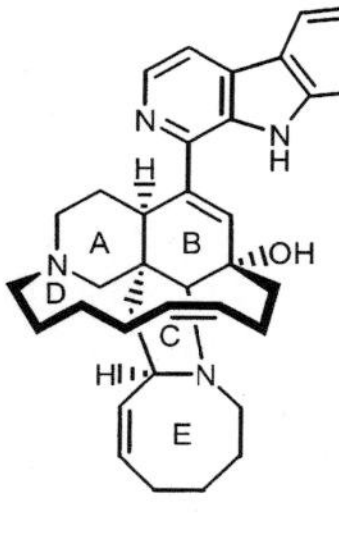

Manzamine A

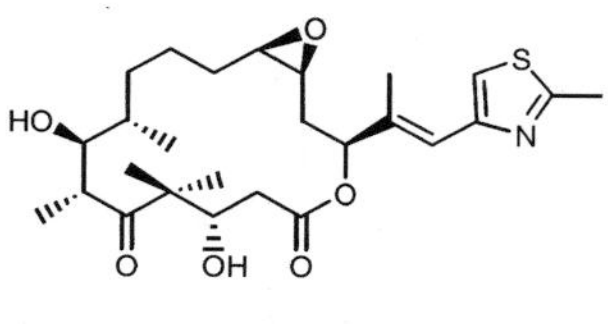

Epothilone A

Manzamine A and epothilone A are two natural products with promising antitumor activities. Manzamine A is isolated from marine sponges, whereas epothilone A is isolated from myxobacteria.

For manzamine A, Martin et al. developed a strategy for the construction of the eight-membered ring E [15], Pandit et al. succeeded in the formation of the 13–membered ring D [18]. The efforts culminated recently in the total synthesis of manzamine A and related manzamine alkaloids, employing the ring-closing metathesis reaction as a key step [16]. Thus, **49** is converted into **50** with a yield of 67 % (Scheme 8).

For epothilone A, among other compounds [17], Nicolaou et al. developed a ring-closing metathesis approach [17c]. Cyclization of diene **51** gave product **52** with an overall yield of 85 %.

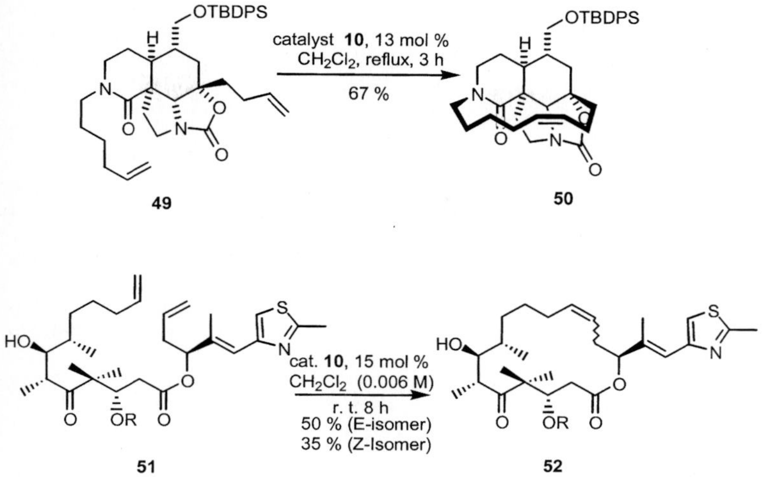

OTBDPS
catalyst **10**, 13 mol %
CH_2Cl_2, reflux, 3 h
67 %
49 **50**

HO
cat. **10**, 15 mol %
CH_2Cl_2 (0.006 M)
r. t. 8 h
50 % (E-isomer)
35 % (Z-Isomer)
51 **52**

Scheme 8. Ring-closing metathesis as the key step in the synthesis of manzamine A and epothilone A.

Indeed, ring-closing metathesis has become an interesting alternative in the formation of many different macrocyclic structures, such as catenanes [19].

Ring-closing metathesis is not limited to olefins. Catalysts promoting ring closure of alkynes [20] have also been developed (Scheme 9). Alkyne **56** can easily be converted into cyclic alkyne **57** with a yield of 73 %. For this purpose, a tungsten catalyst has been used.

Looking at the *E*/*Z* ratio in the formation of macrocyclic structures, for example in the epothilone synthesis (**51** → **52**), a higher *Z* selectivity would be desirable. Here, help may come from ring-closure metathesis of alkynes, leading to cyclic alkynes, which can be reduced selectively to *Z*-olefins.

Ring-closing metathesis has been used successfully in asymmetric synthesis (Scheme 10). Chiral catalysts such as **11** can be employed for the kinetic resolution of racemates [21] (**58** → **59** + **60**). Scheme 10 demonstrates the formation of only one enantiomer **59** with a high level of optical purity. Enantiomer **60** does not undergo

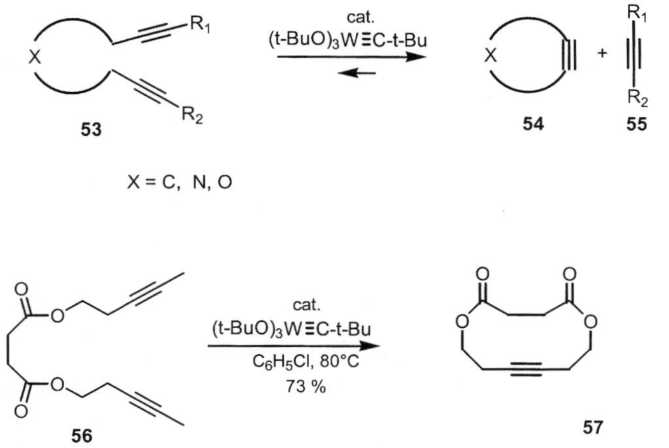

R_1
X
R_2
cat.
$(t\text{-}BuO)_3W{\equiv}C\text{-}t\text{-}Bu$
+
53 **54** **55**
X = C, N, O

cat.
$(t\text{-}BuO)_3W{\equiv}C\text{-}t\text{-}Bu$
C_6H_5Cl, 80°C
73 %
56 **57**

Scheme 9. Ring-closing alkyne metathesis.

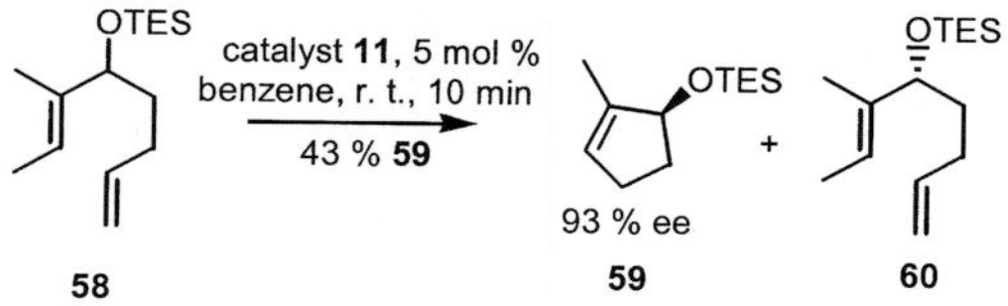

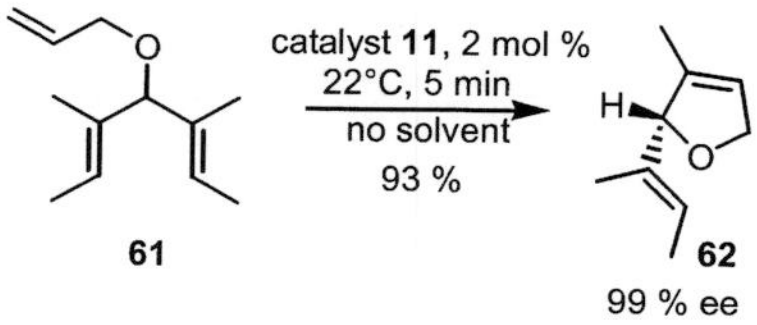

Scheme 10. Asymmetric ring-closing metathesis.

cyclization and can be isolated with a high optical purity.

Another interesting approach in asymmetric ring-closing metathesis is the catalytic enantioselective desymmetrization of achiral starting materials [21] (**61** → **62**). Starting with achiral material **61**, the use of chiral catalyst **11** leads selectively to the enantiomer **62**.

In summary, the development of new powerful catalysts has made ring-closing metathesis a very valuable method for the formation of common, medium-sized and macrocyclic rings. Remarkable features of the reaction are the compatibility of many different functional groups and the good yields, even in synthesis of macrocyclics. For this reason, ring-closing metathesis plays an important role in organic synthesis nowadays, and will contribute substantially in the future.

References

[1] a) K. J. Ivin, *Olefin Metathesis*, Academic Press, London, **1983**; b) V. Dragutan, A T. Balaban, M. Dimonie, *Olefin Metathesis and Ring Opening Polymerization of Cyclo-Olefins*, 2 nd, Ed., Wiley Interscience, New York, **1985**; c) R. H. Grubbs, W. Tumas, *Science* **1989**, *243*, 907; R. H. Schrock, *Acc. Chem. Res.* **1990**, *23*, 158.

[2] R. H. Grubbs, S. H. Pine in *Comprehensive Organic Synthesis* (Ed. B. M. Trost), Pergamon, New York, **1991**, Vol 5, Chapter 9.3.

[3] G. C. Fu, R. H. Grubbs, *J. Am. Chem. Soc.* **1992**, *114*, 5426. For reviews see: R. H. Grubbs, S. Chang, *Tetrahedron* **1998**, *54*, 4413; M. Schuster, S. Blechert, *Angew. Chem.* **1997**, *109*, 2124; M. Schuster, S. Blechert, *Angew. Chem. Int. Ed. Engl.* **1997**, *36*, 2036; H.-G. Schmalz, *Angew. Chem.* **1995**, *107*, 1981; H.-G. Schmalz, *Angew. Chem. Int. Ed. Engl.* **1995**, *34*, 1833.

[4] G. C. Bazan, J. H. Oskam, H.-N. Cho, L. Y. Park, R. R. Schrock, *J. Am. Chem. Soc.* **1991**, *113*, 6899.

[5] W. A. Herrmann, F. E. Kühn, R. W. Fischer, W. R. Thiel, C. C. Romao, *Inorganic Chem.* **1992**, *31*, 4431.

[6] S. T. Nguyen, R. H. Grubbs, J. W. Ziller, *J. Am. Chem. Soc.* **1993**, *115*, 9858.

[7] J. B. Alexander, D. S. La, D. R. Cefalo, A. H. Hoveyda, R. R. Schrock, *J. Am. Chem. Soc.* **1998**, *120*, 4041.

[8] T. A. Kirkland, D. M. Lynn, R. H. Grubbs, *J. Org. Chem.* **1998**, *63*, 9904.

[9] a) A. Fürstner, L. Ackermann, *Chem. Commun.* **1999**, 95; b) A. Fürstner, A. F.Hill, M. Liebl, J. D. E. T. Wilton-Ely, *Chem. Commun.* **1999**, 601.

[10] C. Hinderling, C. Adlhart, P. Chen, *Angew. Chem.* **1998**, *110*, 2831; C. Hinderling, C. Adlhart, P. Chen, *Angew. Chem. Int. Ed. Engl.* **1998**, *37*, 2685.

[11] G. C. Fu, S. T. Nguyen, R. H. Grubbs, *J. Am. Chem. Soc.* **1993**, *115*, 9856.

[12] S. T. Nguyen, L. K. Johnson, R. H. Grubbs, *J. Am. Chem. Soc.* **1992**, *114*, 3974.

[13] C. M. Huwe, S. Blechert, *Tetrahedron Lett.* **1995**, *36*, 1621.

[14] C. M. Huwe, S. Blechert, *Synthesis* **1997**, 61.

[15] S. F. Martin, Y. Liao, H.-J. Chen, M. Pätzel, M. N. Ramser, *Tetrahedron. Lett.* **1994**, *35*, 6005.

[16] S. F. Martin, J. M. Humphrey, A. Ali, M. C. Hillier, *J. Am. Chem. Soc.* **1999**, *21*, 866.

[17] a) D. Schinzer, A. Limberg, A. Bauer, O. M. Böhm, M. Cordes, *Angew. Chem. Int. Ed. Engl.* **1997**, *38*, 523; D. Schinzer, A. Limberg, A. Bauer, O. M. Böhm, M. Cordes, *Angew. Chem.* **1997**, *109*, 543; b) D. Meng, D.-S. Su, A. Balog, P. Bertinato, E. J. Sorenson, S. J. Danishefsky, Y.-H. Zheng, T.-C. Chou, L. He, S. Horwitz, *J. Am. Chem. Soc.* **1997**, *119*, 2733; c) Z. Yang, Y. He, D. Vourloumis, H. Vallberg, K. C. Nicolaou, *Angew. Chem. Int. Ed. Engl.* **1997**, *36*, 166; Z. Yang, Y. He, D. Vourloumis, H. Vallberg, K. C. Nicolaou, *Angew. Chem.* **1997**, *109*, 170.

[18] S. F. Martin, Y. Liao, Y. Wong, T. Rein, *Tetrahedron Lett.* **1993**, *34*, 691.

[19] D. G. Hamilton, N. Feeder, S. J. Teat, J. K. M. Sanders, *New J. Chem.* **1998**, 1019.

[20] A. Fürstner, G. Seidel, *Angew. Chem.* **1998**, *110*, 1758; A. Fürstner, G. Seidel, *Angew. Chem. Int. Ed. Engl.* **1998**, *37*, 1734.

[21] D. S. La, J. B. Alexander, D. R. Cefalo, D. D. Graf, A. H. Hoveyda, R. R. Schrock, *J. Am. Chem. Soc* **1998**, *120*, 9720.

Metal-Catalyzed Hydroformylations

Oliver Reiser

Hydrocarbonylations of C-C multiple bonds belong to the most attractive metal-catalyzed reactions. Inexpensive starting materials can be converted into useful intermediates without generating toxic by-products. Concurrent with the formation of a new C-C bond a versatile and further transformable carboxyl, carbonyl or alcohol functionality is introduced into the substrate. With the development of extremely efficient heterogeneous or homogeneous catalysts, which excel in high turnover numbers and turnover rates, hydrocarbonylations are especially attractive for industrial applications. The drawback of such reactions – the possible formation of several isomeric products – can be minimized in technical processes through effective separation of the products, which are in general *all* useful as intermediates. To arrive at only one hydrocarbonylation product a selective reaction process is of course necessary. In this article the scope and limitation of hydroformylations and related reactions will be discussed.

Hydroformylations of C-C Double Bonds

As early as 1938, Roelen discovered the cobalt-catalyzed hydroformylation of olefins, then known as the oxo reaction, which allowed the synthesis of aldehydes by addition of carbon monoxide and hydrogen to alkenes. Not long after this discovery it was found that cobalt, rhodium, ruthenium and platinum are also suitable as catalysts. However, because of the considerable price advantage for large scale applications in industry, cobalt catalysts are mostly used. Rhodium complexes, however, are generally more reactive and selective and are mainly described in this article.

The processes as well as the potential problems of hydroformylations are depicted in Scheme 1. Catalytically active is the 16-electron complex **4**, which, after coordination of the olefin **1** to the π-complex **5**, forms in a stereospecific *syn* addition the σ-complex **8**. Subsequently, coordination and insertion of CO gives rise to the acyl complex **9**. After addition of H_2 to the Rh(III) complex **7**, a reductive elimination to

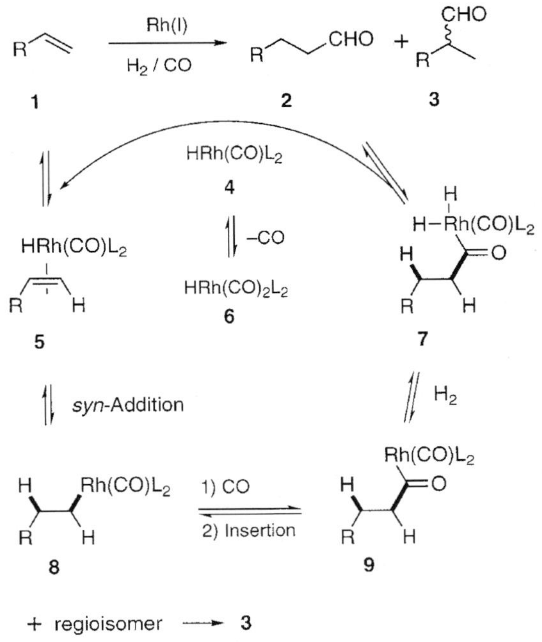

Rh(I)
H_2 / CO
CHO
CHO
1
2
3
$HRh(CO)L_2$
4
–CO
$HRh(CO)_2L_2$
6
$HRh(CO)L_2$
5
H–Rh(CO)L_2
7
syn-Addition
H_2
Rh(CO)L_2
8
1) CO
2) Insertion
Rh(CO)L_2
9
+ regioisomer → 3

Scheme 1. Mechanism of rhodium-catalyzed hydroformylation.

the linear product **2** occurs concurrent with the regeneration of the catalytically active species **4**. Instead of the addition of the rhodium fragment in **5** onto the less-substituted end of the double bond to **8**, the addition can also occur in the opposite way, which overall results in the formation of the branched product **3**. In this case a chiral product is formed, which opens the possibility for an asymmetric reaction process. Further problems can arise from the fact that the combination of H_2/Rh(I) is also a good hydrogenation reagent, which can effect the reduction of the alkene to the corresponding alkane or the reduction of the generated aldehyde to the corresponding alcohol. Since all reaction steps are reversible, migration of double bonds in alkyl-substitued alkenes resulting in even more regioisomeric products, is possible. Only recently has it been possible to develop catalysts through which high regiocontrol (and for branched products high enantiocontrol) has been achieved. The product distribution in hydroformylations is also dependent on the applied pressure, on the CO/H_2 ratio, and on the temperature, but the discussion of these parameters is beyond the scope of this article [1].

Linear or Branched?

Rhodium catalysts that bear phosphorus ligands give rise preferentially to linear hydroformylation products. This can be explained by the greater steric bulk of the catalyst, forcing, in the addition step **5** → **8**, the metal to the less-substituted side of the double bond. According to this simple picture (a more sophisticated analysis is possible by taking different P-Rh-P angles in rhodium phosphine catalysts into account), it can be understood that a rhodium(I)-catalyst developed by Union Carbide [2], which is formed *in situ* from the readily available ligand biphephos (**10**), and rhodium$(CO)_2$(acac) yields with excellent regioselectivity linear hydroformylation products [3]. An interesting application was recently demonstrated by Buchwald with the synthesis of the basic structure of pyrrolizidine alkaloids: starting from the proline derivative **11** one can obtain the linear aldehyde **12** which could be cyclized to **13** upon treatment with HCN (Scheme 2) [4]. The synthesis of pipecolic acid derivatives was accomplished by Ojima by hydroformylation of allylglycine **14** in a single step. Notably, a pressure of only 60 psi was sufficient to achieve

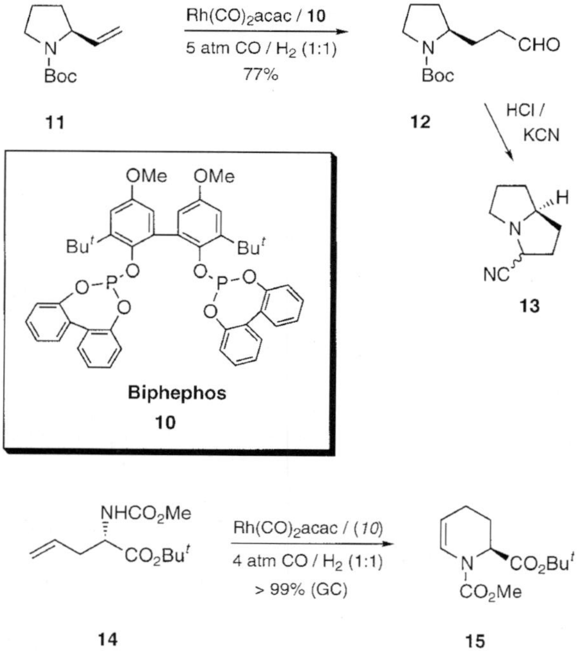

Rh(CO)$_2$acac / **10**
5 atm CO / H$_2$ (1:1)
77%

11 **12** **13**

HCl / KCN

Biphephos **10**

Rh(CO)$_2$acac / (*10*)
4 atm CO / H$_2$ (1:1)
> 99% (GC)

14 **15**

Scheme 2. Regioselective hydroformylation by a Rh(I)-Biphephos catalyst.

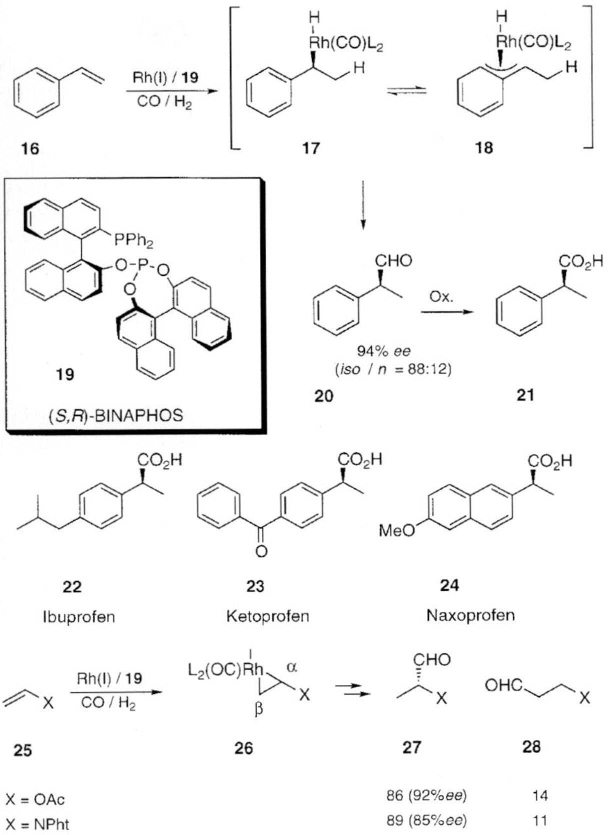

	27	28
X = OAc	86 (92% *ee*)	14
X = NPht	89 (85% *ee*)	11

Scheme 3. Enantioselective hydroformylation by a Rh(I)-BINAPHOS catalyst.

high turnover rates and yields [5]. The primarily formed aldehyde cyclized and subsequently dehydrated spontaneously under the reaction conditions applied.

Multiatomic [6] as well as cationic [7] rhodium catalysts also display a high preference for linear hydroformylation products. However, a catalyst system which generally yields branched hydroformylation products has not yet been found. Vinylarenes, such as styrene (**16**), form preferentially the *iso*-aldehyde **20** and not the *n*-aldehydes. The possibility to form a relatively stable Rh-π-allyl complex **18** is most likely the decisive factor for this result [8]. Subsequent oxidation of **20** leads to 2-arylpropionic acids **21**, of which some derivatives like **22–24** are of great importance as non-steroidal inflammatory drugs (NSID) (Scheme 3) [9]. For their synthesis by the hydroformylation of styrenes, not only a regioselective but also an enantioselective reaction process is necessary. Despite many attempts [10], only in recent times have catalysts been developed which are sufficiently selective for this process. The breakthrough was accomplished by developing mixed phosphine/phosphite-ligands, and the best results so far have been achieved with **19** and its enantiomer [11].

Also, heterosubstituted alkenes **25** lead preferentially to the branched hydroformylation products **27**, which are interesting building blocks for the synthesis of amino acids. An explanation put forward for the observed regiochemistry has been the coordination of rhodium onto the alkene to give **26**. Because of the electron-withdrawing substituent X, the Rh-Cα bond is stronger than the Rh-Cβ bond; therefore, hydride insertion takes place into the Rh-Cβ bond [1c].

An interesting step forward in asymmetric hydroformylation is the development of chiral catalysts on a platinum/tin basis [12], which surpass the previously introduced rhodium catalysts, in part in their enantioselectivity. However, with Pt-Sn systems, more side reactions such as hydrogenation generally occur, and also, with regard to regioselectivity, these catalysts cannot so far compete with the rhodium-phosphine-phosphite systems.

Directed Hydroformylations

The concept of controlling a reaction by intramolecularly directing a metal by a functional group in the substrate [13] has also been successfully applied to hydroformylations in some cases. It is not surprising that phosphine and phosphite groups can exhibit a directing function in a molecule (Scheme 4) [14]. Starting from the cyclohexene derivatives **29** or **30**, the aldehyde **32** is obtained with high regio- and diastereoselectivity *via* the five-membered ring chelate **31**. The example of the hydroformylation of **33** shows in particular that chelation occurs preferentially *via* a five-membered ring instead of a six-membered ring. An intriguing strategy for regio- and diastereoselective hydroformylations of methallylic alcohols **36** was developed by Breit et al. using *ortho*-diphenylphosphinyl benzoate as a removable catalysts-directing group [15]. Polypropionate subunits with *syn* stereochemistry are available in selectivites of up to 96 : 4, which are central building blocks in polyketide natural products.

The directing ability of amide groups (Scheme 5) seems to be preparatively useful. Starting from **38**, formation of **40** is accomplished *via* **39** (six-membered ring chelate instead of seven-membered ring chelate) [16], and starting from **41** the aldehyde **42** is preferentially formed rather than **43** (five-membered ring chelate instead of six-membered ring chelate) [17]. Amines also act as directing groups, however, the amino group coordinates with rhodium so strongly that in such cases stoichiometric amounts of the catalysts have to be used [18].

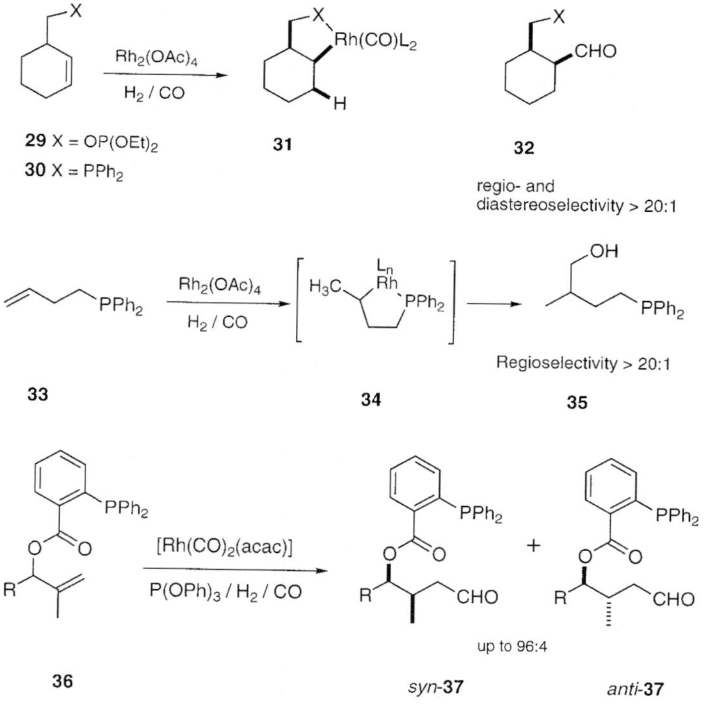

$Rh_2(OAc)_4$, H_2 / CO

29 X = $OP(OEt)_2$
30 X = PPh_2

31 $Rh(CO)L_2$

32 CHO

regio- and diastereoselectivity > 20:1

33 PPh_2 — $Rh_2(OAc)_4$, H_2 / CO → [**34**] → **35**

Regioselectivity > 20:1

36 — $[Rh(CO)_2(acac)]$, $P(OPh)_3$ / H_2 / CO → *syn*-**37** + *anti*-**37**

up to 96:4

Scheme 4. Substrate directed rhodium-catalyzed hydroformylations – phosphorus groups.

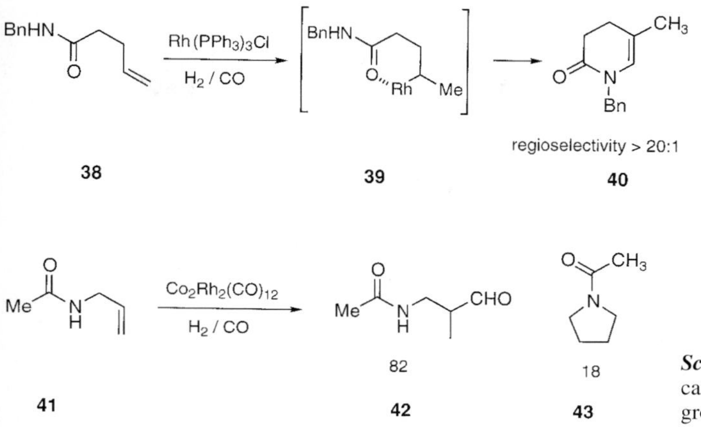

Scheme 5. Substrate directed rhodium-catalyzed hydroformylations – nitrogen groups.

Hydro- and Silylformylation of C-C Triple Bonds

Alkynes can also be hydroformylated, but hydrogenations of the starting materials or of the resulting olefinic products can not usually be suppressed. If one succeeds, however, to trap the primarily formed α,β-unsaturated aldehydes intramolecularly, preparatively useful transformations can be achieved: for example, from β-alkynyl amines **44** one can obtain pyrroles **46** in good yields [19]. It is questionable whether the intermediate **45** is formed, since usually hydroformylation is a stereospecific *cis* addition (cf. Scheme 1), so that in this case an isomerization has to have taken place (Scheme 6). Recently, the hydroformylation of non-terminal alkynes **47** was achieved for the first time in good yields leading to α,β-unsaturated aldehydes **48**, which was accompanied by reduction products only in minor amounts [20]. Again, the ligand **10** was the best choice; however, it was particularly important to lower the H_2/CO-pressure to 15 psi and to stop the reaction at the appropriate time. This method can so far be only applied to symmetrically substituted alkynes; with unsymmetrical substrates one always obtains the two possible regioisomeric products with not very high selectivity.

With excellent chemo- and regioselectivity, alkynes and especially terminal alkynes **49** undergo silylformylations (Scheme 7) [21]. Instead of hydrogen, phenyldimethylsilane is used, which adds, after activation of the Si-H bond by the metal catalyst, exclusively via the silicon-metal and not *via* the hydrogen-metal bond onto the al-

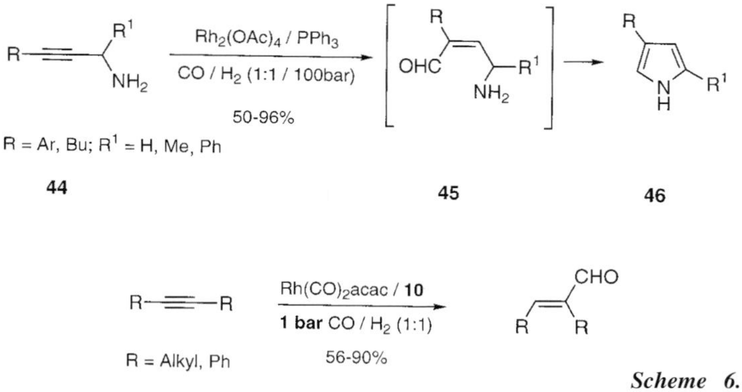

Scheme 6. Rhodium-catalyzed hydroformylations of alkynes.

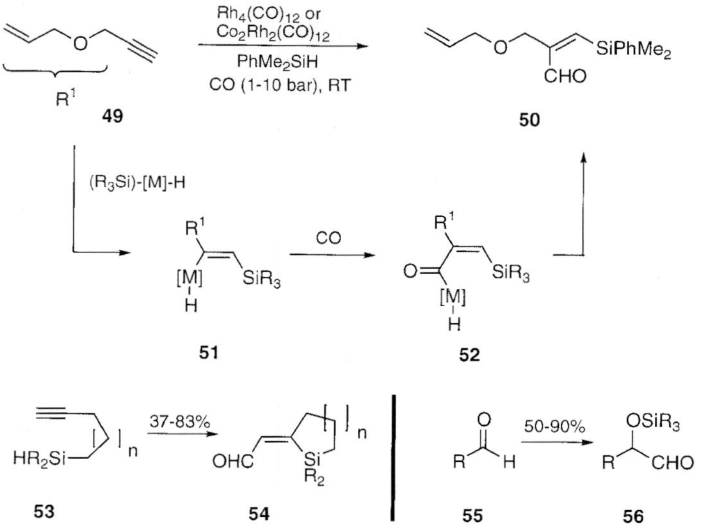

Scheme 7. Silylformylations of alkynes and aldehydes.

kyne to give **51**. The bulky silyl residue orients itself to the less-substituted side of the triple bond. After insertion of CO into the metal-vinyl bond to **52**, a reductive elimination occurs to yield the final silylformylation product **50**. This reaction can also be carried out in an intramolecular fashion, in which the *exo-dig* ring closure products **54** are always formed from the precursors **53** [22]. As demonstrated in the example of **49**, silylformylations do not occur with C-C double bonds. However, C-O double bonds do react, e.g. aldehydes **55** can be transformed to α-siloxyaldehydes **56**.

Domino Processes

The aldehydes **58** which are obtained in hydroformylations can also be directly converted to further products in the course of the reaction. For example, in the presence of a secondary amine, a reductive amination to **59** can be added onto the hydroformylation reaction (aminomethylation, Scheme 8) [23]. Especially elegant seems the possibility to add onto the hydroformylation another carbonylation reaction. If dicobalt octacarbonyl is used as the catalyst and the aldehyde **58** is trapped with a primary amide to give **60**, a second carbonylation occurs, resulting in a

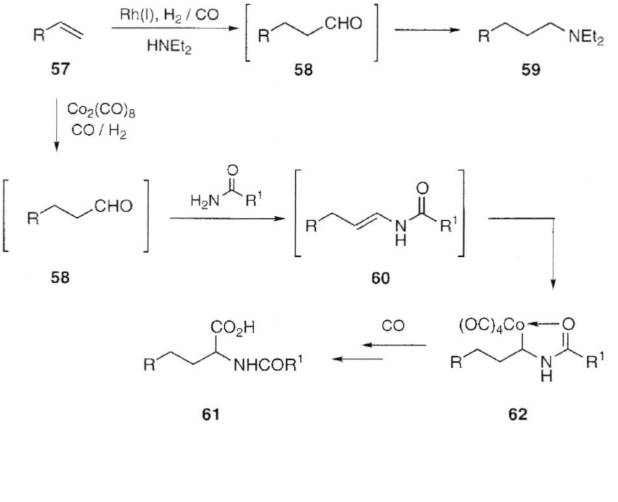

Scheme 8. Domino-carbonylation reactions.

carboxylic acid **61** and not in an aldehyde because of the *N*-acyl group [24]. With this sequence it is therefore possible to transform an alkene into a glycine unit in a single step.

There are many more fascinating aspects and possibilities of carbonylation reactions, which could not be covered within this article. As a result of the continuous discovery of new catalysts, the great synthetic potential of these processes will certainly be further explored.

References

[1] a) B. Cornils, W. A. Herrmann, R. W. Eckl, *J. Mol. Catal. A – Chemical* **1997**, *116*, 27–33. b) M. Beller, B. Cornils, C. D. Frohning, C. W. Kohlpaintner, *J. Mol. Catal. Chem* **1995**, *104*, 17. c) J. K. Stille in *Comprehensive Organic Synthesis Vol. 4*; B. M. Trost, I. Fleming, Ed.; Pergamon Press: Oxford, **1991**; pp. 913. d) I. Ojima, *Chem. Rev.* **1988**, *88*, 1011.

[2] Union Carbide Corporation (E. Billig, A. G. Abatjoglou, D. R. Byant), *Chem. Abstr.* **1987**, *107*, 7392.

[3] G. D. Cuny, S. L. Buchwald, *J. Am. Chem. Soc.* **1993**, *115*, 2066.

[4] G. D. Cuny, S. L. Buchwald, *Synlett* **1995**, 519.

[5] I. Ojima, M. Tzamarioudaki, M. Eguchi, *J. Org. Chem.* **1995**, *60*, 7078.

[6] a) G. G. Stanley, in M. P. Doyle (Ed.): *Advances in Catalytic Processes, Vol. 2*, Jai Press Inc., 55 Old Post Road/ No 2/Greenwich/CT 06836 **1997**, pp. 221–243. b) G. Süss-Fink, *Angew. Chem.* **1994**, *106*, 71.

[7] C. W. Lee, H. Alper, *J. Org. Chem.* **1995**, *60*, 499.

[8] A. van Rooy, E. N. Orij, P. C. J. Kamer, P. W. N. M. van Leeuwen, *Organometallics* **1995**, *14*, 34.

[9] C. Botteghi, S. Paganelli, A. Schionato, M. Marchetti, *Chirality* **1991**, *3*, 355.

[10] F. Agbossou, J. F. Carpentier, A. Mortreaux, *Chem. Rev.* **1995**, *95*, 2485.

[11] a) N. Sakai, K. Nozaki, H. Takaya, *J. Chem. Soc. Chem. Commun.* **1994**, 395. b) N. Sakai, S. Mano, K. Nozaki, H. Takaya, *J. Am. Chem. Soc.* **1993**, *115*, 7033.

[12] S. Gladiali, D. Fabbri, L. Kollar, *J. Organometal. Chem.* **1995**, *491*, 91.

[13] A. Hoveyda, D. A. Evans, G. C. Fu, *Chem. Rev.* **1993**, *93*, 1307.

[14] R. W. Jackson, P. Perlmutter, E. E. Tasdelen, *Aust. J. Chem.* **1991**, *44*, 951.

[15] (a) B. Breit, M. Dauber, K. Harms, *Chem. Eur. J.* **1999**, *5*, 2819–2827. (b) B. Breit, *Eur. J. Org. Chem.* **1998**, 1123–1134.

[16] I. Ojima, A. Korda, W. R. Shay, *J. Org. Chem.* **1991**, *56*, 2024.

[17] I. Ojima, Z. Zhang, *J. Org. Chem.* **1988**, *53*, 4422.

[18] M. E. Krafft, X. Y. Yu, S. E. Milczanowski, K. D. Donnelli, *J. Am. Chem. Soc.* **1992**, *114*, 9215.

[19] E. M. Campi, W. R. Jackson, Y. Nilsson, *Tetrahedron Lett.* **1991**, *32*, 1093.

[20] J. R. Johnson, G. D. Cuny, S. L. Buchwald, *Angew. Chem.* **1995**, *107*, 1877.

[21] a) I. Matsuda, A. Ogiso, S. Sato, Y. Izumi, *J. Am. Chem. Soc.* **1989**, *111*, 2332. b) I. Ojima, R. J. Donovan, M. Eguchi, W. R. Shay, P. Ingallina, A. Korda, Q. Zeng, *Tetrahedron* **1993**, *49*, 5431.

[22] F. Monteil, I. Matsuda, H. Alper, *J. Am. Chem. Soc.* **1995**, *117*, 4419.

[23] T. Baig, J. Molinier, P. Kalck, *J. Organomet. Chem.* **1993**, *455*, 219.

[24] I. Ojima, K. Hirai, M. Fujita, T. Fuchikami, *J. Organomet. Chem.* **1985**, *279*, 203.

Rare Earth Metal Catalysts

Patrick Amrhein and Karola Rück-Braun

Institut für Organische Chemie, Universität Mainz, Germany

The growing demand for efficient chemical transformations and catalysts has inspired a few research groups in recent years to develop rare earth metal catalysts for organic synthesis [1, 2]. Triflates of rare earth metals are strong Lewis acids, which are stable in aqueous solution. Rare earth metal alkoxides on the other hand are of interest as Lewis bases, e.g. in the catalysis of carbonyl reactions, because of the low ionization potentials (5.4–6.4 eV) and electronegativities (1.1–1.3) of the 17 rare earth elements. Rare earth metal-alkali metal complexes in contrast show both Brønsted-basic and Lewis-acidic properties. Impressive applications of such catalysts are presented and discussed here.

Reactions in Aqueous Media

Traditional Lewis acids such as $AlCl_3$ or $BF_3 \cdot OEt_2$ catalyze key steps in many reactions involving carbonyl compounds, leading to carbon-carbon bond formation. Because of their reactivity and instability these catalysts cannot be used in aqueous solution. For this area of application, rare earth metal catalysts open up new perspectives.

For example, aldol reactions of silyl enolates with aldehydes in the presence of $Sc(OTf)_3$ ($Tf = OSO_2CF_3$) or the rare earth metal triflates $Ln(OTf)_3$ (Ln = Yb, Gd, Lu) proceed smoothly in both organic solvents and aqueous solution under mild reaction conditions in high yields [1, 3].

Not only are these triflates highly stable in aqueous media, but in some cases reactions proceed many times faster in the presence of water. Furthermore, these Lewis acids can be recovered easily and completely. In toluene-ethanol-water mixtures, $Cu(OTf)_2$ as well as $Sc(OTf)_3$ and $Sm(OTf)_3$ are effective catalysts, e.g. in catalytic aldol reactions or allylation reactions of aldehydes [4]. Recently, it was found that $Sc(OTf)_3$ catalyzes allylation reactions of aldehydes with tetraallyltin (**2**) solely in water, with sodium dodecyl sulfate (SDS) as an additive, to afford homoallylic alcohols in high yields [5]. For example, 0.25 equiv. of **2** and 2-desoxyribose **1** in the presence of 0.1 equiv. of $Sc(OTf)_3$ and 0.2 equiv. sodium dodecylsulfate at room temperature gave the addition product in 99 % yield within 48 h (Scheme 1).

Based on this and earlier work, Kobayashi et al. in 1998 reported initial investigations of three-component reactions of aldehydes, anilines and allyltributylstannane to give homoallylic amines **4** [6]. With the catalyst $Sc(OTf)_3$ in the presence of sodium dodecyl sulfate for building a micellar system, yields in the range 66–90 % were obtained (Scheme 1). The stability of the catalyst towards oxygen is noteworthy.

Rare earth metal-catalyzed Michael additions of α-nitro esters or β-keto esters with α,β-unsaturated carbonyl compounds were investigated by Feringa and co-workers [7]. According to these authors, the use of $Yb(OTf)_3$ enables addition products **5** to be isolated in higher yields and greater purities than those obtained in organic solvents (Scheme 1).

Until now, the regioselective addition of enolates to imines in the presence of an aldehyde functionality could not be realized with stoichiometric amounts of traditional Lewis acids such as $SnCl_4$, $TiCl_4$ or $BF_3 \cdot OEt_2$. Recently, Kobayashi and coworkers reported that, in the presence of

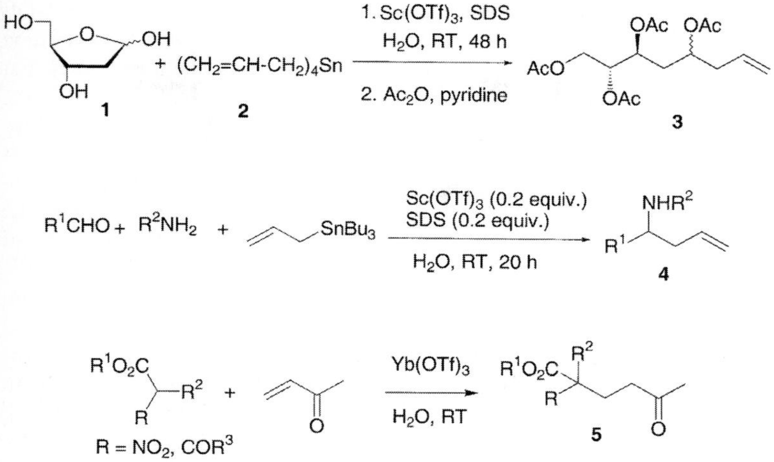

Scheme 1

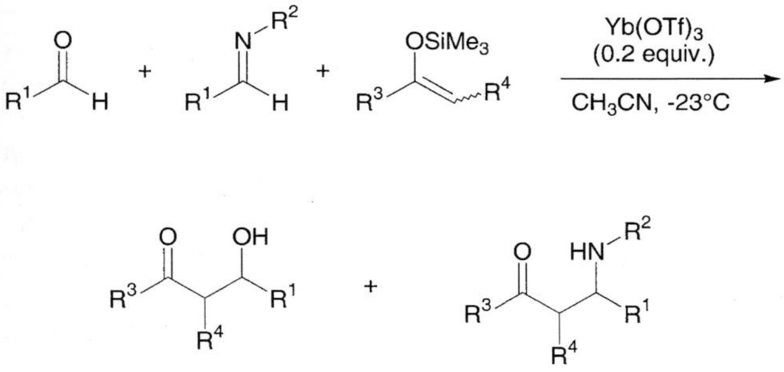

Scheme 2

catalytic amounts of $Yb(OTf)_3$ this type of reaction takes place at low temperature (Scheme 2) [8]. NMR studies prove that the Lewis acid activates the imine functionality and not the aldehyde group by highly selective complex formation.

Kobayashi et al. also published experiments in which they applied polymer-bound rare earth metal triflates [9]. The performance of Lewis acid-catalyzed imino-aldol reactions at the solid phase was realized by linking the silyl enolethers to 5-(4'-chloromethylphenyl)pentyl polystyrene [10].

Friedel-Crafts Catalysts

By use of the catalyst system $Ln(OTf)_3$-$LiClO_4$ and $Sc(OTf)_3$-$LiClO_4$, respectively, Friedel-Crafts acylations are performed in high yields (Scheme 3) [11].

In the absence of $LiClO_4$, the acylations proceed only sluggishly. However, a homogenous solution is obtained and the start of the reaction is observed only after the addition of the above mentioned triflate to a suspension of $LiClO_4$, acetic anhydride and the aromatic compound. Thus, the highly reactive cationic acylation reagent is formed from acetic anhydride and $LiClO_4$ only after the addition of the triflate $M(OTf)_3$ (M = Yb, Sc).

Advantageously, catalytic amounts of rare earth metal triflates are used instead of stoichiometric amounts of aluminum trichloride in catalytic Fries rearrangements of carboxylic acid aryl esters furnishing keto building blocks, e.g. **10**

[12]. The rearrangement of naphthol derivatives is best carried out with 5 mol % of $Sc(OTf)_3$ in toluene at 100 °C (Scheme 3). $Sc(OTf)_3$ is also the catalyst of choice for the direct synthesis of the ketones **10** from naphthol and acid chlorides in toluene/nitromethane (yield > 90 %).

Multi-Component Couplings

Multi-component couplings open up an economic and straightforward route to very different compound libraries. For example, 1,3-dipolar cycloaddition reactions of nitrones with alkenes furnish isoxazolines, which can be transformed reductively to hydroxyketones or *β*-amino alcohols. In 1997, two groups reported on the synthesis of isoxazolines by rare earth metal-catalyzed [3 + 2] cycloadditions (Scheme 4) [13, 14].

Reactions catalyzed by $Yb(OTf)_3$ proceed much more slowly than those catalyzed by $Sc(OTf)_3$ in the presence of molecular sieve (4 Å), but with better *endo*/*exo* selectivity [14]. According to Kobayashi et al. the products **13** are obtained smoothly with high *endo*-selectivity in 31–82 % yield by three-component coupling of aldehydes, hydroxylamines and alkenes catalyzed with $Yb(OTf)_3$ [13]. Enantioselective reaction procedures were carried out with (+)-BINOL (BINOL = 2,2'-dihydroxy-1,1'-binaphthyl) and PyBOX **14** as chiral ligands by Jørgensen et al. for the first time. While $Yb(OTf)_3 \cdot H_2O$ and (+)-BINOL furnished *endo*-**13b** as a racemic mixture, the use of the ligand **14** (PyBOX) enabled enantioselectivities in the range 67–73 % *ee* to be realized. However, for the enantioselective synthesis of *endo*-**13a** [72 %, *endo*:*exo* >99 : 1, 78 % *ee*] starting from **11a** and **12**, (+)-BINOL in combination with 1,2,6-trimethylpiperidine should be used as catalysts according to Kobayashi et al.

Lewis acid-catalyzed tandem Michael iminoaldol reactions enable the one-pot synthesis of *γ*-acyl-*δ*-lactams from *α*,*β*-unsaturated thioesters, silyl enolates and imines [15]. For the initial Michael addition, the combination of $SbCl_5$ with $Sn(OTf)_3$ (5 mol %) proved to be efficient. However, after the addition of the imino compound the iminoaldol product was isolated in moderate yield. For the enhancement of turnover and yield, $Sc(OTf)_3$, once again proved to be the Lewis acid of choice (Scheme 4, **15** : **16** = 81 : 19, 94 %).

Likewise, four-component reactions (silyl enolates, *α*,*β*-unsaturated thio esters, amines and aldehydes) proceed at low temperatures in high yields (65–97 %) with $Sc(OTf)_3$, as catalyst when aromatic aldehydes and aniline building blocks are employed.

With catalytic amounts of rare earth metal triflates, heterocarbonyl compounds, e.g. acylhydrazones, are also successfully activated. From the latter and silyl enolates (Scheme 4), the coupling products are obtained directly or in a one-pot synthesis in the presence of 5 mol% of $Sc(OTf)_3$ or $Yb(OTf)_3$ in 45–96 % yield. For example, compound **17** was isolated in 92 % yield and was subsequently cyclized with base to the corresponding pyrazolone (Scheme 4) [16]. In comparison with typical Lewis acids, such as $SnCl_4$, (10 % yield) and boron trifluoride etherate (42 % yield), $Sc(OTf)_3$ proved to be superior.

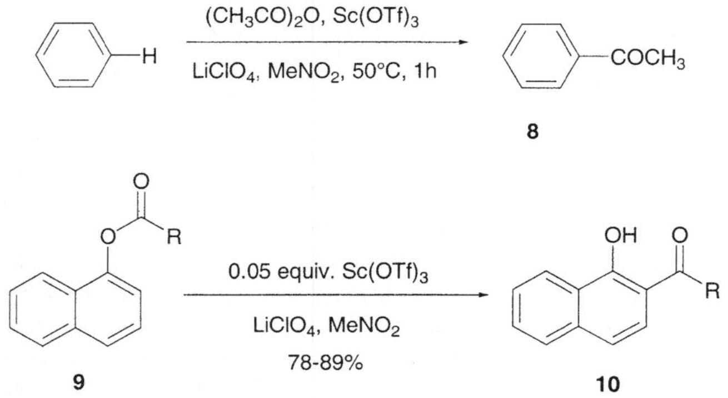

$(CH_3CO)_2O$, $Sc(OTf)_3$
H
$LiClO_4$, $MeNO_2$, 50°C, 1h
$COCH_3$
8

O
O R
0.05 equiv. $Sc(OTf)_3$
$LiClO_4$, $MeNO_2$
78-89%
OH O
R
9
10

Scheme 3

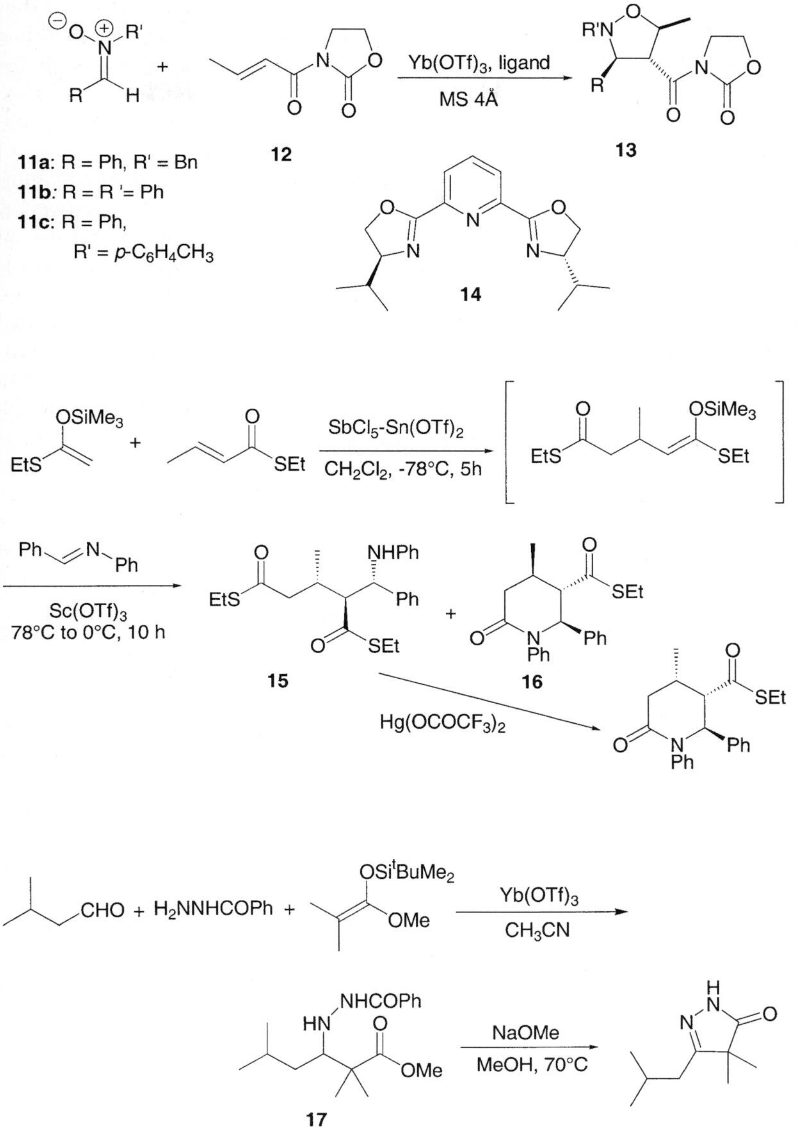

11a: R = Ph, R' = Bn
11b: R = R '= Ph
11c: R = Ph, R' = p-$C_6H_4CH_3$

Yb(OTf)$_3$, ligand
MS 4Å

12 13 14

$SbCl_5$-$Sn(OTf)_2$
CH_2Cl_2, -78°C, 5h

Sc(OTf)$_3$
78°C to 0°C, 10 h

15 16

$Hg(OCOCF_3)_2$

Yb(OTf)$_3$
CH_3CN

NaOMe
MeOH, 70°C

17

Scheme 4

Asymmetric Two-Center Catalysis

During the last decade Shibasaki and co-workers focussed on the application of rare earth metal catalysts with special properties [2]. More recently, impressive studies by this group revealed the broad applicability of chiral heterobimetallic catalysts based on rare earth metal alkoxide complexes in asymmetric catalysis. Whereas initial investigations concentrated on catalytic and asymmetric nitroaldol reactions with nitromethane, hydrophosphonylations of imines [17] and Michael additions [2], more recently the direct catalytic aldol reaction of aldehydes with non-modified ketones [18] and the epoxidation of α,β-unsaturated ketones were reported [19]. By use of the rare earth metal-alkali metal complex $LaNa_3$-tris[(R)-(binaphthoxide)] (10 mol %), the reaction of **18** (Scheme 5) with *tert*-butyl hydroperoxide (TBHP) afforded **21** in 92 % yield with 82 % *ee*.

With the La-BINOL catalysts **19a** and **19b** and the Yb-BINOL catalysts **20a** and **20b**, products with higher optical purity can generally be obtained. Interestingly, the alkali metal-free BINOL complexes **19b** and **20b** (Ln = La, Yb, R = CH_2OH) were found to be the most efficient catalysts. By the latter, epoxidations of differently substituted enones with cumol hydroperoxide (CMHP) or TBHP are catalyzed, providing high enantiofacial selectivity (83–94 % *ee*). These La-BINOL complexes are of oligomeric structure, and the asymmetric catalysis is attributed to the positive interaction of two rare earth metal centers. One metal center, therefore, appears to act as a Lewis acid, activating the enone and controlling the orientation of the carbonyl function. In contrast, a second metal alkoxide unit seems to activate the hydroperoxide because of its Brønsted-basic properties.

For catalytic asymmetric aldol-type reactions, the transformation of the methylene compounds to a silyl enolate or a silyl ketene acetal was at one time necessary. Recently, the aldol reaction of aldehydes with non-modified ketones was realized by use of the lanthanum-Li_3-tris[(*R*)-binaphthoxide] catalyst **22** [18]. According to the proposed catalytic cycle, after abstraction of an *α*-proton from the ketone, the reaction between the lithium-enolate complex and the aldehyde

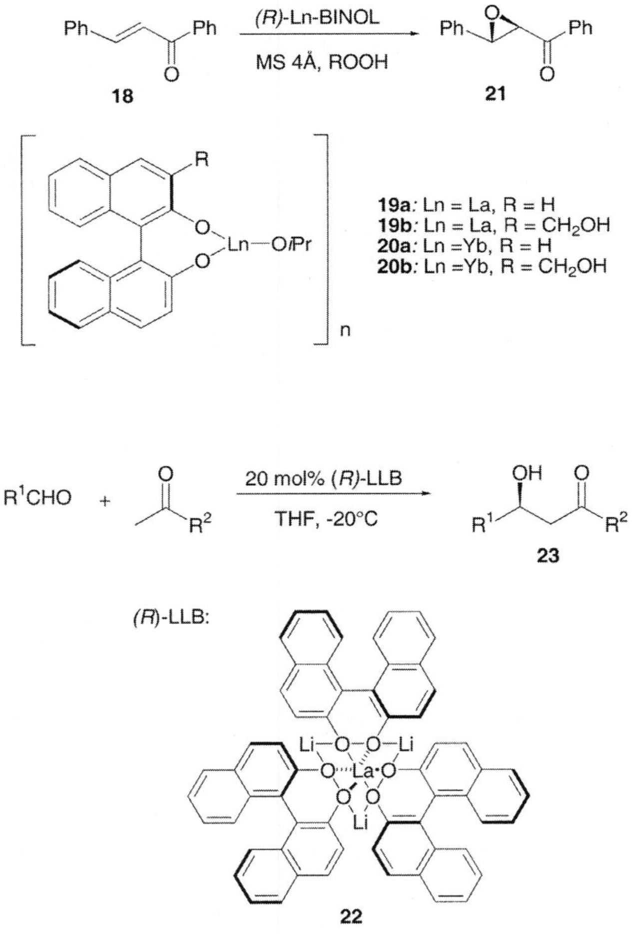

Ph–CH=CH–C(O)–Ph (**18**) → [(*R*)-Ln-BINOL; MS 4Å, ROOH] → epoxide **21**

19a: Ln = La, R = H
19b: Ln = La, R = CH_2OH
20a: Ln =Yb, R = H
20b: Ln =Yb, R = CH_2OH

[BINOL–Ln–O*i*Pr]$_n$

R^1CHO + $CH_3C(O)R^2$ → [20 mol% (*R*)-LLB; THF, -20°C] → **23**

(*R*)-LLB: **22**

Scheme 5

coordinated to the lanthanum center takes place. From the β-keto metal alcoholate obtained by proton exchange, the catalyst is regenerated, and simultaneously the optically active aldol adduct **23** is liberated. The products are obtained in 28–90 % yield with an enantiomeric excess in the range of 44 to 94 %. Self-condensation of the aldehydes applied was only observed with dihydrocinnamic aldehyde.

Only recently, Shibasaki et al. reported on the application of a Sm-Na-(*R*)-BINOL-complex as catalyst in a reaction cascade consisting of an asymmetric Michael addition of thiols to α,β-unsaturated carbonyl compounds followed by an asymmetric enolate protonation [20].

The development and application of these rare earth catalysts excellently demonstrate the fruitful combination of the empirical with the rational approach [2]. In the near future, additional applications of these and other rare earth metal complexes in various enantioselective reactions are expected.

References

[1] Review: S. Kobayashi, *Synlett* **1994**, 689 and literature cited therein.

[2] Review: M. Shibasaki, H. Sasai, T. Arai, *Angew. Chem.* **1997**, *109*, 1290 and literature cited therein.

[3] S. Kobayashi, T. Wakabayashi, S. Nagayama, H. Oyamada, *Tetrahedron Lett.* **1997**, *38*, 4559 and literature cited therein.

[4] S. Kobayashi, S. Nagayama, T. Busujima, *Chem. Lett.* **1997**, 959.

[5] S. Kobayashi, T. Wakabayashi, H. Oyamada, *Chem. Lett.* **1997**, 831.

[6] S. Kobayashi, T. Busujima, S. Nagayama, *J. Chem. Soc. Chem. Commun.* **1998**, 19.

[7] a) E. Keller, B. L. Feringa, *Synlett* **1997**, 842; b) E. Keller, B. L. Feringa, Tetrahedron Lett. **1996**, *37*, 1879.

[8] S. Kobayashi, T. Busujiama, S. Nagayama, *J. Org. Chem.* **1997**, *62*, 232.

[9] a) S. Kobayashi, S. Nagayama, *Synlett* **1997**, 653 and literature cited therein; b) L. Yu, D. Chen, J. Li, P. G. Wang, *J. Org. Chem.* **1997**, *62*, 3575.

[10] S. Kobayashi, M. Moriwaki, *Tetrahedron Lett.* **1997**, *38*, 4251.

[11] A. Kawada, S.Mitamura, S. Kobayashi, *J. Chem. Soc. Comm.* **1996**, 183.

[12] S. Kobayashi, M. Moriwaki, I. Hachiya, *Bull. Chem. Soc. Jpn.* **1997**, *70*, 267.

[13] S. Kobayashi, R. Akiyama, M. Kawamura, H. Ishitani, *Chem. Lett.* **1997**, 1039.

[14] A. I. Sanchez-Blanco, K. V. Gothelf, K. A. Jørgensen, *Tetrahedron Lett.* **1997**, *38*, 7923.

[15] S. Kobayashi, R. Akiyama, M. Moriwaki, *Tetrahedron Lett.* **1997**, *38*, 4819.

[16] H. Oyamada, S. Kobayashi, *Synlett* **1998**, 249.

[17] H. Gröger, Y. Saida, H. Sasai, K. Yamaguchi, J. Martens, M. Shibasaki, *J. Am. Chem. Soc.* **1998**, *120*, 3189.

[18] Y. M. A. Yamada, N. Yoshikawa, H. Sasai, M. Shibasaki, *Angew. Chem.* **1997**, *109*, 1942.

[19] M. Bougauchi, S. Watanabe, T. Arai, H. Sasai, M. Shibasaki, *J. Am. Chem. Soc.* **1997**, *119*, 2329.

[20] E. Emori, T. Arai, H. Sasai, M. Shibasaki, *J. Am. Chem. Soc.* **1998**, *120*, 4043.

Dithioacetals as an Entry to Titanium-Alkylidene Chemistry: New and Efficient Carbonyl Olefination

Bernhard Breit

Organisch-Chemisches Institut der Ruprecht-Karls-Universität Heidelberg, Germany

Reactions in which two halves of a complex molecule are linked by means of a convergent synthetic strategy belong to the most valuable framework-building reactions in organic synthesis. Especially successful are those reactions in which C-C double bonds are formed, such as the Wittig reaction and its variants according to Horner, Wadsworth, and Emmons (referred to here as Wittig-type reactions) [1]. Decisive for the effectiveness of these methods is the simple preparative access to the two individual components – a carbonyl compound: and a Wittig-type reagent, which can be obtained, for instance, by the reaction of an alkyl halide with phosphanes or phosphites followed by deprotonation (Scheme 1).

Only such a simple and general approach enables the synthetic chemist to convert any two complex building blocks into carbonyl and Wittig-reagent components in a late step of the synthetic sequence. The components can be subsequently coupled by either an inter- or intramolecular olefination reaction. However, Wittig-type reactions are also subject to certain limitations. One example is the *cis* selectivity upon use of nonstabilized ylides under salt-free reaction conditions. Fortunately, the Wittig-type reactions can be supplemented by the Julia-Lythgoe olefination (Scheme 2), which is a general method for preparing *trans*-disubstituted olefins [2, 3]. The components required for the Julia-Lythgoe olefination – the carbonyl and sulfone components – also meet the criterion of being readily available, and allow the corresponding functionalization in a late synthetic step.

Another disadvantage of Wittig-type reactions is their limitation to aldehydes and ketones as the carbonyl component; carboxylic acid derivates

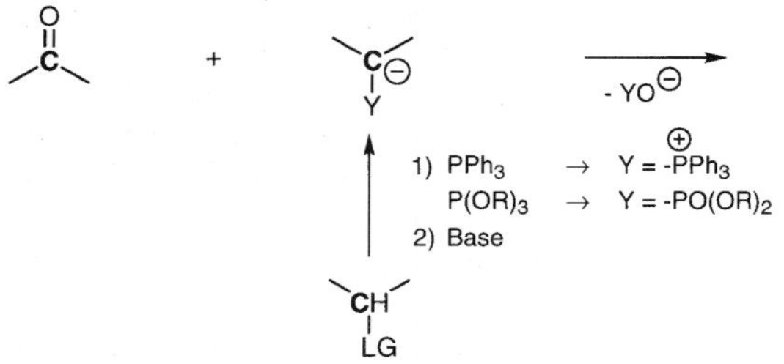

+ $- YO^{\ominus}$

1) PPh_3 → $Y = -\overset{\oplus}{P}Ph_3$

$P(OR)_3$ → $Y = -PO(OR)_2$

2) Base

CH

LG

Scheme 1. Schematic representation of the Wittig and Horner-Wadsworth-Emmons reactions. LG = leaving group.

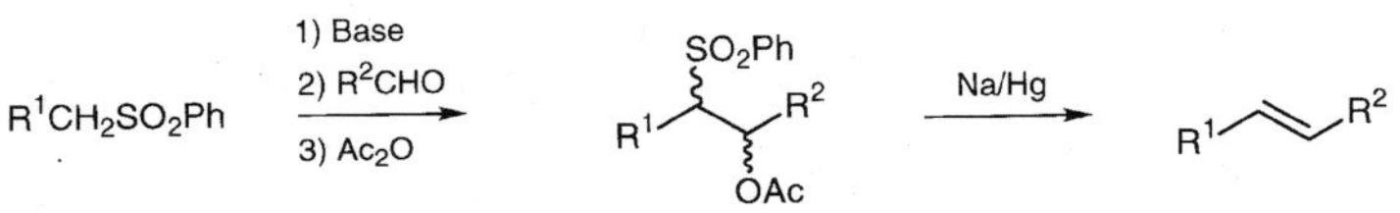

$R^1CH_2SO_2Ph$ 1) Base 2) R^2CHO 3) Ac_2O → SO_2Ph, R^1, R^2, OAc → Na/Hg → R^1, R^2

Scheme 2. Stereoselective synthesis of disubstituted *trans*-olefins according to Julia and Lythgoe.

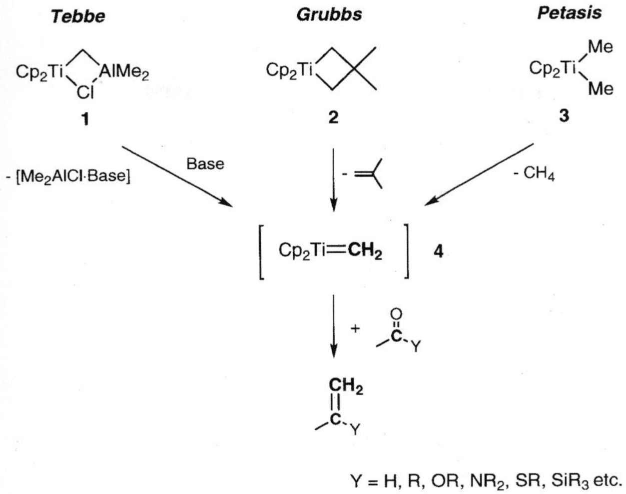

Scheme 3. Carbonyl methylenation with the titanium-methylene species **4** prepared from the Tebbe, Grubbs, or Petasis reagents (**1** – **3**).

are generally inert in this respect. Furthermore, Wittig-type reactions and Julia-Lythgoe olefinations both require a more or less basic reaction medium. Especially in the case of easily enolizable carbonyl compounds, this can lead to undesired side reactions such as elimination and racemization of adjacent stereocenters. The olefination of sterically demanding carbonyl substrates also clearly demonstrates the limitations of the Wittig reaction.

For this reason, considerable efforts have been devoted to finding improved olefination reagents that can overcome these shortcomings of both the Wittig-type reactions and Julia-Lythgoe olefinations. A milestone was reached in 1978 by Tebbe, who recognized the usefulness of the titanium-aluminum complex **1** for carbonyl methylenations [4]. In addition to the Tebbe reagent **1**, the titanacyclobutane **2** reported by Grubbs [5] and the Petasis reagent **3** [6]. are available for efficient methylenation of carbonyl compounds (Scheme 3). These reagents are reactive under neutral to slightly Lewis acidic conditions, which allows easily enolizable carbonyl compounds to be used in methylenation reactions without competing side reactions. Another advantage is the clean methylenation of carboxylic acid derivatives with formation of, for example, preparatively valuable enol ethers and enamines [7].

A plausible intermediate of this olefination is the titanium-methylene species **4**, which is formed from **1** by removal of $AlMe_2Cl$ with a Lewis base, from **2** by fragmentation with elimination of isobutene, and from **3** by α-elimination and release of methane. However, none of these three routes to titanium-carbene complexes of type **4** proved to be generally applicable. Consequently, the use of these reagents in synthesis is essentially limited to the transfer of a methylene unit [8]. From a synthetic viewpoint, a general and easy route to substituted titanium-alkylidene species and their use in carbonyl olefinations would be more desirable.

The first progress was made by Takai and Lombardo, who developed an in situ entry to titanium-alkylidene chemistry starting from the reagent combinations **5** and **6** (Scheme 4) [9]. These reactions proceed via a *gem*-dizinc compound **7** (its formation is catalyzed by traces of lead or lead(II) salts), which is subsequently transmetalated with $TiCl_4$ to the titanium-alkylidene species **8**, the actual olefination reagent. To date, **8** has not been characterized in detail [10]. These in situ reagents exhibit chemoselectivities similar to those of the structurally defined methylenation reagents **1**–**3**.

The advantage of the Takai-Lombardo reagents is the possibility of transferring substituted alky-

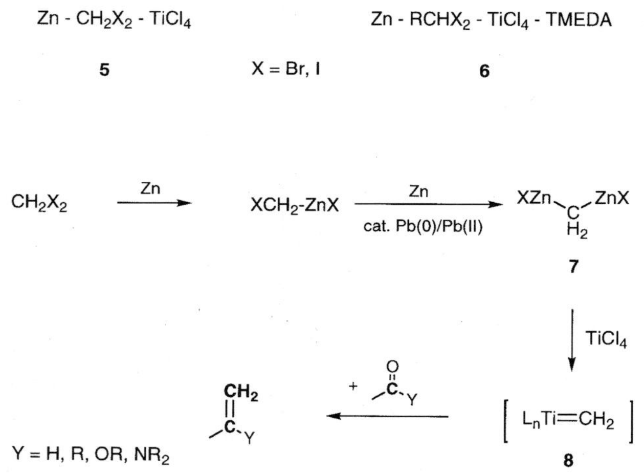

Zn - CH_2X_2 - $TiCl_4$ **5** X = Br, I Zn - $RCHX_2$ - $TiCl_4$ - TMEDA **6**

CH_2X_2 →(Zn) XCH_2-ZnX →(Zn, cat. Pb(0)/Pb(II)) $XZnCH_2ZnX$ **7** →($TiCl_4$) [$L_nTi{=}CH_2$] **8** →(+ Y–C(=O)–) $CH_2{=}C(Y)$–

Y = H, R, OR, NR_2

Scheme 4. Takai-Lombardo reagents; TMEDA= N,N,N',N'-tetramethylethylenediamine.

lidene units in addition to methylene units. The use of the olefination reagent **6** allows the general alkylidenation of carboxylic esters [11]. In this transformation, the (*Z*)-enol ethers **9** are obtained with high stereoselectivity (Scheme 5).

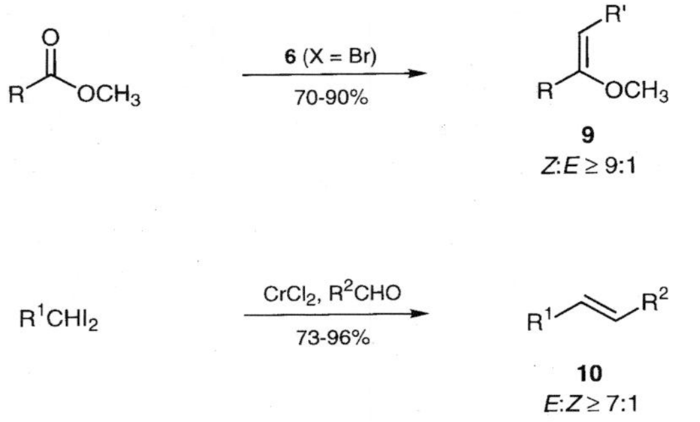

R–C(=O)–OCH_3 →(**6** (X = Br), 70-90%) **9** Z:$E \geq 9$:1

R^1CHI_2 →($CrCl_2$, R^2CHO, 73-96%) $R^1CH{=}CHR^2$ **10** E:$Z \geq 7$:1

Scheme 5. General alkylidenation of esters to enol ethers **9** and synthesis of disubstituted (*E*)-alkenes **10** from aldehydes and chromium(II) reagents according to Takai and Utimoto.

A variation of this reaction was developed in 1986 by Takai and Utimoto, in which geminal dihaloalkanes were added to aldehydes in a reaction mediated by chromium dichloride. This led to the stereoselective formation of the corresponding *trans*-olefins **10** [12]. The major drawback of this method is the rather cumbersome access to the corresponding substituted dihalomethane compounds, which prevents a broad application of this reaction for synthesis.

The solution to the above problem was recently found by Takeda et al., who reported on the desulfurization of dithioacetals as a general and easy entry to titanium-alkylidene chemistry [13]. Dithioacetals, which are easily accessible from carbonyl compounds, are treated with the titanocene source [$Cp_2Ti\{P(OMe)_3\}_2$] (**11**), which was specifically developed for this purpose; the respective titanium-alkylidene species **12** is formed, presumably by desulfurization (Scheme 6). The most important subsequent reac-

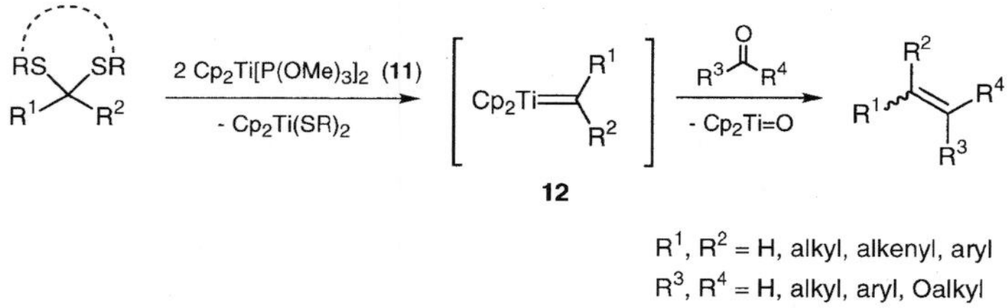

$R^1R^2C(SR)_2$ →(2 $Cp_2Ti[P(OMe)_3]_2$ (**11**), − $Cp_2Ti(SR)_2$) [$Cp_2Ti{=}CR^1R^2$] **12** →(R^3COR^4, − $Cp_2Ti{=}O$) $R^1R^2C{=}CR^3R^4$

R^1, R^2 = H, alkyl, alkenyl, aryl
R^3, R^4 = H, alkyl, aryl, Oalkyl

Scheme 6. Carbonyl olefination with titanium-alkylidene species **12** prepared from dithioacetals according to Takeda et al.

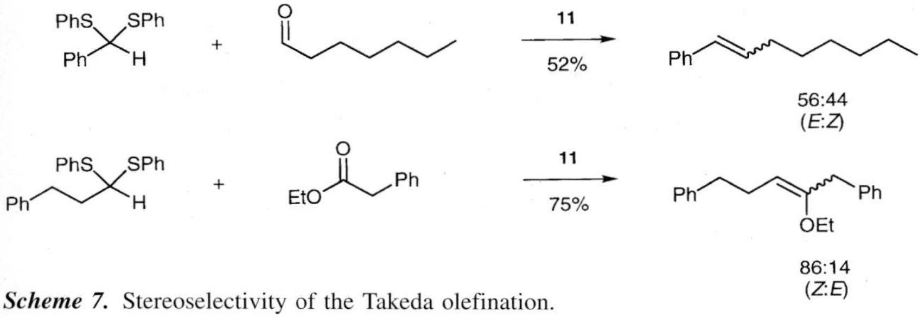

Scheme 7. Stereoselectivity of the Takeda olefination.

tion of this species is carbonyl olefination, which proceeds smoothly with aldehydes, ketones, and esters. The intermediates formed in these reactions exhibit a chemoselectivity spectrum similar to that of the titanium reagents **1**–**3**, **5**, and **6**. No limitations have yet been observed with respect to the structure of the dithioacetals; that is, even 1-substituted dithioacetals with β-hydrogen atoms can be converted. It is still unclear as to which functional groups can be tolerated in this reaction.

A clear disadvantage is the unsatisfactory stereoselectivity observed for the olefination of aldehydes so far. However, one can easily envision that modifying the microenvironment of the reactive titanium center will bring about improvements. Better selectivities have been obtained in the olefination of carboxylic esters (Scheme 7).

The good accessibility of dithioacetals offers the possibility, e.g. by Wittig-type reactions, to convert any two complex fragments into an olefin using the titanium reagent **11** in a late step of the synthetic sequence. This olefination is therefore as valuable as the Wittig-type reactions. Moreover, this reaction offers all the advantages of the titanium-mediated reactions that proceed in Lewis acidic media.

Interestingly, the subsequent reactions of the titanium-alkylidene species **12** obtained from dithioacetals are not limited to carbonyl olefinations. When the carbene complex is prepared in the presence of olefins, the latter are smoothly cyclopropanated (Scheme 8; **13**) [14]. Furthermore, the reaction of symmetrically disubstituted acetylenes with dithioacetals containing a methylene unit provides the corresponding trisubstituted 1,3-dienes **14** in a stereoselective fashion [15].

Interestingly, if an alkene functionality is available intramolecularly (**15**), a ring closing meta-

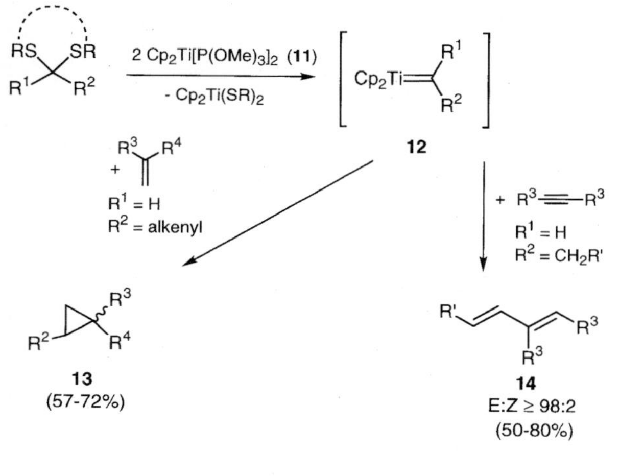

Scheme 8. Reactions of the titanium-alkylidene species **12**, prepared from dithioacetals, with olefins and acetylenes.

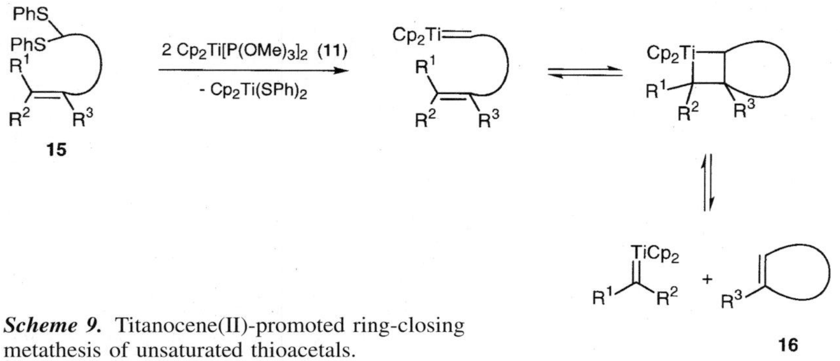

Scheme 9. Titanocene(II)-promoted ring-closing metathesis of unsaturated thioacetals.

thesis towards five-, six-, seven- and eight-membered carbo- and heterocylic ring systems may be realized (Scheme 9; **16**) [16].

Dithioacetals have already proven to be very useful in organic synthesis. For instance, they function as carbonyl protecting groups that can be used orthogonal to *O,O*-acetals [17]. The introduction of a dithioacetal leads to a polarity reversal of the carbonyl group (dithiane method according to Seebach) [18]. As a result, any complex dithioacetal can be obtained by a deprotonation-alkylation sequence. This remarkable multifunctionality of the dithioacetal unit has now been expanded by the work of Takeda et al. with respect to the specific formation of titanium-alkylidene species. Although the subsequent chemistry of this species needs to be established in detail, the intermolecular carbonyl olefination as well as the intramolecular ring-closing metathesis have given a glimpse of the synthetic scope of this reaction strategy. In addition to the Wittig-type reaction and the Julia-Lythgoe olefination, the Takeda olefination has the potential of becoming another general olefination reaction which proceeds under Lewis acidic conditions.

References

[1] a) A. Maercker, *Org. React.* **1965**, *14*, 270–490; b) M. Schlosser, *Top. Stereochem.* **1970**, *5*, 1–30; c) W. S. Wadsworth, *Jr., Org. React.* **1977**, *25*, 73–253; d) B. E. Maryanoff, A. B. Reitz, *Chem. Rev.* **1989**, *89*, 863–927; for the use of the Wittig reaction in the synthesis of complex natural products, see e) H. J. Bestmann, O. Vostrowsky, *Top. Curr. Chem.* **1983**, *109*, 85–163; f) K. C. Nicolaou, M. W. Härter, J. L. Gunzner, A. Nadin, *Liebigs Ann.* **1997**, 1283–1301.

[2] S. E. Kelly in Comprehensive Organic Synthesis, Vol. 1 (Eds.: B. M. Trost, I. Fleming), Pergamon, Oxford, **1991**, p. 792 ff.

[3] Although the Schlosser modification of the Wittig reaction provides access to *trans*-olefins from nonstabilized ylides, the Julia-Lythgoe olefination has proven to be the method of choice for solving this synthetic problem today. a) M. Schlosser, K.-F. Christmann, *Justus Liebigs Ann. Chem.* **1967**, *708*, 1–35; b) M. Schlosser, K.-F. Christmann, A. Piskala, *Chem. Ber.* **1970**, *103*, 2814–2820.

[4] a) F. N. Tebbe, G. W. Parshall, G. S. Reddy, *J. Am. Chem. Soc.* **1978**, *100*, 3611–3613; b) S. H. Pine, R. Zahler, D. A. Evans, R. H. Grubbs, *J. Am. Chem. Soc.* **1980**, *102*, 3270–3272; c) S. H. Pine, R. J. Pettit, G. D. Geib, S. G. Cruz, C. H. Gallego, T. Tijerina, R. D. Pine, *J. Org. Chem.* **1985**, *50*, 1212–1216.

[5] a) T. R. Howard, J. B. Lee, R. H. Grubbs, *J. Am. Chem. Soc.* **1980**, *102*, 6876–6878; b) K. A. Brown-Wensley, S. L. Buchwald, L. Cannizzo, L. Clawson, S. Ho, D. Meinhardt, J. R. Stille, D. Straus, R. H. Grubbs, *Pure Appl. Chem.* **1983**, *55*, 1733–1744.

[6] N. A. Petasis, S.-P. Lu, E. I. Bzowej, D.-K. Fu, J. P. Staszewski, I. Akritopoulou-Zanze, M. A. Patane, Y.-H. Hu, *Pure Appl. Chem.* **1996**, *68*, 667–670.

[7] Review: S. H. Pine, *Org. React.* **1993**, *43*, 1–91.

[8] Reports have appeared on the olefination of carbonyl compounds with the use of dibenzyl-, bis[(trimethylsilyl)methyl]-, and dicyclo- propyltitanocenes: a) N. A. Petasis, E. I. Bzowej, *J. Org. Chem.* **1992**, *57*, 1327–1330; b) N. A. Petasis, I. Akritopoulou, *Synlett* **1992**, 665–667; c) N. A. Petasis, E. I. Bzowej, *Tetrahedron Lett.* **1993**, *34*, 943–946; d) N. A. Petasis, J. P. Staszewski, D.-K. Fu, *ibid.* **1995**, *36*, 3619–3622.

[9] a) K. Takai, Y. Hotta, K. Oshima, H. Nozaki, *Tetrahedron Lett.* **1978**, 2417–2420; b) L. Lombardo, *ibid.* **1982**, *23*, 4293–4296; c) *Org. Synth.* **1987**, *65*, 81–87; J. Hibino, T. Okazoe, K. Takai, H. Nozaki, *Tetrahedron Lett.* **1985**, *26*, 5579–5580; e) T. Okazoe, J. Hibino, K. Takai, H. Nozaki, *ibid.* **1985**, *26*, 5581–5584; f) K. Takai, O. Fujimura, Y. Kataoka, K. Utimoto, *ibid.* **1989**, *30*, 211–214.

[10] K. Takai, T. Kakiuchi, Y.Kataoka, K. Utimoto, *J. Org. Chem.* **1994**, *59*, 2668–2670.

[11] a) T. Okazoe, K. Takai, K.Oshima, K. Utimoto, *J. Org. Chem.* **1987**, *52*, 4410–4412; b) K. Takai, Y. Kataoka, T. Okazoe, K. Utimoto, *Tetrahedron Lett.* **1988**, *29*, 1065–1068.

[12] a) T. Okazoe, K. Takai, K. Utimoto, *J. Am. Chem. Soc.* **1987**, *109*, 951–953; b) K. Takai, K. Nitta, K. Utimoto, *ibid.* **1986**, *108*, 7408–7410; c) K. Takai, Y. Kataoka, T. Okazoe, K. Utimoto, *Tetrahedron Lett.* **1987**, *28*, 1443–1446.

[13] Y. Horikawa, M. Watanabe, T. Fujiwara, T. Takeda, *J. Am. Chem. Soc.* **1997**, *119*, 1127–1128.

[14] Y. Horikawa, T. Nomura, M. Watanabe, T. Fujiwara, T. Takeda, *J. Org. Chem.* **1997**, *62*, 3678–3682.

[15] T. Takeda, H. Shimokawa, Y. Miyachi, T. Fujiwara, *Chem. Commun.* **1997**, 1055–1056.

[16] T. Fujiwara, T. Takeda, *Synlett* **1999**, 354–356.

[17] P. J. Kocienski, *Protecting Groups*, Thieme, Stuttgart, **1994**, p. 171–178.

[18] a) D. Seebach, *Angew. Chem.* **1969**, *81*, 690–700; *Angew. Chem. Int. Ed. Engl.* **1969**, *8*, 639–649; b) B. T. Gröbel, D. Seebach, *Synthesis* **1977**, 357–402; c) D. Seebach, *Angew. Chem.* **1979**, *91*, 259–278; *Angew. Chem. Int. Ed. Engl.* **1979**, *18*, 239–258; d) P. C. Bulman-Page, M. B. van Niel, J. C. Prodger, *Tetrahedron* **1989**, *45*, 7643–7677.

New Developments in the Pauson-Khand Reaction

Oliver Geis and Hans-Günther Schmalz

Institut für Organische Chemie, Universität Köln, Germany

Introduction

Metal mediated and catalyzed reactions have made significant contributions to organic synthesis over the past two decades [1]. One of the earliest and most useful of these is the Pauson-Khand carbon-carbon coupling reaction [2] first reported in 1971. In this reaction, a cyclopentenone is formed from an alkyne and an alkene in the presence of $[Co_2(CO)_8]$ with insertion of carbon monoxide in a formal [2+2+1]-cycloaddition. The exceptional potential of this reaction has been demonstrated in many (mostly intramolecular) syntheses (Scheme 1) [3].

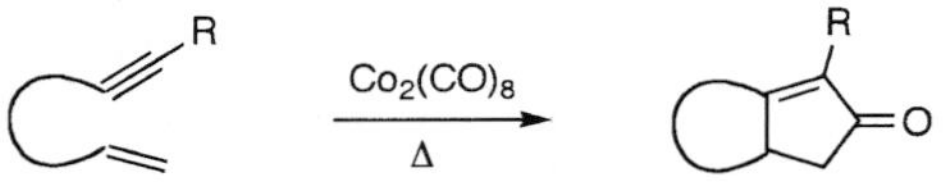

R Co2(CO)8 Δ R O

Scheme 1. The general pattern of the Pauson-Khand reaction.

Although a catalytic approach was discussed in the initial publication [2b], stoichiometric amounts of the metal were usually required to achieve acceptable yields. In most cases, the readily prepared and air-stable alkyne-$Co_2(CO)_6$ complexes (**1**) were heated (60–120 °C) with the alkene (occasionally under CO atmosphere), but long reaction times (often several days) were needed and the yields were frequently unsatisfactory.

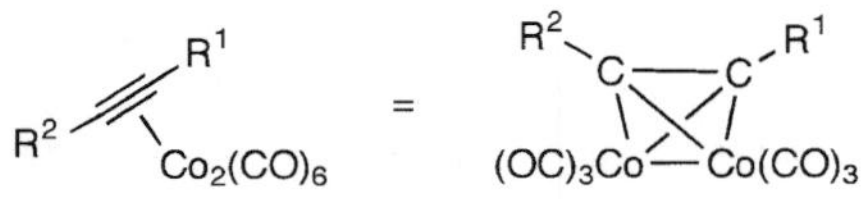

R1 R2 Co2(CO)6 = R2 C C R1 (OC)3Co Co(CO)3

1

Improved Reaction Conditions

Once the high synthetic value of the Pauson-Khand reaction was recognized, considerable efforts were made by several groups in the early 1990s to optimize the reaction conditions. An important improvement came with the use of tertiary amine *N*-oxides that generate free coordination sites at cobalt by oxidative removal of CO ligands. These reactions proceed rapidly at room temperature, often in high yields [4]. A very promising procedure for the stoichiometric Pauson-Khand reaction was recently described by Sugihara et al. [5], who discovered that the use of primary amines as solvent leads to a dramatic increase in reaction rates. Excellent yields are obtained within few minutes with only 3.5 equivalents of cyclohexylamine when the reactions are run in dichloromethane at 83 °C under argon. Alternatively, the reactions can be performed in a 1 : 3 mixture of 1,4-dioxane and 2 M aqueous ammonia at 100 °C (Scheme 2).

Catalytic Variants

Only catalytic Pauson-Khand reactions fulfill the criterion of atom economy [6], and the use of stoichiometrical amounts of the transition metal is not acceptable commercially. It is not surprising, therefore, that several research groups have focused more recently on the development of catalytic variants.

Based on the work of Pauson and Billington [2, 7], Rautenstrauch et al. showed in their 1990's synthesis of the dihydrojasmonate precursor **2** (Scheme 3) that catalytic Pauson-Khand

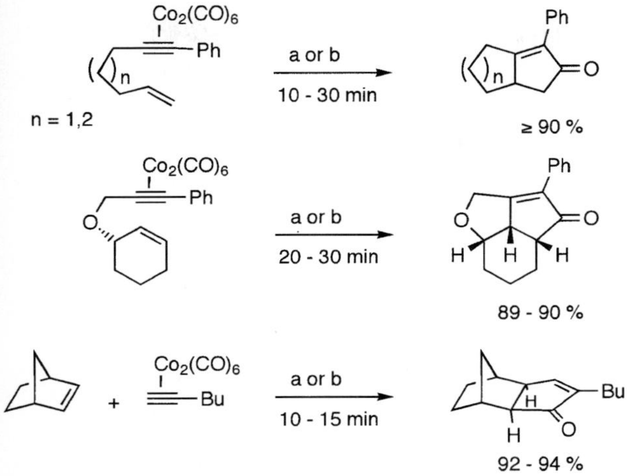

Scheme 2. Stoichiometric Pauson-Khand reactions according to T. Sugihara. a) 3.5 equiv. cyclohexylamine, 1,2-dichloromethane, 83°C; b) 1,4-dioxan/2N NH_3(aq) (1 : 3), 100°C.

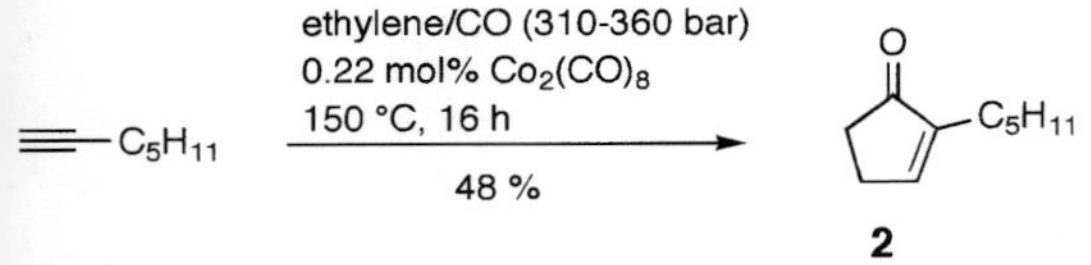

Scheme 3. Catalytic Pauson-Khand reactions according to Rautenstrauch.

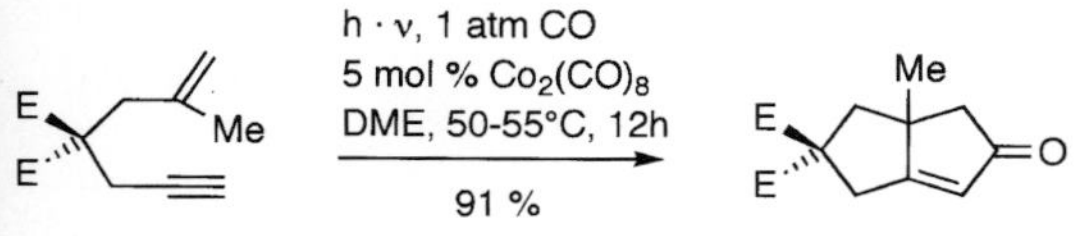

Scheme 4. Photochemically induced, catalytic Pauson-Khand reactions according to Livinghouse.

reactions are possible if high CO pressure and high temperature are used [8]. Korean laboratories have since found that more efficient transformations can be achieved with modified cobalt catalysts such as [$Co_2(CO)_8$/$P(OPh)_3$], [(indenyl)Co(cod)] or [$Co(acac)_2$/$NaBH_4$] under CO pressure or with $Co_2(CO)_8$ in supercritical fluids [9]. A seemingly practicable procedure was published in 1996 by Pagenkopf and Livinghouse, who obtained high yields with photoactivation of [$Co_2(CO)_8$] and low CO pressures (Scheme 4) [10].

Although a whole series of carbonyl complexes of other transition metals (Fe, Mo, W, Ni) could only be used in stoichiometric Pauson-Khand reactions [11], two Japanese laboratories have since independently reported efficient ruthenium-catalyzed (intramolecular) reactions. The desired cyclopentenones are formed in good to excellent yields in dimethylacetamide [12] or dioxane [13] in the presence of 2 mol % of [$Ru_3(CO)_{12}$] at 140–160 °C and 10–15 atm CO pressure.

Pauson-Khand Type Reactions with Metallocenes

Negishi et al. have demonstrated that alkynes react with "zirconocene" generated in situ to give metallacyclopentene species of type **3,** which when treated with carbon monoxide afford typical Pauson-Khand products (Scheme 5) [14]. Tamao et al. have also shown that enynes react with isocyanides to form iminocyclopentenones in the presence of stoichiometric amounts of [$Ni(cod)_2$/n-Bu_3P] [15]. More recently, Buchwald and co-workers have developed a titanocene-based method in which the intermediate titanacyclopentenes are initially captured by isocyanides and the resulting imines are subsequently hydrolyzed to the cyclopentenones [16]. They have also succeeded in performing

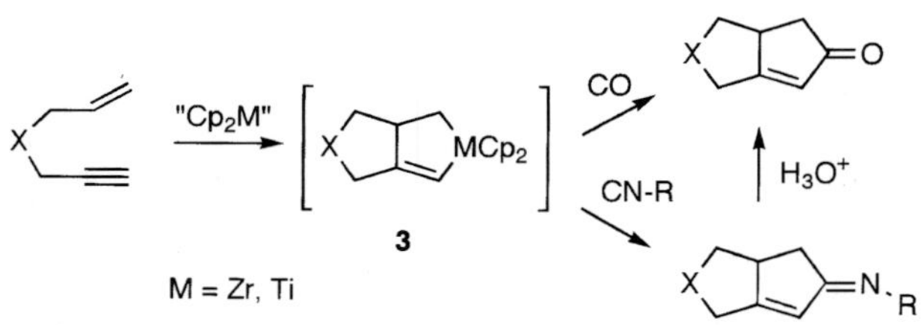

Scheme 5. Metallocene-mediated Pauson-Khand type reactions.

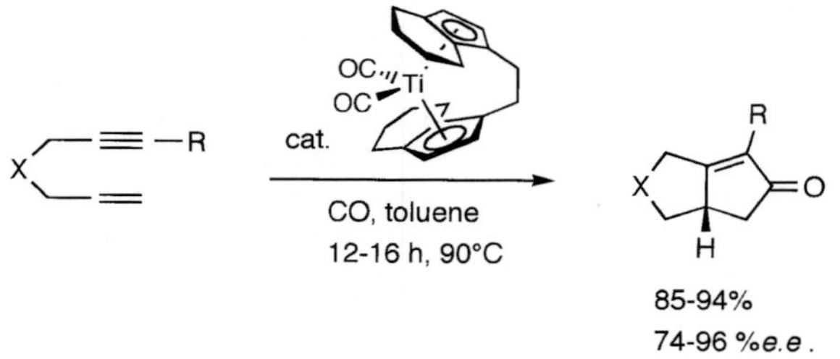

Scheme 6. Catalytic enantioselective bicyclizations according to Buchwald.

the reactions catalytically with trialkylsilylcyanides as the isocyanide source and catalytic amounts of $[Cp_2Ti(PMe_3)_2]$ or, alternatively, a catalyst which is generated in situ from $[Cp_2TiCl_2]$ by addition of two equivalents of *n*-BuLi or EtMgBr [17]. An Ni-based catalyst system ([Ni(cod)$_2$]/BDPEDA) has also proved to be efficient [18].

Recently, Buchwald and co-workers devised an outstanding procedure for the direct, titanocene-catalyzed cyclocarbonylation of enynes [19]. This catalytic method has a number of advantages: it occurs at low CO pressure, tolerates a variety of functional groups including disubstituted alkenes, and gives the cyclopentenones in high yields (>85 %). This industrially attractive process was later modified to proceed enantioselectively. With 5–20 % of [(*S,S*)-(EBTHI)Ti(CO)$_2$] as chiral catalyst [20] the desired cyclopentenones are obtained in high enantiomeric purity (Scheme 6) [21].

Asymmetric Pauson-Khand Reactions

Considerable efforts have been made to develop asymmetrical variants of the classical Pauson-Khand reaction. Initial investigations have shown that compounds derived from cobalt complexes of type **1,** in which a carbonyl ligand is replaced by a chiral phosphane (glyphos), react with high enantioselectivity [22]. However, the procedure is too complex to be of preparative value. The concept of Kerr et al., who achieved significant enantioselectivities (max. 44 % *ee*) in intermolecular Pauson-Khand reactions by

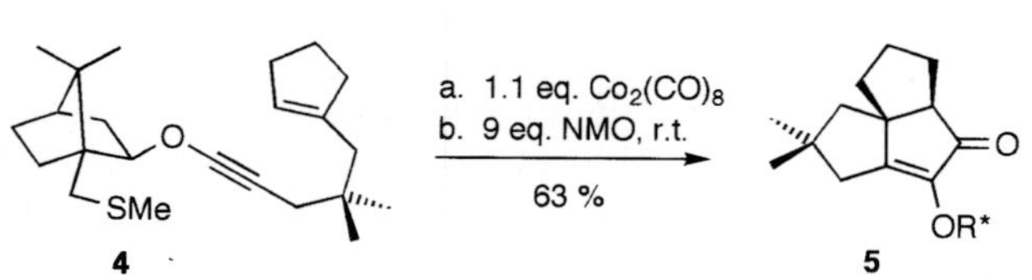

Scheme 7. A diastereoselective Pauson-Khand reaction using a chiral-modified substrate according to M. A. Pericàs.

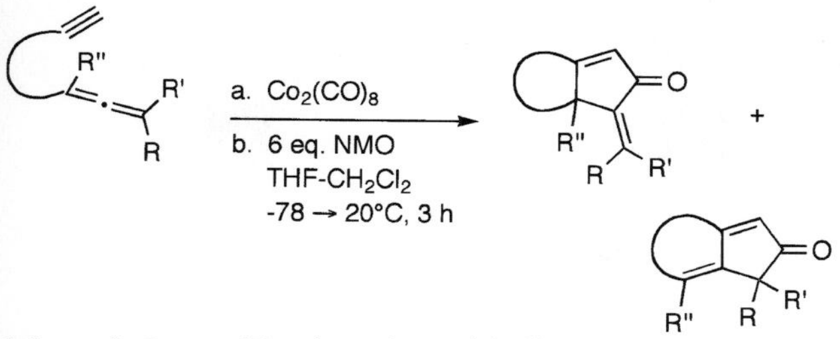

Scheme 8. Pauson-Khand reactions with allenes.

employing chiral amine oxides as promoters, appears more elegant [23]. A different approach was used by Pericàs, Moyano, Riera and Greene, who observed high levels of asymmetric induction in Pauson-Khand reactions when chirally modified substrates were used [24]. The potential of this method was demonstrated in total syntheses of hirsutene [25], brefeldin A [26], and β-cuparenone [27]. The same concept was applied successfully in a recently published synthesis of (+)-15-nor-pentalenene, wherein the key step is conversion of enyne **4** into the tricyclic product **5** (Scheme 7) [28]. As the preparation of chirally modified substrates always requires considerable effort, the attractiveness of the above-mentioned catalytic enantioselective method is obvious, and it would be interesting to test the Buchwald procedure with the achiral analogs of Pericàs' substrates.

Pauson-Khand Reactions with Allenes

Several authors have demonstrated that a Pauson-Khand type formation of methylenecyclopentenones from enynes, allenes and carbon monoxide occurs with stoichiometric ammounts of [$Fe_2(CO)_9$] or [$Mo(CO)_6$], and catalytically with [$Cp_2Ti(CO)_2$] [19, 29]. Cazes et al. recently reported that cobalt-mediated inter- and intramolecular conversions of allenes with enynes are also possible (Scheme 8) [30]. In these reactions, however, the formation of β-methylenecyclopentenones is favoured and di- or trisubstituted allenes must be employed in intramolecular reactions, because less substituted allenes tend to polymerize in the presence of [$Co_2(CO)_8$].

"Hetero-Pauson-Khand Reactions": Synthesis of γ-Butyrolactones

Buchwald and his group have also synthesized γ-butyrolactones successfully by a metallocene mediated cyclization of enones (and ynones) with carbon monoxide in a formal [2+2+1]-addition, and have thus achieved the first "hetero-Pauson-Khand reaction" [31]. The reactions can be conducted in high yields with either stoichiometrical or catalytical amounts of [$Cp_2Ti(PMe_3)_2$] as the example in Scheme 9 shows.

"Interrupted" Pauson-Khand-Reactions

Krafft et al. have found that in the conversions of enyne-$Co_2(CO)_6$ complexes significant amounts of monocyclic by-products are obtained in addition to the desired cyclopentenones (via oxidation

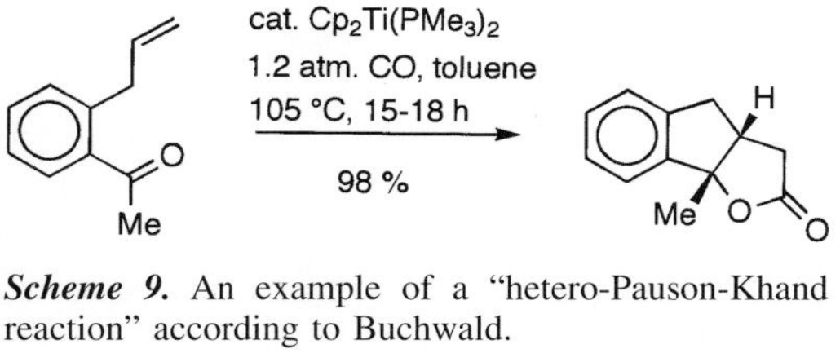

Scheme 9. An example of a "hetero-Pauson-Khand reaction" according to Buchwald.

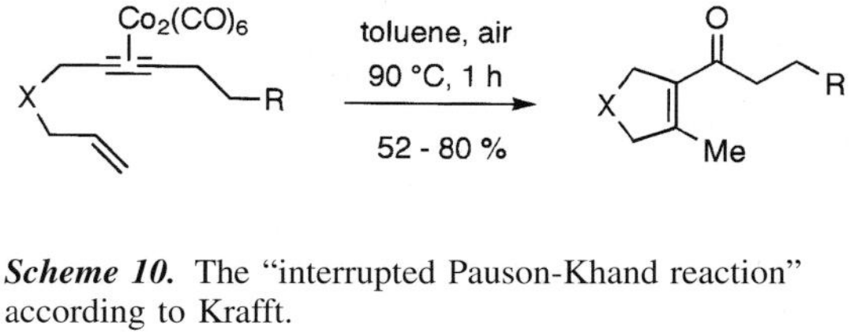

Scheme 10. The "interrupted Pauson-Khand reaction" according to Krafft.

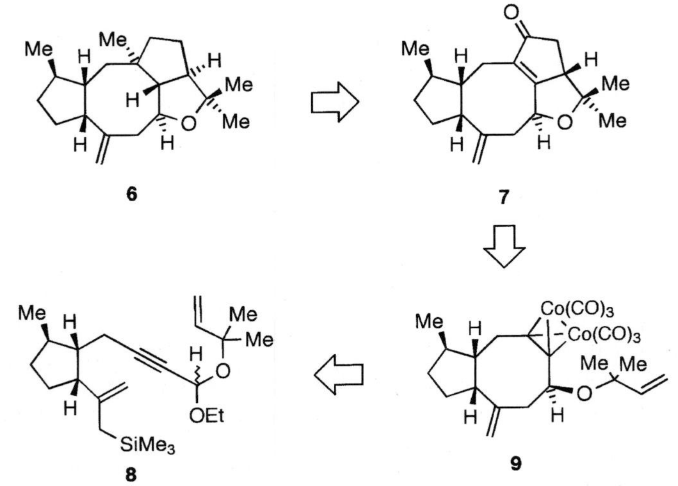

Scheme 11. Strategy of the Schreiber synthesis of (+)-epoxydictymene in a retrosynthetic representation.

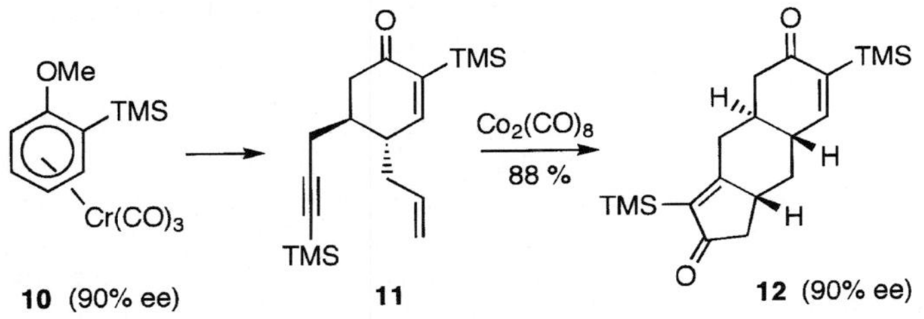

Scheme 12. Combined arene-$Cr(CO)_3$ and Pauson-Khand chemistry according to Kündig.

of the primary enyne cyclization product) [32]. In fact, the conventional Pauson-Khand reaction can be almost totally suppressed if it is conducted in air. An example of an "interrupted Pauson-Khand reaction" is illustrated in Scheme 10.

New Synthetic Applications of the Pauson-Khand Reaction

The high value of the Pauson-Khand reaction in the synthesis of natural products and other complex compounds has been frequently demonstrated [3]. One of the most impressive examples is the synthesis of the marine natural product (+)-epoxydictymene by Schreiber and co-workers [33]. The synthetic strategy (Scheme 11) uses an intermolecular Nicholas reaction [34] (the Lewis acid mediated conversion of the $Co_2(Co)_6$ complex derived from **8**) for the preparation of the actual Pauson-Khand substrate **9,** which is then converted to the epoxydictymene precursor **7**.

A reaction sequence published recently by Kündig and co-workers also deserves notice [35]: in a one-pot reaction the planar chiral arene-$Cr(CO)_3$ complex **10** is first converted (with chirality transfer) to the enyne **11**, which then affords the tricyclic Pauson-Khand product **12** in high yield and completely diastereoselectively (Scheme 12).

Finally, sequential Pauson-Khand reactions (domino reactions) are possible [36, 37]. A particular fascinating application of this concept is the synthesis of a fenestrane by Keese and co-workers (Scheme 13) [36].

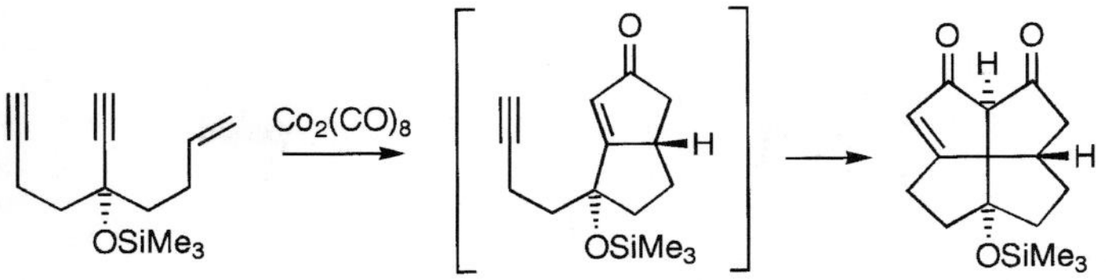

Scheme 13. Synthesis of a fenestrane by domino Pauson-Khand reaction according to Keese.

Conclusions

The Pauson-Khand reaction (together with related metallocene-catalyzed transformations) has established a prominent place in the repertoire of synthetic organic chemists. Its use enables the construction of complex molecules in a convergent and atom economic way starting from structurally simple precursors. High levels of enantioselectivity can be achieved. It is therefore not surprising that an increasing number of research groups is focussing on the further development of this reaction. In the era of combinatorial chemistry the conversion of solid-supported substrates is just one possibility [38].

References

[1] a) L. S. Hegedus, *Transition Metal Organometallics in the Synthesis of Complex Molecules*, 2nd Edn., University Science Books, Sausalito, CA, **1999**; b) M. Beller, C. Bolm (eds.), *Transition Metals for Organic Synthesis,* Wiley-VCH, Weinheim, **1998**; c) R. Noyori, *Asymmetric Catalysis in Organic Synthesis*, Wiley, New York, **1994**.

[2] a) I. U. Khand, G. R. Knox, P. L. Pauson, W. E. Watts, *J. Chem. Soc. Chem. Commun.* **1971**, 36; b) I. U. Khand, G. R. Knox, P. L. Pauson, W. E. Watts, M. I. Foreman, *J. Chem. Soc., Perkin Trans. 1* **1973**, 977; c) P. L. Pauson, I. U. Khand, *Ann. N. Y. Acad. Sci.* **1977**, *295*, 2.

[3] For reviews on the Pauson-Khand reaction, see: a) P. L. Pauson, *Tetrahedron* **1985**, 41, 5855; b) P. L. Pauson in *Organometallics in Organic Synthesis*, A. de Meijere, H. tom Dieck (eds.), Springer-Verlag, Berlin **1987**, p. 233; c) N. E. Schore, *Org. React.* **1991**, *40*, 1; d) N. E. Schore in *Comprehensive Organic Synthesis*, *Vol 5*; B. M. Trost (ed.), Pergamon, Oxford, **1991**, p. 1037; e) N. E. Schore in *Comprehensive Organometallic Chemistry II*, *Vol. 12*, E. W. Abel, F. G. A. Stone, G. Wilkinson (eds.), Pergamon, Oxford, **1995**, p. 703; f) N. Jeong in: *Transition Metals for Organic Synthesis, Vol. 1*, M. Beller, C. Bolm (eds.), Wiley-VCH, Weinheim, **1998**, p. 561; for a recent review on metal mediated cycloadditions in general, see: g) H.-W. Frühauf, *Chem. Rev.* **1997**, *97*, 523.

[4] a) S. Shambayati, W. E. Crowe, S. L. Schreiber, *Tetrahedron Lett.* **1990**, *31*, 5289; b) N. Jeong, Y. K. Chung, B. Y. Lee, S. H. Lee, S.-E. Yoo, *Synlett* **1991**, 204; c) A. R. Gordon, C. Johnstone, W. J. Kerr, *Synlett* **1996**, 1083.

[5] T. Sugihara, M. Yamada, H. Ban, M. Yamaguchi, C. Kaneko, *Angew. Chem.* **1997**, *109*, 2884; *Angew. Chem. Int. Ed. Engl.* **1997**, *36*, 2801; (for a related paper which also describes a method for the in situ complexation of alkynes with $CoBr_2$, Zn and CO, see: T. Rajesh, M. Periasamy, *Tetrahedron Lett.* **1998**, *39*, 117).

[6] a) B. M. Trost, *Science* **1991**, *254*, 1471; b) B. M. Trost, *Angew. Chem.* **1995**, 107, 285; *Angew. Chem. Int. Ed. Engl.* **1995**, *34*, 259.

[7] D. C. Billington, W. J. Kerr, P. L. Pauson, C. F. Farnocchi, *J. Organomet. Chem.* **1988**, *356*, 213.

[8] V. Rautenstrauch, P. Mégard, J. Conesa, W. Küster, *Angew. Chem.* **1990**, *102*, 1441; *Angew. Chem. Int. Ed. Engl.* **1990**, *29*, 1413.

[9] a) N. Jeong, S. H. Hwang, Y. Lee, Y. K. Chung, *J. Am. Chem. Soc.* **1994**, *116*, 3159; b) B. Y. Lee, Y. K. Chung, N. Jeong, Y. Lee, S. H. Hwang, *J. Am. Chem. Soc.* **1994**, *116*, 8793; c) N. Y. Lee, Y. K. Chung, *Tetrahedron Lett.* **1996**, *37*, 3145; d) N. Jeong, S. H. Hwang, Y. W. Lee, Y. S. Lim, *J. Am. Chem. Soc.* **1997**, *119*, 10549.

[10] a) B. L. Pagenkopf, T. Livinghouse, *J. Am. Chem. Soc.* **1996**, *118*, 2285; for newer work on catalytic Pauson-Khand reactions, see: b) M. Hayashi, Y. Hashimoto, Y. Yamamoto, J. Usuki, K. Saigo, *Angew. Chem.* **2000**, *112*, 645; *Angew. Chem. Int. Ed. Engl.* **2000**, *39*, 631; and refs. cited therein.

[11] a) A. J. Pearson, R. A. Dubbert, *J. Chem. Soc., Chem. Commun.* **1991**, 202; b) A. J. Pearson, R. A. Dubbert, *Organometallics* **1994**, *13*, 1656; c) T. R. Hoye, J. A. Suriano, *Organometallics* **1992**, *11*, 2044; d) T. R. Hoye, J. A. Suriano, *J. Am. Chem. Soc.* **1993**, *115*, 1154; e) C. Mukai, M. Uchiyama, M. Hanaoka, *J. Chem. Soc. Chem. Commun.* **1992**, 1014; f) N. Jeong, S. J. Lee, *Tetrahedron Lett.* **1993**, *34*, 4027; g) L. Pagès, A. Llebaria, F. Camps, E. Molins, C. Miravitlles, J. M. Moretó, *J. Am. Chem. Soc.* **1992**, *114*, 10449.

[12] T. Kondo, N. Suzuki, T. Okada, T.-a. Mitsudo, *J. Am. Chem. Soc.* **1997**, *119*, 6187.
[13] T. Morimoto, N. Chatani, Y. Fukumoto, S. Murai, *J. Org. Chem.* **1997**, *62*, 3762.
[14] a) E.-i. Negishi, S. J. Holmes, J. M. Tour, J. A. Miller, *J. Am. Chem. Soc.* **1985**, *107*, 2568; b) E.-i. Negishi, F. E. Cederbaum, T. Takahashi, *Tetrahedron Lett.* **1986**, *27*, 2829; c) E.-i. Negishi, S. J. Holmes, J. M. Tour, J. A. Miller, F. E. Cederbaum, D. R. Swanson, T. Takahashi, *J. Am. Chem. Soc.* **1989**, *11*, 3336.
[15] a) K. Tamao, K. Kobayashi, Y. Ito, *J. Am. Chem. Soc.* **1988**, *110*, 1286; b) K. Tamao, K. Kobayashi, Y. Ito, *Synlett* **1992**, 539.
[16] R. B. Grossmann, S. L. Buchwald, *J. Org. Chem.* **1992**, *57*, 5803.
[17] a) S. C. Berk, R. B. Grossmann, S. L. Buchwald, *J. Am. Chem. Soc.* **1993**, *115*, 4912; b) S. C. Berk, R. B. Grossmann, S. L. Buchwald, *J. Am. Chem. Soc.* **1994**, *116*, 8593; c) F. A. Hicks, S. C. Berk, S. L. Buchwald, *J. Org. Chem.* **1996**, *61*, 2713.
[18] M. Zhang, S. L. Buchwald, *J. Org. Chem.* **1996**, *61*, 4498; BDPEDA = *N,N'*-bis(diphenylmethylene) ethylenediamin.
[19] a) F. A. Hicks, N. M. Kablaoui, S. L. Buchwald, *J. Am. Chem. Soc.* **1996**, *118*, 9450; b) F. A. Hicks, N. M. Kablaoui, S. L. Buchwald, *J. Am. Chem. Soc.* **1999**, *121*, 5881.
[20] For the use of EBTHI complexes in enantioselective synthesis, see: A. H. Hoveyda, J. P. Morken, *Angew. Chem.* **1996**, *108*, 1378; *Angew. Chem. Int. Ed. Engl.* **1996**, *35*, 1262.
[21] F. A. Hicks, S. L. Buchwald, *J. Am. Chem. Soc.* **1996**, *118*, 11688.
[22] a) P. Bladon, P. L. Pauson, H. Brunner, R. Eder, *J. Organomet. Chem.* **1988**, *355*, 449; b) H. Brunner, A. Niederhuber, *Tetrahedron Asym.* **1990**, *1*, 711; c) A. M. Hay, W. J. Kerr, G. G. Kirk, D. Middlemiss, *Organometallics* **1995**, *14*, 4986.
[23] a) W. J. Kerr, G. G. Kirk, D. Middlemiss, *Synlett* **1995**, 1085; b) W. J. Kerr, D. M. Lindsay, J. S. Scott, S. Watson, *OMCOS 9 Conference*, Göttingen **1997** (Poster Nr.318).
[24] a) J. Castro, A. Moyano, M. A. Pericàs, A. Riera, A. E. Greene, *Tetrahedron Asym.* **1994**, *5*, 307; b) V. Bernardes, X. Verdaguer, N. Kardos, A. Riera, A. Moyano, M. A. Pericàs, A. E. Greene, *Tetrahedron Lett.* **1994**, *35*, 575; c) X. Verdaguer, A. Moyano, M. A. Pericàs, A. Riera, V. Bernardes, A. E. Greene, A. Alvarez-Larena, J. F. Piniella, *J. Am. Chem. Soc.* **1994**, *116*, 2153; d) S. Fonquerna, A. Moyano, M. A. Pericàs, A. Riera, *Tetrahedron* **1995**, *51*, 4239; e) S. Fonquerna, A. Moyano, M. A. Pericàs, A. Riera, *J. Am. Chem. Soc.* **1997**, *119*, 10225; f) E. Montenegro, M. Poch, A. Moyano, M. A. Pericàs, A. Riera, *Tetrahedron Lett.* **1998**, *39*, 335; g) S. Fonquerna, R. Rios, A. Moyano, M. A. Pericàs, A. Riera, *Eur. J. Org. Chem.* **1999**, 3459.
[25] J. Castro, H. Sörensen, A. Riera, C. Morin, A. Moyano, M. A. Pericàs, A. E. Greene, *J. Am. Chem. Soc.* **1990**, *112*, 9388.
[26] V. Bernardes, N. Kann, A. Riera, A. Moyano, M. A. Pericàs, A. E. Greene, *J. Org. Chem.* **1995**, *60*, 6670.
[27] J. Castro, A. Moyano, M. A. Pericàs, A. Riera, A. E. Greene, A. Alvarez-Larena, J. F. Piniella, *J. Org. Chem.* **1996**, *61*, 9016.
[28] J. Tormo, A. Moyano, M. A. Pericàs, A. Riera, *J. Org. Chem.* **1997**, *62*, 4851.
[29] a) R. Aumann, H.-J. Weidenhaupt, *Chem. Ber.* **1987**, *120*, 23; b) J. L. Kent, H. Wan, K. M. Brummond, *Tetrahedron Lett.* **1995**, *36*, 2407.
[30] a) M. Ahmar, F. Antras, B. Cazes, *Tetrahedron Lett.* **1995**, *36*, 4417; b) M. Ahmar, O. Chabanis, J. Gauthier, B. Cazes, *Tetrahedron Lett.* **1997**, *38*, 5277; c) M. Ahmar, C. Locatelli, D. Colombier, B. Cazes, *Tetrahedron Lett.* **1997**, *38*, 5281.
[31] a) N. M. Kablaoui, F. A. Hicks, S. L. Buchwald, *J. Am. Chem. Soc.* **1996**, *118*, 5818; b) N. M. Kablaoui, F. A. Hicks, S. L. Buchwald, *J. Am. Chem. Soc.* **1997**, *119*, 4424.
[32] M. E. Krafft, A. M. Wilson, O. A. Dasse, B. Shao, Y. Y. Chung, Z. Fu, L. V. R. Bonaga, M. K. Mollmann, *J. Am. Chem. Soc.* **1996**, *118*, 6080.
[33] a) T. F. Jamison, S. Shambayati, W. E. Crowe, S. L. Schreiber, *J. Am. Chem. Soc.* **1994**, *116*, 5505; b) T. F. Jamison, S. Shambayati, W. E. Crowe, S. L. Schreiber, *J. Am. Chem. Soc.* **1997**, *119*, 4353.
[34] K. M. Nicholas, *Acc. Chem. Res.* **1987**, *20*, 207.
[35] A. Quattropani, G. Anderson, G. Bernardinelli, E. P. Kündig, *J. Am. Chem. Soc.* **1997**, *119*, 4773.
[36] M. Thommen, R. Keese, *Synlett* **1997**, 231.
[37] S. G. Van Ornum, J. M. Cook, *Tetrahedron Lett.* **1997**, *38*, 3657.
[38] a) G. L. Bolton, *Tetrahedron Lett.* **1996**, *37*, 3433; b) J. L. Spitzer, M J. Kurth, N. E. Shore, S. D. Najdi, *Tetrahedron* **1997**, *53*, 6791; c) G. L. Bolton, J. C. Hodges, J. R. Rubin, *Tetrahedron* **1997**, *53*, 6611.

Multicomponent Catalysis for Reductive Bond Formations

Alois Fürstner

Max-Planck-Institut für Kohlenforschung, Mülheim/Ruhr, Germany

Catalysis in general and asymmetric catalysis in particular are at the forefront of chemical research [1]. Their impact on industrial production can hardly be overestimated and is likely to increase further [2]. However, the high degree of sophistication reached in many respects may hide the simple notion that there still remain fairly large domains in preparative organic chemistry in which no catalytic alternatives to well-established stoichiometric transformations yet exist. The following account is intended to put into perspective some pioneering studies which address this problem and try to develop new concepts for metal-catalyzed reductive bond formations [3].

Catalytic Processes Mediated by Chlorosilanes

Titanium. The high reducing ability and the pronounced oxophilicity of early transition metals in low oxidation states act jointly as a formidable driving force in many transformations. However, such processes are usually hampered by the fact that the metal oxides or alkoxides formed as the inorganic by-products usually resist attempted re-reductions to the active species and thus render catalysis a difficult task.

A prototype example is the famous McMurry coupling of carbonyl compounds to alkenes (Scheme 1) [4]. The very high stability of the accumulating titanium oxides constitutes the thermodynamic sink which drives the conversion but demands the use of stoichiometric or excess amounts of the low-valent titanium reagent [Ti]. Only recently has it been possible to elaborate a procedure that for the first time enables us to perform intramolecular carbonyl coupling reactions catalyzed by titanium species [5].

This development was triggered by some earlier insights into the essentials of this transformation: it has been shown that a large number of low-valent titanium species, [Ti], differing in their formal oxidation states, ligand spheres and solubilities promote carbonyl coupling processes with comparable ease. Because this fact refutes previous assumptions that metallic titanium was essential, the strong and aggressive reducing agents required for its preparation can be avoided. As a consequence it was possible to develop a particularly convenient – but still stoichiometric – "instant method" for performing carbonyl coupling reactions based on the formation of [Ti] from $TiCl_3$ and Zn *in the presence of the substrate* [6].

Since the latter feature meets a fundamental requirement for catalysis, this set-up paved the

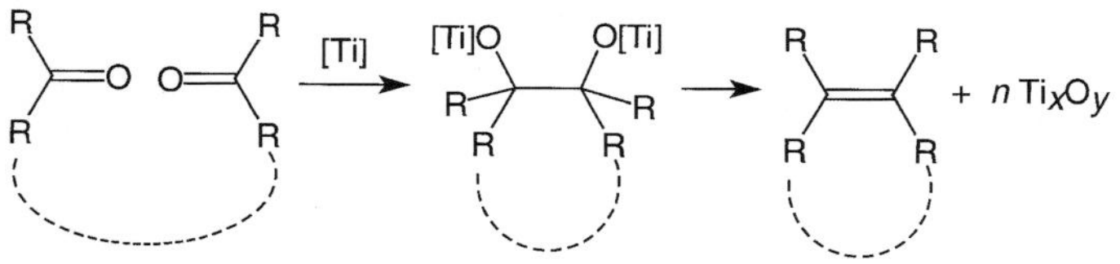

Scheme 1. The McMurry coupling of carbonyl compounds to alkenes.

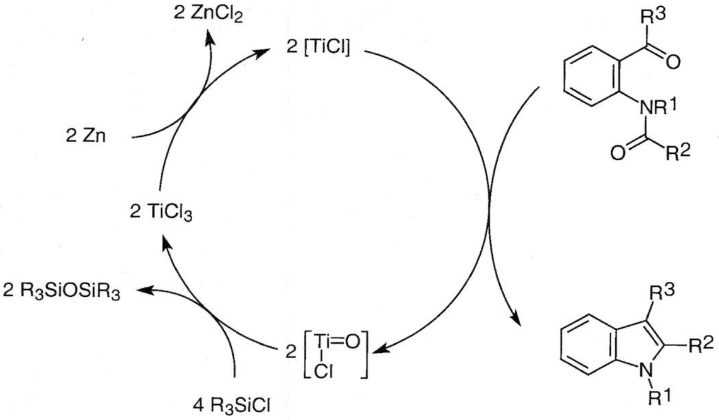

Scheme 2. Catalytic cycle for McMurry-type couplings.

way for a truly catalytic process. However, Zn dust cannot re-reduce the titanium oxides or oxychlorides initially formed into any active low-valent [Ti] species. Therefore an indirect way to complete a catalytic cycle was devised which relies on a ligand exchange between the titanium oxides or oxychlorides and a chlorosilane (Scheme 2). In fact, a multicomponent redox system consisting of $TiCl_3$ cat., Zn and a chlorosilane accounts for the first titanium-catalyzed carbonyl coupling reactions. The efficiency of this method can be tuned to some extent by choosing the R_3SiCl additive. The examples compiled in Table 1 show that this new catalytic procedure compares favorably with the existing stoichiometric precedent in terms of yields and reaction rates. These studies have been carried out using oxo-amide derivatives as substrates, which ex-

Table 1. Titanium-induced indole syntheses: comparison of the catalytic with the stoichiometric "instant" procedure.

Product	Catalytic procedure $TiCl_3$ cat., Zn, TMSCl, 0.5 h	"Instant method" $TiCl_3$ (2–4 equiv), Zn, 1–4 h
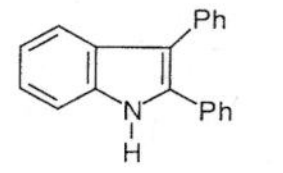 2,3-diphenylindole (N–H)	80 %	98 %
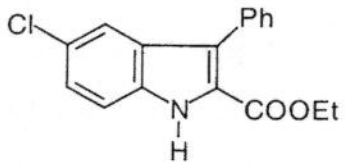 5-Cl, 3-Ph, 2-COOEt indole (N–H)	79 %	87 %
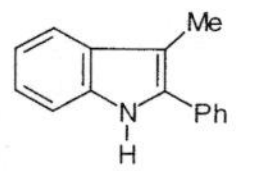 3-Me, 2-Ph indole (N–H)	82 % [a]	76 %
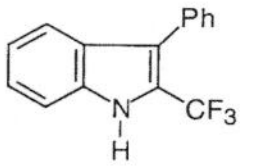 3-Ph, 2-CF_3 indole (N–H)	88 %	82 %
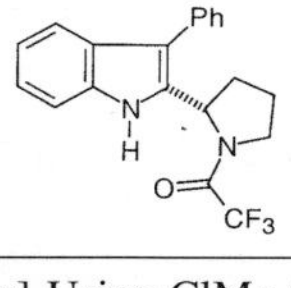 3-Ph, 2-(N-COCF$_3$-pyrrolidin-2-yl) indole (N–H)	67 %	90 %

[a] Using $ClMe_2Si(CH_2)_3CN$ as additive instead of TMSCl.

hibit a pronounced tendency to cyclize to indole derivatives on treatment with [Ti] [5–7].

It was clear at the outset that the basic principle of this catalytic scenario may apply to other transformations as well. An obvious extension concerns the pinacol coupling since any McMurry reaction probably passes through the 1,2-diolate stage (*cf.* Scheme 1) [4]. In fact, several titanium-catalyzed procedures have been reported which rely on chlorosilane additives for the liberation of the product and the simultaneous regeneration of the $TiCl_x$ salt. They involve either [Cp_2TiCl_2] cat., Zn, chlorosilane [8], or [Cp_2TiCl_2] cat., Mn, chlorosilane [9], or [$TiCl_3(THF)_3$] cat., Zn, TMSCl, *t*-BuOH [10], or [$Cp_2Ti(Ph)Cl$] cat., Zn, chlorosilane [11]. The use of Cp_2TiCl_2 in this context deserves particular emphasis, because titanium sources of this type open up new vistas for stereocontrol if *ansa*-titanocene derivatives are used to transfer chiral information from the ligand to the diol [12].

Chromium. Similar chlorosilane-mediated catalytic processes can be envisaged with many other early transition metals. The development of the first Nozaki-Hiyama-Kishi reactions catalyzed by chromium species [13] illustrates how to avoid the use of an excess of a physiologically suspect and rather expensive salt without compromising the efficiency, practicality and scope of the reaction. The tentative catalytic cycle is depicted in Scheme 3.

In this case, the silylation of the metal alkoxide initially formed represents the key step of the overall process which releases the chromium salt from the organic product. The other crucial parameter is the use of the stoichiometric reducing agent for the regeneration of the active Cr^{II} species. Commercial Mn turned out to be particularly well suited, as it is very cheap, its salts are essentially non-toxic and rather weak Lewis acids, and the electrochemical data suggest that it will form an efficient redox couple with Cr^{III}. Moreover, the very low propensity of commercial Mn to insert on its own into organic halides guarantees that the system does not deviate from the desired chemo- and diastereoselective chromium path. Thus, a mixture of CrX_n ($n = 2, 3$) cat., TMSCl and Mn accounts for the first Nozaki reactions catalytic in chromium [13].

This method applies to aryl, alkenyl, allyl and alkynyl halides as well as to alkenyl triflates and exhibits the same selectivity profile as its stoichiometric precedent (Scheme 4). Moreover, it does not matter if the catalytic cycle is started at the Cr^{II} or Cr^{III} stage as implied by Scheme 3. Therefore it is possible to substitute cheap and stable $CrCl_3$ for the expensive and air-sensitive $CrCl_2$ previously used for Nozaki reactions. In some cases other chromium templates such as [Cp_2Cr] or [$CpCrCl_2$] can be employed, improving the total turnover number of this transformation even further [13, 14].

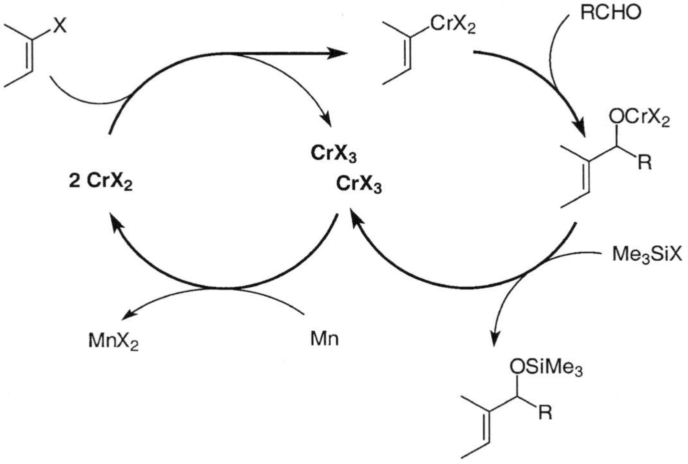

Scheme 3. Proposed catalytic cycle for the first Nozaki-Hiyama-Kishi reactions catalyzed by chromium species.

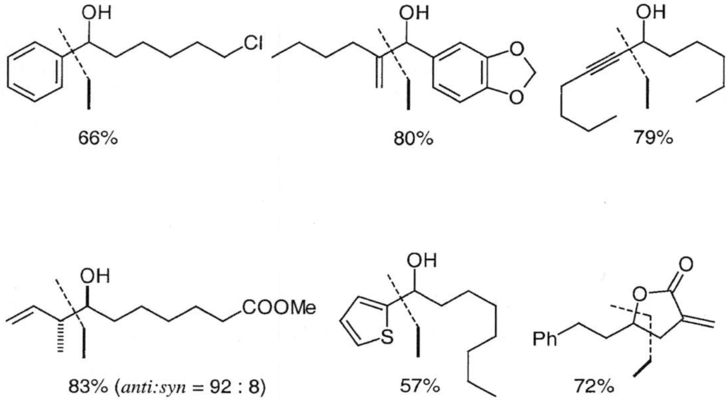

66%

80%

79%

83% (*anti:syn* = 92 : 8)

57%

72%

Scheme 4. The scope of Nozaki reactions catalyzed by chromium resembles that of the stoichiometric version.

Electrochemically driven Nozaki-Hiyama-Kishi reactions constitute an attractive modification of this basic concept (Scheme 5) [15]. Although the current density turned out to be a critical parameter and must be carefully controlled, the authors show that in this case the $LiClO_4$ used as supporting electrolyte also acts as the oxophilic mediator instead of TMSCl. They also used a palladium cocatalyst in order to form more highly nucleophilic "chromium ate" complexes as the actual intermediates. Encouraged by this precedent, further studies using electrons as the ultimate reducing agent are likely to appear in the near future.

Other Transition Metals. The rather general validity of the chlorosilane-based catalysis concept is further substantiated by some recent examples of pinacol coupling processes catalyzed either by low-valent vanadium ($[CpV(CO)_4]$ cat., Zn, chlorosilane) [16] or low-valent samarium (SmI_2 cat., Mg, chlorosilane) [17]. Likewise, a report from Coreys group on samarium iodide catalyzed additions of carbonyl compounds to acrylates deserves mention; these follow essentially the same rationale (SmI_2 cat., Zn(Hg), TMSOTf, LiI) [18]. In view of the extensive use of SmI_2 in stoichiometric transformations, the possible impact of catalytic alternatives is easy to imagine.

Catalytic Processes Based on Other Mediators

Although chlorosilanes are an obvious choice as mediators for catalysis on account of their high affinity for oxygen, low price and lack of toxicity, several other additives can also be envisaged. The recent publication on the electrochemical version of the Nozaki reaction mentioned above simply employs the Li cations of the supporting electrolyte for this very purpose [15], whereas another titanium-catalyzed pinacol

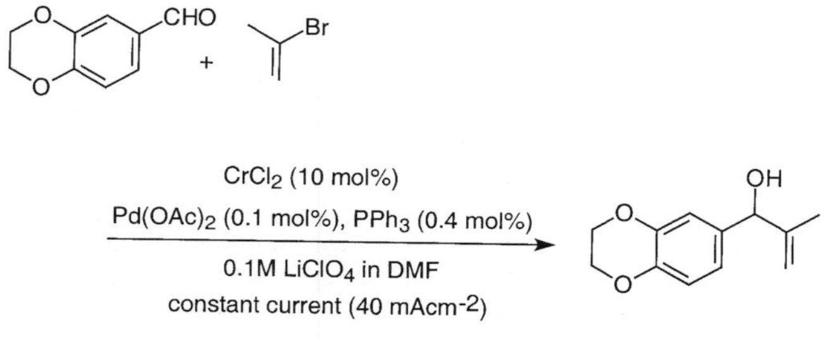

$CrCl_2$ (10 mol%)
$Pd(OAc)_2$ (0.1 mol%), PPh_3 (0.4 mol%)
0.1M $LiClO_4$ in DMF
constant current (40 $mAcm^{-2}$)

Scheme 5. An electrochemically driven Nozaki-Hiyama-Kishi reaction.

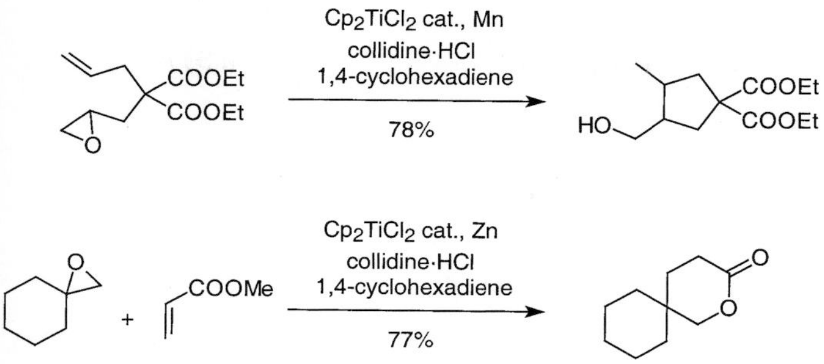

Scheme 6. Inter- and intramolecular C-C coupling reactions by the method of Gansäuer et al.

coupling reaction [$TiCl_4$ cat., Li(Hg), $AlCl_3$] is based on the oxophilicity of Al^{III} [19].

An even more interesting development concerns the use of protons. Thus, Gansäuer et al. were able to achieve epoxide ring-openings and pinacol coupling reactions with catalytic amounts of Cp_2TiCl_2, simply by using pyridinium hydrochlorides as scavengers for the product and Zn or, preferably, Mn as the stoichiometric reducing agent [20, 21]. The pK_a of the pyridinium salt is properly adjusted, and the protic medium does not interfere with the radical intermediates prior to product formation. This method applies to inter- and intramolecular C-C coupling reactions (Scheme 6) as well as to simple reductions and turned out to be compatible with various sensitive functional groups in the oxirane substrates.

Another approach to multicomponent redox catalysis employs silanes (R_3SiH) as the additives. This allows the stoichiometric reducing agent and the oxophilic reaction partner to be merged into a single component. Two independent reports from Buchwald [22] and Crowe [23] on the cyclization of unsaturated carbonyl compounds based on the turnover of a "Cp_2Ti" template rely on this principle (Scheme 7) [24].

A similar idea allows the well-known Barton-McCombie deoxygenation of alcohols to proceed for the first time with catalytic rather than stoichiometric or excess amounts of tributylstannane (Scheme 8) [25]. As shown in the proposed catalytic cycle, this exceptionally versatile but highly toxic reagent is regenerated from the otherwise accumulating "dead-end" product $Bu_3Sn(OPh)$ by means of polymethylhydrosiloxane (PMHS). Once again it is the affinity for oxygen in combination with the reducing ability of this inexpensive, non-toxic and easily handled silicon hydride which qualifies PMHS as an an-

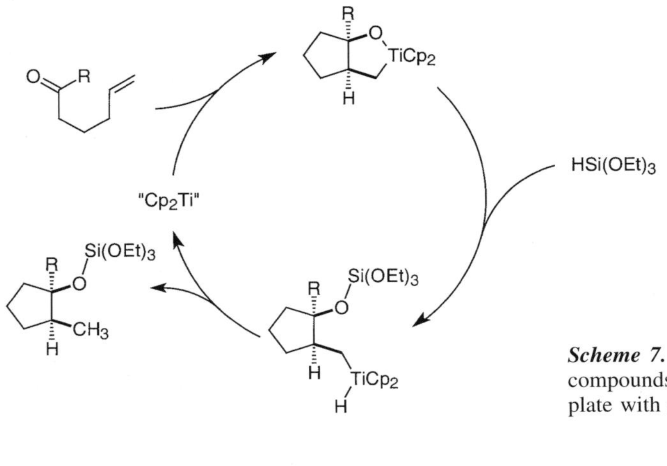

Scheme 7. The cyclization of unsaturated carbonyl compounds based on the turnover of a Cp_2Ti template with a silane as the additive.

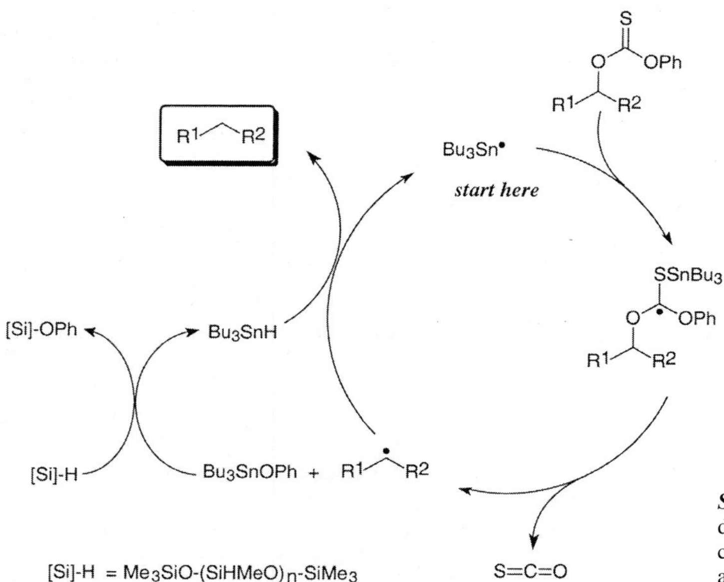

Scheme 8. The Barton-McCombie deoxygenation of alcohols with catalytic rather than stoichiometric amounts of tributylstannane.

cillary component for catalysis. The authors show that the addition of *n*-BuOH facilitates the regeneration of the tin hydride, improves the key step of the catalytic process and makes the reaction as efficient as the stoichiometric version (Table 2).

These and related examples rival – and may well replace – their stoichiometric counterparts. Although none of them is "atom economical" [26] in the pure sense, they do at least permit economy in the key component. If the latter is expensive, difficult to handle and/or of physiological concern, such multicomponent catalyst systems up-grade established transformations to a significant extent, quite apart from the heuristic lessons in and the stimulus for catalysis research which they provide [27].

Table 2. Barton-McCombie deoxygenation of phenyl thionocarbonate esters: comparison of the catalytic with the stoichiometric reaction [25].

Substrate	Product	Catalytic	Stoichiometric
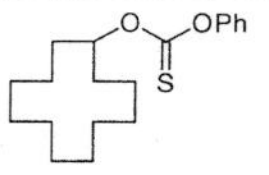	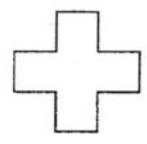	66 %	68 %
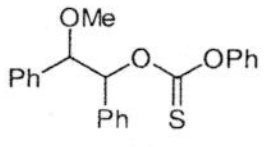	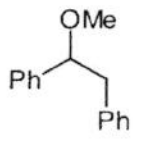	70 %	65 %
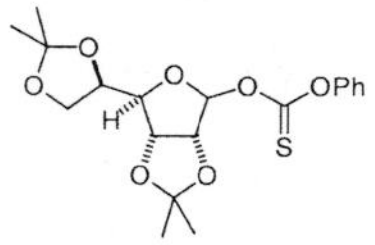	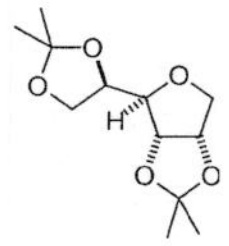	63 %	72 %
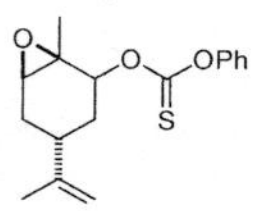	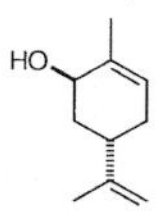	68 %	61 %

References

[1] a) R. Noyori, *Asymmetric Catalysis in Organic Synthesis*, Wiley, New York, **1994**; b) I. Ojima (Ed.), *Catalytic Asymmetric Synthesis*, VCH, New York, **1993** and ref. cited.

[2] For a leading reference see: B. Cornils, W. A. Herrmann, *Applied Homogeneous Catalysis with Organometallic Compounds*, VCH, Weinheim, **1996**.

[3] a) This article is an up-dated version of the following account: A. Fürstner, *Chem. Eur. J.* **1998**, *4*, 567–570. b) See also: A. Fürstner, *Pure Appl. Chem.* **1998**, *70*, 1071–1076.

[4] a) For a recent review see: A. Fürstner, B. Bogdanovic, *Angew. Chem.* **1996**, *108*, 2583–2609; *Angew. Chem. Int. Ed. Engl.* **1996**, *35*, 2442–2469; b) J. E. McMurry, *Chem. Rev.* **1989**, *89*, 1513–1524. c) A. Fürstner in *Transition Metals for Organic Synthesis* (M. Beller, C. Bolm, Eds.), Wiley-VCH, Weinheim, **1998**, Vol. 1, 381–401.

[5] A. Fürstner, A. Hupperts, *J. Am. Chem. Soc.* **1995**, *117*, 4468–4475.

[6] A. Fürstner, A. Hupperts, A. Ptock, E. Janssen, *J. Org. Chem.* **1994**, *59*, 5215–5229.

[7] For applications see: a) A. Fürstner, A. Ptock, H. Weintritt, R. Goddard, C. Krüger, *Angew. Chem.* **1995**, *107*, 725–728; *Angew. Chem. Int. Ed. Engl.* **1995**, *34*, 678–681; b) A. Fürstner, A. Ernst, *Tetrahedron* **1995**, *51*, 773–786; c) A. Fürstner, A. Ernst, H. Krause, A. Ptock, *Tetrahedron* **1996**, *52*, 7329–7344; d) A. Fürstner, D. N. Jumbam, G. Seidel, *Chem. Ber.* **1994**, *127*, 1125–1130; e) A. Fürstner, A. Hupperts, G. Seidel, *Org. Synth.* **1998**, *76*, 142–150 and ref. cited.

[8] a) A. Gansäuer, *J. Chem. Soc. Chem. Commun.* **1997**, 457–458; b) A. Gansäuer, M. Moschioni, D. Bauer, *Eur. J. Org. Chem.* **1998**, 1923–1927; c) T. Hirao, B. Hatano, M. Asahara, Y. Muguruma, A. Ogawa, *Tetrahedron Lett.* **1998**, *39*, 5247–5248.

[9] M. S. Dunlap, K. M. Nicholas, *Synth. Commun.* **1999**, *29*, 1097–1106.

[10] T. A. Lipski, M. A. Hilfiker, S. G. Nelson, *J. Org. Chem.* **1997**, *62*, 4566–4567.

[11] Y. Yamamoto, R. Hattori, K. Itoh, *Chem. Commun.* **1999**, 825–826.

[12] a) A. Gansäuer, *Synlett* **1997**, 363–364; b) For a recent complementary approach towards diastereoselective pinacol coupling using Schiff bases as ligands to $TiCl_4(THF)_2$ cat. in combination with Mn and TMSCl see: M. Bandini, P. G. Cozzi, S. Morganti, A. Umani-Ronchi, *Tetrahedron Lett.* **1999**, *40*, 1997–2000.

[13] a) A. Fürstner, N. Shi, *J. Am. Chem. Soc.* **1996**, *118*, 2533–2534; b) A. Fürstner, N. Shi, *J. Am. Chem. Soc.* **1996**, *118*, 12349–12357; c) For a recent comprehensive review on stoichiometric and catalytic Nozaki-Hiyama-Kishi reactions see: A. Fürstner, *Chem. Rev.* **1999**, *99*, 991–1045.

[14] For recent applications of this multicomponent system to other Nozaki-Hiyama-Kishi reactions catalytic in chromium see: a) With acrolein acetals: R. K. Boeckman, R. A. Hudack, *J. Org. Chem.* **1998**, *63*, 3524–3525; b) With trichloroethane: J. R. Falck, D. K. Barma, C. Mioskowski, T. Schlama, *Tetrahedron Lett.* **1999**, *40*, 2091–2094; c) $CrCl_2$ cat., TMSCl, $NiCl_2$ cat., Al-powder: M. Kuroboshi, M. Tanaka, S. Kishimoto, K. Goto, H. Tanaka, S. Torii, *Tetrahedron Lett.* **1999**, *40*, 2785–2788.

[15] R. Grigg, B. Putnikovic, C. J. Urch, *Tetrahedron Lett.* **1997**, *38*, 6307–6308.

[16] a) T. Hirao, T. Hasegawa, Y. Mugumura, I. Ikeda, *J. Org. Chem.* **1996**, *61*, 366–367; b) T. Hirao, M. Asahara, Y. Muguruma, A. Ogawa, *J. Org. Chem.* **1998**, *63*, 2812–2813; c) For an application of the same multicomponent system to the reductive dimerization of aldimines see: B. Hatano, A. Ogawa, T. Hirao, *J. Org. Chem.* **1998**, *63*, 9421–9424.

[17] R. Nomura, T. Matsuno, T. Endo, *J. Am. Chem. Soc.* **1996**, *118*, 11666–11667.

[18] E. J. Corey, G. Z. Zheng, *Tetrahedron Lett.* **1997**, *38*, 2045–2048.

[19] O. Maury, C. Villiers, M. Ephritikhine, *New. J. Chem.* **1997**, *21*, 137–139.

[20] Epoxide openings: a) A. Gansäuer, M. Pierobon, H. Bluhm, *Angew. Chem.* **1998**, *110*, 107–109; *Angew. Chem. Int. Ed. Engl.* **1998**, *37*, 101–103; b) A. Gansäuer, H. Bluhm, *Chem. Commun.* **1998**, 2143–2144; c) A. Gansäuer, H. Bluhm, M. Pierobon, *J. Am. Chem. Soc.* **1998**, *120*, 12849–12859.

[21] Pinacol couplings by Cp_2TiCl_2 cat. and Mn under buffered protic conditions: a) A. Gansäuer, D. Bauer, *J. Org. Chem.* **1998**, *63*, 2070–2071; b) A. Gansäuer, D. Bauer, *Eur. J. Org. Chem.* **1998**, 2673–2676.

[22] a) N. M. Kablaoui, S. L. Buchwald, *J. Am. Chem. Soc.* **1995**, *117*, 6785–6786; b) N. M. Kablaoui, S. L. Buchwald, *J. Am. Chem. Soc.* **1996**, *118*, 3182–3191; c) See also: N. M. Kablaoui, F. A. Hicks, S. L. Buchwald, *J. Am. Chem. Soc.* **1996**, *118*, 5818–5819.

[23] W. E. Crowe, M. J. Rachita, *J. Am. Chem. Soc.* **1995**, *117*, 6787–6788.

[24] Closely related are reports from the same authors in which isocyanates or CO insert into titanaoxacycles in order to release the catalytically active "Cp_2Ti" species, cf: a) F. A. Hicks, S. C. Berk, S. L. Buchwald, *J. Org. Chem.* **1996**, *61*, 2713–2718; b) W. E. Crowe, A. T. Vu, *J. Am. Chem. Soc.* **1996**, *118*, 1557–1558 and ref. cited.

[25] R. M. Lopez, D. S. Hays, G. C. Fu, *J. Am. Chem. Soc.* **1997**, *119*, 6949–6950.

[26] B. M. Trost, *Angew. Chem.* **1995**, *107*, 285–307; *Angew. Chem. Int. Ed. Engl.* **1995**, *34*, 259.

[27] For recent accounts on similar topics see also: a) T. Hirao, *Synlett* **1999**, 175–181. b) A. Gansäuer, *Synlett* **1998**, 801–809.

Natural Product Synthesis by Rh-mediated Intramolecular C–H Insertion

Douglass F. Taber

Department of Chemistry and Biochemistry, University of Delaware, USA

Salah-E. Stiriba

Department of Chemistry, Texas A & M University, USA

Since the observation that Rh(II) carboxylates are superior catalysts for the generation of transient electrophilic metal carbenoids from α-diazocarbonyls compounds, intramolecular carbenoid insertion reactions have assumed strategic importance for C–C bond construction in organic synthesis [1]. Rhodium(II) compounds catalyze the remote functionalization of carbon-hydrogen bonds to form carbon-carbon bonds with good yield and selectivity. These reactions have been particularly useful in the intramolecular sense to produce preferentially five-membered rings. We summarize here the applications to natural product synthesis of this method for ring construction.

Although most of the work with Rh-mediated intramolecular C–H insertion has focussed on five-membered ring construction, the first application to natural product synthesis, by Cane, involved establishment of a six-membered ring. Thus, on exposure to $Rh_2(OAc)_4$, diazoketone **2** was cyclized to the tricyclic lactone **3** [2]. This product had previously been transformed by Paquette into **pentalenolactone E** (**4**) [3].

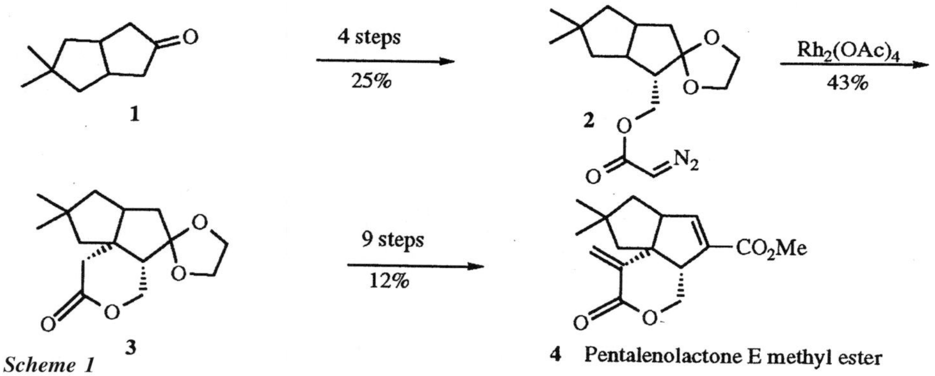

1 → 4 steps, 25% → **2** → $Rh_2(OAc)_4$, 43% → **3** → 9 steps, 12% → **4** Pentalenolactone E methyl ester

Scheme 1

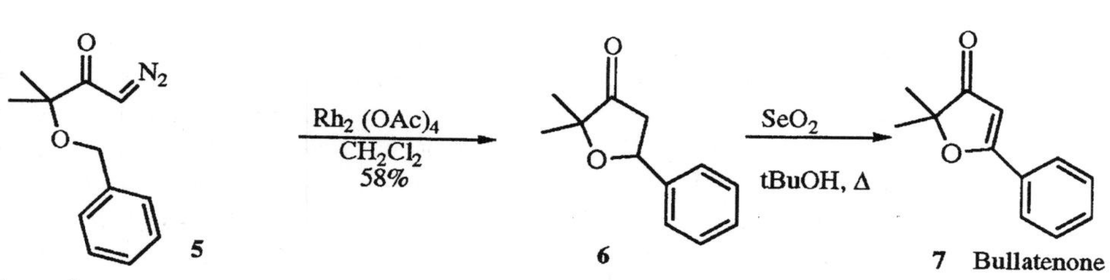

5 → $Rh_2(OAc)_4$, CH_2Cl_2, 58% → **6** → SeO_2, tBuOH, Δ → **7** Bullatenone

Scheme 2

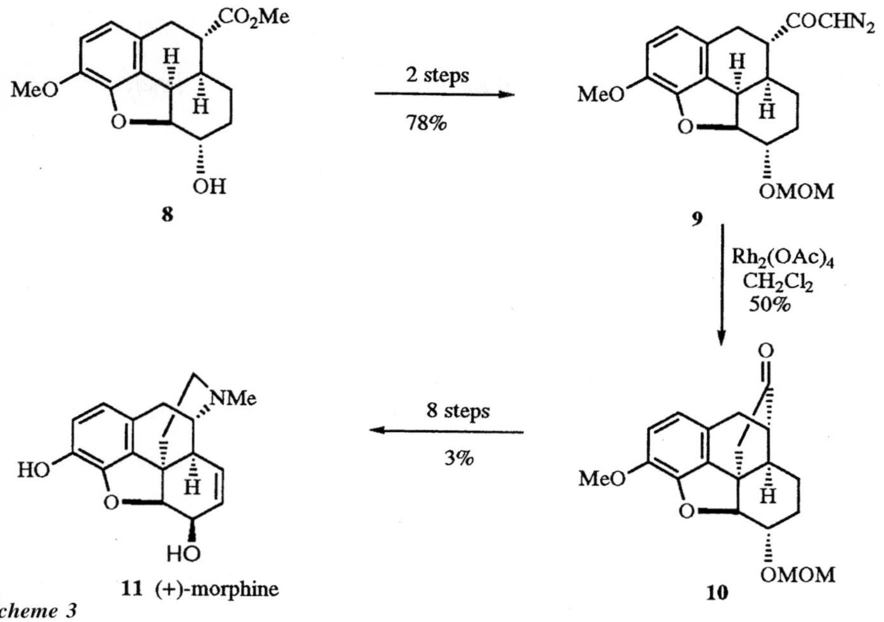

Scheme 3

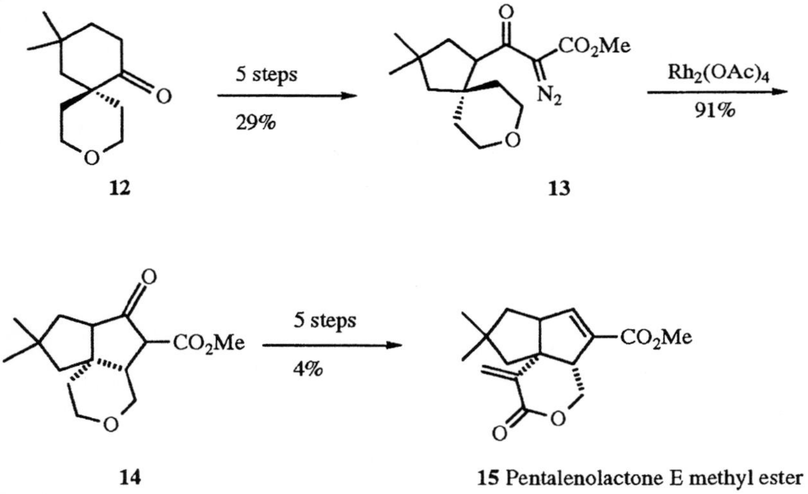

Scheme 4

The utility of this approach to five-membered *heterocycles* is illustrated by the synthesis of bullatenone (**7**) by Adams. Rhodium acetate-mediated insertion is especially preferred adjacent to ether oxygen, as illustrated by the cyclization of **5** to **6** [4]. Oxidation of furanone (**6**) with SeO_2 according to the procedure of Smith provided **bullatenone** (**7**) [5].

Rh-mediated C–H insertion is also useful for *carbocyclic* construction, as illustrated by the new asymmetric route to (+)-**morphine** (**11**) recently reported by White [6]. Cyclopentane formation is used to fashion a pentacyclic skeleton (**10**) from which the piperidine ring of **11** is evolved at a later stage.

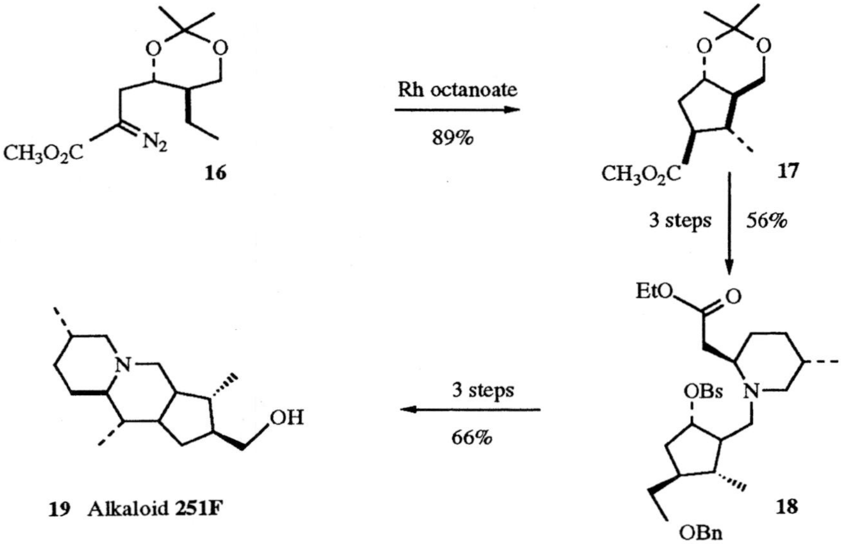

Scheme 5

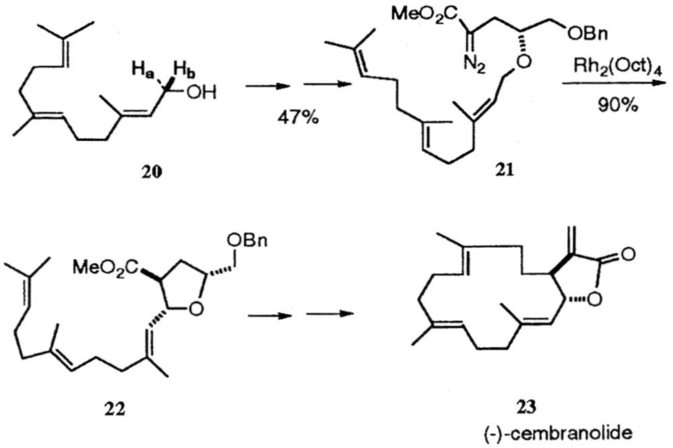

Scheme 6

Intramolecular C–H insertion is, essentially, a method for the specific remote functionalization of hydrocarbons. An important implication of this for synthetic strategy is that the C–H insertion process can dissolve symmetry, thus leading from a simple precursor to a much more complex product. An alternative route to **pentalenolactone E** (**15**) takes advantage of this idea [7]. In the key step, β-keto ester **13**, which has a single stereogenic center, is transformed into the tricycle **14**, which has four stereogenic centers.

A simple route to the Dendrobatid alkaloid **251F** (**19**) nicely illustrates the synthetic utility of Rh-mediated C–H insertion [8]. The excellent diastereoselectivity observed in the cyclization of **16** to **17** was in fact predicted computationally [9].

A single stereogenic center on the bridge between the target C–H bond and the diazocarbonyl can be sufficient to induce high diastereoselectivity. This is illustrated [10] by the cyclization of **21**, prepared from farnesol (**20**), which has two enantiotropic H atoms on the alcohol methy-

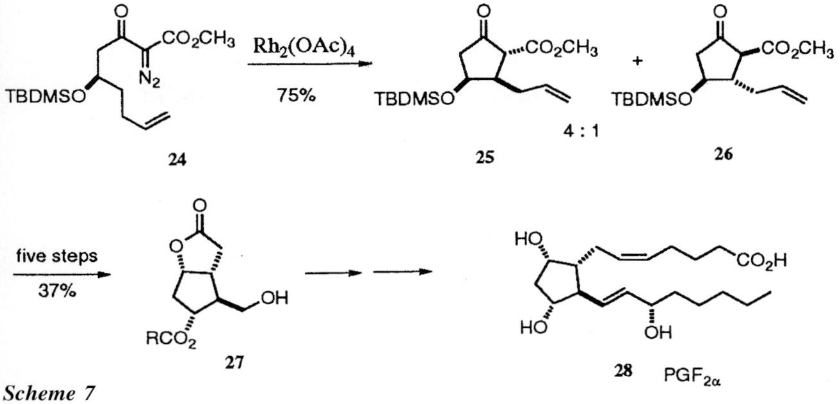

Scheme 7

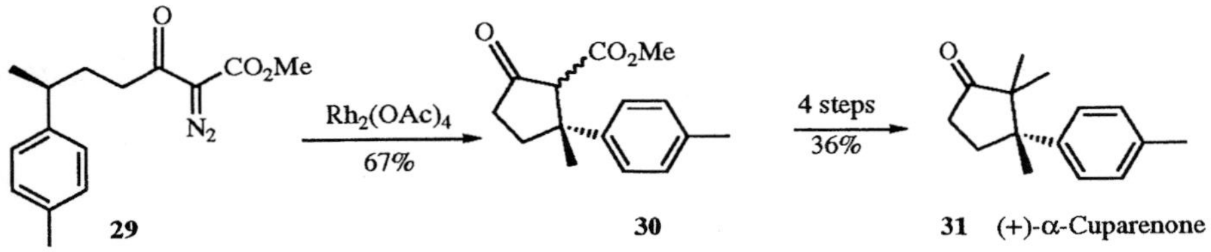

Scheme 8

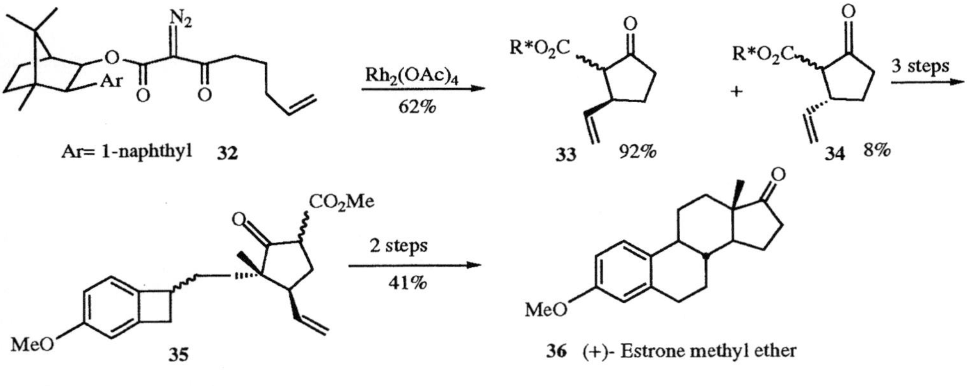

Scheme 9

lene. Rh-mediated insertion occurred with high selectivity for H_a to give **22**. Ester **22** was carried on over several steps to the marine natural product (−)-**cembranolide** (**23**).

A single stereogenic center can also induce substantial diastereoselectivity in the course of Rh-mediated carbocyclization. Diazo ester **24**, for instance [11], cyclized with a 4 : 1 preference for **25**. Ester **25** was carried on to the Corey lactone (**27**), the starting point for the total synthesis of the prostaglandins, exemplified by **PGF**$_{2\alpha}$ (**28**).

A key feature of intramolecular C–H insertion is the inherent ability to transform an acyclic tertiary stereogenic center into a cyclic quaternary stereogenic center, with retention of absolute configuration [12]. This was first demonstrated by rhodium-mediated cyclization of **29** to **30**, leading to (+)-α-cuparenone (**31**) [13].

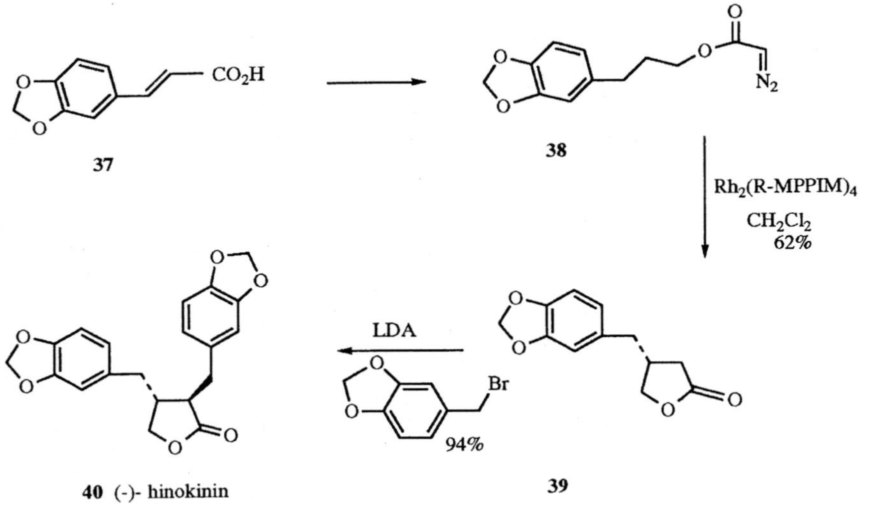

37 → 38

38 → 39: Rh_2(R-MPPIM)$_4$, CH_2Cl_2, 62%

39 → 40 (-)- hinokinin: LDA, ArCH$_2$Br, 94%

Scheme 10

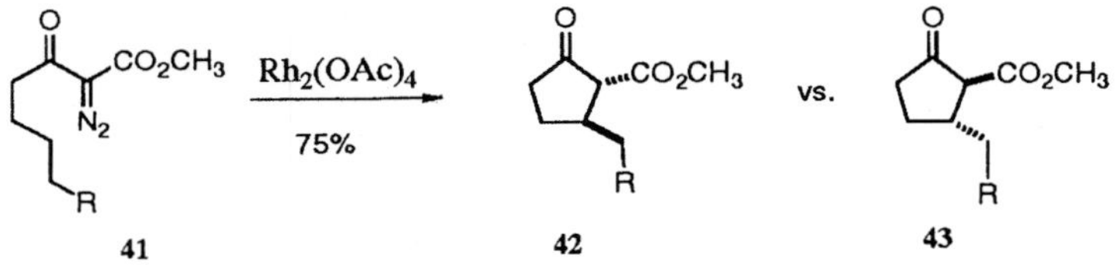

41 → 42 vs. 43: $Rh_2(OAc)_4$, 75%

Scheme 11

The synthesis of (+)-estrone methyl ether (**36**) illustrates the enantioselective construction of a polycyclic target by the use of chiral auxiliary control to establish the first cyclic stereogenic center [14]. In this case, the specific design of the naphthyldiazoester **32** directed Rh-mediated intramolecular C–H insertion selectively toward one of the two diastereotopic C–H bonds on the target methylene. The new ternary center so created then biased the formation of the adjacent quaternary center in the course of the alkylation. The chiral skew in the product cyclopentanone (**35**) controlled the relative and absolute course of the intramolecular cycloaddition, to give the steroid **(+)-estrone methyl ether** (**36**).

The high point in the development to date of Rh-mediated C–H insertion has been the design by Doyle of enantiomerically pure Rh(II) complexes that direct the absolute sense of the cyclization of α-diazo acetates. The applicability of such cyclizations to natural product synthesis has been demonstrated by Doyle with the construction of chiral lignane lactones such as **(–)–hinokinin** (**40**) [15].

The α-diazocarbonyl derivatives used in these studies are easily prepared, and the rhodium-mediated cyclizations proceed rapidly, with high catalyst turnover (ca. 100–1000). The catalysts are stable at room temperature for years, and are not air sensitive. The reactions work best in inert solvents such as dichloromethane or benzene, and the solvent must be dry. Slow addition of the diazocarbonyl compound to the catalyst is not usually necessary.

There are many aspects of these Rh-mediated cyclizations that are yet to be explored. What factors, for instance, govern the ratio of **25** to **26** (Scheme 7)? Would an Rh catalyst that was more readily polarizable and so more sensitive to electronic effects give a higher proportion of **25**? The enantioselective lactone cyclizations of Doyle [15] are particularly intriguing. Attempts toward enantioselective *carbocyclization* using a chiral rhodium catalyst have to date [16] not yielded preparatively useful enantiomeric excess. If, for instance, a generally useful catalyst for the selective transformation of **41** specifically to **42** or to **43** could be developed, it would have widespread utility.

As the factors governing regio-, chemo-, diastereo- and enantioselectivity come to be better understood, the Rh-mediated cyclization of an α-diazocarbonyl derivative will come to be a powerful tool for natural product synthesis.

References

[1] Reviews: a) For an excellent recent overview of stereoselection in metal-mediated intramolecular C–H insertion, see G.A. Sulikowski, K.L.Cha, M.M. Sulikowski, *Tetrahedron: Asymm.* **1998**, 9, 3145. For other reviews, see b) A. Padwa, D. J. Austin, *Angew. Chem. Int. Ed. Engl.* **1994**, 106, 1881–1899; c) M. A. McKervey, T. Ye, *Chem. Rev.* **1994**, 94, 1091–1160; M. A. McKervey, M. P. Doyle, *J. Chem. Soc. Chem. Commun.* **1997**, 983–1072. d) M. P. Doyle, *In Homogeneous Transition Metal Catalysts in Organic Synthesis*; (Eds.: Moser, W. R., Slocum, D. W.), ACS Advanced Chemistry Series 230; American Chemical Society, Washington, DC, **1992**; Chapter 30. e) D. F. Taber in *Comprehensive Organic Synthesis, Vol. 3*, (Ed.: B.M. Trost), Pergamon Press, Oxford, **1991**, p. 1045. f) M. P. Doyle, A. B. Dyatkin, G. H. P. Roos, F. Ganas, D. A. Pierson, A. van Basten, P. Muller, P. Polleux, *J. Am. Chem. Soc.* **1994**, 116, 4507. g) P. Wang, J. Adams, *J. Am. Chem. Soc.* **1994**, 116, 3296. h) D. F. Taber in *Methods of Organic Chemistry, (Houben-Weyl) Vol. E21*; (Eds.: G. Helmchen, R. W. Hoffmann, J. Mulzer, E. Schaumann), George Thieme Verlag, Stuttgart, **1995**, p. 1127.i) For a recent review of intramolecular C–H insertion by *alkenylidenes*, see W. Kirmse, *Angew. Chem. Int. Ed. Engl.* **1997**, *36*, 1164.

[2] D. E. Cane, P. J. Thomas, *J. Am. Chem. Soc.* **1984**, 106, 5295.

[3] L. A. Paquette, G. D. Annis, H. Schostarez, J. F. Blount, *J. Org. Chem.* **1981**, 46, 3768.

[4] J. Adams, M.-A. Poupart, L. G. Chris, *Tetrahedron Lett.* **1989**, 30, 1749–1752.

[5] A. B. Smith III, P. Jerris, *Syn. Commun.* **1978**, 8, 421.

[6] J. D. White, P. Hrnciar, F. Stappenbeck, *J. Org. Chem.* **1997**, 62, 5250–5251.

[7] D. F. Taber, J. L. Schuchardt, *J. Am. Chem.* Soc. **1985**, 107, 5289.

[8] D. F. Taber, K. K. You, *J. Am. Chem. Soc.* **1995**, 117, 5757.

[9] D. F. Taber, K. K. You, A. L. Rheingold, *J. Am. Chem. Soc.* **1996**, 118, 547.

[10] D.F. Taber, D.F.; Y. Song, *J. Org. Chem.* **1997**, *62*, 6603.

[11] Y. Takayuki, S. Yamada, M. Azuma, A. Ueki, M. Ikeda *Synthesis* **1998**, 973.

[12] J. C. Gilbert, D. H. Giamalva, M. E. Baze, *J. Org. Chem.* **1985**, 50, 2557.

[13] D. F. Taber, E. M. Petty, K. Raman, *J. Am. Chem.* Soc. **1985**, 107, 196.

[14] D. F. Taber, K. Raman, M. D. Gaul, *J. Org. Chem.* **1987**, 52, 28.

[15] J. W. Bode, M. P. Doyle, M. N. Protopopova, Q.-L. Zhou, *J. Org. Chem.* **1996**, 61, 9146.

[16] S.-l. Hashimoto, N. Watanabe, T. Sato, M. Shiro, S. Ikegami, *Tetrahedron Lett.* **1993**, *33,* 5109.

C. Enantioselective Catalysis

Enantioselective Heck Reactions

Markus Jachmann and Hans-Günther Schmalz

Institut für Organische Chemie, Universität Köln, Germany

Chirogenic reactions, i.e. reactions which lead from achiral starting materials to chiral products [1], deserve particular attention, as only such transformations offer the possibility to be enantioselectively catalyzed.

Among the non-enzymatic, catalytic-asymmetric reactions, homogeneous transition metal catalyzed processes play a predominant role [2]. This is mainly due to the fact that by means of chiral ligands it is comparatively facile to transfer absolute stereochemical information to a catalytically active metal center. However, the success of some of these reactions (e.g. the Sharpless asymmetric epoxidation or the Noyori hydrogenation) must not hide the fact that the number of powerful transition metal-catalyzed *C–C coupling reactions,* which proceed reliably with high enantioselectivity, is still rather small.

The Heck reaction, i.e. the palladium(0) catalyzed vinylation of aryl- or vinylhalides (or the corresponding triflates), belongs undoubtedly to the most important metal-catalyzed C–C coupling reactions [3, 4]. Accordingly, it enjoys increasing application as a key reaction in total synthesis [5].

The basic pattern of the Heck reaction in its classical form is depicted in Scheme 1. It involves, for instance, the reaction of an aryl halide (**1**) with an olefin in the presence of a base and a catalytic amount of a palladium complex to give a styrene derivative (**2**) under elimination of HX.

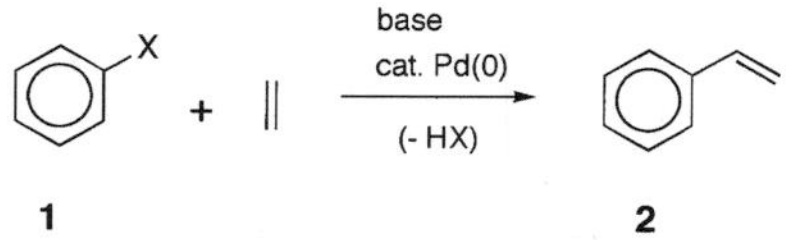

base
cat. Pd(0)
(- HX)
1
2

Scheme 1

Mechanistically, the Heck reaction can be rationalized as follows (Scheme 2): First, a Pd(II) complex of type **3** is formed by oxidative addition of the halide **1** to a $L_2Pd(0)$ species. Complex **3** then reacts with an olefin via the π-complex **4** to give the β-insertion product, i.e. the alkyl complex **5**. After β-H elimination, the product **2** is released and the active catalyst is regenerated by base-assisted reductive elimination of HX.

The Heck reaction in its original form is *not* a chirogenic reaction. However, the situation changes if cyclic alkenes are employed as a coupling component, as was initially shown by Larock et al. [6]. In such cases, non-conjugated, chiral products of type **6** are formed because only one *syn*-H atom is available for β-H elimination in the intermediates of type **7** (Scheme 3). While racemic mixtures are obtained with achiral catalysts, such transformations of course afford the possibility of achieving enantioselective Heck reactions.

Base-HX
Base
$L_2Pd(0)$
1
L_2Pd X H
L_2Pd X
3
2
L_2Pd X H
5
L_2Pd X
4

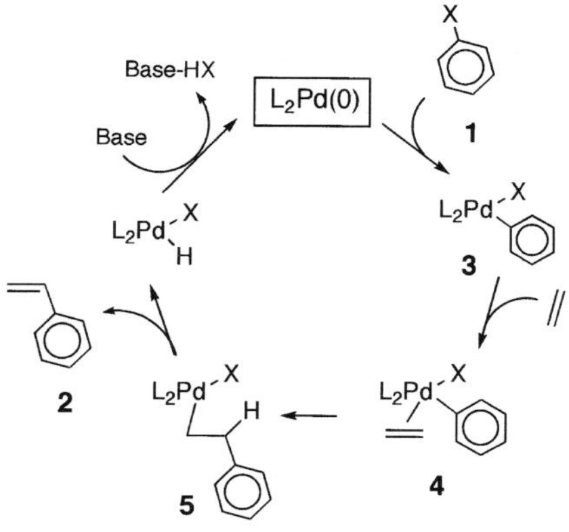

Scheme 2 Mechanism of the Heck reaction.

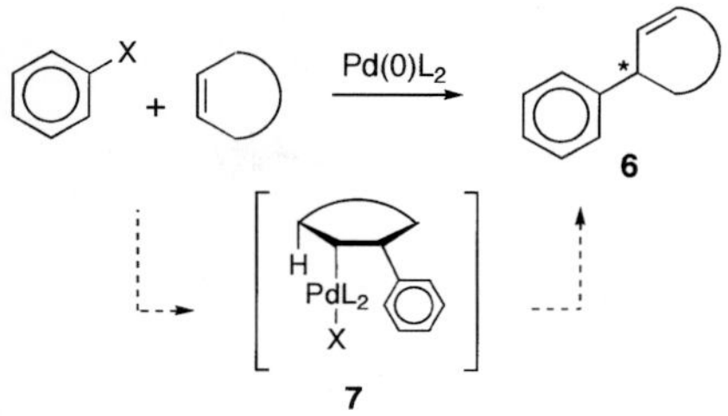

Scheme 3

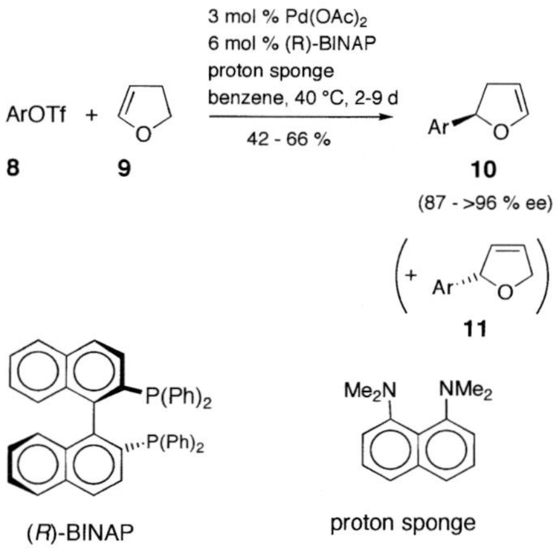

Intermolecular Reactions

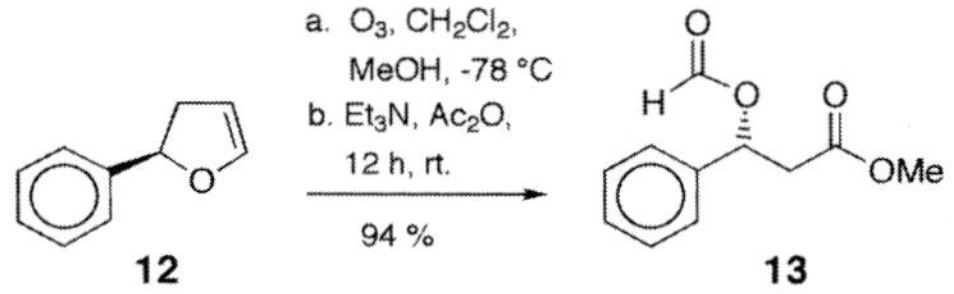

While several groups had been working on asymmetric intermolecular Heck reactions [7], Hayashi and Ozawa et al. were the first to report synthetically useful selectivities [8, 9]. The reaction of various aryl triflates **8** with 2,3–dihydrofuran (**9**) proceeds under optimized conditions [9d] with high enantioselectivity (>96 % *ee*) and leads to (*R*)-2-aryl-2,3-dihydrofurans **10** with acceptable yields (Scheme 4). As catalyst, an in situ generated Pd-(*R*)-BINAP complex is used in combination with 1,8-bis-(dimethylamino)-naphthalene (proton sponge), which has been established as the base of choice. The (*S*)-configured regioisomers **11** are often observed as by-products. They can be separated by chromatographic methods and exhibit significantly lower enantiomeric purity than the major products **10** (6–53 % *ee*). Obviously, the isomerization of the double bond under the reaction conditions (via β-H-insertion/β-H elimination) is accompanied by an additional kinetic resolution.

Scheme 4

Scheme 5

The enolethers of type **10** represent useful intermediates for further synthetic use. For example, Reiser and co-workers showed a way to transform the chiral 2,3–dihydrofuranderivative **12** by ozonolysis directly to the synthetically valuable β-hydroxyester **13** (Scheme 5) [10].

The power of their methodology was demonstrated by Hayashi and Ozawa in a remarkably short synthesis of **18**, an antagonist of the plate-

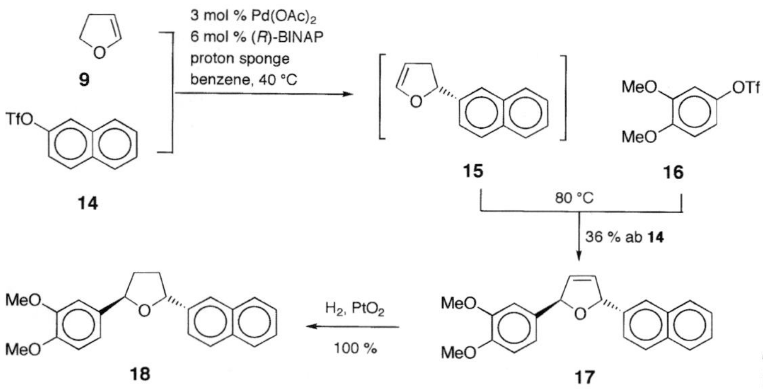

Scheme 6 Total synthesis of a PAF inhibitor according to T. Hayashi.

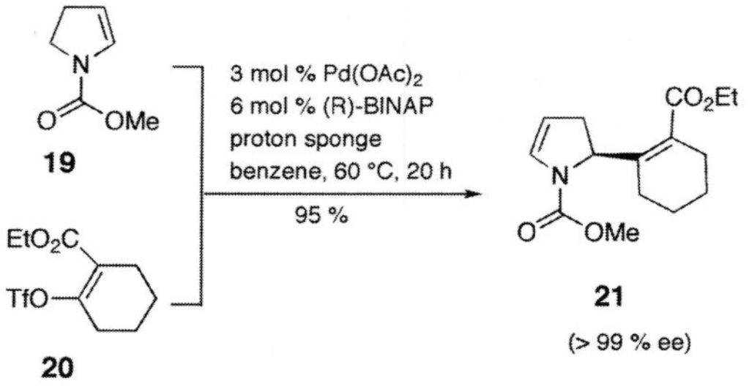

Scheme 7

let-activating factors (Scheme 6). Initially, the 2,3-dihydrofuran **9** is reacted under the established conditions with the β-naphthyl triflate **14** to give **15,** which is not isolated but directly reacted in a second Heck reaction with triflate **16** at elevated reaction temperatures. Finally, the double arylated product **17** is hydrogenated to afford the target molecule **18**.

The method of Hayashi and Ozawa is neither restricted to aryl triflates nor to 2,3-dihydrofuranes. Vinyl triflates can also be employed, and *N*-(methoxycarbonyl)-pyrroline **19** as the olefin component gives even better results. Thus, the reaction of **19** with the enol triflate **20** results in the formation of **21** with excellent yield and almost complete enantioselectivity (Scheme 7) [9e].

Intramolecular Reactions

The enantioselective cyclization of prochiral substrates of type **22** to bicyclic products **23** was examined by Shibasaki and co-workers [11–14]. In the presence of BINAP as a chiral ligand the two enantiotopic double bonds of **22** are differentiated, and two new chiral centers are generated simultaneously in a single step (Scheme 8).

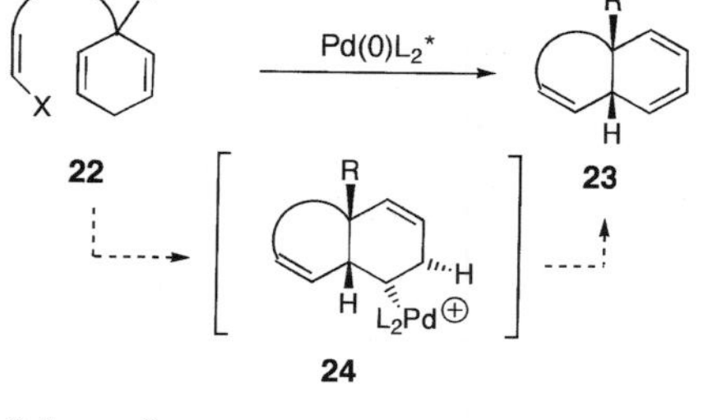

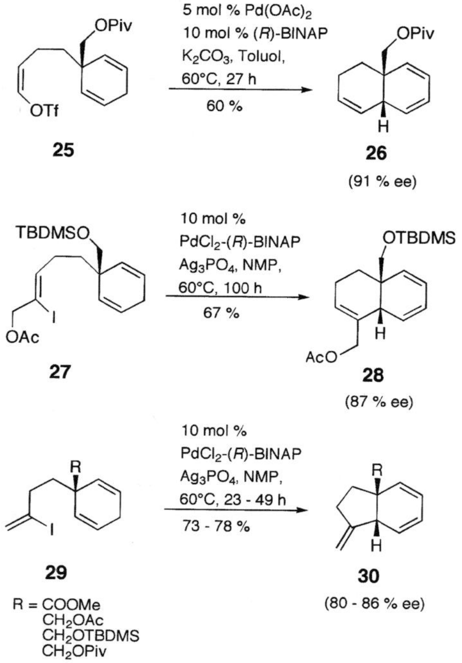

Scheme 8

Scheme 9

In order to obtain high enantioselectivities it seems to be important that the reactions proceed via cationic intermediates (e.g. **24**) to disfavor the partial dissociation of the chiral ligand. For this reason, silver salts are added to reactions of vinyliodides. These reactions are best performed in *N*-methylpyrrolidone (NMP) as a solvent. As the examples shown in Scheme 9 demonstrate, both *cis*-decalin [12] and *cis*-hydrindane [13] derivatives can be obtained in useful yields and enantiomeric purities.

In their search for suitable synthetic applications of their methodology, Shibasaki and co-workers spared no efforts and carried out an 18-step synthesis of lactone **35**, which represents an early intermediate of Danichefsky's synthesis of (+)-vernolepin (Scheme 10) [14]. First, the ester **31** is transformed via **32** into the allylic alcohol **33**, which is then cyclized with good enantioselectivities to yield the enone **34** (which is initially formed as an enol by β-H-elimination).

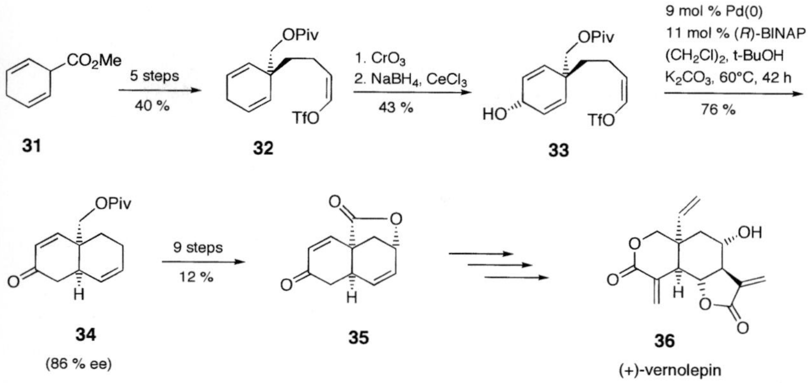

Scheme 10 Formal total synthesis of (+)-vernolepine according to M. Shibasaki.

This (formal) total synthesis of vernolepin deserves attention because of the elegance of the key step, which generates **34** in non-racemic form. However, the long and inefficient overall sequence impairs the competitiveness of the synthesis.

Three new chirality centers are formed with high enantio- and complete diastereoselectivity in the course of the reaction of the enol triflate **37** to the bicyclo [3.3.0]octane derivative **38** (Scheme 11) [15]. In this transformation, the intermediate **39**, formed by oxidative addition, leads to the cationic palladium-π-allyl complex **40**, which is finally converted to the isolated product **38** by regio- and diastereoselective nucleophilic addition of an acetate anion. The bicyclic product **38** is of interest as a building block for the synthesis of capnellene sesquiterpenes.

By converting the enol triflate **41** to the spirotricyclic dienone **42**, Overman and co-workers had already shown in 1989 that the direct enantioselective formation of quaternary chiral carbon centers can be carried out through an intramolecular Heck reaction. While the enantioselectivities were only moderate at the beginning [16], the same authors later succeeded in achieving the Pd(0)-BINAP-catalyzed cyclization of substrates of type **43** to spiro-oxindoles **44** with up to 95 % *ee* (Scheme 12) [17].

Subsequently, Overman and co-workers reported an application of their method in an enantioselective total synthesis of the alkaloid (−)-physostigmine (**50**), which as an effective acetylcholine esterase inhibitor is of interest for

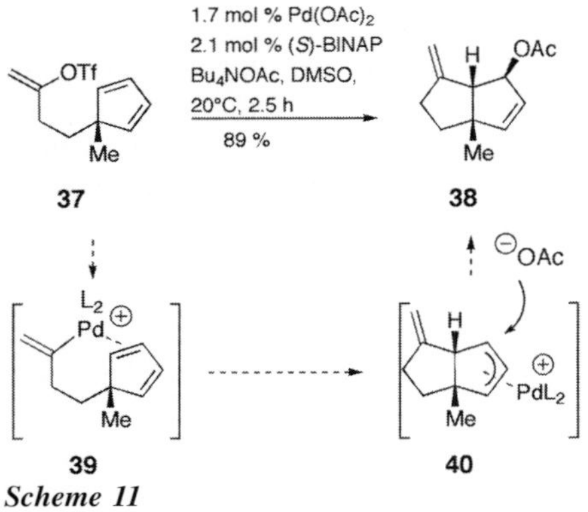

Scheme 11

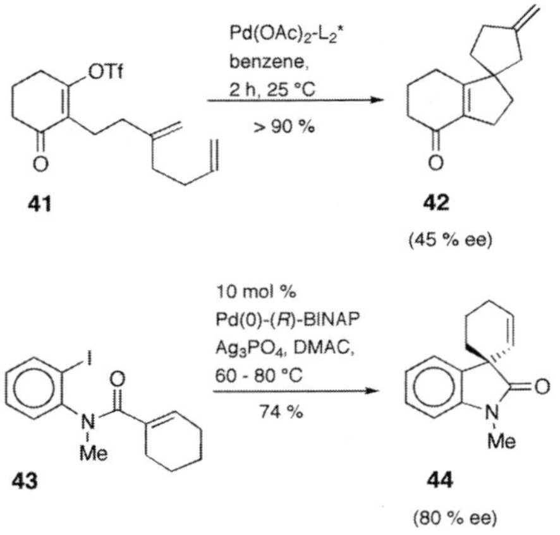

Scheme 12

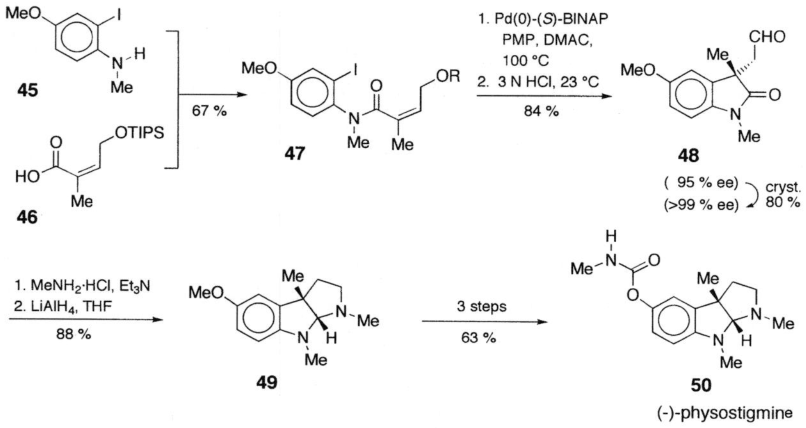

1. Pd(0)-(*S*)-BINAP PMP, DMAC, 100 °C 2. 3 N HCl, 23 °C 84 %

(95 % ee) cryst. 80 % (>99 % ee)

1. $MeNH_2 \cdot HCl$, Et_3N 2. $LiAlH_4$, THF 88 %

3 steps 63 %

(-)-physostigmine

Scheme 13 Total synthesis of (−)-physostigmin according to L. E. Overman.

the treatment of glaucoma, myastenia gravis and Alzheimers desease [18]. The synthesis (Scheme 13) starts with the coupling of the two readily accessible components **45** and **46** to give the amide **47**. The enantioselective Heck cyclization of **47** was then performed using a Pd(0)-(*S*)-BINAP catalyst in the presence of the bulky base 1,2,2,6,6-pentamethylpiperidine (PMP) in dimethylacetamide (DMAC) as solvent. After acidic hydrolysis of the initially formed enol ether to the corresponding aldehyde **48**, the enantioselection was completed by a single recrystallization. The conversion of **48** [via (−)-esermethol **49**] into the target molecule **50** was accomplished in only a few steps.

All in all, the synthesis is remarkably short and efficient, and impressively demonstrates the power of the asymmetric Heck cyclization. It should be pointed out that only the *Z*-configured substrates of type **51** react with good selectivities (80 to 90 % *ee*), while the *E*-isomers **52** lead to products of much lower enantiomeric purity (< 45 % *ee*).

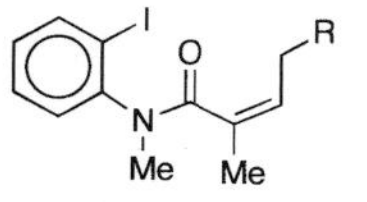

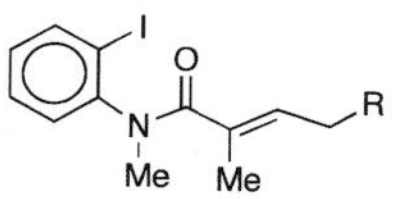

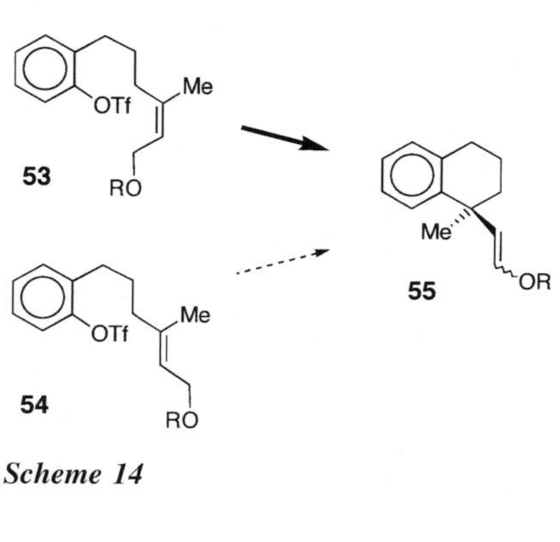

Scheme 14

As Shibasaki and co-workers demonstrated, tetralin derivatives of type **55** (with a quaternary benzylic carbon center) can also be enantioselectively prepared by an intramolecular Heck reaction (Scheme 14) [19].

Again, the geometry of the double bond proved to be of central importance: while only moderate selectivities were obtained with *E*-configured substrates **54**, Pd(0)-BINAP-catalyzed reactions of the *Z*-configured substrates **53** provided the

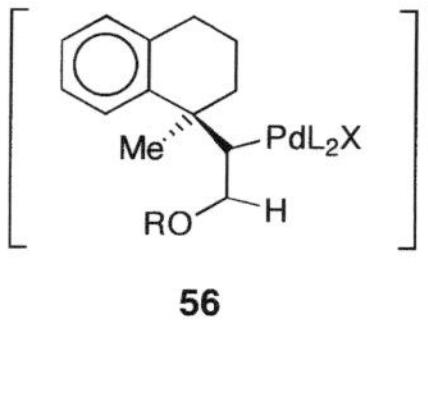

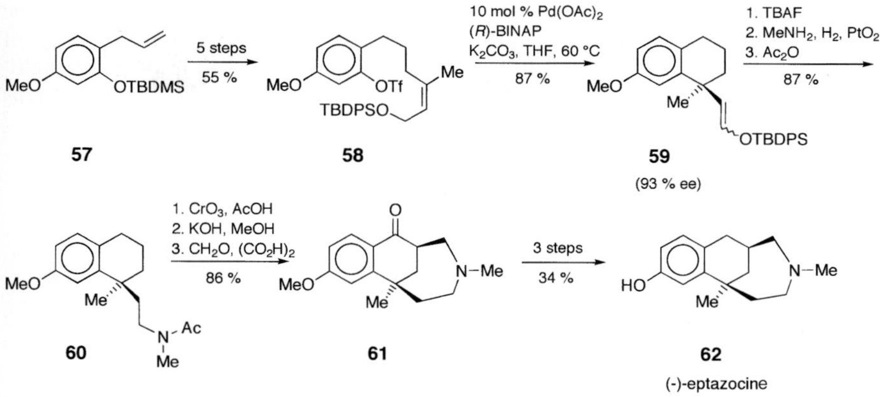

Scheme 15 Total synthesis of (–)-eptazocine according to M. Shibasaki.

products **55** with good yield (79 % to 85 %) and enantioselectivity (87 % to 91 %). In these reactions, a quaternary chirality center is generated in the insertion step. As a consequence, the β-H elimination of the intermediate **56** can only proceed to one side. This avoids the formation of any undesirable regioisomers.

The methodology discussed above was applied by Shibasaki in an efficient total synthesis of (–)-eptazocine, an analgetic compound which contains (like other analgesics such as morphine) a 1,1–disubstituted tetralin as a substructure (Scheme 15) [19]. Starting from the trisubstituted benzene derivative **57**, the prochiral cyclization precursor **58** is prepared in only five steps. In the chirogenic key reaction, **58** is then cyclized to **59** with high enantioselectivity by treatment with a catalyst formed in situ from $Pd(OAc)_2$ and (*R*)-BINAP. After cleavage of the enol ether, reductive amination and N-acetylation, the resulting intermediate **60** is converted to the tricyclic pre-target compound **61** by benzylic oxidation, amide hydrolysis and Mannich cyclization.

The above discussed early work of Overman and Shibasaki had shown that products of type **64** with a quaternary carbon atom in the benzylic position ($R^1 \neq H$) can be prepared by Pd-BINAP-catalyzed Heck cyclization with high enantioselectivity (Scheme 16). It is important to keep in mind that the configuration of the double bond in the starting material **63** severely influences the selectivity of the reaction and that the absolute configuration of the products has to be established in all individual cases.

An important further development came from the group of Tietze, who introduced the concept of silyl termination in order to control the regioselectivity of the final β-H elimation step in the case of substrates (**63**) with $R^1 = H$ [20]. An example is the transformation shown in Scheme 17, a key step in an enantioselective synthesis of a cytostatic nor-sesquiterpene [20c].

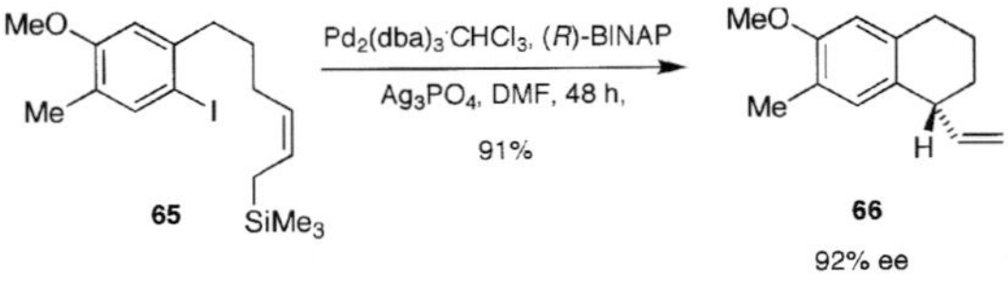

Scheme 17

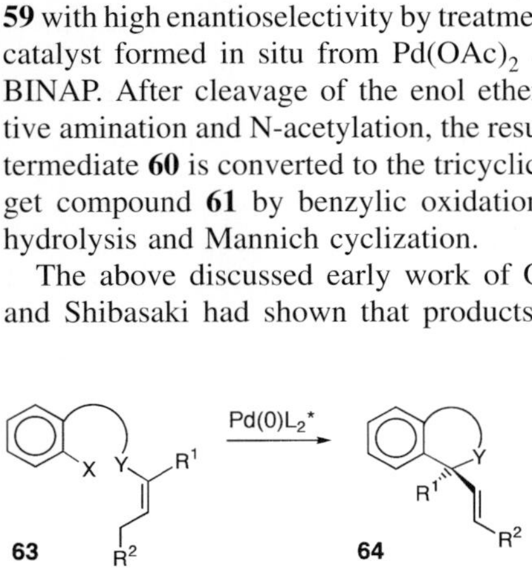

Scheme 16

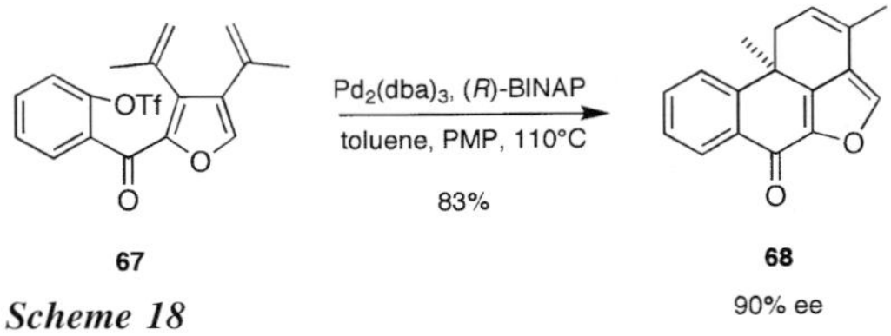

Scheme 18

Another field was opened by Keay et al., who reported interesting enantioselective tandem cyclizations of polyenes, for instance, the reaction shown in Scheme 18. This chemistry was successfully employed in a total synthesis of (+)-xestoquinone [21].

While BINAP had been used as the sole ligand of choice in the early days of asymmetric Heck chemistry, the introduction of the chiral P,N-ligand **69** by Pfaltz and co-workes has led to another great improvement of the general methodology [22].

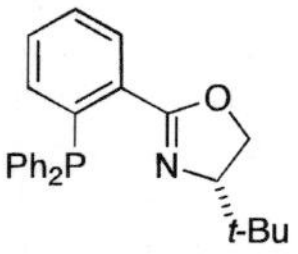

Ph$_2$P N O *t*-Bu

69

The few selected examples shown in Scheme 19 impressively demonstrate the efficiency of the catalyst system derived from **69** and a (halogen-free !!) Pd source. Interestingly, the reactions employing the Pfaltz catalyst proceed in most cases without concomitant double bond migration.

Conclusion

The enantioselective Heck reaction has matured into a powerful method for asymmetric C–C bond formation and has proven its value in several total syntheses. One can expect that it will find many more applications in the future, even in industry. There is still an extensive space for chemists to design new suitable substrates and to search for new effective chiral ligands [23].

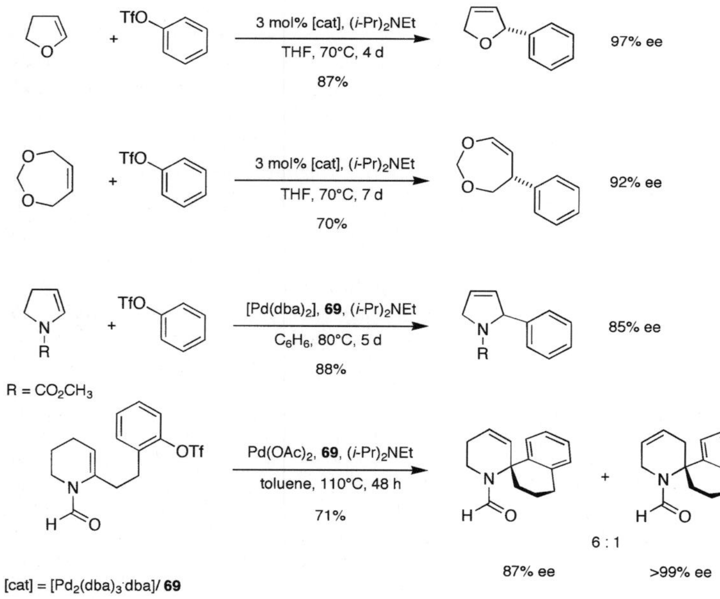

TfO + ; 3 mol% [cat], (i-Pr)$_2$NEt; THF, 70°C, 4 d; 87%; 97% ee

TfO + ; 3 mol% [cat], (i-Pr)$_2$NEt; THF, 70°C, 7 d; 70%; 92% ee

N R + TfO ; [Pd(dba)$_2$], **69**, (i-Pr)$_2$NEt; C_6H_6, 80°C, 5 d; 88%; N R; 85% ee

R = CO_2CH_3

OTf N H O ; Pd(OAc)$_2$, **69**, (i-Pr)$_2$NEt; toluene, 110°C, 48 h; 71%; N H O + N H O; 6 : 1; 87% ee; >99% ee

[cat] = [Pd$_2$(dba)$_3$·dba]/ **69**

Scheme 19

References

[1] S. Drenkard, J. Ferris, A. Eschenmoser, *Helv. Chim. Acta* **1990**, 1373.

[2] See, for instance: a) I. Ojima (Ed.) *Catalytic Asymmetric Synthesis*, VCH, New York, **1993**; b) R. Noyori, Asymmetric Catalysis in Organic Synthesis, Wiley, New York, **1994**; c) L. S. Hegedus, *Transition Metals in the Synthesis of Complex Organic Molecules* (2nd ed.), University Science Books, Sausalito, **1999**.

[3] Reviews: a) R. F. Heck, *Organic Reactions*, **1982**, *27*, 345; b) R. F. Heck in *Comprehensive Organic Synthesis*, (Eds.: B. M. Trost, I. Fleming) Pergamon, Oxford, **1991**; Vol. 4, chapter 4.3., p. 833; c) A. de Meijere, F. Meyer, *Angew. Chem.* **1994**, *106*, 2437; *Angew. Chem. Int. Ed. Engl.* **1994**, *33*, 2379; d) M. Beller, T. H. Riermeier, G. Stark in: *Transition Metals for Organic Synthesis, Vol. 1* (M. Beller, C. Bolm; eds.), Wiley-VCH, Weinheim **1998**, chapter 2.13; and refs. cited therein.

[4] K. Ritter, *Synthesis* **1993**, 735.

[5] Selected work: a) C. Y. Hong, N. Kado, L. E. Overman, *J. Am. Chem. Soc.* **1993**, *115*, 11028; b) J. J. Masters, D. K. Jung, W. G. Bornmann, S. J. Danishefsky, *Tetrahedron Lett.* **1993**, *34*, 7253; c) D. J. Kucera, S. J. OConnor, L. E. Overman, *J. Org. Chem.* **1993**, *58*, 5304; d) A.O. King, E. G. Corley, R. K. Anderson, R. D. Larsen, T. R. Verhoeven, P. J. Reider, Y. B. Xiang, M. Belley, Y. Leblanc, M. Labelle, P. Prasit, R. J. Zamboni, *J. Org. Chem.* **1993**, *58*, 3731.

[6] R. C. Larock, W. H. Gong, *J. Org. Chem.* **1989**, *54*, 2047; and refs. cited therein.

[7] a) H. Brunner, K. Kramler, *Synthesis* **1991**, 1121; b)T. Sakamoto, Y. Kondo, H. Yamanaka, *Tetrahedron Lett.* **1992**, *33*, 6845.

[8] Review: T. Hayashi, A. Kubo, F. Ozawa, *Pure. Appl. Chem.* **1992**, *64*, 421.

[9] a) F. Ozawa, A. Kubo, T. Hayashi, *J. Am. Chem. Soc.* **1991**, *113*, 1417; b) F. Ozawa, T. Hayashi, *J. Organomet. Chem.* **1992**, *428*, 267; c) F. Ozawa, A. Kubo, T. Hayashi, *Chem. Lett.* **1992**, 2177; d) F. Ozawa, A. Kubo, T. Hayashi, *Tetrahedron Lett.* **1992**, *33*, 1485; e) F. Ozawa, Y. Kobatake, T. Hayashi, *Tetrahedron Lett.* **1993**, *34*, 2505.

[10] S. Hillers, A. Niklaus, O. Reiser, *J. Org. Chem.* **1993**, *58*, 3169.

[11] Overviews: a) M. Shibasaki, C. D. J. Boden, A. Kojima, *Tetrahedron* **1997**, *53*, 7371; b) M. Shibasaki, E. M. Vogl, *J. Organomet. Chem.* **1999**, *576*, 1.

[12] a) Y. Sato, M. Sodeoka, M. Shibasaki, *J. Org. Chem.* **1989**, *54*, 4738; b) Y. Sato, M. Sodeoka, M. Shibasaki, *Chem. Lett.* **1990**, 1953; c) Y. Sato, S. Watanabe, M. Shibasaki, *Tetrahedron Lett.* **1992**, *33*, 2589.

[13] Y. Sato, T. Honda, M. Shibasaki, *Tetrahedron Lett.* **1992**, *33*, 2593.

[14] a) K. Kondo, M. Sodeoka, M. Mori, M. Shibasaki, *Synthesis* **1993**, 920; b) K. Kondo, M. Sodeoka, M. Mori, M. Shibasaki, *Tetrahedron Lett.* **1993**, *34*, 4219; c) K. Ohrai, K. Kondo, M. Sodeoka, M. Shibasaki, *J. Am. Chem. Soc.* **1994**, *116*, 11737.

[15] a) K. Kagechika, M. Shibasaki, *J. Org. Chem.* **1991**, *56*, 4093; b) T. Ohshima, K. Kagechika, M. Adachi, M. Sodeoka, M. Shibasaki, *J. Am. Chem. Soc.* **1996**, *118*, 7108.

[16] N. E. Carpenter, D. J. Kucera, L. E. Overman, *J. Org. Chem.* **1989**, *54*, 5846.

[17] a) A. Ashimori, L. E. Overman, D. J. Poon, *J. Org. Chem.* **1992**, *57*, 4571; b) A. Ashimori, B. Bachand, L. E. Overman, D. J. Poon, *J. Am. Chem.* Soc. **1998**, *120*, 6477; c) A. Ashimori, B. Bachand, M. A. Calter, S. P. Govek, L. E. Overman, D. J. Poon, *J. Am. Chem.* Soc. **1998**, *120*, 6488.

[18] a) A. Ashimori, T.Matsuura, L. E. Overman, D. J. Poon, *J. Org. Chem.* **1993**, *58*, 6949; b) L. E. Overman, *Pure Appl. Chem.* **1994**, *66*, 1423; c) T. Matsuura, L. E. Overman, D. J. Poon, *J. Am. Chem. Soc.* **1998**, *120*, 6500.

[19] T. Takemoto, M. Sodeoka, H. Sasai, M. Shibasaki, *J. Am. Chem. Soc.* **1993**, *115*, 8477.

[20] a) L. F. Tietze, R. Schimpf, *Angew. Chem.* **1994**, *106*, 1138; *Angew. Chem. Int. Ed. Engl.* **1994**, *33*, 1089; b) L. F. Tietze, T. Raschke, *Synlett* **1995**, 597; c) L. F. Tietze, T. Raschke, *Liebigs. Ann.* **1996**, 1981.

[21] a) B. A. Keay, S. P. Maddaford, W. A. Cristofoli, N. G. Andersen, M. S. Passafaro, N. S. Wilson, J. A. Nieman, *Can. J. Chem.* **1997**, *75*, 1163; b) S. Y. W. Lau, B. A. Keay, *Synlett* **1999**, 605.

[22] a) O. Loiseleur, P. Meier, A. Pfaltz, *Angew. Chem.* **1996**, *108*, 218; *Angew. Chem. Int. Ed. Engl.* **1996**, *35*, 200; b) O. Loiseleur, M. Hayashi, N. Schmees, A. Pfaltz, *Synthesis* **1997**, 1338; c) O. Loiseleur, M. Hayashi, M. Keenan, N. Schmees, A. Pfaltz, *J. Organomet. Chem.* **1999**, *576*, 16; see also: d) L. Ripa, A. Hallberg, *J. Org. Chem.* **1997**, *62*, 595.

[23] a) M. Shibasaki, C. D. J. Boden, A. Kojima, *Tetrahedron* **1997**, *53*, 7371; b) F. Miyazaki, K. Uotsu, M. Shibasaki, *Tetrahedron Lett.* **1997**, *38*, 3459; c) S. Y. Cho, M. Shibasaki, *Tetrahedron Lett.* **1998**, *39*, 1773; d) L. F. Tietze, K. Thede, F. Sannicolo, *Chem. Commun.* **1999**, 1811; e) M. Tschoerner, P. S. Pregosin, A. Albinati, *Organometallics* **1999**, *18*, 670.

Catalytic Asymmetric Aldol Reactions

Rolf Krauss and Ulrich Koert

Institut für Chemie, Humboldt-Universität, Germany

Carbon-carbon coupling reactions belong to the most important and often hardest steps in organic synthesis. Nowadays, stereoselective C–C bond formations using covalently bound chiral auxiliaries are well known and established [1]. Because of economical and ecological requirements, the use of chiral catalysts rather than covalently bound auxiliaries is an urgent necessity [2a]. Apart from oxidation reactions (Sharpless-Jacobsen epoxidation, Sharpless dihydroxylation [2b-d]) and reductions (CBS reduction, catalytic hydrogenation [2e-f]), the number of truly useful and widely applicable enantioselective catalysts is still limited [3]. Therefore asymmetric catalysis for C–C linkage will be important in the future. Especially in the case of the aldol reaction important progress has been made recently [4].

Catalytic asymmetric aldol additions can at present be divided into

- the Mukaiyama-type reactions, which can be catalyzed by Lewis acids and Lewis bases, and
- the direct methods, where no preconversion of the ketone moiety into a more reactive species is necessary (Scheme 1).

In the Mukaiyama aldol reaction an aldehyde (**1**) reacts with a silyl enol ether (**3**) under Lewis-acid catalysis to yield the aldol adduct (**4**). The use of a chiral Lewis acid (**L***) offers the opportunity to perform the reaction in an asymmetric manner (Scheme 1) [5].

If the right reaction conditions are chosen, only a small amount of catalyst is needed. The catalytic cycle is demonstrated in Scheme 2. The aldehyde **1** is activated by coordination of the Lewis acid

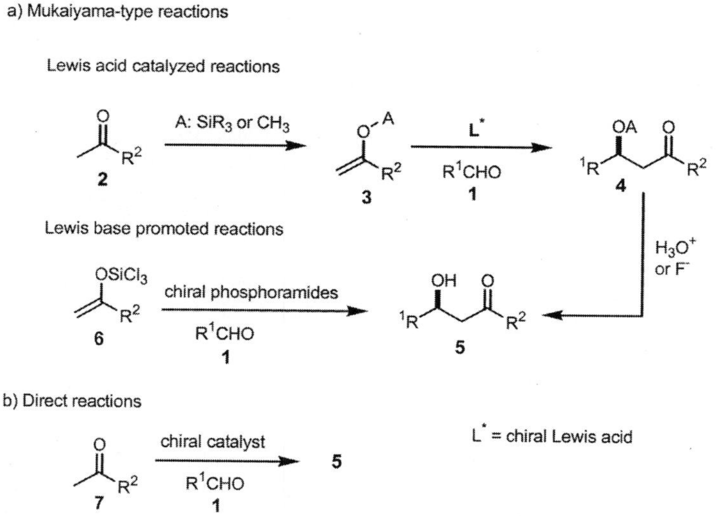

Scheme 1 Different types of catalytic asymmetric aldol reactions.

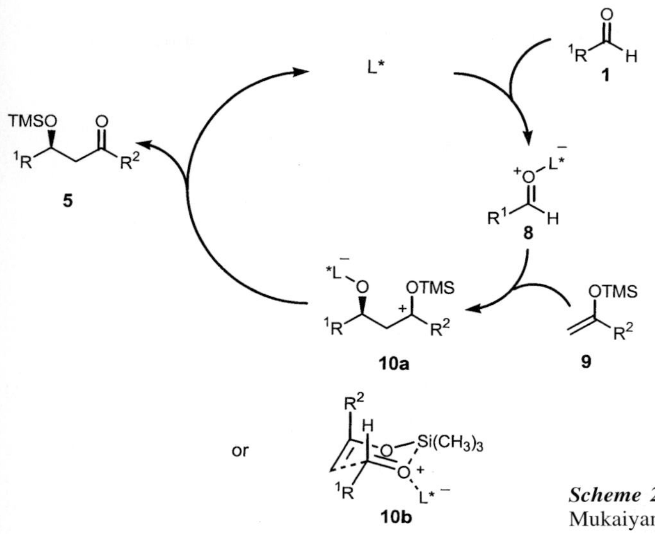

Scheme 2. Mechanism of the Lewis acid-catalyzed Mukaiyama aldol reaction.

L* forming intermediate **8**. Trapping **8** with a silyl enol ether **9** leads to the aldol addition product **5**. The open-chain structure **10a** and the cyclic structure **10b** are discussed as transition states in the literature. Release of the Lewis acid **L*** is the final step of the catalytic cycle.

As one of the first groups, Reetz et al. [6] reported a catalytic Mukaiyama aldol reaction (Scheme 3a) with the chiral aluminum complex **13**. However, low yields and a low level of enantioselectivity made this reaction not generally applicable.

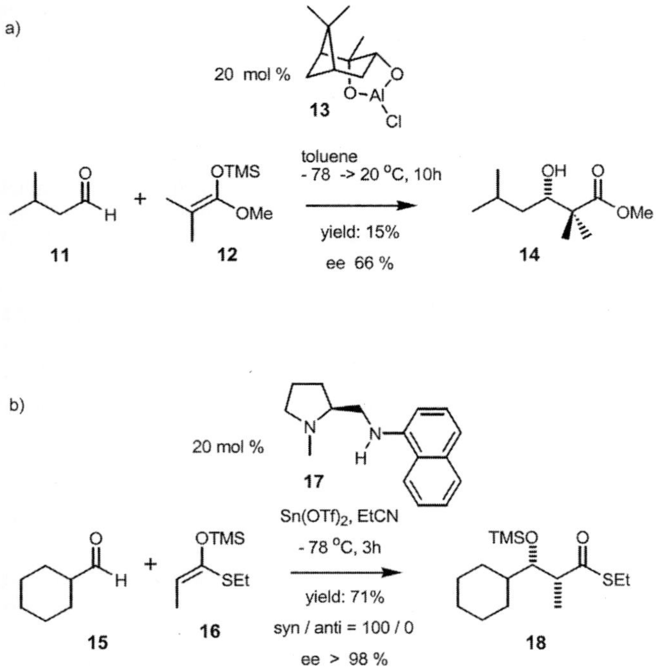

Scheme 3. Catalytic asymmetric aldol reaction with aluminum- and tin-containing chiral Lewis acids: a) Reetz and b) Mukaiyama.

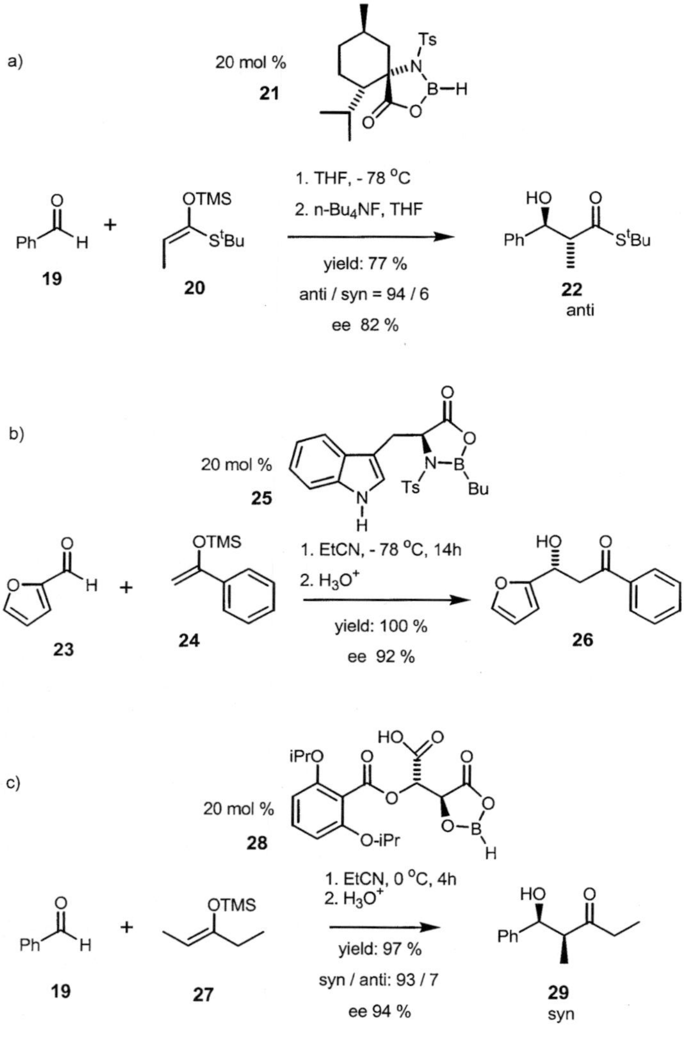

a)
20 mol % **21**
19 + **20** → 1. THF, - 78 °C; 2. n-Bu_4NF, THF; yield: 77 %; anti / syn = 94 / 6; ee 82 % → **22** anti

b)
20 mol % **25**
23 + **24** → 1. EtCN, - 78 °C, 14h; 2. H_3O^+; yield: 100 %; ee 92 % → **26**

c)
20 mol % **28**
19 + **27** → 1. EtCN, 0 °C, 4h; 2. H_3O^+; yield: 97 %; syn / anti: 93 / 7; ee 94 % → **29** syn

Scheme 4. Catalytic, asymmetric aldol reaction with boron-containing Lewis acids: a) Masamune, b) Corey and c) Yamamoto.

A chiral Lewis acid derived from $Sn(OTf)_2$ and the proline derivative **17** has proven to catalyze the aldol reaction effectively. As Mukaiyama et al. [7] demonstrated, a high degree of enantioselectivity was achieved (Scheme 3b).

Boron compounds formed the next generation of chiral catalysts (Scheme 4). Masamune [8] was able to use a thioketeneacetal (**20**) and the chiral boron-based Lewis acid **21** in the stereoselective formation of the *anti* aldol product **22**. Unfortunately, the demand for 20 mol % catalyst was still very high. Corey's group [9] reported the successful use of the tryptophane-derived catalyst **25** in an asymmetric Mukaiyama aldol reaction (**23** + **24** → **26**). Tartaric acid is the starting point of the synthesis of acyloxyborane complex **28**, introduced by Yamamoto et al. [10]. With this catalyst the silyl enol ether **27** was converted to *syn*-α-methyl-β-hydroxyketone **29** with high enantio- and diastereoselectivities (Scheme 4c). Worth mentioning is the practicable reaction temperature (0 °C) in Yamamoto's example.

Chiral titanium Lewis acids belong today to a very promising class of catalysts. In contrast to boron(III), titanium(IV) has the advantageous ability to expand its coordination sphere from tet-

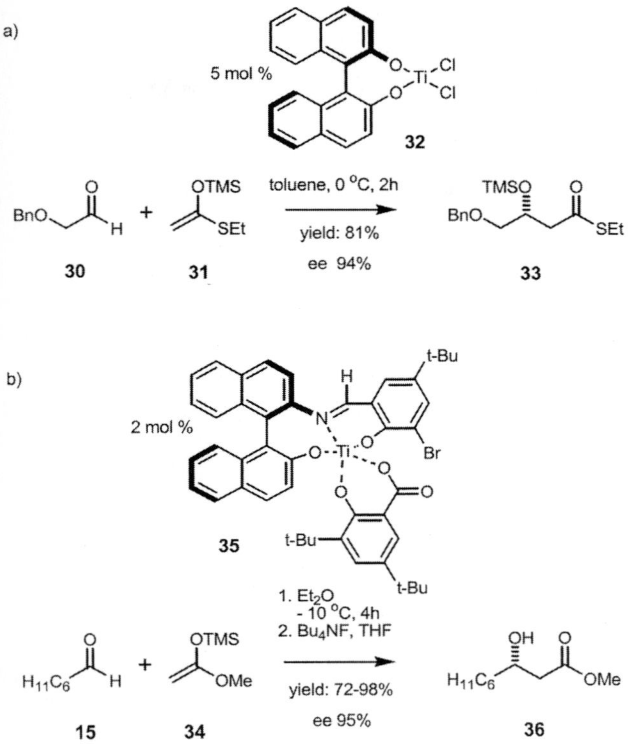

a)

5 mol % **32**

30 + **31** → (toluene, 0 °C, 2h; yield: 81%; ee 94%) → **33**

b)

2 mol % **35**

15 + **34** → (1. Et_2O, - 10 °C, 4h; 2. Bu_4NF, THF; yield: 72-98%; ee 95%) → **36**

Scheme 5. Catalytic, asymmetric aldol reaction with titanium containing chiral Lewis acids: a) Mikami and b) Carreira.

rahedral to trigonal-bipyramidal or even octahedral (Scheme 5).

The amount of catalyst can be reduced to 5 mol % using the binaphthyl-derived Ti-complex **32** [11]. Only 2 mol% of Carreira's [12] catalyst **35** is necessary to obtain good yields and high *ee* values. However, the chiral binaphthyl ligand for **35** is not commercially available.

Lewis acid catalysts activate the aldehyde by coordination to the carbonyl oxygen. Shibasaki et al. [13] were able to demonstrate that the activation of the enol ether is possible too. The reaction of the aldehyde **37** with the silyl enol ether **38** in the presence of the catalyst **39** proceeds with good, but still not excellent enantioselectivity to yield the aldol adduct **40**. Only 5 mol % of the chiral palladium(II) complex **39** was used (Scheme 6a). Activation of the Pd(II)-BINAP complex **39** by AgOTf is necessary. Therefore, addition of a small amount of water is important.

Better results were achieved by the cationic palladium(II) complex **41** (Scheme 6b) [14]. The reaction was performed at 0 °C in 1,1,3,3-tetramethylurea without any activation of the palladium(II) complex.

NMR-spectroscopic analysis led to the postulation of a mechanism [13] (Scheme 7) involving the formation of a palladium enolate.

The neutral palladium(II) compound **43** is transformed by addition of AgOTf into the cationic complex **44**. In the presence of water an exchange of the triflate anion to hydroxide occurs (**44** → **45**). Finally, the palladium enolate **46** is formed from the palladium complex and the silyl enol ether.

C_2-Symmetric tridentate bis(oxazolinyl)pyridine-Cu(II) complexes, introduced by Evans et al., can function as effective chiral Lewis acid catalysts in the Diels-Alder reaction [15a]. When applied to catalytic asymmetric aldol reactions [15], remarkable results were achieved (Scheme 8) [15a]. Only 0.5 mol % of catalyst **48** was needed for the reaction of **30** with the silylketene thioacetal **47** to yield after deprotection **49** in 99 % chemical yield. The *ee* values were determined to be 99 %. Today, catalysts

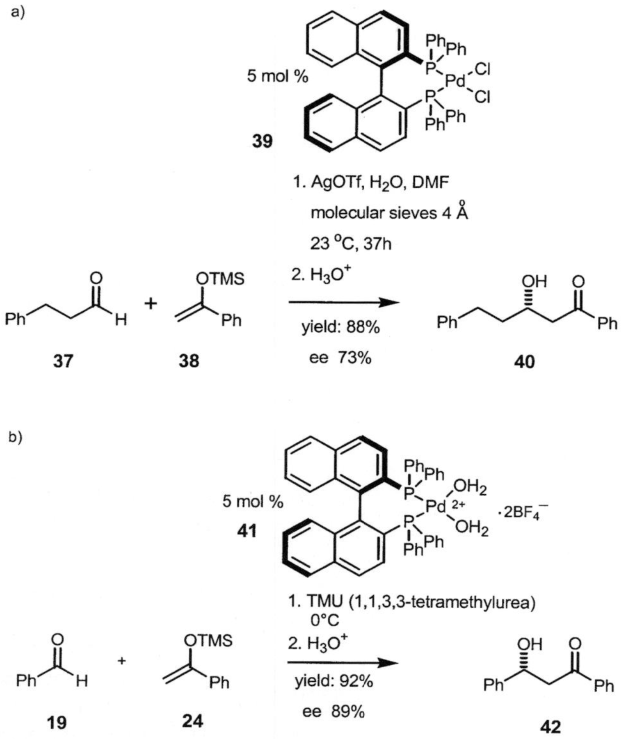

a)

5 mol % **39**

1. AgOTf, H_2O, DMF molecular sieves 4 Å 23 °C, 37h
2. H_3O^+

37 + **38** → **40**

yield: 88%
ee 73%

b)

5 mol % **41**

$\cdot 2BF_4^-$

1. TMU (1,1,3,3-tetramethylurea) 0°C
2. H_3O^+

19 + **24** → **42**

yield: 92%
ee 89%

Scheme 6. Palladium-catalyzed Mukaiyama aldol reactions according to Shibasaki.

of type **48** rank in the top position for asymmetric aldol reactions.

Compared to the great variety of Lewis acid catalysts for the catalytic asymmetric aldol reaction the field of nucleophilic (Lewis base) catalysts is less explored. This strategy involves the transient activation of the latent enolate equivalent via Lewis base coordination to the silyl enol ether (Scheme 9) [3]. For instance the trichlorosilyl enol ether **50** is able to expand its valency at the silicon atom from four to five and six. It reacts with an aldehyde (**51**), proceeding through a closed Zimmerman-Traxler-like transition state (**54**), to give **53** after quenching with saturated aqueous $NaHCO_3$ [16].

A useful synthetic alternative to the Mukaiyama aldol addition is the carbonyl-ene reaction [17]. This reaction of an aldehyde **51** with an enol ether **55**, bearing at least one hydrogen atom in the allylic position, under Lewis-acid catalysis, yields a *β*-hydroxy-enol ether of type **56** (Scheme 10). By use of a chiral Lewis acid (**L***) enantioselectivity can be achieved. For the carbonyl-ene reaction a cyclic transition state **57** is postulated [18].

The primary product **56** may be further transformed either into the ketone **58**, under acidic conditions, or oxidized to the ester **59** by ozone. The compounds **58** and **59** are the aldol products of the aldehyde **51** with acetone or

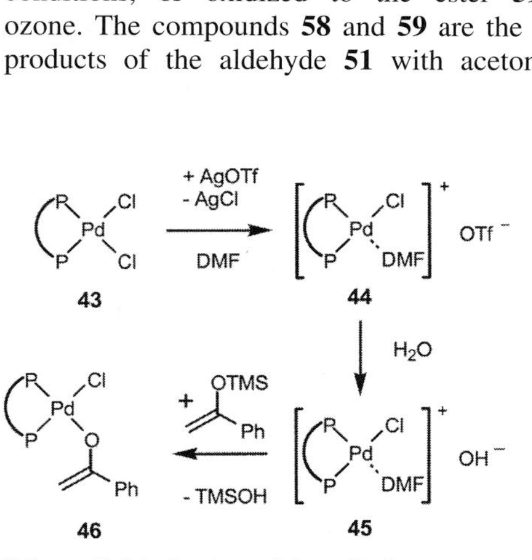

43 → (+ AgOTf, - AgCl, DMF) → **44** OTf^- → (H_2O) → **45** OH^- → (+ OTMS/Ph, - TMSOH) → **46**

Scheme 7. Mechanism of the palladium-catalyzed Mukaiyama aldol reaction.

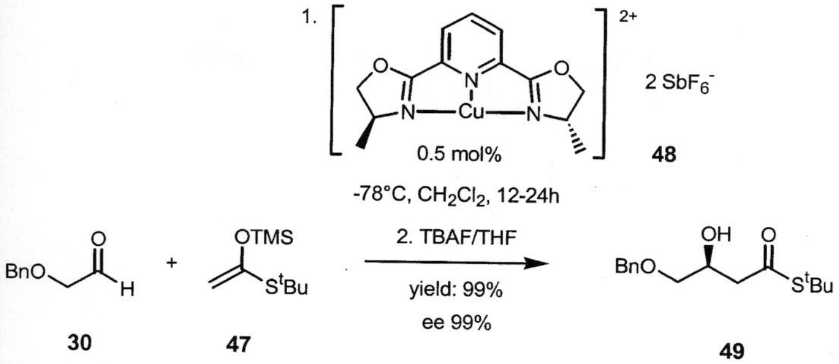

1. 0.5 mol% **48** ($2\ SbF_6^-$), $-78°C$, CH_2Cl_2, 12-24h
2. TBAF/THF

yield: 99%
ee 99%

30 + **47** → **49**

Scheme 8. Copper(II)-catalyzed aldol reaction according to Evans.

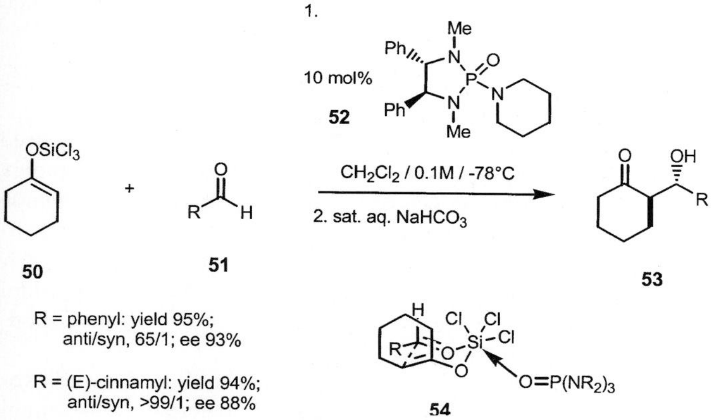

1. 10 mol% **52**, CH_2Cl_2 / 0.1M / -78°C
2. sat. aq. $NaHCO_3$

50 + **51** → **53**

R = phenyl: yield 95%;
anti/syn, 65/1; ee 93%

R = (E)-cinnamyl: yield 94%;
anti/syn, >99/1; ee 88%

54

Scheme 9. Lewis base-promoted catalytic asymmetric aldol reaction according to Denmark.

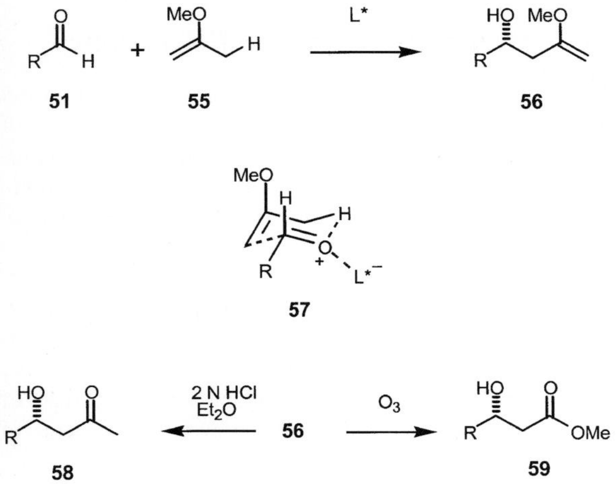

51 + **55** —L*→ **56**

57

58 ←(2 N HCl, Et_2O)— **56** —(O_3)→ **59**

Scheme 10. The carbonyl-ene-reaction: a synthetic alternative to the Mukaiyama aldol reaction.

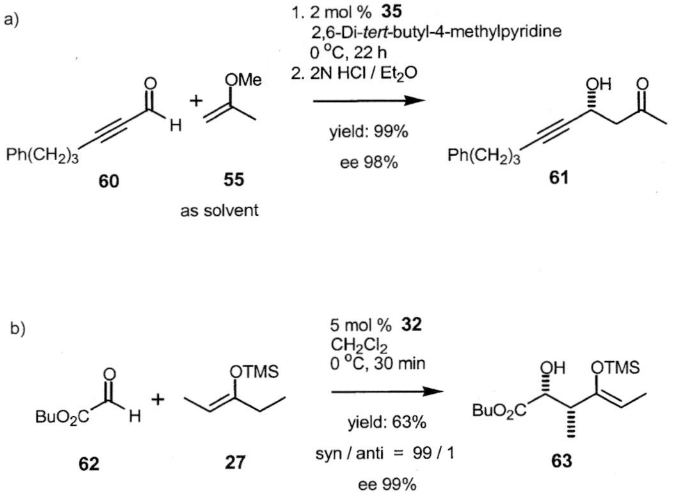

a)

1. 2 mol % **35**
2,6-Di-*tert*-butyl-4-methylpyridine
0 °C, 22 h
2. 2N HCl / Et_2O

yield: 99%
ee 98%

$Ph(CH_2)_3$ **60** + **55** as solvent → $Ph(CH_2)_3$ **61**

b)

5 mol % **32**
CH_2Cl_2
0 °C, 30 min

yield: 63%
syn / anti = 99 / 1
ee 99%

BuO_2C **62** + **27** → BuO_2C **63**

Scheme 11. Catalytic, asymmetric carbonyl-ene reaction with titanium-containing chiral Lewis acids: a) Carreira and b) Mikami.

methyl acetate respectively, illustrating the similarity of the Mukaiyama aldol and the carbonyl-ene reactions.

In 1995 Carreira et al. [19] reported a catalytic variant of the asymmetric carbonyl-ene reaction (Scheme 11a). By treatment of the aldehyde **60** with 2 mol % of titanium catalyst **35**, already used in the Mukaiyama aldol reaction, the β-hydroxyketone **61** is formed in quantitative yield and with an excellent *ee* value. Here, the ene-compound, 2–methoxypropene, is used simultaneously as solvent in a large excess. The high enantioselectivity is still limited to aldehydes similar to **60**; benzaldehyde for instance is converted with an *ee* of only 66 %.

Mikami et al. [18] demonstrated that under Lewis acid catalysis silyl enol ethers, bearing at least one hydrogen atom in the allylic position, form carbonyl-ene products. They succeeded in using the titanium catalyst **32** for the asymmetric catalysis of this reaction (Scheme 11b). If the aldehyde contains an activating substituent, as in the case of the glyoxolate **62**, an excess of the enecompound is not necessary. For example, the reaction of **62** with the silyl enol ether **27** to the carbonyl-ene adduct **63** still proceedes with good stereoselectivity, but yields drop to a moderate value.

Besides the Mukaiyama aldol and the carbonyl-ene reactions another successful application of asymmetric catalysis is the nitro-aldol reaction (Scheme 12) [20a]. Shibasaki et al. [20b] used a chiral in situ generated lanthanide complex (**64**) as catalyst. The optically active lanthanide complex **66** is postulated as the basic intermediate, activating the nitromethane as shown in **67**. However, in the case of the Mukaiyama aldol addition, lanthanide Lewis acids still give moderate *ee* values.

Worth mentioning are chiral gold complexes [20d, e] as well as chiral quaternary ammonium fluorides [21], which are used successfully as catalysts in the asymmetric aldol reaction.

Direct Catalytic Aldol Reactions

Recent success was achieved in carrying out direct catalytic asymmetric aldol reactions of aldehydes with unmodified ketones [22]. No pre-conversion of the ketone moiety to a more reactive species such as an enol silyl ether or enol methyl ether is necessary.

As Scheme 13 shows, reaction of the tertiary aldehyde **68** with methyl-phenyl ketone **69** under catalysis by the barium complex **70** gave compound **71** in a smooth reaction and in quantitative yield [23]. Only 5 mol % of catalyst and 2.0 eq. of the ketone are needed. However, the *ee* value of 70 % is only moderate.

Scheme 14 gives a mechanistic rationale for the role of the barium complex **70**. After substitution

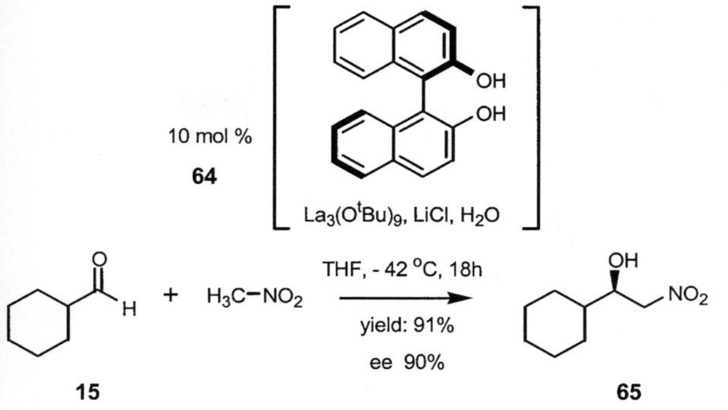

10 mol %
64
$La_3(O^tBu)_9$, LiCl, H_2O
+ H_3C-NO_2
THF, - 42 °C, 18h
yield: 91%
ee 90%
15 **65**
66 **67**

Scheme 12. Chiral lanthanum complex as catalyst in the asymmetric nitro-aldol reaction according to Shibasaki.

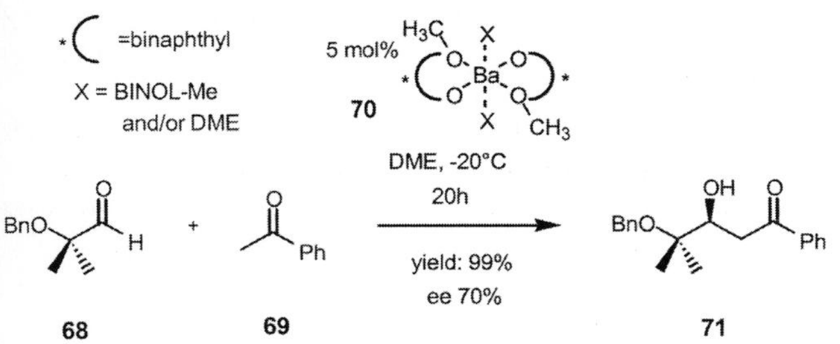

=binaphthyl
X = BINOL-Me and/or DME
5 mol%
70
DME, -20°C
20h
yield: 99%
ee 70%
68 **69** **71**

Scheme 13. Barium-catalyzed direct asymmetric aldol reaction.

of **X** by the aldehyde **51**, complex **72** is formed. Here, the barium center acts as a Lewis acid and activates the aldehyde. In the following, the addition of the ketone **69** occurs and **73** is generated. The coordination sphere of the barium atom is expanded from six to seven. Then a Brønsted base unit of **73** deprotonates an α-proton of methyl phenyl ketone to yield the intermediate **74**. Now the reaction of the aldehyde and the barium enolate takes place in a chelation controlled manner to deliver the aldol product **75**.

Inspired by enzyme chemistry, Shibasaki et al. [24] developed several heterobimetallic asymmetric catalysts [25], displaying both Lewis acidity and Brønsted basicity. Best results so far were achieved with a chiral lanthanum-lithium binaphthoxide complex **78** [24]. Reaction of **76** with ethyl methyl ketone (**77**) under catalysis by 20 mol % **78** gave **79** with an *ee* value of 94 % and in 71 % chemical yield.

The mechanism is similar to that of the barium-catalyzed direct aldol reaction (Scheme 16). The reaction commences with deprotonation of the ketone (**2**) by the Brønsted base unit of the catalyst under generation of the enolate **81**. After addition of the aldehyde **1** the Lewis acid-base adduct **82** is formed. Then the reaction of the aldehyde and the enolate occurs (**82** → **83**). After release of the aldol product **5** the catalyst **80** is recovered ready for the next cycle.

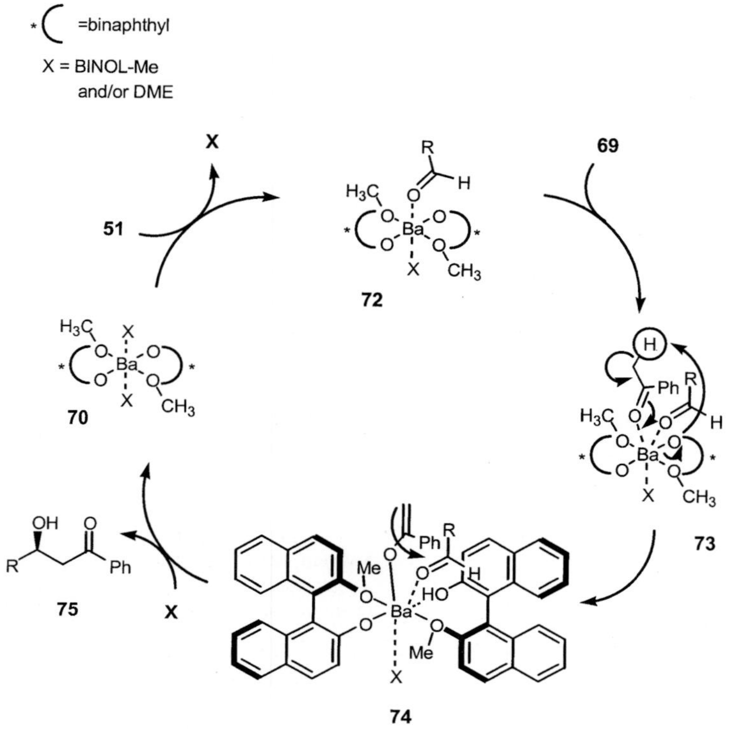

Scheme 14. Mechanism of the barium-catalyzed direct aldol reaction.

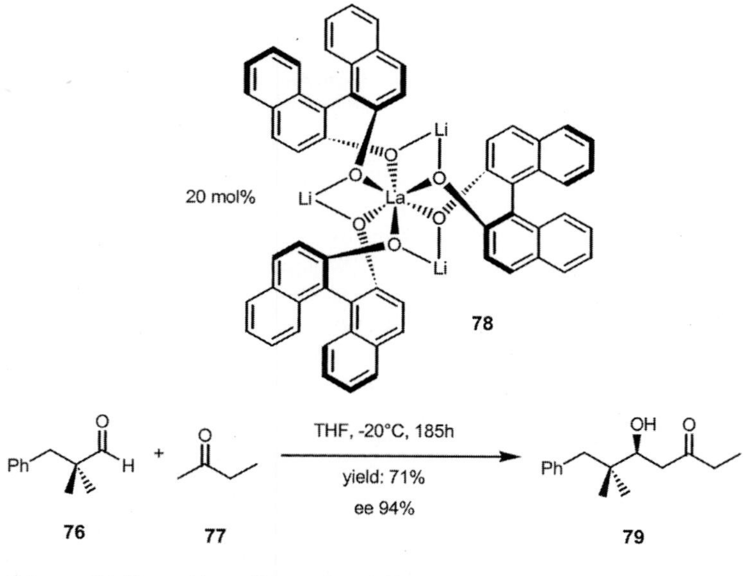

Scheme 15. Heterobimetallic-catalyzed direct asymmetric aldol reaction.

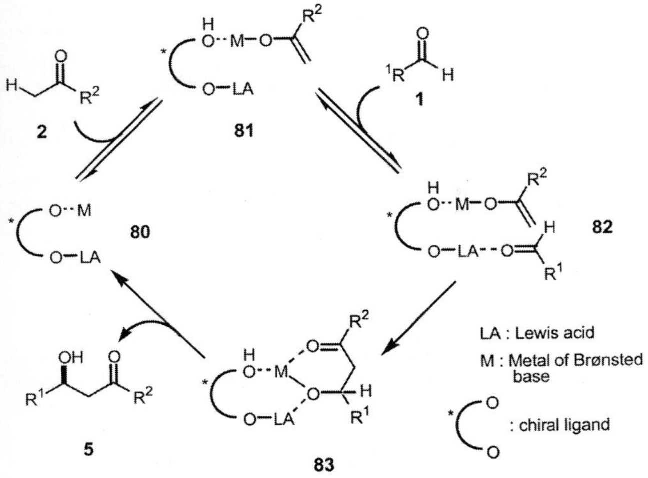

LA : Lewis acid
M : Metal of Brønsted base
: chiral ligand

Scheme 16. Catalytic cycle of the direct heterobimetallic asymmetric aldol reaction.

Direct methods mentioned above are generally only practicable with tertiary aldehydes. In the case of secondary and primary aldehydes *ee* values achieved are still too low. Nevertheless, direct methods are important supplementary alternatives to the Mukaiyama-type reactions.

Last but not least, the synthetic power of enzymes in this field is noteworthy [26]. Aldolases are chiral catalysts optimized during evolution, and are able to work in aqueous systems. Only a minimum amount of catalyst is necessary. One example is [26c] the reaction of **29** and **84** with D-fructose-1,6-bisphosphate aldolase **85** as catalyst with formation of **86** (Scheme 17).

Compared to nonbiological catalysts, enzymes often provide products of higher enantiomeric purity. The required amount of catalyst is much lower. Synthetic catalysts however, are often much more widely applicable; their substrate specifity is not as limited as those of their biological counterparts. With regard to their use in complex syntheses, both synthetic and biological catalysts will be indispensable in the future.

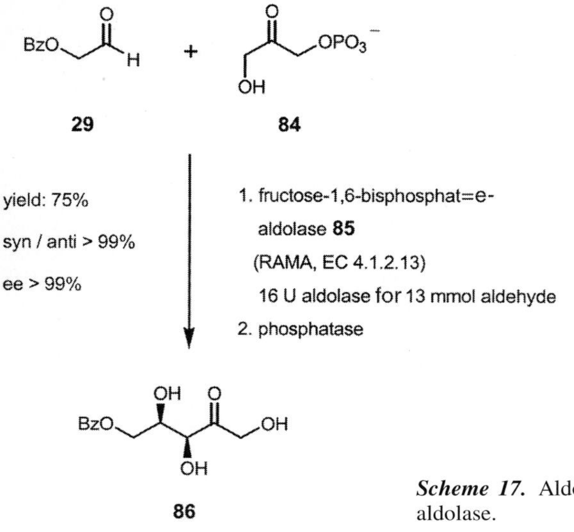

yield: 75%
syn / anti > 99%
ee > 99%

1. fructose-1,6-bisphosphat=e-aldolase **85**
(RAMA, EC 4.1.2.13)
16 U aldolase for 13 mmol aldehyde
2. phosphatase

Scheme 17. Aldol reaction catalyzed by fructose-1,6-bisphosphate-aldolase.

References

[1] *Methods of Organic Chemistry*, (Houben-Weyl) Additional and Suppl. Vol. of the 4th Edn., Vol E 21b, *Stereoselective Synthesis* (Eds.: G. Helmchen, R. W. Hoffmann, J. Mulzer, E. Schaumann), Stuttgart, **1995**.

[2] a) *Catalytic Asymmetric Synthesis* (Ed.: I. Ojima), VCH, Weinheim, **1993**; b) R. A. Johnson, K. B. Sharpless in a) P. 103; c) E. N. Jacobsen in a) p. 159; d) R. A. Johnson, K. B. Sharpless in a) p. 227; e) E. J. Corey, R. K. Bakshi, B. Shibata, *J. Org. Chem.* **1988**, *53*, 2861; f) H. Takaya, T. Ohta, R. Noyori in a) p. 1.

[3] a) M. Sawamura, Y. Ito in 2a) p. 367; b) K. Maruoka, H. Yamamoto in 2a) p. 413; c) P. Knochel, R. D. Singer, *Chem. Rev.* **1993**, *93*, 2117; d) T. Hayashi in 2a) . 325.

[4] For a comprehensive review of catalytic enantioselective aldol reactions: S. G Nelson, *Tetrahedron: Asymmetry* **1998**, *9*, 357.

[5] a) C. H. Heathcock in *Modern Synthetic Methods 1992* (Ed.: R. Scheffold) VHCA, Basel, **1992**, p. 1. b) K. Narasaka, *Synthesis* **1991**, 1; c) T. Bach, *Angew. Chem.* **1994**, *106*, 433; *Angew. Chem. Int. Ed. Engl.* **1994**, *33*, 417; d) T. K. Hollis, B. Bosnich, *J. Am. Chem. Soc.* **1995**, *117*, 4570.

[6] M. T. Reetz, S.-H. Kyung, C. Bolm, T. Zierke, *Chem. Ind.* **1986**, 824.

[7] S. Kobayashi, Y. Fujishita, T. Mukaiyama, *Chem. Lett.* **1990**, 1455.

[8] E. R. Parmee, Y. Hong, O. Tempkin, S. Masamune, *Tetrahedron Lett.* **1992**, *33*, 1729; E. R. Parmee, O. Tempkin, S. Masamune, A. Abiko, *J. Am. Chem. Soc.* **1991**, *113*, 9365.

[9] E. J. Corey, C. L. Cywin, T. D. Roper, *Tetrahedron Lett.* **1992**, *33*, 6907.

[10] a) K. Furuta, T. Maruyama, H. Yamamoto, *J. Am. Chem. Soc.* **1991**, *113*, 1041; b) K. Furuta, T. Maruyama, H. Yamamoto, *Synlett* **1991**, 439.

[11] a) K. Mikami, S. Matsukawa, *J. Am. Chem. Soc.* **1994**, *116*, 4077. b) G. E. Keck, D. Krishnamurthy, *J. Am. Chem. Soc.* **1995**, *117*, 2363.

[12] E. M. Carreira, R. A. Singer, W. Lee, *J. Am. Chem. Soc.* **1994**, *116*, 8837.

[13] M. Sodeoka, K. Ohrai, M. Shibasaki, *J. Org. Chem.* **1995**, *60*, 2648.

[14] M. Sodeoka and M. Shibasaki, *Pure Appl. Chem.* **1998**, *70*, 411.

[15] a) D. A. Evans, , J. A. Murry, C. S. Burgey, K. R. Campos, B. T. Connell, and R. J. Staples, *J. Am. Chem. Soc.* **1999**, *121*, 669; b) D. A. Evans, C. S. Burgey, M. C. Kozlowski, and S. T. Tregay, *J. Am. Chem. Soc.* **1999**, *121*, 686; c) D. A. Evans, T. Rovis, M. C. Kozlowski, and J. S. Tedrow, *J. Am. Chem. Soc.* **1999**, *121*, 1994; d) D. A. Evans, D. W. C. MacMillan, and K. R. Campos, *J. Am. Chem. Soc.* **1997**, *119*, 10859; e) D. A. Evans, M. C. Kozlowski, C. S. Burgey, and D. W. C. MacMillan, *J. Am. Chem. Soc.* **1997**, *119*, 7893; f) D. A. Evans, J. A. Murry, and M. C. Kozlowski, *J. Am. Chem. Soc.* **1996**, *118*, 5814 and references therein.

[16] S. E. Denmark, S. B. D. Winter, X. Su, and K.-T. Wong, *J. Am. Chem. Soc.* **1996**, *118*, 7404.

[17] D. J. Berrisford, C. Bolm, *Angew. Chem.* **1995**, *107*, 1862; *Angew. Chem. Int. Ed. Engl.* **1995**, *34*, 1717.

[18] K. Mikami, S. Matsukawa, *J. Am. Chem. Soc.* **1993**, *115*, 7039.

[19] E. M. Carreira, W. Lee, R. A. Singer, *J. Am. Chem. Soc.* **1995**, *117,* 3649.

[20] a) For a review, see: M. Shibasaki, H. Sasai, T. Arai, *Angew. Chem.* **1997**, *109*, 1290; *Angew. Chem. Int. Ed. Engl.* **1997**, *36*, 1236; b) H. Sasai, T. Suzuki, S. Arai, T. Arai, M. Shibasaki, *J. Am. Chem.Soc.* **1992**, *114,* 4418; c) K. Uotsu, H. Sasai, M. Shibasaki, *Tetrahedron: Asymmetry* **1995**, *6*, 71; d) Y. Ito, M. Sawamura, T. Hayashi, *J. Am. Chem. Soc.* **1986**, *108,* 6405; e) A. Togni, S. D. Pastor, *J. Org. Chem.* **1990**, *55*, 1649.

[21] T. Shioiri, A. Bohsako, and A. Ando, *Heterocycles* **1996**, *42*, 93.

[22] For a review on new catalytic concepts for the asymmetric aldol reaction, see: H. Gröger, E. M. Vogl, and M. Shibasaki, *Chem. Eur. J.* **1998**, *4*, 1137.

[23] Y. M. A. Yamada, and M. Shibasaki, *Tetrahedron Let.* **1998**, *39*, 5561

[24] N. Yoshikawa, Y. M. A. Yamada, J. Das, H. Sasai, and M. Shibasaki, *J. Am. Chem. Soc.* **1999**, *121*, 4168 and references therein.

[25] For reviews on heterobimetallic catalysts, see: a) Shibasaki et al. in footnote 20a; b) H. Steinhagen, G. Helmchen, *Angew. Chem.* **1996**, *108*, 2489; *Angew. Chem. Int. Ed. Engl.* **1996**, *35*, 2339; c) M. Shibasaki, T. Iida, Y. M. A. Yamada, *J. Synth. Org. Chem. Jpn.* **1998**, *56*, 344.

[26] a) W.-D. Fessner in 1) Chapter 1.3.4.6, p. 1736; b) H. Waldmann, *Nachrichten Chem. Techn. Lab.* **1991**, 1408; c) M. D. Bednarski, E. S. Simon, N. Bischofberger, W.-D. Fessner, M.J. Kim, W. Lees, T. Saito, H. Waldmann, G. M. Whitesides, *J. Am. Chem. Soc.* **1989**, *111*, 627; d) W.-D. Fessner, A. Schneider, H. Held, G. Sinerius, C. Walter, M. Hixon, J. V. Schloss, *Angew. Chem.* **1996**, *108*, 2366; *Angew. Chem. Int. Ed. Engl.* **1996**, *35*, 2219; e) G. Zhong, D. Shabad, B. List, J. Anderson, S. C. Sinha, R. A. Lerner, C. F. Barbas III, *Angew. Chem.* **1998**, *110*, 2609; *Angew. Chem. Int. Ed. Engl.* **1998**, *37*, 2481.

Binaphthyls: Universal Ligands for Catalysis

Tobias Wabnitz and Oliver Reiser

Institut für Organische Chemie, Universität Regensburg, Germany

In stereoselective synthesis, the use of chiral catalysts – consisting of metals and chiral ligands – is of central significance. While the periodic table imposes natural boundaries upon the number of available metals, the structural variety of the ligands is only limited by the fantasy of the research scientist. Thus, in 1992 there were already known more than 2000 different chiral ligands [1], and their number has been constantly growing ever since.

Why does a need for so many different ligands exist? Given a certain metal-catalyzed reaction, in general there exists only one suitable metal, and only in a limited number of cases a number of metals can be used as catalysts. Normally, the catalyst found works about equally well with the whole range of substrates of the reaction concerned. However, within a class of reactions brought about by the chiral ligands, the induced diastereo- or enantioselectivity is found to vary considerably with minor structural alterations of the substrates. Especially research groups that resort to asymmetric catalysis only occasionally and just as a means to an end have to limit their pool to a few chiral ligands, particularly since the latter need to be synthesized or acquired for non-negligible amounts of money.

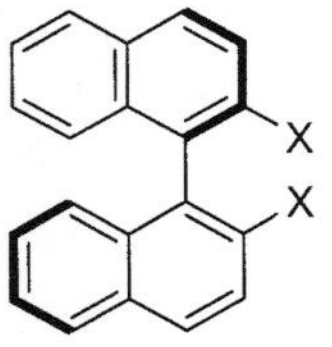

(*R*)-**1a**: X = OH

(*R*)-**1b**: X = PPh_2

(*R*)-**1c**: X = $OPPh_2$

(*R*)-**1d**: X = $P(O)Ph_2$

(*R*)-**1e**: X = NH_2

(*R*)-**1f**: X = $NHPPh_2$

(*R*)-**1g**: X = N=CHAr

(*R*)-**1h**: X = $AsPh_2$

Figure 1

It would be required of an „ideal" chiral ligand to form chiral catalysts with as many metals as possible, which in turn are able to catalyze a great number of reactions and allow of broad substrate variability without loss of selectivity. There are no such universal ligands yet, but binaphthyls **1** come close to this ideal. The C_2 axially-symmetric biaryl frame is an excellent transmitter of chiral information, and the possibility of varying the coordination sites X (e. g. **1a–h**) [2] enables the use of a wide range of metals.

Synthesis and Structure of Binaphthyl Compounds

The most frequently applied ligands are 2,2'-binaphthol (BINOL **1a**) and 2,2'-bis(diphenylphosphino)binaphthyl (BINAP **1b**). Both antipodes of these compounds are commercially available in enantiopure form, though not at little cost [3]. It is rewarding, therefore, to become acquainted with the syntheses of these compounds, which have been described in detail [4] and have been simplified substantially on the basis of recent publications [5]. The iron(III)-catalyzed dimerization of 2-naphthol (**2**) to give racemic BINOL (*rac*)-**1a** succeeds smoothly and on a large scale (Scheme 1). Its resolution can be achieved easily with *N*-benzylcinchoidinium chloride (**3**) and yields both (*R*)- and (*S*)-**1a** in high enantiomeric excesses. After conversion into the ditriflate **4**, enantiopure **1a** can be coupled with diphenylphosphine (or, in lower yield, with cheaper chlorodiphenylphosphine) in a nickel-catalyzed

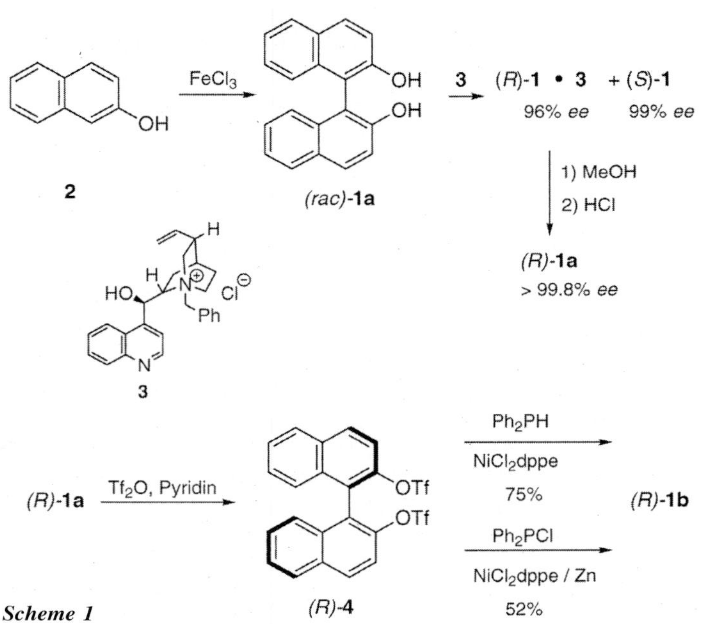

2 $\xrightarrow{FeCl_3}$ (*rac*)-**1a** $\xrightarrow{3}$ (*R*)-**1** • **3** + (*S*)-**1**

96% *ee* 99% *ee*

1) MeOH

2) HCl

(*R*)-**1a**

> 99.8% *ee*

HO, H, N⊕, Cl⊖, Ph, N, H

3

(*R*)-**1a** $\xrightarrow{Tf_2O, \text{Pyridin}}$ (*R*)-**4** (OTf, OTf)

Ph_2PH, $NiCl_2dppe$, 75% → (*R*)-**1b**

Ph_2PCl, $NiCl_2dppe$ / Zn, 52% → (*R*)-**1b**

Scheme 1

reaction which directly affords **1b**. Hence both the formerly necessary, difficult bromination of (*rac*)-**1a** and the costly optical resolution at a later stage of the previously employed synthetic procedure [4] can be dispensed with.

What is the distinction of the binaphthyls **1a** and **1b** as outstanding ligands for catalysis? On the basis of their pliant structure due to the possibility of rotation about the biaryl axis, a great number of differently sized metal ions can coordinate without significant increase in torsional strain of the resulting complexes. The oxophilic early main-group (e. g. B, Al, Sn) and early transition metals (e. g. Ti, Zr) as well as the lanthanoids (e.g. La, Yb) are especially suited to coordinate BINOL **1a,** whereas **1b** is generally used to complex transition metals of the Group VIII (e.g. Pd, Rh, Ru). Binaphthyl complexes can normally be prepared by simply mixing the ligand and a metal salt. However, it is not easy to understand the principles upon which these complexes perform as such excellent chiral multiplicators. Difficulties arise since the exact structures of the catalytically active species are unknown in most cases and the depiction of the complexes as [metal-ligand] monomers often represents an unrealistic simplification.

In every case the coordination of a binaphthyl-type ligand by a metal atom results in the formation of a seven-membered chelate ring. Due to its backbone of sp2-hybridized carbon atoms, this ring is rigid in conformation and skewed in an unambiguous way, as determined by the chiral binaphthyl unit. In these complexes, the ligand transfers its chirality effectively to the central metal and to other ligands or substrates coordinated by the same center. For example octahedral complexes containing two additional bidentate ligands as in *(S)*-**5** always form only one diastereomer, which exhibits Λ-configuration.

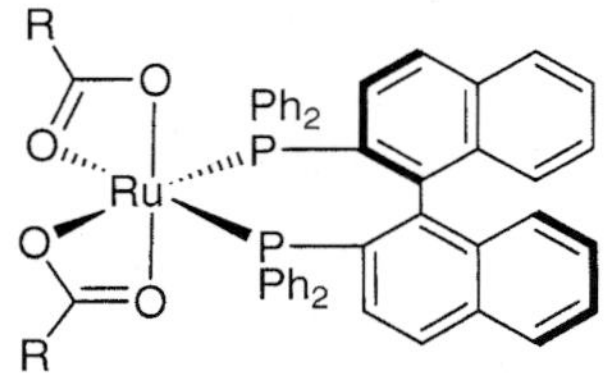

R, O, O, Ru, Ph_2, P, P, Ph_2, O, O, R

(*S*)-**5**

Figure 2

BINAP Catalysts

The Ru(II) complex **5** has proved to be an extremely versatile catalyst in asymmetric reductions of C-C and C-O double-bonds (Scheme 2) [6]. Suitable substrates generally seem to be those containing a hetero atom for coordination, thus a chelate ring accommodating the ruthenium atom and the moiety that is to be reduced can be formed. The reduction of α,β- and β,γ-unsaturated carboxylic acids to their saturated, optically active analog can be achieved in this way. Especially amino acids can be prepared in high optical purities, e. g. the reduction of **9** affords the phenylalanine derivative **6** with a selectivity of up to 92 % *ee*. The hydrogenation of geraniol (**11**) with *(R)*-**5** distinctly demonstrates the necessity of pre-coordination via two functional groups of the substrate: the exclusive product citronellol (**8**) is obtained in excellent enantiomeric excess and the double bond between C6 and C7 remains completely intact. Allyl- and homoallyl alcohols generally can be reduced using *(R)*-**5**, but if another methylene group is inserted, the distance between the double bond and the hydroxyl group becomes too large and no hydrogenation occurs.

The reduction of β-ketoesters to aldols is one of the most important applications of Ru(II)-BINAP catalysts [7]. As a special bonus, the chirally labile C2 stereogenic center can be exploited in a dynamic kinetic resolution such that racemic reactants yield only one of the four conceivable stereoisomers in high diastereomeric and enantiomeric excess. This strategy has been extended to the reduction of β-ketophosphonates **10**. The 3-hydroxyphosphonic acids **7** which are accessible by this route constitute promising starting materials for the synthesis of peptide analog and antibiotics [8].

Whereas **5** only reacts with bifunctional substrates, a new modification enables the catalytic hydrogenation of simple ketones **13**. The ternary catalyst consisting of [Ru(II)-BINAP] and a chiral 1,2-diamine in an alcoholic solution of KOH exhibits an activity more than 1000 times that of [Ru(II)-BINAP] alone. The products have been obtained with optical purities greater than 99 % *ee* by choosing a suitable chiral diamine

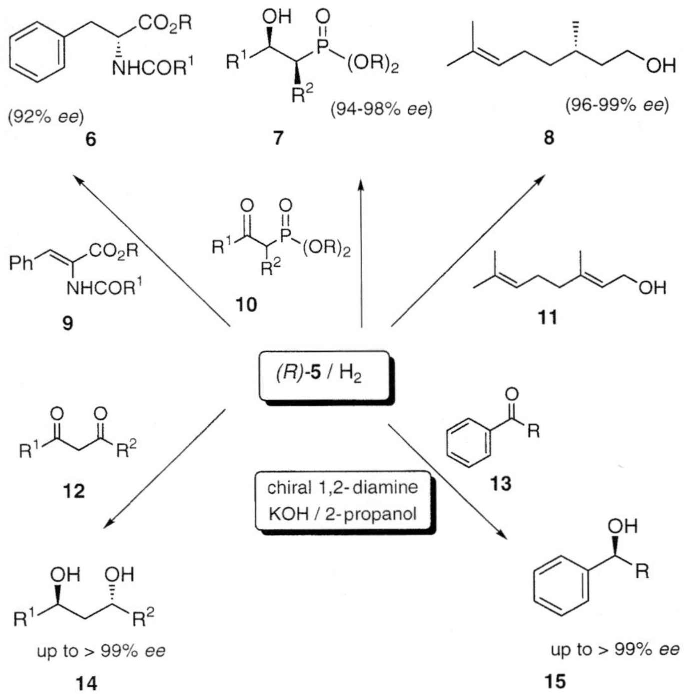

(92% *ee*) **6**
(94-98% *ee*) **7**
(96-99% *ee*) **8**
9 **10** **11**
(R)-**5** / H_2
12 **13**
chiral 1,2-diamine
KOH / 2-propanol
up to > 99% *ee* **14**
up to > 99% *ee* **15**

Scheme 2

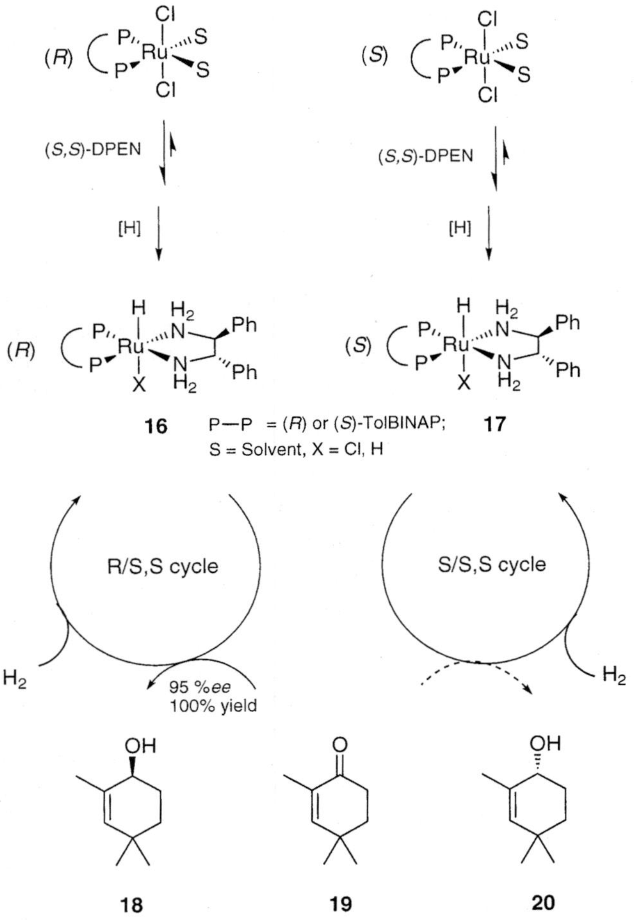

(R) (S)
(S,S)-DPEN (S,S)-DPEN
[H] [H]
(R) (S)
16 P—P = (R) or (S)-TolBINAP; 17
S = Solvent, X = Cl, H
R/S,S cycle S/S,S cycle
H_2 H_2
95 %ee
100% yield
18 19 20

Scheme 3

in order to increase the enantioselectivity in a "*matched case*" situation. Probably the most exciting recent discovery has been that as well as hydrogen, isopropanol can be used as the reducing agent [9].

The concept of asymmetric activation [10] has been transferred successfully to [Ru(II)-BINAP] catalysts and chiral diamine auxiliaries (Scheme 3). *α,β*-Unsaturated ketones such as **19** have been hydrogenated in 100 % yield and 95 % *ee* using Ru(II) complexes comprising optically pure *(S,S)*-diphenylethylenediamine and racemic BINAP analog s [11]. This surprising result can be explained by the formation of *diastereomeric* complexes [Ru(BINAP)(diamine)] **16** and **17** exhibiting *R/S,S*- or *S/S,S*-configuration with respect to the BINAP and diamine ligand. Under the hydrogenation conditions employed, only one of these coexisting catalysts was significantly activated by the diamine; hence high selectivities are attainable.

A highly interesting variation of this concept is the application of achiral, conformatively flexible 2,2–bis(diphenylphosphino)biphenyl-type ligands (BIPHEP) instead of BINAP. Here, the configuration of BIPHEP only becomes fixed when the ligand is incorporated into [Ru(BIPHEP)(diamine)] complexes **21** or **22**. In the reduction of 1-acetonaphthone, selectivities up to 92 % *ee* have been attained exploiting the different catalytic activities of the coexisting diastereomeric complexes thus formed [12].

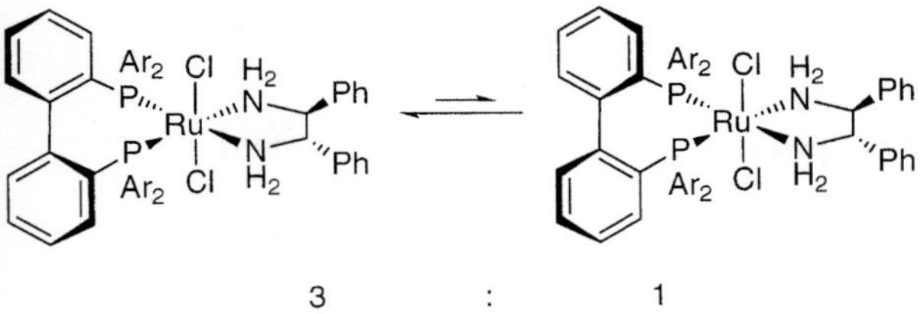

Scheme 4

Chirality Transfer in Metal-BINAP Complexes

The high asymmetric inductions that can be attained in these reductions are above all to be attributed to steric interactions of the substrate with the phosphorus-bound phenyl groups of the [Ru-(*R*)-BINAP] complex (*R*)-**5** (Scheme 5). It can be assumed that hydride, being the smallest ligand, coordinates *cis* to both phosphine ligands. The P-bound phenyl groups occupy equatorial (parallel to the substrate coordination plane) and axial positions (perpendicular to the substrate coordination plane) with respect to the metal-ligand cycle. The skewed BINAP skeleton requires the equatorial phenyl substituents to protrude directly into two of four coordination quadrants that are accessible to the substrate.

The complexation of the substrate occurs in the way that entails the minimal steric interactions with these occupied quadrants. Compared to other C_2-symmetric biphosphines, the differentiation of the four quadrants is particularly distinct with BINAP due to the bond angles and the rigidity of the composite system. Therefore, a high chiral recognition of prochiral substrates can be achieved (Fig. 3).

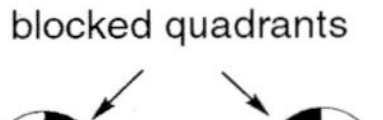

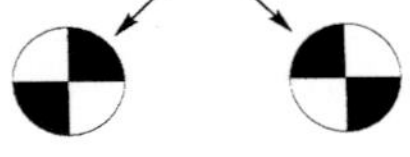

Figure 3

Apart from a few exceptions [13], the quadrant model can be employed to predict the asymmetric induction of BINAP complexes. For example, the model can be referred to when considering the reduction of *β*-dicarbonyl compounds (Scheme 5). In **24**, repulsion between the equatorial phenyl moiety and R^1 occurs when the substrate adapts the orientation where the keto function is complexed coparallel to the Ru–H bond, which is

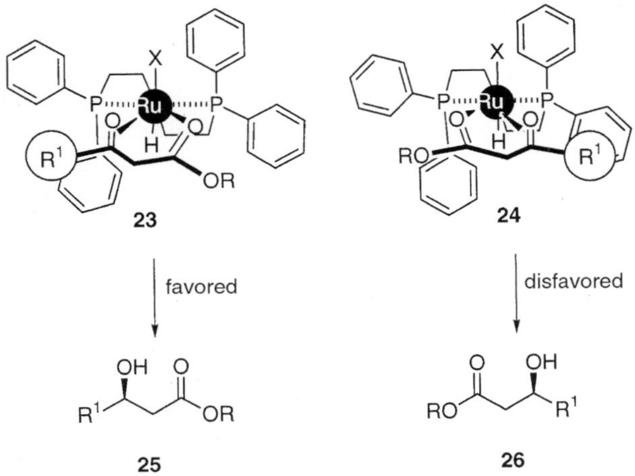

Scheme 5

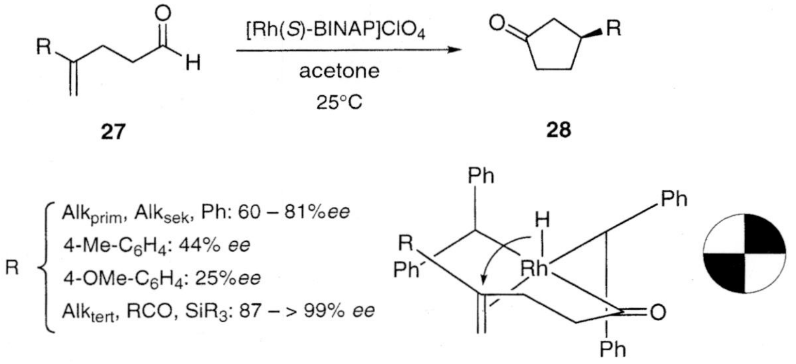

[Rh(*S*)-BINAP]ClO_4, acetone, 25°C: **27** → **28**

R: Alk$_{prim}$, Alk$_{sek}$, Ph: 60 – 81% *ee*; 4-Me-C_6H_4: 44% *ee*; 4-OMe-C_6H_4: 25% *ee*; Alk$_{tert}$, RCO, SiR_3: 87 – > 99% *ee*

Scheme 6

necessary for the transfer of hydrogen. On the other hand, the coordination of the ester group in **23** is realized via a lone electron pair of the carbonyl group so that the substituent OR does not experience steric hindrance by the phenyl moiety of the BINAP ligand. If the priority is defined as O > large substituent (corresponds to R^1 in this case) > small substituent, the carbonyl group is always attacked from the *re* face when using *(R)*-**5**. The relative stereochemistry between C-2 and C-3 is determined by the structure of a substituent attached to C-2. It can be memorized as a rule of thumb that acyclic β-ketoesters are reduced to *syn*-aldols and cyclic β-ketoesters are reduced to *anti*-aldols.

This model can also be employed successfully for explaining why the double reduction of 1,3-diketones **12** with **5** affords *anti*-configured 1,3-diols in high diastereomeric and enantiomeric excess [14].

The quadrant model which has been described for [Ru-BINAP] can normally be applied analogously with [Pd-BINAP] [15] and [Rh-BINAP] complexes. A new example for the latter is the hydroacylation of 4-substituted 4-pentenal **27** to 3-substituted cyclopentenones **28** using the [Rh-BINAP]ClO_4 catalyst (Scheme 6). However, the strong dependence of the selectivity upon the nature of the substituents R suggests that stereoelectronic factors have to be taken into account as well [16].

Binaphthyls in Cyclic Transition States

Many reactions for forming C–C bonds such as the aldol reaction, the carbonyl-ene reaction or the allylation of aldehydes require the participation of a Lewis acid and proceed through a chair-like transition state in accordance with the Zimmermann-Traxler model. This model can explain the relative configuration of newly formed stereogenic centers, but in the absence of chiral information there can always exist two enantiomeric chair-like structures which govern the absolute product configuration. These transition state structures can be differentiated effectively if binaphthol **1a** participates in the reaction, e.g. via complexation of a Lewis acid such as titanium, aluminum or boron which coordinates the ligand tetrahedrally. The following model can be used for predicting the asymmetric induction if the metal is fitted *directly* into the chair-like transition state. If analyzing the metal atom that bears the *(S)*-**1a** ligand from its front face, it can be seen that both the axial and the equatorial position of the neighboring atom on the *left-hand* side experience strong steric interactions by the binaphthyl skeleton. Less strong, but nevertheless perceptible, steric hindrance is experienced by the equatorial position of the neighboring atom on the *right-hand* side. The axial position of the latter atom is the least hindered site, but it is subjected, of course, to the 1,3-diaxial interactions of the chair configuration.

The reduction of ketones **29**, where S signifies small substituent and L signifies large substituent, by stoichiometric amounts of *(S)*-BINAL-H **30**

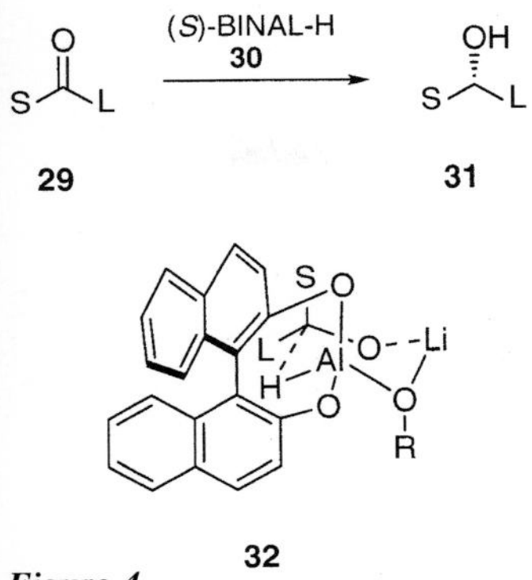

Figure 4

demonstrates that this model can be referred to successfully for prediction of the asymmetric induction [17]. In the complex **32**, the small substituent H is oriented towards the left, and the larger OR appendage to the right of the Al atom. The small and the large substituents of the ketone are positioned axially and equatorially, respectively. It must be noted that the definition of small and large cannot be based solely on steric arguments.

On the contrary, the arrangement is dominated by electronic effects because of π-π-interactions between the axial O–Al and the substituents of the ketone. Repulsive interactions that occur with π-systems of unsaturated groups such as phenyl, alkenyl or alkynyl make these groups large substituents. The presence of electronic effects becomes obvious with the formation of **34** and **36**. Only in the case of very strong π-acceptors such as 4–cyclopent-1-ene-3-one the repulsive (nπ) interaction is surpassed by the attractive (nπ^*) interaction. Hence the formation of **37** occurs with the opposite absolute configuration to that of **33**–**36** (Scheme 7).

If the metal-binaphthyl complex is *not* fitted *directly* into the cyclic transition state, it becomes difficult to explain the asymmetric inductions observed. The following rule seems to be generally valid for both BINOL and BINAP complexes: *The complexation of carbonyl or imine moieties by (R)-binaphthyl-metal complexes results in a shielding of the si face, the reaction proceeds from the re face.* Correspondingly, the opposite principle applies when *(S)*-binaphthyl complexes are used. All aldol reactions and carbonyl-ene reactions which are catalyzed by binaphthyl complexes abide by this rule [18], and the scheme can also be applied to the addition of ketene-silyl-acetals to imines with boron-BINOL catalysts [19].

Diels-Alder reactions have been very successfully subjected to asymmetric catalysis by binaphthyl complexes. Accordingly, the synthesis of tetrahydropyranes **41** and **42** can be realized by reaction of glyoxylic esters **39** with methoxydienes **38** (Scheme 8) [20]. Some of these reactions take place with excellent *endo*-control and

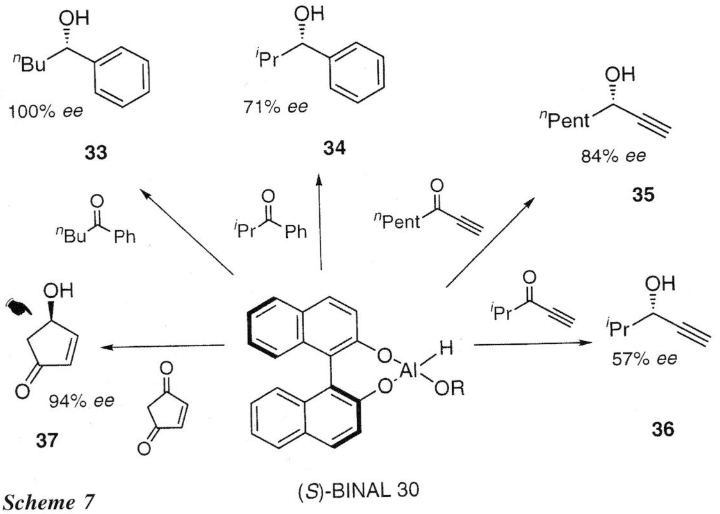

Scheme 7

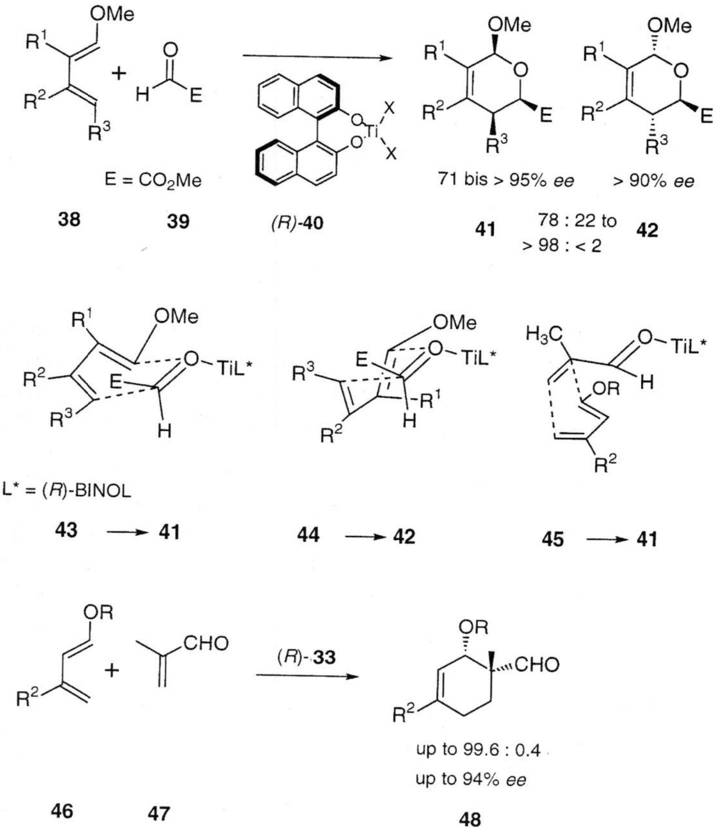

Scheme 8

enantioselectivity. In complete analogy to the aldol and ene reactions mentioned earlier, catalysis with *(R)*-**40** yields both the *endo*-product **41** via **43** and the *exo*-product **42** via **44**, in each case the aldehyde **39** is attacked from the *re* face. In addition to its ability of shielding the *si* face of the carbonyl group, the *(R)*-BINOL complex *(R)*-**40** is also capable of shielding a vicinal double bond from the *same* face. Methacrolein **47** is expected to adapt *s-trans*-configuration and *anti*-complexation by the titanium catalyst before reacting with the diene **46**. In fact, the cyclohexene **48** is obtained with high selectivity. This can be explained on the basis of the transition state **45**. The selectivities of the cyloadditions between cyclopentadiene and 2–bromoacrolein with boron-BINOL complexes as catalysts, which have been described by Yamamoto, can also be understood using this model [21].

Further Applications of Binaphthyls in Homogenous Catalysis

The new binaphthyl ligands **49** [22] and **50** [23] have led to a breakthrough in the copper catalyzed 1,4-addition of organozinc reagents to α,β-unsaturated ketones, a reaction which has proved to be difficult if carried out enantioselectively. Selectivities up to 96 % *ee* have been reached with cyclic substrates, while acyclic substrates still leave room for further improvement. **50** has also been successfully employed in palladium-catalyzed alkylations of allyl acetates [24].

Finally, a fascinating development in the field of lanthanum-BINOL complexes remains to be mentioned [25]. These compounds so far have proved to catalyze enantioselectively hydrophosphonylations of imines [26], nitroaldol reactions [27], Michael additions [28] and epoxidations of

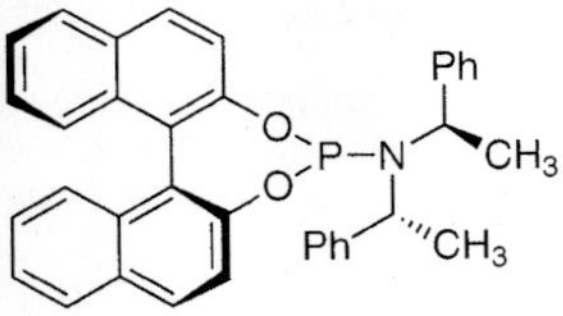

49 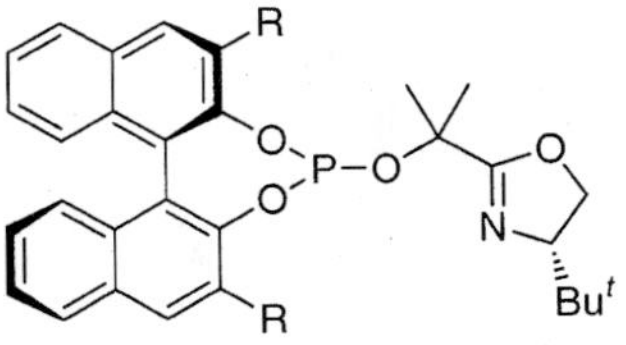 **50**

Scheme 9

enones [29]. In every case, the direction of asymmetric induction is identical to the titanium-catalyzed aldol and Diels-Alder reactions: if *(R)*-BINOL is used as ligand, the addition to the imine, to the aldehyde or to the α,β-unsaturated carbonyl compound always occurs from the *re* face.

The hypothesis that the structure of these catalysts is not always in accordance with a monomeric structure has been expressed for both the titanium-BINOL complexes [30] and the corresponding lanthanum species. In the latter case, this hypothesis has been corroborated impressively. The complexes, which are prepared in situ from alkali-BINOL compounds and lanthanum alkoxides, are in fact bimetallic compounds of the type La-(alkali)$_3$[BINOL]$_3$. The complexation of the lanthanum atom by the three BINOL ligands leads exclusively to the formation of the Δ-diastereomer. Every pair of BINOL ligands is linked via a bridging alkali metal atom, which coordinates the BINOL oxygen atoms. The alkali atoms play an important role in catalytic activity and selectivity. For example, it is assumed that in Michael addition the sodium atom coordinates the

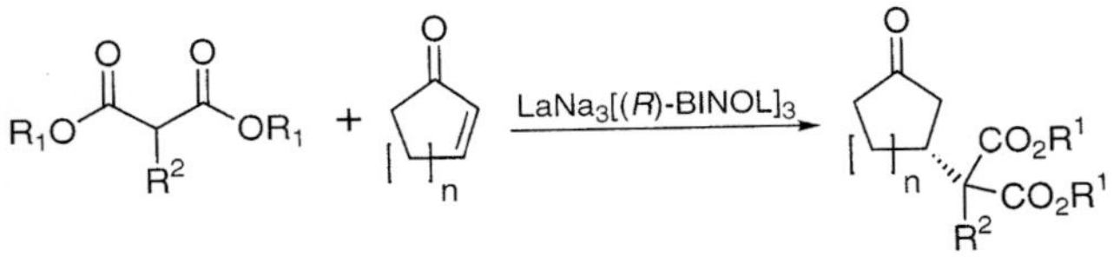

LaNa$_3$[(*R*)-BINOL]$_3$

up to 92% *ee*

42 **43** **44**

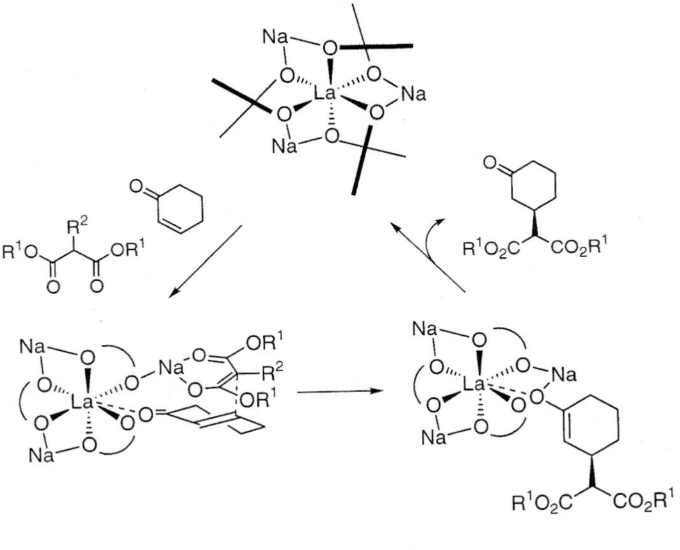

Scheme 10

nucleophile and the central lanthanum atom coordinates the Michael acceptor (Scheme 10). The positions of both substrates are unambiguously determined by the screw-like structure of the complex; therefore, the reactions can proceed with high selectivity. A marvelous, multicolored graphical illustration of these facts can be seen in the original paper [28].

These complexes are the first examples of multifunctional catalysts and demonstrate impressively the opportunities that can reside with the as yet hardly investigated bimetallic catalysis. The concept described here is not limited to lanthanides but has been further extended to main group metals such as gallium [31] or aluminum [32]. In addition, this work should be an incentive for the investigation of other metal-binaphthyl complexes to find out whether polynuclear species play a role in catalytic processes there as well. For example, the preparation of titanium-BINOL complexes takes place in the presence of alkali metals [molecular sieve (!)]. A leading contribution in this direction has been made by Kaufmann et al. as early as 1990 [33]. It was proven that the reaction of *(S)*-**1a** with monobromoborane dimethyl sulfide leads exclusively to a binuclear, propeller-like borate compound. This compound was found to catalyze the Diels-Alder reaction of cyclopentadiene and methacrolein with excellent *exo*-stereoselectivity and enantioselectivity in accordance with the empirical rule for carbonyl compounds which has been presented earlier.

Acknowledgement: The authors would like to acknowledge support from the Fonds der Chemischen Industrie and the Studienstiftung des Deutschen Volkes (fellowship T.W.)

References

[1] H. Brunner, W. Zettlmeier, *Handbook of Enantioselective Catalysis with Transition metal compounds - Ligands - References*, VCH, Weinheim, **1993,** 359.

[2] *Recent developments*: **1a**: F.-Y. Zhang, C.-C. Pai, A. S. C. Chan, *J. Am. Chem. Soc.* **1998**, *120*, 5808. **1b**: H. Suga, T. Fudo, T. Ibata, *Synlett* **1998**, 933. **1c**: A. Kojima, C. D. J. Boden, M. Shibasaki, *Tetrahedron Lett.* **1997**, *38*, 3459.

[3] Approx. prices per mmol - BINOL: 35 Euro, BINAP: 300 Euro.

[4] a) J. Jaques, C. Fouquey in *Organic Synthesis* (ed.: B. E. Smart), Organic Synthesis, Inc., **1988,** *67,* 1; b) H. Takaya, S. Akutagawa, R. Noyori, *ibid.* **1988,** *67,* 20; c) L. K. Truesdale, *ibid.* **1988,** *67,* 13.

[5] a) D. Cai, B. L. Huges, T. R. Verhoeven, P. J. Reider, *Tetrahedron Lett.* **1995,** *44,* 7991; b) D. Cai, J. F. Payack, D. R. Bender, D. L. Hughes, T. R. Verhoeven, P. J. Reider, *J. Org. Chem* **1994,** *59,* 7180; c) D. J. Ager, M. B. East, M. Eisenstadt, S. A. Laneman, *Chem. Commun.* **1997,** 2359.

[6] R. Noyori, H. Takaya, *Acc. Chem. Res.* **1990,** *23,* 345.

[7] R. Noyori, M. Tokunaga, M. Kitamura, *Bull. Chem. Soc. Jpn.* **1995,** *68,* 36.

[8] K. M. Tokunaga, R. Noyori, *J. Am. Chem. Soc.* **1995,** *117,* 2675.

[9] a) T. Ohkuma, H. Ooka, S. Hashigushi, T. Ikariya, R. Noyori, *J. Am. Chem. Soc.* **1995,** *117,* 2675; b) T. Ohkuma, H. Ooka, M. Yamakawa, T. Ikariya, R. Noyori, *J. Org. Chem.* **1996,** *61,* 4872; c) K.-J. Haack, S. Hashiguchi, A. Fujii, T. Ikariya, R. Noyori, *Angew. Chem. Int. Ed.* **1997**, *36*, 285; d) T. Okuma, M. Koizumi, H. Doucet, T. Pham, M. Kozawa, K. Kunihiko, Y. Eijii, T. Ikariya, R. Noyori, *J. Am. Chem. Soc.* **1998**, *120*, 13529.

[10] a) J. W. Faller, J. Parr, *J. Am. Chem. Soc.* **1993,** *115,* 804; b) K. Mikami, S. Matukawa, *Nature* **1997,** *385,* 613.

[11] T. Ohkuma, H. Doucet, T. Pham, K. Mikami, T. Korenaga, M. Terada, R. Noyori, *J. Am. Chem. Soc.* **1998,** *120,* 1086.

[12] K. Mikami, T. Korenaga, M. Terada, T. Ohkuma, T. Pham, R. Noyori, *Angew. Chem. Int. Ed.* **1999,** *38,* 495.

[13] A. Miyshita, A. Yasuda, T. Souchi, R. Noyori, *Tetrahedron* **1984,** *117,* 2931.

[14] M. Kitamura, T. Ohkuma, S. Inoue, N. Sayo, H. Kumobayashi, S. Akutagawa, T. Ohta, H. Takaya, R. Noyori, *J. Am. Chem. Soc.* **1988,** *110,* 629.

[15] a) T. Hayashi in *Organic Synthesis in Japan - Past, Present, and Future* (ed.: R. Noyori), Tokyo Kagaku Dozin, Tokyo, **1992,** 105; b) K. Ohrai, K. Kondo, M. Sodeoka, M. Shibasaki, *J. Am. Chem. Soc.* **1994,** *116,* 11737.

[16] R. W. Barnhart, X. Wang, P. Noheda, S. H. Bergens, J. Whelan, B. Bosnich, *J. Am. Chem. Soc.* **1994,** *116,* 1821.

[17] a) R. Noyori, I. Tomino, M. Yamada, M. Nishizawa, *J. Am. Chem. Soc.* **1984,** *106,* 6709; b) R. Noyori, I. Tomino, M. Yamada, M. Nishizawa, *ibid.* **1984,** *106,* 6717.

[18] U. Koert, *Nachr. Chem. Tech. Lab.* **1995,** *43,* 1068.

[19] K. Ishihara, Y. Kuroki, H. Yamamoto, *Synlett* **1995,** 41.

[20] K. Mikami, Y. Motoyama, M. Terada, *J. Am. Chem. Soc.* **1994,** *116,* 2812.
[21] K. Ishihara, H. Yamamoto, *J. Am. Chem. Soc.* **1994,** *116,* 1561.
[22] B. L. Feringa, M. Pineschi, L. A. Arnold, R. Imbos, A. H. M. de Vries, *Angew. Chem. Int. Ed.* **1997**, *36*, 2620.
[23] A. K. H. Knöbel, I. H. Escher, A. Pfaltz, *Synlett* **1997**, 1429.
[24] R. Pretot, A. Pfaltz, Angew. Chem. Int. Ed. **1998**, *37*, 323.
[25] M. Shibasaki, H. Sasai, T. Arai, *Angew. Chem. Int. Ed.* **1997,** *36,* 1237.
[26] H. Sasai, S. Arai, Y. Tahara, M. Shibasaki, *J. Org. Chem.* **1995,** *60,* 6656.
[27] a) H. Sasai, T. Tokunaga, S. Watanabe, T. Suzuki, N. Itoh, M. Shibasaki, *J. Org. Chem.* **1995,** *60,* 7388; b) H. Sasai, T. Suzuki, S. Arai, M. Shibasaki, *J. Am. Chem. Soc.* **1992,** *114,* 4418.
[28] a) H. Sasai, T. Arai, Y. Satow, K. N. Houk, M. Shibasaki, *J. Am. Chem. Soc.* **1995,** *117,* 6194; b) M. Shibasaki, H. Sasai, T. Arai, T. Iida, *Pure Appl. Chem.* **1998**, *70*, 1027.
[29] M. Bougachi, S. Watanabe, T. Arai, H: Sasai, M. Shibasaki, *J. Am. Chem. Soc.* **1997,** *119,* 2329.
[30] G. E. Keck, D. Krishnamurthy, *J. Am. Chem. Soc.* **1995,** *117,* 2363.
[31] a) E. M. Vogl, S. Matsunaga, M. Kanai, T. Iida, M. Shibasaki, *Tetrahedron Lett.* **1998**, *39*, 7917. b) T. Iida, N. Yamamoto, S. Matsunaga, H.-G. Woo, M. Shibasaki, *Angew, Chem. Int. Ed.* **1998**, *37*, 2223.
[32] a) T. Arai, H. Sasai, K. Yamaguchi, M. Shibasaki, *J. Am. Chem. Soc.* **1998**, *120*, 441; b) S. Yamasaki, T. Iida, M. Shibasaki, *Tetrahedron Lett.* **1999**, *40*, 307.

Fluorotitanium Compounds – Novel Catalysts for the Addition of Nucleophiles to Aldehydes

Rudolf O. Duthaler

Pharma Research, Novartis Pharma AG, Basel Switzerland

Andreas Hafner

Consumer Care Division CIBA SPECIALTY CHEMICALS, Basel, Switzerland

Transformations involving chiral catalysts most efficiently lead to optically active products. The degree of enantioselectivity rather than the efficiency of the catalytic cycle has up to now been in the center of interest. Compared to hydrogenations, catalytic oxidations or C–C bond formations are much more complex processes and still under development. In the case of catalytic additions of dialkyl zinc compounds[1], allylstannanes [2], allyl silanes [3] , and silyl enolethers [4] to aldehydes, the degree of asymmetric induction is less of a problem than the turnover number and substrate tolerance. Chiral Lewis acids for the enantioselective Mukaiyama reaction have been known for some time [4a – 4c], and recently the binaphthol-titanium complexes **1** [2c – 2e, 2j] and **2** [2b, 2i] have been found to catalyze the addition of allyl stannanes to aldehydes quite efficiently. It has been reported recently that a more active catalyst results upon addition of $Me_3SiS(i\text{-}Pr)$ [2k] or $Et_2BS(i\text{-}Pr)$ [2l, 2m] to binaphthol-Ti(IV) preparations.

Most desirable, however, would be chiral catalysts for the addition of the more readily available and less toxic allyl silanes, but so far the efforts towards an enantioselective variant of the Sakurai-Hosomi reaction have been less successful. Some time ago Ketter and Herrmann [3a] obtained 24 % *ee* for the addition of allyl silane to aldehydes catalyzed by the dichlorotitanate **1**.

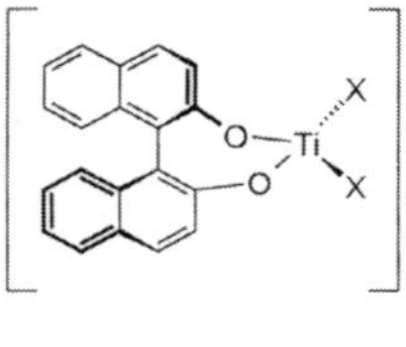

1	X = i-PrO	Ti-O-Ti bridged oligomer
2	X = Cl	
3	X = F	Ti-F-Ti bridged oligomer?

Better results (80 % *ee*) have been reported by Mikami, Nakai and co-workers [3c] for the addition of crotyl silane also catalyzed by complex **1**. Yamamoto and co-workers [3b] used chiral acyloxy boranes to catalyze the Sakurai-Hosomi-reaction. While an excellent 96 % *ee* was obtained for the addition of 2,3–disubstituted allyl groups, the conversion with parent allyl silane was low (46 %) and the asymmetric induction mediocre (55 % ee). Gauthier and Carreira [5] then made a big leap forward by using the difluorotitanium-binaphthol complex **3**. The catalyst **3** is prepared in situ via the TiF_4-binaphthol adduct **4** and formal HF elimination mediated by allyl silane **5**. The addition of **5** to aldehydes **6** (→ **7**) catalyzed by 10 % of **3** proceeds with 61–94 % *ee* and good yields (69–93 %), the best results being observed for aldehydes with tertiary alkyl residues (Scheme 1).

The prerequisites for enantioselective catalysis of the Mukaiyama reaction with chiral metal complexes have been thoroughly analyzed by Hollis and Bosnich [4a] (see also [4b,c]). They are depicted in Scheme 2 for the Sakurai-Hosomi reaction with a titanium catalyst: (1) the Lewis acid should form a complex **8** with the aldehyde, rather than being transformed to an allyltitanium compound by reaction with allyl silane **5**. (2) The Ti–O bond of intermediate **9** should be comparatively weak, thus enabling direct conversion to the *O*-silylated adduct **7** with regeneration of catalyst **10** according to the variant **"A"**. In the case of pathway **"B"** a titanated product **11** is formed together with Me_3SiX **12**. For $X = CF_3SO_3$ **12** is an excellent catalyst for the allylsilylation, proceeding to racemic product **7**. In most cases catalysis by **12** is indeed much faster than the me-

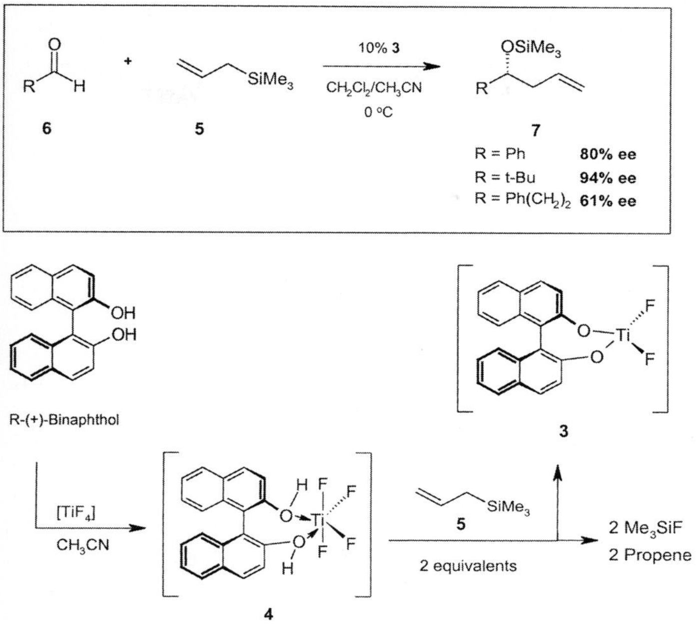

Scheme 1

tathesis of **11** and **12** to product **7** and chiral catalyst **10**. We have been able to document such a metathesis by treating the primary product **13** of the allyltitanation of benzaldehyde by **14** either with Me_3SiBr or Me_3SiI in toluene, whereby the silylether **15** and the titanium halogenide **16** are formed. In a stepwise manner **16** (or chloride **17**) can be converted back to the allyltitanium reagent **14** by reaction with allyl-Grignard, allowing the one-pot conversion of a further batch of benzaldehyde [6]. Interesting in this context are recent reports on the Mukaiyama reaction catalyzed by copper complexes. By working in hydroxylic solvents the achiral pathway can be suppressed, as any trimethylsilyl triflate is hydrolyzed [4d]. Carreira and co-workers, on the other hand, postulate formation of a Cu enolate in a copper-catalyzed addition of a silyl dienolate. The catalytic cycle would have to include silylation of the Cu-aldolate by the silylenol ether, thereby regenerating the Cu-enolate [4e, 4f].

As put forward by Gauthier and Carreira [5], the special nature of the titanium fluoride bond is decisive for the catalytic properties of complex **3**. Because of the extreme electronegativity of fluorine the Ti–F bond is highly polar and, hence, also very strong. A characteristic feature of fluoride ligands, as opposed to chloride and oxido substituents, is the high tendency to bridge metal centers. This has been well documented by Roesky, Noltemeyer, and co-workers [7] with numerous crystal structures. Quite analogous to the allylstannane addition catalyzed by alkoxide **1** [2c – 2e, 2j], the asymmetric induction observed with complex **3** exceeds the optical purity of the binaphthol ligand [5]. This is explained by aggregate formation (see also [2d, 2i]) and the crystal structure of **1,** which reveals oligomer formation by Ti-O-Ti bridging [8]. Based on the many structures of titanium fluorides [7] it would, on the contrary, be astonishing if the fluoride **3** was not aggregating via Ti-F-Ti bridging. One could therefore envisage, that, starting from dimer **18**, a ternary transition state **19** is responsible for the catalytic Sakurai-Hosomi reaction, whereby the electrophilic titanium center activates the aldehyde, while at the same time the nucleophilic fluoride bridges to silicon and thereby increases the reactivity of the allylsilane. It is known that pentacoordinated, negatively charged allyl siliconates add without Lewis acid to aldehydes [9]. The adduct **20** as the primary product could then fragment to the silyl ether **7**, and catalyst **18** would be regenerated. Direct catalysis by

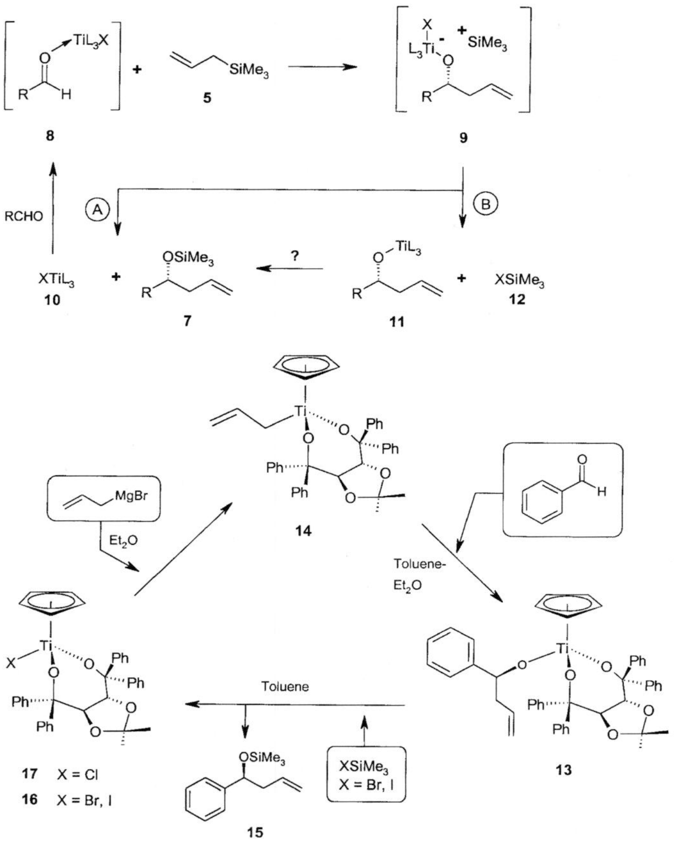

Scheme 2

dimer **18** or higher oligomers is also possible (*cf.* discussion in [2d, 2i]).

We speculated earlier whether a fluoride ligand could add catalytic properties to our chiral dialkoxy cyclopentadienyltitanium complexes. Chlorotitanium-TADDOLate **17** – an excellent template for controlling the stereochemistry of stoichiometric additions also to chiral aldehydes [10] – could easily be transformed to the comparably inert methyltitanium derivative **21**, isopropoxide **22**, and trifluoromethylsulfonate **23**. However, the preparation of fluoride **24** turned out to be rather tricky [11]. We finally succeeded when a new method developed by Roesky and coworkers [7a,b] appeared, allowing a clean conversion of chloride **17** to **24** with Me_3SnF (Scheme 4) [11]. An X-ray analysis revealed that the fluoride **24** is monomeric in the crystal and also nearly congruent with chloride **17**, except for the Ti–F bond [182.5(5) pm], which as expected is shorter than the Ti–Cl bond of **17** [228.4(3) pm] [12]. We were of course interested in the catalytic properties of **24** in comparison with triflate **23**. In the presence of **23,** allylsilane **5** adds smoothly to benzaldehyde; the product is, however, racemic. Most probably the catalyst is trimethylsilyl triflate, which is formed according to pathway **"B"** (Scheme 2, *cf.* [4a]). On the contrary, no reaction between benzaldehyde and allyl silane **5** is observed in the presence of fluoride **24**. As we observed earlier, all cyclopentadienyltitanium TADDOLates studied catalyzed the conversion of benzaldehyde **25** with salt-free $(i\text{-PrO})_3TiCH_3$ **26** to 2-phenylethanol **27** [11]. In this case the titanium fluoride effect was indeed very pronounced, as at –78 °C 0,5 % of **24** afforded 60 % of product **27** in 17 h with 78 % *ee*; with 2 % of **24** the induction is 93 % *ee*.

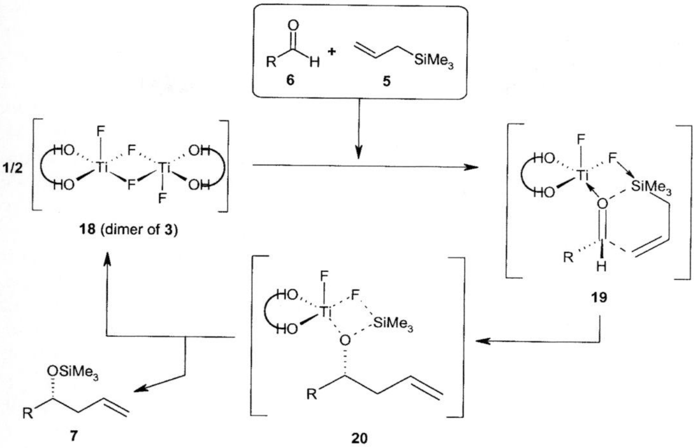

6 + 5 ($SiMe_3$)

1/2 [18 (dimer of 3)] → [19] → [20] → 7 ($OSiMe_3$)

Scheme 3

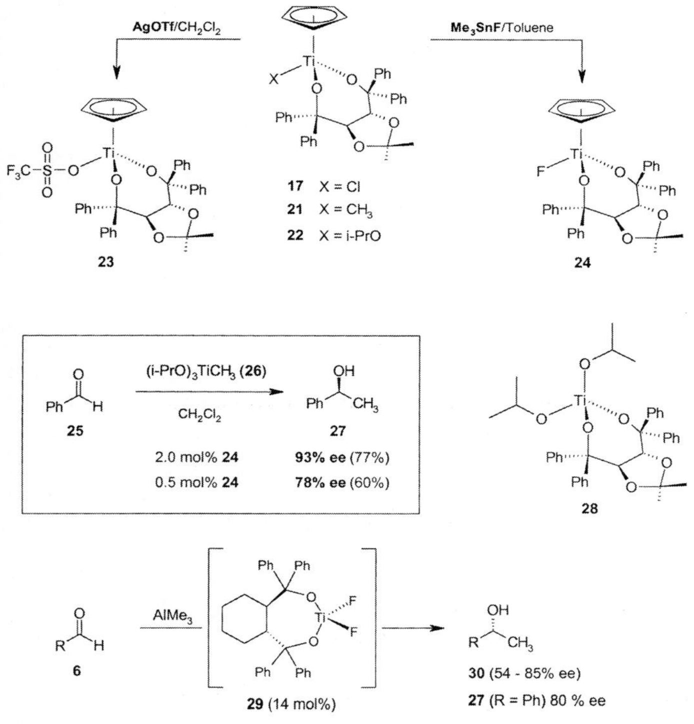

17 X = Cl
21 X = CH_3
22 X = i-PrO

$AgOTf/CH_2Cl_2$ → 23

Me_3SnF/Toluene → 24

25 → 27 with $(i\text{-}PrO)_3TiCH_3$ (26), CH_2Cl_2

2.0 mol% **24**: **93% ee** (77%)
0.5 mol% **24**: **78% ee** (60%)

28

6 → [29 (14 mol%)] with $AlMe_3$ → 30 (54 - 85% ee); 27 (R = Ph) 80 % ee

Scheme 4

Fluoride **24** thus by far outperforms the analogous complexes, as **27** of only 40–60 % *ee* is obtained in the presence of 10 % of **17**, and also **21**–**23** [11]. The reaction of **26** with aldehydes can also be catalyzed by titanate **28** lacking the cyclopentadienyl ligand. In this case, however, 20 % or even stoichiometric amounts of catalyst were used [13]. For comparison we have tested complex **28** under comparable conditions. With 0.5 % of **28** only 6 % of benzaldehyde **25** was converted to product **27** [11]. Originally we hoped for high versatility and substrate tolerance of fluoride **24**. However, with complex aldehydes such as glyceraldehyde acetonide no conversion with $(i\text{-PrO})_3TiCH_3$ was observed.

Recently Pagenkopf and Carreira reported that the addition of trimethyl aluminum to aldehydes can be catalyzed by the putative difluorotitanium-TADDOLate **29**, giving secondary alcohols **30** of moderate enantiomerical purity (Scheme 4) [14]. In accordance with structural studies by Roesky, Noltemeyer, and co-workers [15] a fluoride bridging titanium and aluminum is postulated for the transition state.

As exemplified by the reactions of Schemes 1 and 4, fluorotitanium compounds could open new possibilities for metal-catalyzed processes. Their fascinating structural diversity [7] as well as further catalytic possibilities in the field of olefin polymerizations [7i, 16] have been put forward by the pioneering work of Roesky, Noltemeyer and co-workers. Similar properties were also exhibited by the analogous zirconium and hafnium compounds [7b,i]. A Zr binaphtholate has already been successfully applied for the enantioselective allylstannylation of aldehydes [2f]. Buchwald and co-workers successfully used a chiral titanocene difluoride as precursor for the corresponding Ti(III) hydride, a very efficient catalyst for the enantioselective hydrosilylation of imines [17].

References:

[1] K. Soai, S. Niwa, *Chem. Rev.* **1992**, *92*, 833–856.

[2] a) J.A. Marshall, Y. Tang, *Synlett* **1992**, 653 - 654; b) A.L. Costa, M.G. Piazza, E. Tagliavini, C. Trombini, A. Umani-Ronchi, *J. Am. Chem. Soc.* **1993**, *115*, 7001 - 7002; c) G.E. Keck, K.H. Tarbet, L.S. Geraci, *ibid.* **1993**, *115*, 8467 - 8468; d) J.W. Faller, D.W.I. Sams, X. Liu, *ibid.* **1996**, *118*, 1217 - 1218; e) R. Brückner, St. Weigand, *Chemistry, A European Journal* **1996**, *2*, 1077 - 1084; f) P. Bedeschi, S. Casolari, A.L. Costa, E. Tagliavini, A. Umani-Ronchi, *Tetrahedron Lett.* **1995**, *36*, 7897 - 7900. g) A. Yanagisawa, H. Nakashima, A. Ishiba, H. Yamamoto, *J. Am. Chem. Soc.* **1996**, *118*, 4723 – 4724. h) A. Yanagisawa, A. Ishiba, H. Nakashima, H. Yamamoto, *Synlett* **1997**, 88 – 90. i) P.G. Cozzi, E. Tagliavini, A. Umani-Ronchi, *Gazz. Chim. Ital.* **1997**, *127*, 247 – 254. j) G.E. Keck. T. Yu, *Org. Lett.* **1999**, *1*, 289 – 291. k) Ch.-M. Yu, H.-S. Choi, W.-H. Jung, S.-S. Lee, *Tetrahedron Lett.* **1996**, *37*, 7095 – 7098. l) Ch.-M. Yu, H.-S. Choi, W.-H. Jung, H.-J. Kim, J. Shin, *Chem. Commun.* **1997**, 761 – 762. m) Ch.-M. Yu, S.-K. Yoon, H.-S. Choi, K. Baek, *ibid.*, 763 – 764.

[3] a) A. Ketter, R. Herrmann, *Z. Naturforsch B, Chem. Sci.* **1990**, *45*, 1684–1688; b) K. Ishihara, M. Mouri, Q. Gao, T. Maruyama, K. Furuta, H. Yamamoto, *J. Am. Chem. Soc.* **1993**, *115*, 11490 - 11495; c) S. Aoki, K. Mikami, M. Terada, T. Nakai, *Tetrahedron* **1993**, *49*, 1783 - 1792.

[4] a) The most important catalysts for the *Mukaiyama*-reaction are cited in: T.K. Hollis, B. Bosnich, *J. Am. Chem. Soc.* **1995**, *117*, 4570 - 4581; b) E.M. Carreira, R.A. Singer, *Tetrahedron Lett.* **1994**, *35*, 4323 - 4326; c) S.E. Denmark, Ch.-T. Chen, *ibid.* **1994**, *35*, 4327 - 4330. d) Sh. Kobayashi, S. Nagayama, T. Busujima, *Tetrahedron* **1999**, *55*, 8739 – 8746. e) J. Krüger, E.M. Carreira, *J. Amer. Chem. Soc.* **1998**, *120*, 837 – 838. f) B.L. Pagenkopf, J. Krüger, A. Stojanovich, E.M. Carreira, *Angew. Chem.* **1998**, *110*, 3312 – 3314; *Angew. Chem., Int. Ed. Engl. 37*, 3124.

[5] D.R. Gauthier, Jr., E.M. Carreira, *Angew. Chem.* **1996**, *108*, 2521–2523; *Angew. Chem., Int. Ed. Engl.* **1996**, *35*, 2363–2365.

[6] A. Hafner, unpublished results.

[7] a) A. Herzog, F.-Q. Liu, H.W. Roesky, A. Demsar, K. Keller, M. Noltemeyer, F. Pauer, *Organometallics* **1994**, *13*, 1251–1256; b) E.F. Murphy, P. Yu, St. Dietrich, H.W. Roesky, E. Parisini, M. Noltemeyer, *J. Chem. Soc., Dalton Trans* **1996**, 1983 - 1987; c) H.W. Roesky, M. Satoodeh, M. Noltemeyer, *Angew. Chem.* **1992**, *104*, 869 - 870; *Angew. Chem., Int. Ed. Engl.* **1992**, *31*, 864 - 866; d) F.-Q. Liu, H. Gornitzka, D. Stalke, H.W. Roesky, *ibid.* **1993**, *105*, 447 - 448; *Int. Ed. Engl.* **1993**, *32*, 442 - 444; e) H.W. Roesky, I. Leichtweis, M. Noltemeyer, *Inorg. Chem.* **1993**, *32*, 5102 - 5104; f) F.-Q. Liu, A. Kuhn, R. Herbst-Irmer, D. Stalke, H.W. Roesky, *Angew. Chem.* **1994**, *106*, 577 - 578; *Angew. Chem., Int. Ed. Engl.* **1994**, *33*, 555 - 556; g) F.-Q. Liu, D. Stalke, H.W. Roesky, *ibid.* **1995**, *107*, 2004 - 2006; *Int. Ed. Engl.* **1995**, *34*, 1872 - 1874; h) A. Künzel, H.W. Roesky, M. Noltemeyer, H.-G. Schmidt, *J. Chem. Soc., Chem. Commun.* **1995**, 2145–2146; i) S.A.A. Shah, H. Dorn, A. Voigt, H.W. Roesky, E. Parisini, H.-G. Schmidt, M. Noltemeyer, *Organometallics* **1996**, *15*, 3176 - 3181. j) E.F. Murphy, R. Murugavel, H.W. Roesky *Chem. Rev.* **1997**, *97*, 3425 – 3468. k) H.W. Roesky, I. Haiduc *J. Chem. Soc., Dalton Trans.* ***1999,*** 2249 – 2264.

[8] a) C.A. Martin, Ph.D. Thesis, Massachusetts Institute of Technology, 1988; b) R.O. Duthaler, A. Hafner, *Chem. Rev.* **1992**, *92*, 807 - 832 (Chart 11).

[9] a) T. Hayashi, Y. Matsumoto, T. Kiyoi, Y. Ito, Sh. Kohra, Y. Tominaga, A. Hosomi, *Tetrahedron Lett.* **1988**, *29*, 5667–5670; b) M. Kira, K. Sato, H. Sa-

kurai, *J. Am. Chem. Soc.* **1990**, *112*, 257 - 260; c) M. Kira, K. Sato, H. Sakurai, M. Hada, M. Izawa, J. Ushio, *Chem. Lett.* **1991**, 387 - 390.

[10] R.O. Duthaler, A. Hafner, P.L. Alsters, P. Rothe-Streit, G. Rihs, *Pure Appl. Chem.* **1992**, *64*, 1897–1910.

[11] R.O. Duthaler, A. Hafner, P.L. Alsters, M. Tinkl, unpublished results, partially presented *e.g.* at the *Seventh IUPAC Symposium on Organo-Metallic Chemistry directed towards Organic Synthesis*, Sept. 19 – 23, 1993, Kobe (Japan).

[12] G. Rihs, unpublished results.

[13] a) D. Seebach, D.A. Plattner, A.K. Beck, Y.M. Wang, P. Hunziker, W. Petter, *Helv. Chim. Acta* **1992**, *75*, 2171 - 2209 (Scheme 3); b) B. Weber, D. Seebach, *Tetrahedron* **1994**, *50*, 7473 - 7484; c) Y.N. Ito, X. Ariza, A.K. Beck, A. Bohác, C. Ganter, R.E. Gawley, F.N.M. Kühnle, J. Tuleja, Y.M. Wang, D. Seebach, *Helv. Chim. Acta* **1994**, *77*, 2071 - 2110 (Scheme 7a).

[14] B.L. Pagenkopf, E.M. Carreira *Tetrahedron Lett.* **1998**, *39*, 9593 – 9596.

[15] a) P. Yu, H.W. Roesky, A. Demsar, Th. Albers, H.-G. Schmidt, M. Noltemeyer, *Angew. Chem* **1997**, *109*, 1846 – 1847; *Angew. Chem., Int. Ed. Engl.* *36*, 1766. b) H. Wessel, M.L. Montero, C. Rennekamp, H.W. Roesky, P. Yu, I. Usón *Angew. Chem.* **1998**, *110*, 862 – 864; *Angew. Chem., Int. Ed. Engl.* *37*, 843.

[16] a) W. Kaminsky, S. Lenk, V. Scholz, H.W. Roesky, A. Herzog, *Macromolecules* **1997**, *30*, 7647 – 7650. b) P. Yu, P. Müller, M.A. Said, H.W. Roesky, I. Usón, G. Bai, M. Noltemeyer, *Organometallics* **1999**, *18*, 1669 – 1674.

[17] X. Verdaguer, U.E.W. Lange, M.T. Reding, St.L. Buchwald, *J. Amer. Chem. Soc.* **1996**, *118*, 6784 – 6785.

Enzymes and Transition Metal Complexes in Tandem – a New Concept for Dynamic Kinetic Resolution

Rainer Stürmer

BASF AG, Ludwigshafen, Germany

The resolution of racemic compounds mediated by enzymes has become a valuable tool for the synthesis of chiral intermediates. In most cases, however, only one enantiomer of the intermediate is required for the next step in the synthesis; thus, the unwanted isomer must be either discarded or racemized for reuse in the enzymatic resolution process. Dynamic kinetic resolution is one way of avoiding this problem: the unwanted enantiomer is racemized during the selective enzymatic process and serves as fresh starting material in the resolution.

This concept has been known for a long time in pure enzymatic synthesis, e.g. amino acid synthesis via hydantoins [1] or oxazolidinones [2]. Cyanohydrins [3] and lactols [4] are prone to in situ racemization as well and may serve as substrates in kinetic resolutions.

Dynamic kinetic resolution is well known in pure chemical synthesis, as illustrated by work by Noyori et al. on the asymmetric hydrogenation of α-substituted β-ketoesters. Noyori et al. [5], Ward [6] and Caddick et al. [7] have reviewed the chemical syntheses, and biocatalytic routes have been discussed by Faber et al. [8].

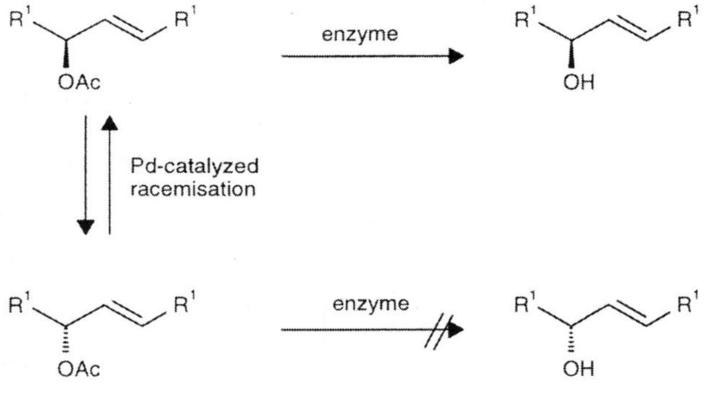

Scheme 1. Pd-catalyzed racemization of an allylic acetate [8].

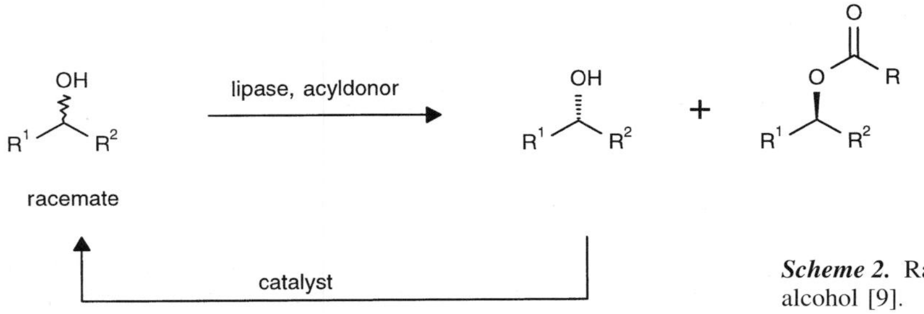

Scheme 2. Racemization of a secondary alcohol [9].

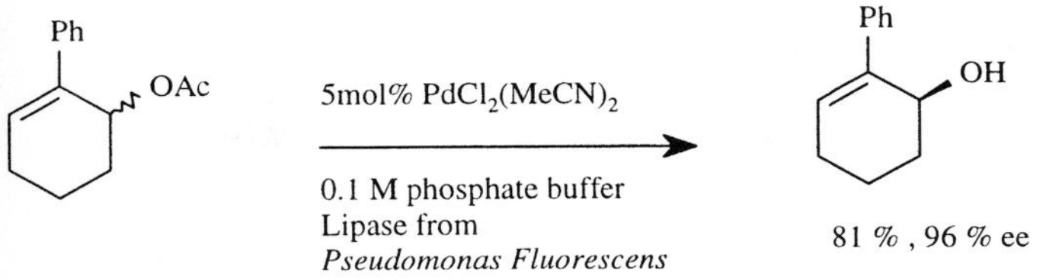

Ph OAc → 5mol% $PdCl_2(MeCN)_2$ / 0.1 M phosphate buffer / Lipase from *Pseudomonas Fluorescens* → Ph OH

81 % , 96 % ee

Scheme 3. Specific example of the racemization shown in Scheme 1.

Williams et al. have now demonstrated the compatibility of enzyme and transition metal complexes in one reaction vessel. Two examples illustrate the concept: first a palladium-catalyzed racemization [9] of an allylic acetate in the presence of a hydrolase (Scheme 1), and second the racemization of a secondary alcohol [10] through Oppenauer oxidation / Meerwein-Ponndorf-Verley reduction with concomitant acylation of one enantiomer with a lipase from *Pseudomonas fluorescens* (Scheme 2).

To utilize these reactions, a few conditions must be met. A selective enzyme is crucial and the metal-organic catalyst must facilitate a fast racemization of the substrate. Last but not least the catalyst should not influence the enzyme in terms of selectivity and reactivity. In the ideal case the enzyme hydrolyzes one enantiomer of the allylic acetate, giving rise to the allylic alcohol, which itself is not susceptible to Pd-catalyzed racemization.

Scheme 3 provides a specific example of the enzymatic hydrolysis outlined in Scheme 1. The substituted allylic acetate is hydrolyzed with a lipase from *Pseudomonas fluorescens* in a phosphate buffer in the presence of 5 mol% palladium complex.

After 19 days the conversion is 96 % (yield is 81 %), and the enantiomeric excess reaches 96 %. The rate-determining step is most likely the slow Pd-catalyzed step, since 50 % conversion is achieved after only two days. A more active metal complex has yet to be identified for a more practical reaction.

The reaction time in the second example (Scheme 4) is rather long as well; the rate-determining step is again the metal-catalyzed racemization. After 6 days the conversion is 76 % with 3 mol % $[Rh(COD)Cl]_2$ and the enantiomeric excess reaches 80 %.

Theoretically, in a simple kinetic resolution the *ee* value should not exceed 32 % at this specific conversion. In addition to the rhodium complex, this reaction requires acetophenone as stoichiometric hydride acceptor, phenanthroline as coligand and potassium hydroxide as base. An *ee* value of 98 % at 60 % conversion (theoretical value 67 %)is achieved with $[Rh_2(OAc)_4]$ without an added base after 3 days. Surprisingly, the enzyme tolerates potassium hydroxide in amounts up to 20 mol% at elevated temperatures; however, the enantiomeric excesses are somewhat lower than those obtained in an ordinary kinetic resolution. Unselective, base- or metal-catalyzed acylation might be the reason for the somewhat lower *ee* value.

Bäckvall et al. [10] have recently reported substantial improvements in the described process. A range of 1–phenylethanol derivatives can be synthesized from the racemates in excellent yields and >99 % *ee* by using a binuclear ruthenium complex combined with an immobilized lipase and a specifically designed acyl donor (4-Cl-PhOAc). Even aliphatic alcohols and diols

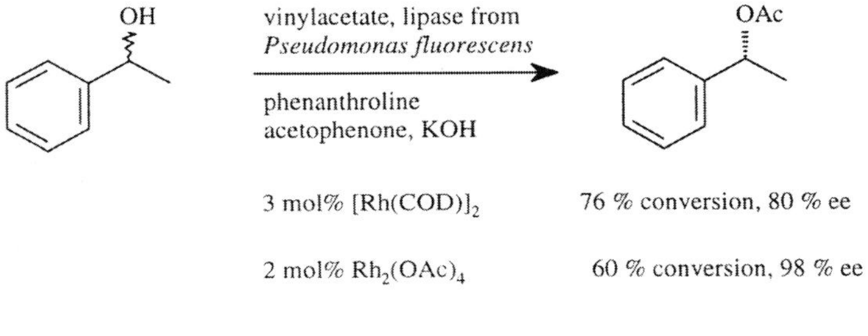

OH → vinylacetate, lipase from *Pseudomonas fluorescens* / phenanthroline / acetophenone, KOH → OAc

3 mol% $[Rh(COD)]_2$ — 76 % conversion, 80 % ee

2 mol% $Rh_2(OAc)_4$ — 60 % conversion, 98 % ee

Scheme 4. Specific example of the racemization shown in Scheme 2.

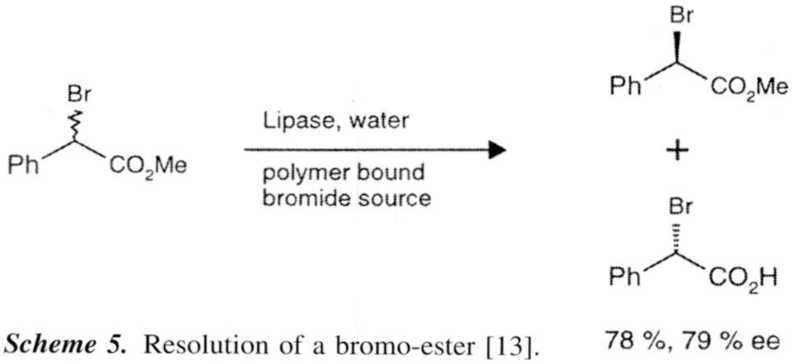

Scheme 5. Resolution of a bromo-ester [13].

can be used in this optimized protocol. Furthermore, no additional base is necessary and the reaction is considerably faster.

This concept was also recently extended by Reetz et al. to the resolution of phenylethylamine [12]. In this case, an immobilised lipase and ethyl acetate as acyl donor are used; the non-acylated (*S*)-enantiomer of the amine is racemized in situ by palladium on charcoal. After 8 days – the metal catalyzed racemization is again likely to be the rate-determining step – (*R*)-*N*-acetyl-phenylethylamine is isolated in 64 % yield and 99 % enantiomeric excess.

Williams et al.[13] used a combination of a polymer-bound phosphonium bromide and a cross-linked lipase from *Candida rugosa* for a resolution process of an α-bromo ester (Scheme 5).

The corresponding acid was obtained in 79 % enantiomeric excess and in 78 % yield.

In summary, a new concept for dynamic kinetic resolution has been demonstrated. It is still a long way to practical systems, but the first steps to team up enzymes with classical metal-organic catalysts have been successfully achieved.

References:

[1] S.Takahashi in *Progress in Industrial Microbiology Vol.24* (Eds. K. Aida, I. Chibata, K. Nakayama, T. Takanami and H. Yamada). Elsevier, Amsterdam, **1986**, S. 269–279.

[2] C.J. Sih et al. in *Stereocontrolled Organic Synthesis*, (Ed. B.M. Trost). Blackwell Scientific Publications. Oxford, **1994**, PP. 401–404.

[3] H. v.d. Deen, A.D. Cuiper, R.P. Hof, A. v. Oeveren, B.L. Feringa, R.M. Kellog, *J. Am. Chem. Soc.* **1996,** *118*, 3801–3803.

[4] M. Inagaki, J. Hiratake, T. Nishioka, J. Oda, J. Org. Chem. **1992,** *57*, 5643–5649.

[5] a) R. Noyori, *Science* **1990**, *248*, 1194–1199; b) R. Noyori, H. Takaya, *Acc. Chem. Res.* **1990**, *23*, 345–350.

[6] R.S. Ward, *Tetrahedron:Asymmetry* **1995**, *6*, 1475–1490.

[7] S. Caddick, K. Jenkins, *Chem.Soc.Rev.* **1996**, *25*, 447–456.

[8] a) H. Stecher, K. Faber *Synthesis* **1997**, 1–17; b) U.T. Strauss, U. Felfer, K. Faber, *Tetrahedron:Asymmetry* **1999**, *10*, 107–117.

[8] P.M. Dinh, J.A. Howarth, A.R. Hudnott, J.M.J. Williams, W. Harris, *Tetrahedron Lett.* **1996**, *37*, 7623–7626.

[10] J.V. Allen, J.M.J. Williams *Tetrahedon Lett.* **1996**, *37*, 1859–1862.

[11] a) A.L.E. Larsson, B.A. Persson, J.-E. Bäckvall, *Angew. Chem.* **1997**, *109*, 1256–1258; *Angew. Chem. Int. Ed. Engl.* **1997**, *36*, 1211–1212 b) B.A. Persson, A.L.E. Larsson, M. Le Ray, J.-E. Bäckvall, *J. Am. Chem. Soc.* **1999,** *121*, 1645–1650.

[12] M.T. Reetz, K. Schimossek, *Chimia* **1996**, *50*, 668–669

[13] M.M. Jones, J.M.J. Williams, *Chem. Commun.* **1998**, 2519–2520.

Non-Enzymatic Kinetic Resolution of Secondary Alcohols

Peter Somfai

Organic Chemistry, Royal Institute of Technology, Stockholm, Sweden

Secondary alcohols represent an important class of readily derivatizable compounds that can be incorporated into a variety of synthetic strategies. The kinetic resolution of these compounds, or more often of their corresponding acetates, has traditionally been done by using esterases and often with excellent results [1]. Recent developments of this approach, which use metal catalysts in order to racemize the substrate in combination with appropriate enzymes, have resulted in efficient dynamic kinetic resolutions [2]. However, the non-enzymatic kinetic resolution of racemic alcohols has proven to be more difficult, the only prominent example being the Sharpless resolution of allylic alcohols [3], but recently several impressive results have been documented. In principle, four conceptually different strategies have been pursued with varying degrees of success. In the most straightforward one, a preformed enantiomerically pure acylating agent, e.g. an acyl halide [4] or *N*-benzoyl oxazolidinone **1** [5], is reacted with a racemic secondary alcohol (**2**) to give the corresponding acylated product (**3**) and recovered starting material (Scheme 1). The inherent drawbacks with this method are that a stoichiometric amount of the chiral reagent is required and, in the former case, that the acylated product will be formed as a mixture of diastereomers unless the resolution proceeds with complete selectivity.

Alternatively, advantage can be taken of the fact that the acylation of secondary alcohols is subjected to nucleophilic catalysis. In a pioneering study more than 60 years ago Wegler investigated the acylation of several different racemic secondary alcohols mediated by brucine and strychnine [6]. For example, when racemic **2a** was reacted with acetic anhydride in the presence of stoichiometric amounts of brucine, (*S*)-*α*-methylbenzyl acetate [(*S*)-**4**] was formed, although its optical purity cannot be ascertained from the data given. More recently the efficiency of other tertiary amines as acylation promoters has been investigated, the best result for the acylation of **2a** being obtained with amine **5** (**2a**:AcCl:**5** = 2 : 1:1) yielding (*S*)-**4** with an optical purity of 68 % *ee* [7].

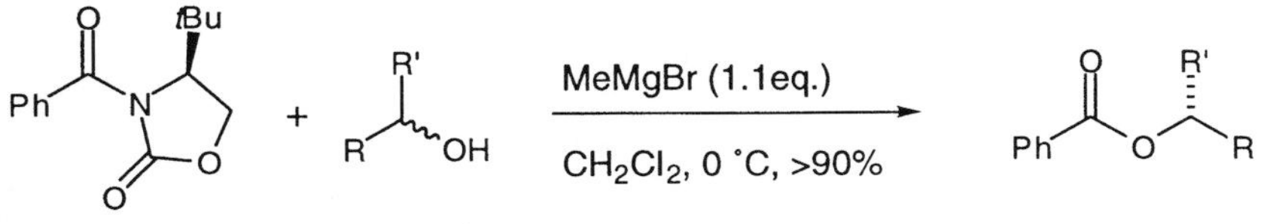

MeMgBr (1.1eq.); CH_2Cl_2, 0 °C, >90%

1 (1 eq.) **2** (10 eq.)
a R=Ph, R'=Me
b R=Ph, R'=*i*Pr
c R, R'=*c*-C_6H_{11}, Me

3a ee=95%
b ee=65%
c ee=5%

Scheme 1. Kinetic resolution of **2** using oxazolidinone **1** [5].

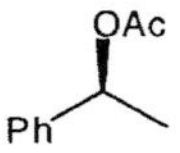

(*S*)-4 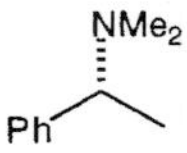 5 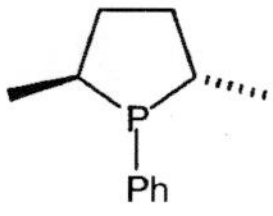 6

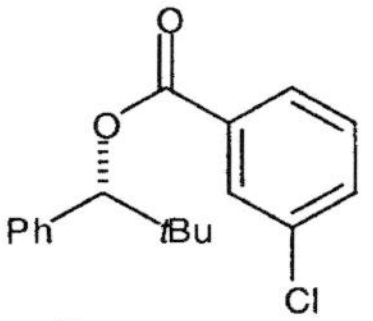

7 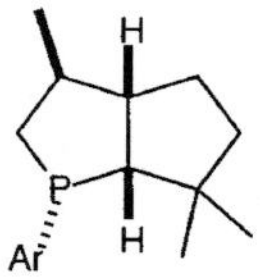 8 Ar=3,5-di(*t*Bu)Ph

The first example of a chiral nucleophilic promoter that is also capable of catalytic turnover was described by Vedejs et al.[8], who showed that acylation of **2a** with Ac_2O in the presence of the C_2-symmetric phospholane **6** (**2a**:Ac_2O:**6** = 1 : 2.5 : 0.16) gave **4** with $ee = 34\ \%$ (44 % conversion, $s = 2.7$, $s = k_{fast}/k_{slow}$), while racemic 1–phenyl-2,2-dimethylpropanol was acylated with *m*-chlorobenzoic anhydride under the same reaction conditions to yield the corresponding ester **7** with $ee = 81\ \%$ (25 % conversion, $s = 12–15$). In a continuation of these studies the bicyclic phosphine **8** was prepared and evaluated (Scheme 2) [9]. It was found that **8** is an efficient catalyst for the resolution of several secondary aryl-alkyl carbinols, resulting in unprecedentedly high *s*-values. The choice of anhydride is important for obtaining optimal selectivity, with isobutyric anhydride performing well for a variety of alcohols, and the resolutions using **8** can be conducted at low temperatures, with concomitant improvements in selectivity due to the its favorable activity.

In a parallel investigation the enantiomerically pure DMAP-analog **10** (*ee*›99 %) was prepared by the same group and converted into the pyridinium salt **11** (Scheme 3), which, for steric reasons, is not sufficiently reactive to promote acyl transfers to alcohol **2** by itself [10]. However, in the presence of Lewis acids ($ZnBr_2$ or $MgBr_2$) and Et_3N, neither of which can be omitted, a slow acylation ensues to give the mixed carbonates **12** in high *ee* and with good *s* values.

The inherent problem with these resolutions is that there is a continuous enrichment of the slower reacting isomer of the starting material as the reaction proceeds, thus requiring exceptional rate differences between the enantiomers in order to obtain high *ee* of the product as well as the recovered starting material *and* high conversion (e. g. at 50 % conversion $s = 200$, $ee = 96\ \%$; $s = 500$, $ee = 98\ \%$). In an attempt to overcome this an ingenious protocol was developed by Vedejs and Chen in which two competing processes, with

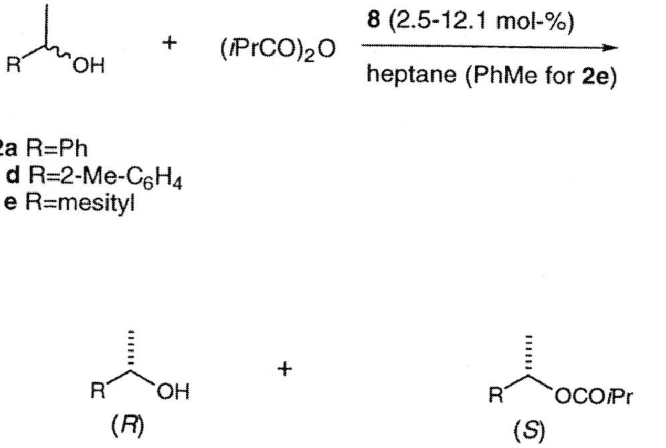

R–CH(OH)–CH3 + $(iPrCO)_2O$ —[**8** (2.5-12.1 mol-%); heptane (PhMe for **2e**)]→ (*R*) + (*S*) R–CH(OCO*i*Pr)–CH3

2a R=Ph
d R=2-Me-C_6H_4
e R=mesityl

(*R*):
2a conv.=29.2%, ee=38.4%, -20 °C
a conv.=42.4%, ee=61.9%, rt
d conv.= 50.1%, ee=95.3%, -40°C
e conv.=44.4%, ee=78.8%, -40°C

(*S*):
9a ee=93.3%, s=42
a ee=84.0%, s=22
d ee=94.9%, s=145
e ee=98.7%, s=369

Scheme 2. Kinetic resolution of **2** using phosphine **8** [9].

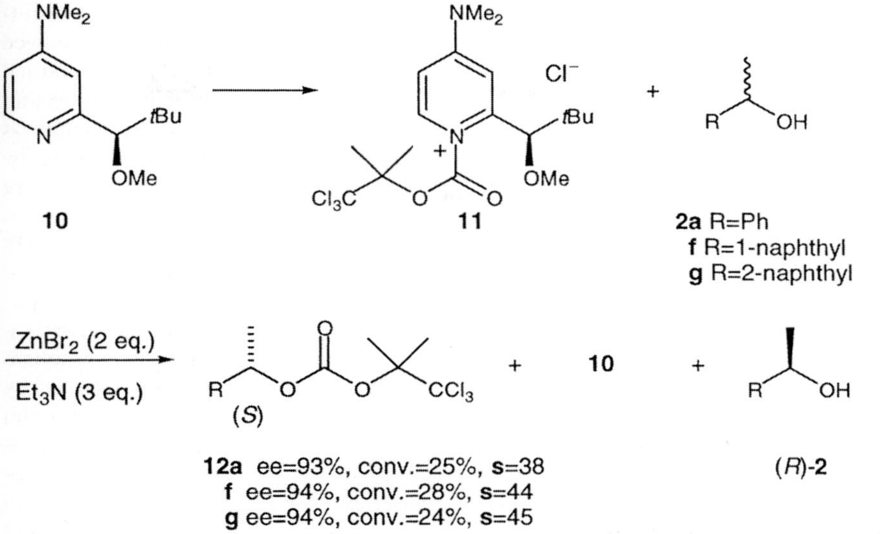
NMe2
tBu
OMe
10
NMe2
Cl⁻
tBu
OMe
Cl3C
11
R
OH
2a R=Ph
f R=1-naphthyl
g R=2-naphthyl
ZnBr2 (2 eq.)
Et3N (3 eq.)
R
CCl3
(S)
10
R
OH
12a ee=93%, conv.=25%, s=38
f ee=94%, conv.=28%, s=44
g ee=94%, conv.=24%, s=45
(R)-2

Scheme 3. Kinetic resolution of **2** using pyridinium salt **9** [10].

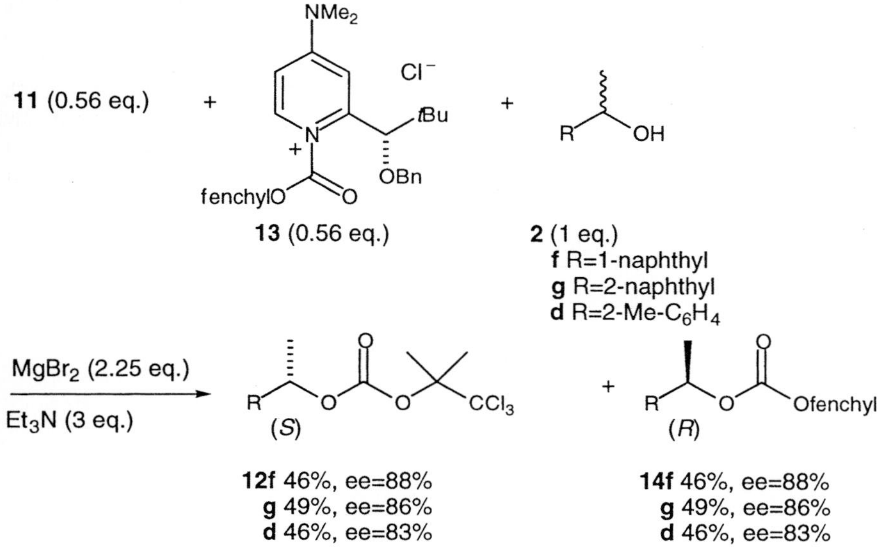
11 (0.56 eq.)
NMe2
Cl⁻
tBu
OBn
fenchylO
13 (0.56 eq.)
R
OH
2 (1 eq.)
f R=1-naphthyl
g R=2-naphthyl
d R=2-Me-C6H4
MgBr2 (2.25 eq.)
Et3N (3 eq.)
R
CCl3
(S)
R
Ofenchyl
(R)
12f 46%, ee=88%
g 49%, ee=86%
d 46%, ee=83%
14f 46%, ee=88%
g 49%, ee=86%
d 46%, ee=83%

Scheme 4. Parallel kinetic resolution of **2** [11].

similar rates and selectivities but for opposite enantiomers, are run in parallel (parallel kinetic resolution, PKR), the advantage being that the optimal 1 : 1 ratio between the stereoisomers is maintained throughout the experiment [11]. The advantages can be appreciated by considering a PKR experiment involving two simultaneous reactions with selectivities for opposite isomers, each reaction with $s = 49$ (100 % conversion) and which, ideally, would give both products in

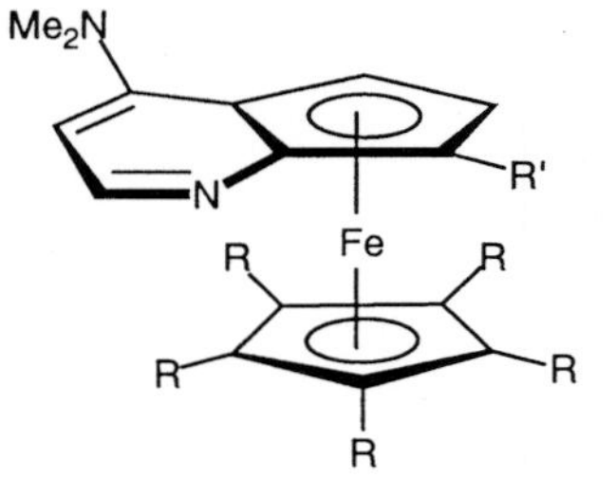

15a R=Me, R'=H
b R=Me, R'=Me, $SiMe_3$
c R=Ph, R'=H

ee = 96 %. In comparison, a simple kinetic resolution would need to have $s = 200$ (50 % conversion) in order to equal the former result. Reagents **11** and **13** were selected with the expectation that they would behave as quasi-enantiomers and acylate opposite enantiomers of racemic secondary alcohols to give products that could be readily separated. In the event, when **11** and **13** were reacted with three different aromatic secondary alcohols the corresponding carbonates were formed in high yield and *ee*, clearly demonstrating the potential of the method (Scheme 4).

Two other groups have described useful nucleophilic catalysts for the kinetic resolution of secondary alcohols. Fu and coworkers have prepared a family of novel chiral π-complexes of heterocycles with iron, e. g. **15**, and showed them to be potent nucleophilic catalysts for a variety of transformations [12]. This design allows for an asymmetric environment close to the planar sp^2-hybridized nitrogen atom that is quite different from that found in **10**. Kinetic resolution of **2a** using catalyst **15a** resulted in only a low enantioselection, and introduction of a methyl or TMS group in the 7 position (**15b**) did not improve the situation. Instead, replacing the η^5-C_5Me_5 group in **15a** with a η^5-C_5Ph_5 moiety gave derivatives **15c**, which have been shown to be excellent catalysts for the kinetic resolution of secondary aryl-alkyl and alkynyl-alkyl carbinols (Schemes 5 and 6). Several aspects of these resolutions are noteworthy. It was shown that the enantioselective acylation of racemic 1-phenylethanol using **15c** and acetic anhydride was dependent on the choice of solvent, the best result being obtained in *tert*-amyl alcohol, which itself apparently is not acetylated under the reaction conditions. Under the optimized reaction conditions, exceptional levels of resolution can be obtained using only 1 mol% of **15c** at 0 °C, and the reaction is not sensitive to small amounts of oxygen, moisture or adventitious impurities [12c]. In the kinetic resolution of propargylic alcohols, it

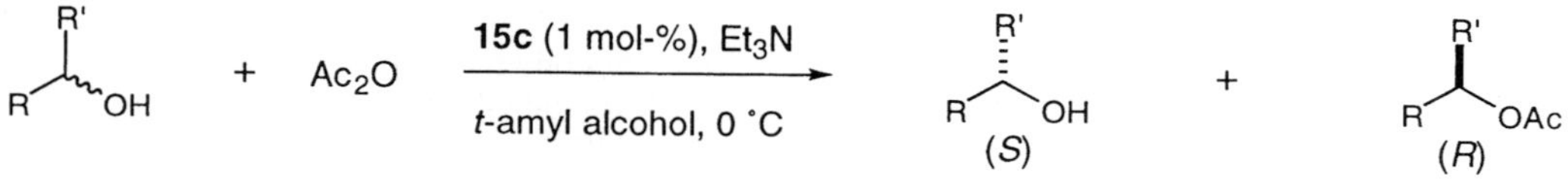

15c (1 mol-%), Et_3N, *t*-amyl alcohol, 0 °C; + Ac_2O → (*S*) alcohol + (*R*) acetate

2a R=Ph, R'=CH_3
d R=2-Me-C_6H_4, R'=CH_3
h R=Ph, R'=*t*Bu

2a conv.=55%, ee=99%, s=43 — **15a** ee=79%
d conv.= 53%, ee=99%, s=71 — **d** ee=87%
h conv.=51%, ee=96%, s=95 — **h** ee=92%

Scheme 5. Kinetic resolution of **2** using **15c** [12c].

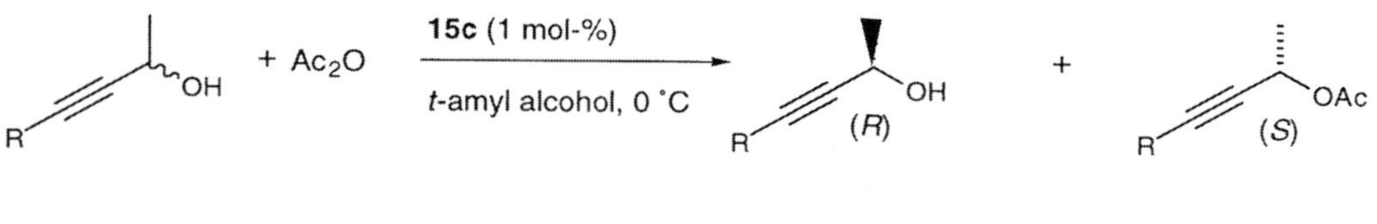

15c (1 mol-%), *t*-amyl alcohol, 0 °C; + Ac_2O → (*R*) alcohol + (*S*) acetate

16a R=Ph
b R=4-MeO-C_6H_4

16a conv.=58%, ee=96%, s=20 — **17a** ee=69%
b conv.= 60%, ee=94%, s=14 — **b** ee=63%

Scheme 6. Kinetic resolution of **16** using **15c** [12d].

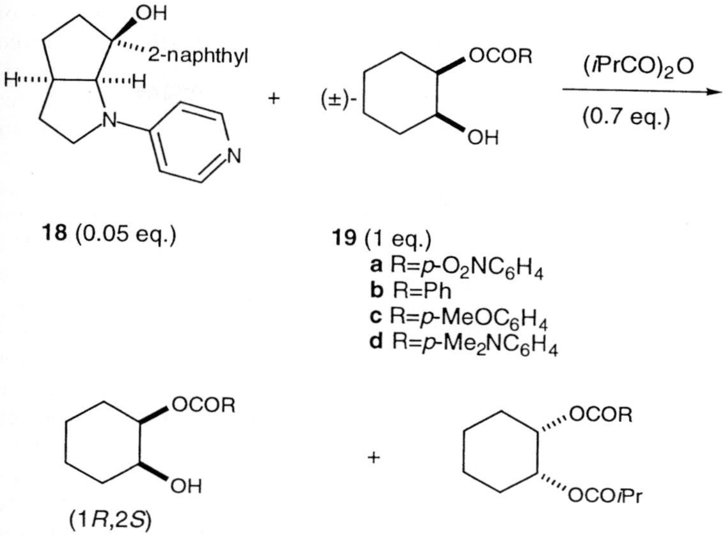

18 (0.05 eq.) + (±)-**19** (1 eq.) $\xrightarrow[(0.7\ \text{eq.})]{(i\text{PrCO})_2\text{O}}$ (1*R*,2*S*) +

19 (1 eq.)
a R=*p*-$O_2NC_6H_4$
b R=Ph
c R=*p*-$MeOC_6H_4$
d R=*p*-$Me_2NC_6H_4$

19a conv.=73%, ee=54%, **s**=2.4
b conv.=71%, ee=81%, **s**=4.5
c conv.=70%, ee=>85%, **s**=5.3
d conv.=72%, ee=>99%, **s**>10.1

Scheme 7. Kinetic resolution of **19** using pyridine derivative **18** [14].

was found that external bases resulted in a decreased enantioselectivity, optimal results being obtained by simply omitting the addition of a base (Scheme 6) [12d]. These workers also succeeded in isolating the acetylated form of **15c** (**15c**$Ac^+ \cdot SbF_6^-$), which is the actual acylating species, and to confirm the structure by X-ray crystallography [12d]. In this acetylpyridinium ion, the NMe_2 moiety, the pyridine ring and the acetyl group reside in a common plane, consistent with an extended delocalization of the positive charge, while the two cyclopentadienyl rings deviate from coplanarity by about 8°. The last observation might be due to steric interactions between the η^5-C_5Ph_5 moiety and the pyridine ring and also results in an efficient blocking of one face of the acetylpyridinium complex. In an interesting extension of this study, Fu and co-workers prepared the ruthenium analog of **15c**; its catalytic activity in the acetylation of 1–phenylethanol was slightly higher than that of **15c,** while its enantioselectivity in the kinetic resolution of the same substrate was markedly lower [13].

A completely different approach was pursued by Fuji and coworkers [14]. The enantiomerically pure pyridine derivative **18** was prepared, the rationale being that the reduced steric hindrance in the proximity of the nitrogen nucleus should result in efficient catalytic turnover and that chiral recognition might be possible by remote asymmetric induction in a process similar to the "induced fit" mechanism in enzymes. When **18** was reacted with racemic **19b** and isobutyric anhydride (1*R*,2*S*)-**19b** was recovered with *ee* = 81 % (71 % conversion, *s* = 4.5, Scheme 7). Interestingly, the optical purity of the recovered alcohol is dependent on the electron donating ability of the aromatic nuclei in the substrate, possibly indicating that $\pi-\pi$ stacking plays a pivotal role in the enantiodifferentiating event. It was also shown that the pyridinium ion derived from **18** and isobutyric anhydride adopts a "closed conformation", in contrast to **18**, in which the naphthyl moiety is situated over the pyridine ring, thus shielding the *si* face of the carbonyl group and supporting the "induced fit" concept.

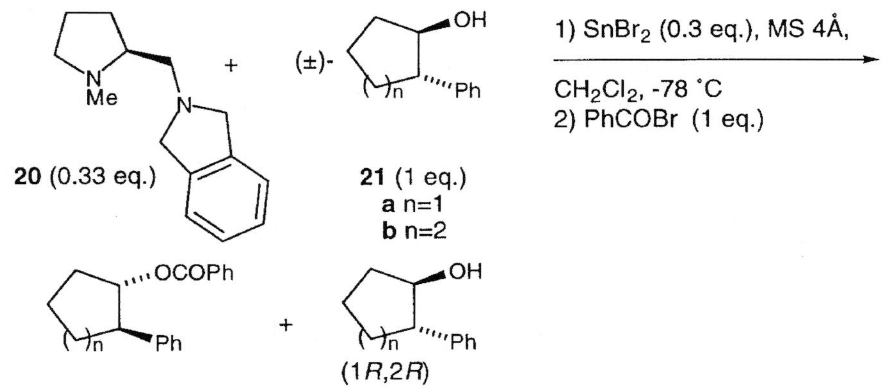

20 (0.33 eq.) + (±)-21 (1 eq.) a n=1 b n=2

1) $SnBr_2$ (0.3 eq.), MS 4Å, CH_2Cl_2, -78 °C
2) PhCOBr (1 eq.)

22a n=1; 44%, ee=86%
b n=2: 44%, ee=97%

(1*R*,2*R*)
21a 46%, ee=86%, s=27
b 49%, ee=84%, s>100

Scheme 8. Lewis acid catalyzed kinetic resolution of **21** [15].

The kinetic resolution of secondary alcohols is also subjected to asymmetric Lewis acid catalysis, as described in a recent elegant study by Oriyama et al [15]. It was shown by these authors that the complex formed by mixing $SnBr_2$ and amine **20**, derived from proline, in the presence of molecular sieves was able to catalyze the benzoylation of racemic **21** to give benzoate **22** along with recovered (1*R*,2*R*)-**21** in exceptional yields and selectivities (Scheme 8). Several other alcohols were also investigated, the cyclic species generally performing better than the acyclic ones.

Finally, Noyori and coworkers have shown that chiral Ru^{II}-diamine complexes are efficient catalysts for the enantioselective transfer hydrogenation of several prochiral ketones, using 2–propanol as the hydrogen source, to give the corresponding alcohols in high *ee* [16a]. The reaction is microscopically reversible and the more rapidly formed stereoisomer of the product is also the one that is more readily oxidized in the reverse process. Consequently, secondary alcohols with a high reduction potential should be eligible for kinetic resolution by using these types of catalysts [16a,b]. That this is indeed the case was elegantly demonstrated using the Ru^{II} complex **23a** or the corresponding mesitylene derivative **23b** (Scheme 9). Several alkyl-aryl carbinols could be successfully resolved by using **23**, with excellent recovery of the unreacted substrate in high *ee*. Two cyclic allylic alcohols, 2-cyclohexen-1-ol and 3-methyl-2-cyclohexen-1-ol, were also shown to be suitable substrates in this resolution, both of which are particularly valuable substrates in organic synthesis. Although this transformation superficially bears a resemblance

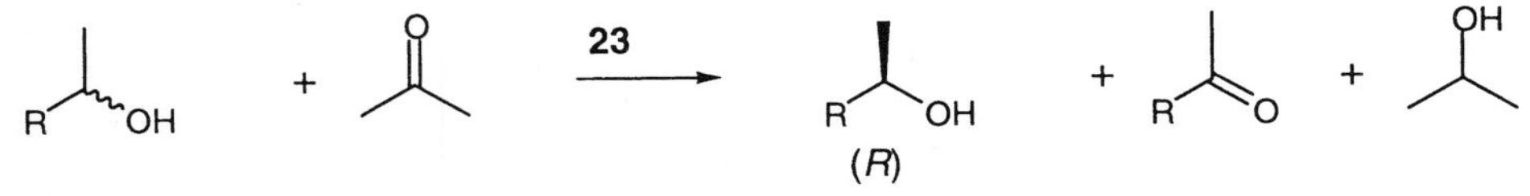

2 a R=Ph
i R=2-$MeOC_6H_4$
j R=2-$Me_2NC_6H_4$

23 → (*R*)

2 a 50%, ee=92%, s>80
i 47%, ee=92%, s>30
j 44%, ee=98%, s>30

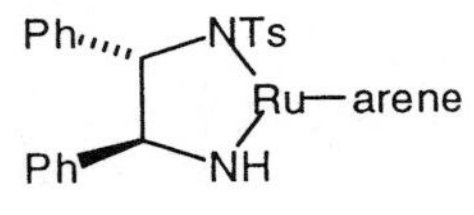

23a arene=*p*-cymene
b arene=mesitylene

Scheme 9. Ru^{II}-Catalyzed kinetic resolution [16b].

to the Oppenauer oxidation, it has been unequivocally shown that the Ru^{II}-catalyzed kinetic resolution proceeds by an 18–electron metal hydride species and not by a metal alkoxide intermediate, as is the case for the latter transformation [16c].

In summary, scalemic secondary alcohols can be obtained by non-enzymatic kinetic resolution and in some cases with excellent selectivities. Although these results in general fall short when compared with those obtained with the esterase-promoted resolutions, it is expected that a more thorough understanding of the enantiodiscriminating event in these processes will result in the development of even more efficient catalysts.

References

[1] For some recent reviews on esterase-promoted resolutions, see: a) C. J. Sih, S.-H. Wu, *Top. Stereochem.* **1989**, *19*, 63; b) C.-S. Chen, C. J. Sih, *Angew. Chem.* **1989**, *101*, 711; *Angew. Chem. Int. Ed. Engl.* **1989**, *28*, 695; c) A. M. Klibanov, *Acc. Chem. Res.* **1990**, *23*, 114; d) S. C. Ward, *Chem. Rev.* **1990**, *90*, 1; e) D. G. Drueckhammer, W. J. Hennen, R. L. Pederson, C. F. Barbas III, C. M. Gautheron, T. Krach, C.-H. Wong, *Synthesis* **1991**, 499; f) S. M. Roberts, *Chimia* **1993**, *47*, 85; for the use of designed peptides, see: g) S. J. Miller, G. T. Copeland, N. Papaioannou, T. E. Horstmann, E. M. Ruel, *J. Am. Chem. Soc.* **1998**, *120*, 1629; h) G. T. Copeland, E. R. Jarvo, S. J. Miller, *J. Org. Chem.* **1998**, *63*, 6784.

[2] R. Stürmer, *Angew. Chem.* **1997**, *109*, 1221; *Angew. Chem. Int. Ed. Engl.* **1997**, *36*, 1173.

[3] Y. Gao, R. M. Hanson, J. M. Klunder, S. Y. Ko, H. Masamune, K. B. Sharpless, *J. Am. Chem. Soc.* **1987**, *109*, 5765.

[4] A. Mazón, C. Nájera, M. Yus, A. Heumann, *Tetrahedron: Asymmetry* **1992**, *3*, 1455; K. Ishihara, M. Kubota, H. Yamamoto, *Synlett* **1994**, 611.

[5] D. A. Evans, J. C. Anderson, M. K. Taylor, *Tetrahedron Lett.* **1993**, *34*, 5563.

[6] R. Wegler, *Liebigs Ann. Chem.* **1932**, *498*, 62; R. Wegler, *ibid.* **1933**, *506*, 77; R. Wegler, *ibid.* **1934**, *510*, 72.

[7] P. J. Weidert, E. Geyer, L. Horner, *Liebigs Ann. Chem.* **1989**, 533.

[8] E. Vedejs, O. Daugulis, S. T. Diver, *J. Org. Chem.* **1996**, *61*, 430.

[9] E. Vedejs, O. Daugulis, *J. Am. Chem. Soc.* **1999**, *121*, 5813.

[10] E. Vedejs, X. Chen, *J. Am. Chem. Soc.* **1996**, *118*, 1809.

[11] E. Vedejs, X. Chen, *J. Am. Chem. Soc.* **1997**, *119*, 2584.

[12] a) J. C. Ruble, G. C. Fu, *J. Org. Chem.* **1996**, *61*, 7230; b) J. C. Ruble, H. A. Latham, G. C. Fu, *J. Am. Chem. Soc.* **1997**, *119*, 1492; c) J. C. Ruble, J. Tweddell, G. C. Fu, *J. Org. Chem.* **1998**, *63*, 2794; d) B. Tao, J. C. Ruble, D. A. Hoic, G. C. Fu, *J. Am. Chem. Soc.* **1999**, *121*, 5091.

[13] C. E. Garrett, G. C. Fu, *J. Am. Chem. Soc.* **1998**, *120*, 7479.

[14] T. Kawabata, M. Nagoto, K. Takasu, K. Fuji, *J. Am. Chem. Soc.* **1997**, *119*, 3169.

[15] T. Oriyama, Y. Hori, K. Imai, R. Sasaki, *Tetrahedron Lett.* **1996**, *37*, 8543.

[16] a) R. Noyori, S. Hashiguchi, *Acc. Chem Res.* **1997**, *30*, 97; b) S. Hashiguchi, A. Fujii, K.-J. Haack, K. Matsumura, T. Ikariya, R. Noyori, *Angew. Chem.* **1997**, *109*, 300; *Angew. Chem. Int. Ed. Engl.* **1997**, *36*, 288; c) K.-J. Haack, S. Hashiguchi, A. Fujii, T. Ikariya, R. Noyori, *Angew. Chem.* **1997**, *109*, 297; *Angew. Chem. Int. Ed. Engl.* **1997**, *36*, 285.

Copper-Catalyzed Enantioselective Michael Additions: Recent Progress with New Phosphorus Ligands

Norbert Krause

Lehrstuhl für Organische Chemie II, Universität Dortmund, Germany

The enantioselective Michael addition of a chiral organometallic reagent to a prochiral substrate is an attractive method for creating a center of chirality in an organic molecule [1]. For this purpose, chirally modified organocopper compounds of the composition RCu(L*)Li can be used; the chiral nontransferable ligand L* controls the stereochemical course of the transfer of the group R to the substrate **1**. By using stoichiometric amounts of these "chiral cuprates", the research groups of Bertz, Corey, Dieter, Rossiter and Tanaka obtained the 1,4-adducts **2** with good enantioselectivities (over 90 % *ee* in some cases). Naturally occurring alcohols and amines were used as chiral ligand L* (e.g., ephedrine and proline derivatives) [1]. However, these investigations also revealed two fundamental problems of enantioselective Michael additions:

1. In solution, organocopper compounds show a dynamic behavior, with equilibria between several species. If this leads to the formation of achiral, but more reactive cuprates, a loss of enantioselectivity is unavoidable. Therefore, it is crucial to develop chiral reagents which react so rapidly with the substrate that undesired competing reactions are suppressed.

2. Many chiral organocopper reagents exhibit a high substrate specificity; that is, they give good stereoselectivities with only one or very few Michael acceptors.

Both problems may be solved by taking advantage of the concept of *ligand-accelerated catalysis* [2], which involves catalytic reactions characterized by dynamic ligand exchange processes. The presence of a suitable ligand can lead to the formation of a highly reactive and selective catalyst by self-assembly. If a chiral ligand is used, a stereoselective reaction may be favored over a nonselective one. Other advantages of catalytic transformations compared to their stoichiometric counterparts are the more efficient use of the metal and the chiral ligand, as well as the minimization of waste production.

Michael additions of organolithium, Grignard, and diorganozinc reagents to enones and other α,β-unsaturated carbonyl compounds are catalyzed inter alia by copper, nickel and cobalt salts. The best results are obtained with copper(I) catalysts, especially those in which copper is bound to a "soft", readily polarizable center (sulfur or phosphorus).

The first reaction of this type was reported by Lippard et al. in 1988: the reaction of 2-cyclohexenone (**3**) with Grignard reagents in the presence of the chiral aminotroponeimine copper complex **5** as catalyst gave the 1,4-adducts **4** with 4–14 % *ee* [3a]. The selectivity was increased to 74 % *ee* by addition of hexamethylphosphoric triamide (HMPA) and silyl halides [3b].

RCu(L*)Li

or R–M / CuX (cat.) / L* (cat.)

1 **2**

Formula 1

RMgX or R_2Zn

Cat*

3 → 4

Cat* =

5 (4-14% ee [3a])
(max. 74% ee; + $tBu_2PhSiCl$ / HMPA[3b])

6 (max. 60% ee[4])

7 (60-72% ee; + HMPA[5])

8 (32% ee[6])

9 (67-90% ee[7])

10 (63% ee[8])

Formula 2

Spescha et al. [4] used the copper complex **6**, which was obtained from a thioglucofuranose derivative, as catalyst for 1,4–additions of Grignard reagents to **3**, and observed enantioselectivities of up to 60 % *ee*. The dihydrooxazolylthiophenolato copper complex **7** was employed by Pfaltz et al. [5] for the enantioselective catalysis of Michael additions to cyclic enones; the best results were obtained with tetrahydrofuran as solvent and HMPA as additive. There was a pronounced dependence of the stereoselectivity on the ring size of the substrate: 16–37 % *ee* for 2-cyclopentenone, 60–72 % *ee* for **3**, and 83–87 % *ee* for 2–cycloheptenone. Alexakis et al. [6] used the heterocycle **8**, which is readily accessible from ephedrine, as chiral ligand for the copper(I)-catalyzed addition of diethylzinc to **3**, and obtained ketone **4** (R = Et) with 32 % *ee*. Another neutral phosphorus ligand is the proline derivative **9**, with which Kanai and Tomioka [7] obtained 67–90 % *ee* in the Cu(I)-catalyzed 1,4–addition of Grignard reagents to 2-cyclohexenone (**3**). Recently, Feringa et al. [8] reported the use of phosphorus amidites of type **10** in Michael additions of diethylzinc to enones. These reactions are catalyzed by copper(I) salts and by copper(II) triflate; when **3** was used as substrate, stereoselectivities of 60 % *ee* (with CuOTf) and 63 % *ee* [with $Cu(OTf)_2$] were obtained (a much higher value of 81 % *ee* resulted with 4,4–dimethyl-2-

R^2MgX or $R^2{}_2Zn$, Cat*: **11** → **12**

Cat* =

13 ($R^1 = R^2$ = Me: 76% ee[5])

10 (R^1 = Ph, R^2 = Et: 87% ee[8]) / $Cu(OTf)_2$

Formula 3

cyclohexenone in the presence of **10** and $Cu(OTf)_2$). Even when $Cu(OTf)_2$ is employed, the actual chiral catalyst is probably a copper(I) species which is formed by in situ reduction of the copper(II) complex. In all these cases, the regioselectivities (1,4- vs 1,2-addition) and chemical yields are acceptable or good.

As these examples show, cyclic enones are normally used as substrates for copper-catalyzed enantioselective Michael additions. In some cases, however, good stereoselectivities were also attained with acyclic enones of type **11**. Thus, van Klaveren, van Koten et al. [9] employed the copper arenethiolate **13** as catalyst for the 1,4–addition of methylmagnesium iodide to benzylideneacetone and obtained adduct **12** ($R^1 = R^2$ = Me) with 76 % *ee*. Interestingly, this catalyst is not suitable for analogous additions to cyclic enones; similarly, the structurally related copper thiolate **7** does not catalyze enantioselective Michael additions of Grignard reagents to acyclic enones of type **11**. With **7** and **13**, complicated nonlinear relationships between the enantiomeric excesses of the catalyst and the product were observed, indicating that the product is formed via different organocopper intermediates in several reaction pathways.

Progress with regard to this undesired substrate specificity was achieved with the phosphorus amidite **10**, which catalyzes not only Michael additions of organozinc reagents to cyclic enones but also to chalcone (**11**; R^1 = Ph) and related acyclic substrates. In the case of the addition of diethylzinc to chalcone, a good enantioselectivity of 87 % *ee* was observed [8].

The last example shows that neutral binaphthol phosphorus ligands are particularly well suited for enantioselective Michael additions. On the basis of this structural feature, Feringa et al. [10] achieved a breakthrough. By combining the C_2-symmetrical axially chiral binaphthol with the C_2-symmetrical bis(1–phenylethyl)amine through a phosphorus center, they obtained the new ligand **15**, which can be used in highly enantioselective copper-catalyzed Michael additions of diorganozinc reagents to numerous cyclic enones. Here, the steric properties of the substrate and the reagent are unimportant, since the transfer of methyl, ethyl and isopropyl groups to 2-cyclohexenone takes place with high enantioselectivities ($\geq$ 94 % *ee*) and good chemical yields (72–95 %), as does the transfer of an ethyl group to 4,4-diphenyl-2-cyclohexenone. Additional functionalities can be introduced into the addition product in several ways without affecting the stereochemical course of the reaction. Thus, the addition of diethylzinc to 4,4–dimethoxy-2,5-cyclohexadienone (**17**), catalyzed by **15** and $Cu(OTf)_2$, gives adduct **18** with 94 % *ee*. The functionalized zinc reagent $[AcO(CH_2)_5]_2Zn$, prepared by hydroboration of the corresponding alkene and transmetalation according to Knochel's method, can be added to 2-cyclohexenone with excellent selectivity of 95 % *ee*. Finally, the zinc enolates formed in these transformations can also be treated with more complex electrophiles instead of protons. For example, the three-component coupling of 2–cyclohexenone, diethylzinc and benzaldehyde in the presence of **15** and $Cu(OTf)_2$ fur-

R²₂Zn, Cu(OTf)₂ (cat.)

14 15 (cat.) 16

> 98% ee > 98% ee 94% ee > 98% ee

95% ee 95% ee

Et_2Zn

$Cu(OTf)_2$ (cat.), **15** (cat.)

17 18 94% ee

Formula 4

$R^2{}_2Zn$, $Cu(OTf)_2$ (cat.)

3 19 (cat.) 4

R = Me: 95% ee

R = Et: 90% ee

Formula 5

nishes the expected hydroxyketone with 95 % *ee*; however, the diastereoselectivities of these transformations are not yet satisfactory [10].

With the combination of phosphorus amidite **15** and $Cu(OTf)_2$, a catalyst has been developed for the first time which can be applied with confidence to almost any combination of Michael acceptor and organozinc reagent. This system apparently exploits the principle of ligand-accelerated catalysis. However, an important limitation has to be taken into account: only enones with six-membered rings give high stereoselectivities (the analogous reactions of diethylzinc with 2-cyclopentenone and 2-cycloheptenone gave the adducts with 10 and 53 % *ee*, respectively). Another recent report by Pfaltz et al. [11] underlines the fact that the bridging of binaphthol and a chiral amine through a phosphorus center is a general feature of catalysts which are suitable for highly enantioselective Michael additions. In the presence of the phosphite **19** with oxazoline structure, the 1,4-additions of dimethyl- and diethylzinc to **3** also give enantioselectivities of 95 and 90 % *ee*, respectively. Hopefully, this major progress will trigger further investigations leading to new chiral copper catalysts which can be applied even more generally to cyclic Michael acceptors with different ring sizes as well as to acyclic substrates [12].

References

[1] Reviews: a) B.E. Rossiter, N.M. Swingle, *Chem. Rev.* **1992**, *92*, 771–806; b) N. Krause, *Kontakte (Darmstadt)* **1993**, (1), 3–13; c) N. Krause, A. Gerold, *Angew. Chem.* **1997**, *109*, 194–213; *Angew. Chem. Int. Ed. Engl.* **1997**, *36*, 186–204.

[2] D.J. Berrisford, C. Bolm, K.B. Sharpless, *Angew. Chem.* **1995**, *107*, 1159–1171; *Angew. Chem. Int. Ed. Engl.* **1995**, *34*, 1050–1064.

[3] a) G.M. Villacorta, C.P. Rao, S.J. Lippard, *J. Am. Chem. Soc.* **1988**, *110*, 3175–3182; b) K.H. Ahn, R.B. Klassen, S.J. Lippard, *Organometallics* **1990**, *9*, 3178–3181.

[4] M. Spescha, G. Rihs, *Helv. Chim. Acta* **1993**, *76*, 1219- 1230.

[5] a) Q.-L. Zhou, A. Pfaltz, *Tetrahedron Lett.* **1993**, *34*, 7725–7728; b) Q.-L. Zhou, A. Pfaltz, *Tetrahedron* **1994**, *50*, 4467–4478; cf.: c) Y. Takemoto, S. Kuraoka, N. Hamaue, C. Iwata, *Tetrahedron: Asymmetry* **1996**, *7*, 993–996; d) Y. Takemoto, S. Kuraoka, N. Hamaue, K. Aoe, H. Hiramatsu, C. Iwata, *Tetrahedron* **1996**, *52*, 14177–14188.

[6] A. Alexakis, J. Frutos, P. Mangeney, *Tetrahedron: Asymmetry* **1993**, *4*, 2427–2430.

[7] M. Kanai, K. Tomioka, *Tetrahedron Lett.* **1995**, *36*, 4275–4278.

[8] A.H.M. de Vries, A. Meetsma, B.L. Feringa, *Angew. Chem.* **1996**, *108*, 2526–2528; *Angew. Chem. Int. Ed. Engl.* **1996**, *35*, 2374–2376.

[9] a) F. Lambert, D.M. Knotter, M.D. Janssen, M. van Klaveren, J. Boersman, G. van Koten, *Tetrahedron: Asymmetry* **1991**, *2*, 1097–1100; b) G. van Koten, *Pure Appl. Chem.* **1994**, *66*, 1455–1462; c) M. van Klaveren, F. Lambert, D.J.F.M. Eijkelkamp, D.M. Grove, G. van Koten, *Tetrahedron Lett.* **1994**, *35*, 6135–6138.

[10] B.L. Feringa, M. Pineschi, L.A. Arnold, R. Imbos, A.H.M. de Vries, *Angew. Chem.* **1997**, *109*, 2733–2736; *Angew. Chem. Int. Ed. Engl.* **1997**, *36*, 2620–2623.

[11] A.K.H. Knöbel, I.H. Escher, A. Pfaltz, *Synlett* **1997**, 1429–1431.

[12] Recent reports on copper-catalyzed enantioselective Michael additions: a) V. Wendisch, N. Sewald, *Tetrahedron: Asymmetry* **1997**, *8*, 1253–1257; b) N. Sewald, V. Wendisch, *Tetrahedron: Asymmetry* **1998**, *9*, 1341–1344; c) A.H.M. De Vries, R.B. Hof, D. Staal, R.M. Kellogg, B.L. Feringa, *Tetrahedron: Asymmetry* **1997**, *8*, 1539–1543; d) E. Keller, J. Maurer, R. Naasz, T. Schader, A. Meetsma, B.L. Feringa, *Tetrahedron: Asymmetry* **1998**, *9*, 2409–2413; e) R. Naasz, L.A. Arnold, M. Pineschi, E. Keller, B.L. Feringa, *J. Am. Chem. Soc.* **1999**, *121*, 1104–1105; f) D. Seebach, G. Jaeschke, A. Pichota, L. Audergon, *Helv. Chim. Acta* **1997**, *80*, 2515–2519; g) A. Alexakis, J. Vastra, J. Burton, P. Mangeney, *Tetrahedron: Asymmetry* **1997**, *8*, 3193–3196; h) A. Alexakis, J. Burton, J. Vastra, P. Mangeney, *Tetrahedron: Asymmetry* **1997**, *8*, 3987–3990; i) A. Alexakis, J. Vastra, J. Burton, C. Benhaim, P. Mangeney, *Tetrahedron Lett.* **1998**, *39*, 7869–7872; j) E.L. Stangeland, T. Sammakia, *Tetrahedron* **1997**, *53*, 16503–16510; k) F.-Y. Zhang, A.S.C. Chan, *Tetrahedron: Asymmetry* **1998**, *9*, 1179–1182; l) M. Yan, L.-W. Yang, K.-Y. Wong, A.S.C. Chan, *Chem. Commun.* **1999**, 11–12. m) T. Mori, K. Kosaka, Y. Nakagawa, Y. Nagaoka, K. Tomioka, *Tetrahedron: Asymmetry* **1998**, *9*, 3175–3178; n) Y. Nakagawa, M. Kanai, Y. Nagaoka, K. Tomioka, *Tetrahedron* **1998**, *54*, 10295–10307; o) M. Kanai, Y. Nakagawa, K. Tomoika, *Tetrahedron* **1999**, *55*, 3843–3854; p) S.M.W. Bennett, S.M. Brown, J.P. Muxworthy, S. Woodward, *Tetrahedron Lett.* **1999**, *40*, 1767–1770; q) A. Alexakis, C. Benhaim, X. Fournioux, A. van der Heuvel, J.-M. Leveque, S. March, S. Rosset, *Synlett* **1999**, 1811–1813; r) A. Alexakis, *Chima* **2000**, *54*, 55–56; s) O. Pamies, G. Net, A. Ruiz, C. Claver, *Tetrahedron: Asymmetry* **1999**, *10*, 2007–2014; t) M. Yan A. S. C. Chen, *Tetrahedron Lett.* **1999**, *40*, 6645–6648; u) X. Hu, H. Chen, X. Zhang, *Angew. Chem.* **1999**, *111*, 3720–3723; *Angew. Chem. Int. Ed.* **1999**, *38*, 3518–3521; v) J. P. G. Versleijen, A. M. van Leusen, B. L. Feringa, *Tetrahedron Lett.* **1999**, *40*, 5803–5806; w) R. Imbos, M. H. G. Brilman, M. Pineschi, B. L. Feringa, *Org. Lett.* **1999**, *1*, 623–625.

C_3-Symmetric Ligands for Catalysis

Mark Mikulás and Karola Rück-Braun

Institut für Organische Chemie, Universität Mainz, Germany

Metalloenzymes with copper, iron or manganese in the active center are involved in biological oxidations and oxygenations as well as in other enzymatic transformations [1]. For the better understanding of the function of these biological metal catalysts numerous model complexes with suitable ligands were synthesized and their physical and chemical properties were investigated. Tridentate amido ligands have been employed successfully for years to simulate the enzymatic processes of non-heme metalloproteins [1–4]. Recently, an increasing interest in such ligands and their chiral analogs has arisen in the area of catalytic asymmetric synthesis. Therefore, a series of tripodal ligands and their current applications are presented here.

Auxiliary Ligands in Bioinorganic Chemistry

Nitrogen-containing, acyclic and cyclic C_3-symmetric ligands, such as hydrotris(pyrazolyl) borates [1], tris[(2-pyridyl)methyl] amines [4] or 1,4,7-triazacyclononanes [5] are well established tools in bioinorganic chemistry for the preparation of biomimetic model compounds (see Schemes 1 and 2). The oxygen-carrying protein hemocyanin from mollusks (e.g. squids) and arthropods (e.g. spiders) contains in the deoxy form two neighboring Cu(I) centers [1]. Model compound **1** bearing hydrotris(pyrazolyl)borate ligands has been developed to clarify the fast and reversible dioxygen binding by this metalloenzyme.

Protein crystallographic data give evidence for two Cu(II) centers, which are bridged by the side-

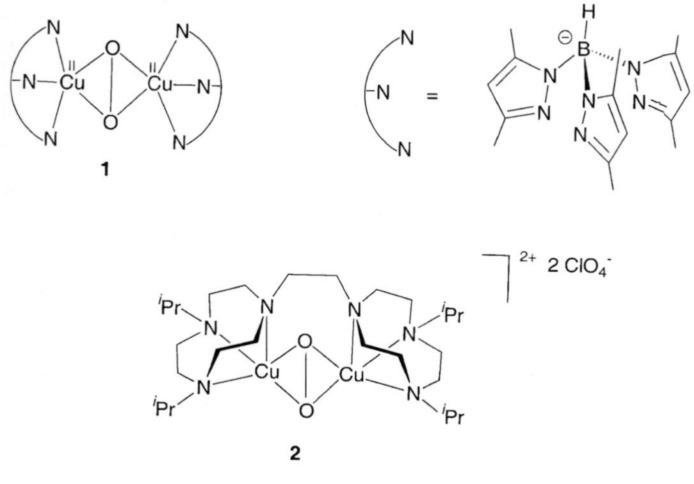

Scheme 1

on coordinated peroxide ion (O_2^{2-}) with μ-η^2:η^2 bonds in oxy-hemocyanin as well as in model compound **1**. Copper-dioxygen complexes display a vast variability of the Cu-O bonding geometry. Binuclear copper(I) complexes of bis(triazacyclononane) ligands and their dioxygen adducts, e.g. **2**, appropriate mononuclear compounds and iron-containing analogs are further efficient model systems for the examination of the assimilation and activation of oxygen by metalloproteins such as hemocyanin, tyrosinase, ribonucleotide reductase or methane monooxygenase [1]. The structure and reactivity of these catalytically active systems depends primarily on the properties of the ligands used. Significant structural differences are already caused by the presence or absence of an ethylene bridge between tridentate triazacyclononane ligands of binuclear copper complexes (see compound **2**) [5]. Therefore, structure-reactivity relationships remain largely unexplained.

Methane monooxygenase consists of a catalytically active diiron center. In the presence of oxygen this enzyme oxidizes methane and other hydrocarbons, i.e. molecules without any anchoring group for the metal center, with a certain chemo- and regiospecifity [6]. Because of the importance of the controlled functionalization of non-activated aliphatic carbon-hydrogen bonds, it is not surprising that nowadays biomimetic catalyst systems are finding increasing attention.

Biomimetic Bleaching Agents and Oxidants

Some of the first catalytic model systems for the simulation of the function of methane monooxygenase comprise monomeric as well as dimeric iron-containing model complexes bearing hydrotris(pyrazolyl)borate ligands [6]. These complexes, e.g. **3**, catalyze the oxidation of aromatic and aliphatic carbon-hydrogen bonds in the presence of oxygen (1 atm), acetic acid and zinc powder at room temperature (Scheme 2).

The conversion of cyclohexane using this catalyst system yields cyclohexanol highly selectively, as well as cyclohexanone (22 : 1). There is also a great interest in the design of biomimetic metal catalysts for the hydroxylation of al-

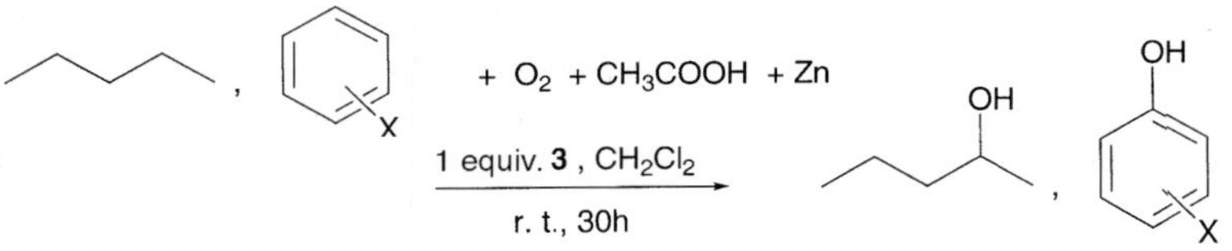

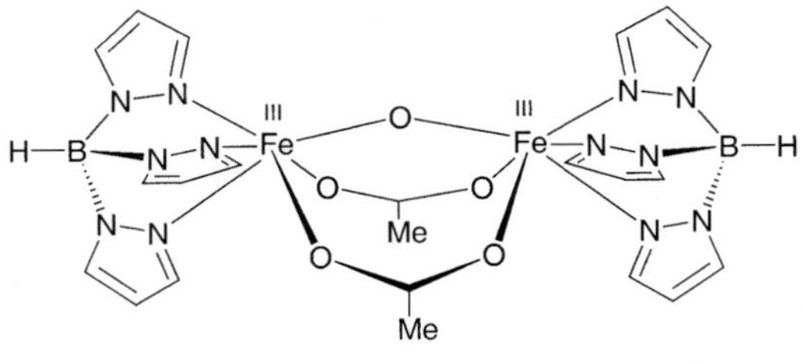

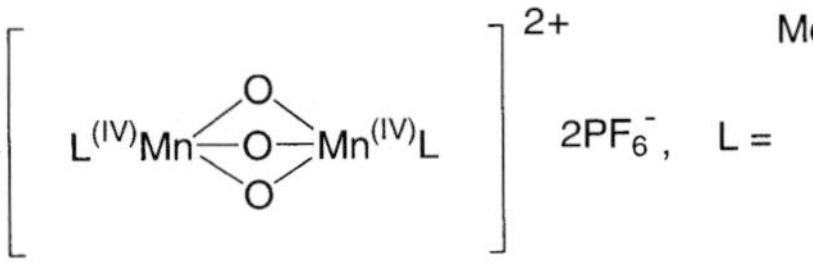

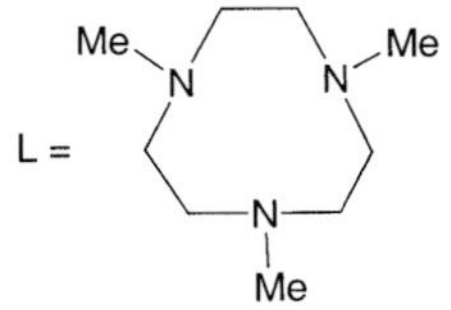

Scheme 2

kanes, alkenes and arenes in the presence of peroxides such as hydrogen peroxide or *tert*-butyl hydroperoxide. In this context the development of new bleaching agents with high activity at low temperature and good environmental compatibility is of great commercial interest, too. Recently, ecologically tolerable iron and manganese complexes have been synthesized and tested for their bleaching activity with hydrogen peroxide [7]. During these industrial studies, the binuclear manganese complex **4** (Scheme 2) proved to be an efficient catalyst (pH 9–10). By oxidation of styrene and 4–vinylbenzoic acid with complex **4** in the presence of hydrogen peroxide in water or acetone (pH 9) at room temperature, the corresponding epoxides were successfully obtained within 5 h or 2 h, respectively. Allylic oxidation products were not observed. In this, as well as in other oxidations, achiral triple-alkylated 1,4,7-triazacyclononane ligands such as **5** have been tested successfully [7–10].

Triazacyclononane ligands containing polyfluoroalkyl side chains at the nitrogen atom were prepared based on the fluorous biphasic catalysis concept (FBC) developed by Horváth and Rábai for catalysis in homogenous systems, and were employed in oxidations of alkanes and alkenes with tert-butyl hydroperoxide and oxygen, in order to achieve an effective separation of the mangenese catalyst from the product [8]. Recently, immobilized ligands were utilized for the manganese-mediated oxidation of alkenes to epoxides with hydrogen peroxide [9]. The application of complex **4** (Scheme 2) for the oxidation of benzylic alcohols to benzaldehydes has been described by Feringa and co-workers [10]. In this case, high turnover numbers in the range 280–1000 and high selectivities were obtained.

These few examples already demonstrate that polyfunctional ligands with C_3 symmetry permit unusual complex geometries with metals in a multitude of oxidation states. As a rule, extremely stable complexes are obtained, especially of early transition metals and metals in low oxidation states.

Enantioselective Oxidations

For quite some time now, chiral, enantiomerically pure tripodal ligands bearing C_3 symmetry have been actively developed [2]. C_2-symmetric optically active compounds are well-established ligands and auxiliaries in asymmetric synthesis. Numerous examples prove the high stereoselectivities achieved with chiral C_2-symmetric ligands in transition metal-catalyzed reactions [2,3]. The efficiency of appropriate catalyst systems is explained by the reduced number of possible diastereomeric transition states. With a bidentate C_2-symmetric ligand, a square-planar metal complex has two remaining identical (homotopic) coordination sites. An octahedral complex bearing a chiral bidentate C_2-symmetric

Mn(OAc)$_2$, **7**, R = CH$_2$CH(OH)Me
H$_2$O$_2$, MeOH, 0°C
6 → **8** 40% *ee*

Cu(OTf)$_2$, **9**
tert-butyl perbenzoate, 0°C, 92h
44%, 84% *ee*

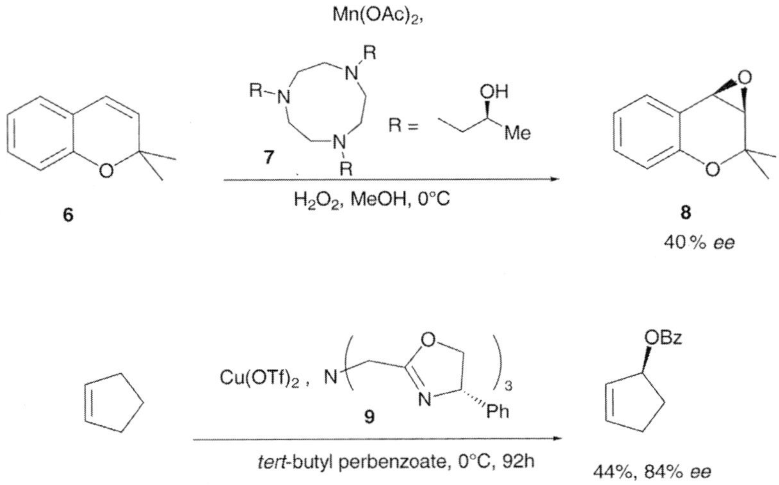

Scheme 3

ligand offers two inequivalent (diastereotopic) coordination sites (axial and equatorial) for substrate binding. However, with C_3-symmetric ligands in octahedral environment, three equivalent (homotopic) positions are obtained. Because of the three fold symmetry, an effective steric shielding should lead to high enantioselectivities in asymmetric synthesis. Thus, a precondition for the rational application of C_3-symmetric ligands is precise knowledge of the complex geometry in the selectivity-dependent reaction step. Recently, chiral C_3-symmetric 1,4,7–triazacyclononanes have been employed in catalytic enantioselective epoxidations for the first time [11]. Oxidation of the chromene **6** with excess 30 % aqueous hydrogen peroxide was achieved in methanol using manganese(II) acetate and ligand **7** (3 mol%, 1 : 1.5) yielding the epoxide **8** with (3*R*,4*R*)-configuration after 15 h and 50 % conversion with 40 % *ee* (Scheme 3).

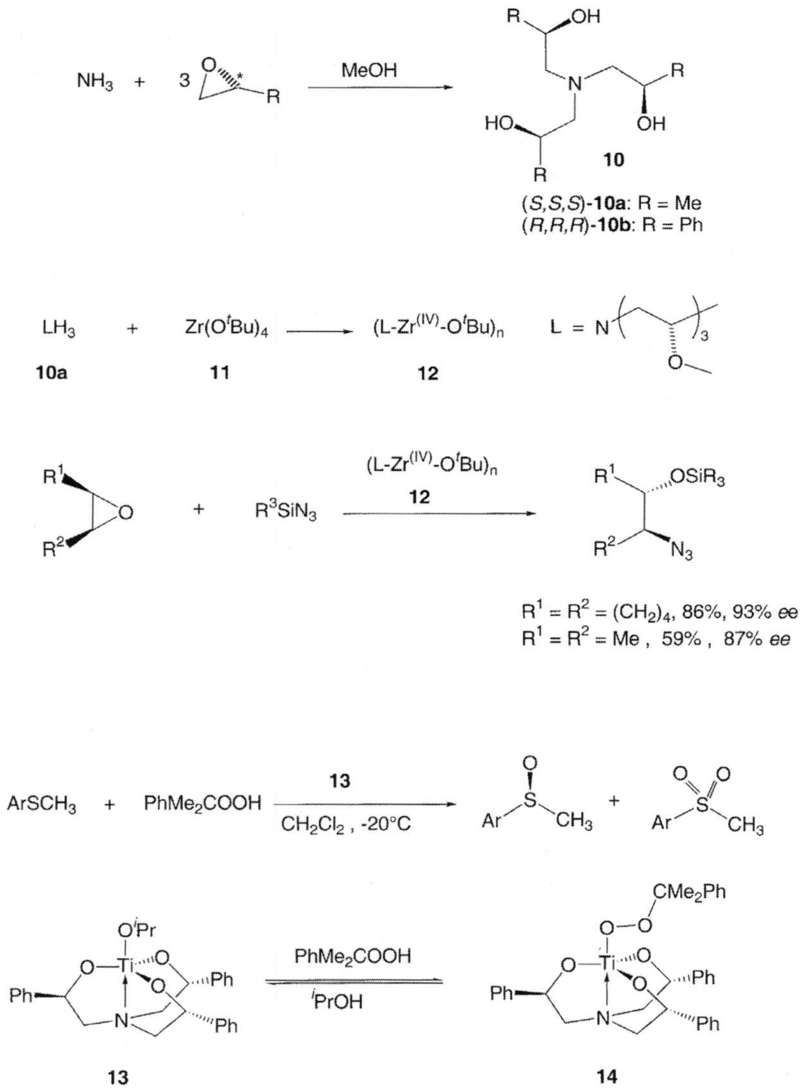

Scheme 4

From $Cu(OTf)_2$ or Cu(OTf) and the chiral C_3-symmetric tris(oxazoline) ligand **9**, copper complexes are obtained that are capable of catalyzing the allylic oxidation of cyclopentene by *tert*-butyl perbenzoate in up to 84 % *ee* [12]. Even today, for most oxidations with chiral or achiral ligand systems, the structures of the real active metal catalysts are unknown. Because of this it is difficult to give a scientific rationale for the selectivities and inductions observed.

Chiral trialkanolamines **10** represent yet another class of chelating polyfunctional chiral ligands for use in asymmetric synthesis (Scheme 4) [13–18].

These tetradentate C_3-symmetric ligands allow the synthesis of stable alkoxy complexes of the early transition metals. Ligands **10** can be prepared by the reaction of ammonia with chiral epoxides in yields ranging from 40 to 97 % [14,17]. Chiral trialkanolamine complexes with titanium(IV) as well as zirconium(IV) centers have been employed in asymmetric catalysis so far. The ring opening of *meso* epoxides was successfully accomplished at 0–25 °C with silyl azides in high enantiomeric excess (83–93 %), when a pre-prepared oligomeric complex obtained from **10a** and **12** was used [13,17,18]. A two-centered zirconium(IV) complex is discussed as the catalytically active species.

In contrast, titanium(IV) isopropoxide and ligand **10a** form the monomeric highly reactive ti-

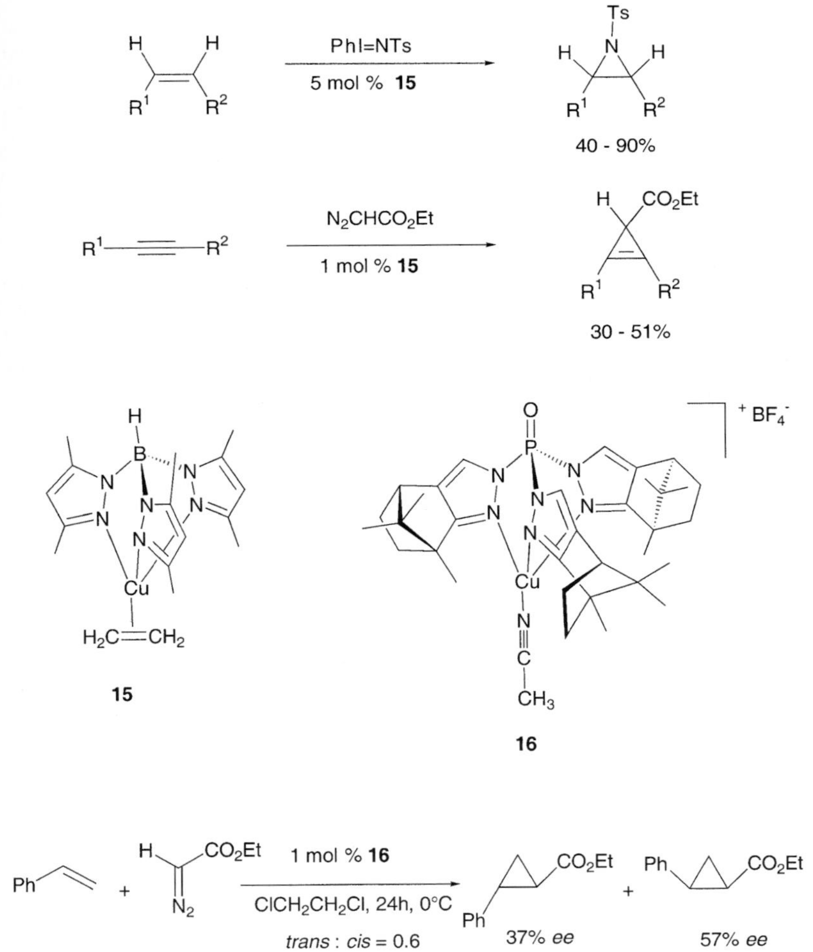

PhI=NTs, 5 mol % **15** — 40 - 90%

N_2CHCO_2Et, 1 mol % **15** — 30 - 51%

15 **16** $^{+}BF_4^{-}$

1 mol % **16**, $ClCH_2CH_2Cl$, 24h, 0°C; *trans* : *cis* = 0.6 — 37% *ee* + 57% *ee*

Scheme 5

tanium(IV) complex **13**, which has been employed successfully in the catalytic enantioselective oxidation of sulfides to sulfoxides (Scheme 4) [15–17]. On account of the distinct electrophilicity of the peroxotitanium(IV) complex, a high selectivity is achieved. Therefore, sulfones could only be detected gas-chromatographically. The best selectivities were obtained for **10b** and Ar = MeC_6H_4 (40 % *ee*). This stereoselection is caused by two processes. After the preferred oxidation furnishing the chiral (*S*)-configured sulfoxide a kinetic resolution during further oxidation to the sulfone leads to an additional accumulation of the (*S*)-configured product by removing more of the (*R*)-configured sulfoxide.

Tris(pyrazolyl) Ligands

Metal catalysts with tripodal ligands are in no way restricted to oxidations in their application.

For instance, the reaction of hydrotris(pyrazolyl)borate-copper complex **15** (Scheme 5) with alkenes and ethyl diazoacetate results in the formation of cyclopropanes in yields in the range 63–78 % [19].

Starting from alkynes, cyclopropenes were obtained under mild reaction conditions at room temperature (30–51 %). Aziridines were isolated in 40–90 % yield with catalyst **15** using [(*p*-tolylsulfonyl)imino]phenyliodinane, PhI = NTs. The first chiral tris(pyrazolyl) ligand was prepared as early as 1992 by Tolman and co-workers starting from a camphor-pyrazole derivative and $POCl_3$ [20]. With $[Cu(CH_3CN)_4]BF_4$ this compound forms the cationic monomeric copper(I) complex **16**. The latter catalyzes the cyclopropanation of styrene with ethyl diazoacetate in up to 60 % *ee*.

In recent years the synthesis of chiral and achiral tripodal phosphines and their application in homogeneous catalysis has been studied in more detail [2]. Enantiomerically pure tripodal ligands were synthesized from the corresponding trichloro compounds and chiral, cyclic lithiophosphanes, e.g. **17**, (Scheme 6) [21,22]. Using a rhodium(I) complex of ligand **18**, an enantiomeric excess of 89 % was obtained in the asymmetric hydrogenation reaction of methyl acetamidocinnamate (**19**).

With dimethyl itaconate (**20**) an enantiomeric excess of 94 % was achieved, whereas with the corresponding C_2-symmetric bidentate ligand lower enantiomeric excesses were observed. However, the reactions with catalyst **18** proceed at higher temperature and also require longer reaction times [22].

Numerous achiral ligands containing C_3 symmetry are known today [2, 3]. The application of their metal complexes in organic synthesis will undoubtedly lead to many surprises. Suitably, the first applications display biomimetic characteristics. The results obtained should lead to a more general application of achiral and chiral tripodal ligands in catalysis.

C_3-symmetry has the potential to be a key element in achieving high stereoselectivities in enantioselective transformations.

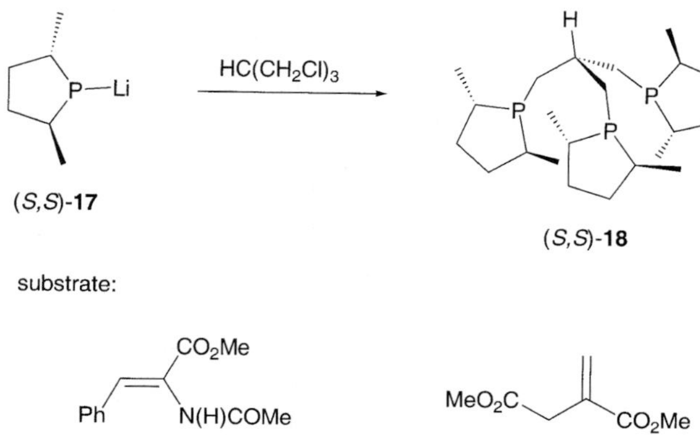

P—Li

(*S,S*)-**17**

$HC(CH_2Cl)_3$

H

P P P

(*S,S*)-**18**

substrate:

CO_2Me

Ph N(H)COMe

19

MeO_2C CO_2Me

20

Scheme 6

References

[1] Review: Y. Moro-oka, K. Fujisawa, N. Kitajima, *Pure Appl. Chem.* **1995**, *67*, 241–248, and references cited.

[2] Review: M. C. Keyes, W. B. Tolman in *Advances in Catalytic Processes* (ed.: M. P. Doyle), JAI Press Inc., Greenwich **1997**, 189, and references cited therein.

[3] Review: C. Moberg, *Angew. Chem.* **1998**, *110*, 260–281; *Angew. Chem. Int. Ed. Engl.* **1998**, *37*, 248–268, and references cited therein.

[4] C. Kim, K. Chen, J. Kim, L. Que, *J. Am. Chem. Soc.* **1997**, *119*, 5964–5965, and references cited therein.

[5] S. Mahapatra, V. G. Young, S. Kaderli, A. D. Zuberbühler, W. B. Tolman, *Angew. Chem.* **1997**, *109*, 125–127; *Angew. Chem. Int. Ed. Engl.* **1997**, *36*, 130–133, and references cited therein.

[6] N. Kitajima, M. Ito, H. Fukui, Y. Moro-oka, *J. Chem. Soc. Chem. Commun.* **1991**, 102–104, and references cited therein.

[7] R. Hage, J. E. Iburg, J. Kerschner, J. H. Koek, E. L. M. Lempers, R. J. Martens, U. S. Racherla, S. W. Russell, T. Swarthoff, M. R. P. van Vliet, J. B. Warnaar, L. van der Wolf, B. Krijnen, *Nature* **1994**, *369*, 637–639, and references cited therein.

[8] J.-M. Vincent, A. Rabion, V. K. Yachandra, R. H. Fish, *Angew. Chem.* **1997**, *109*, 2438–2440; *Angew. Chem. Int. Ed. Engl.* **1997**, *36*, 2346–2349, and references cited therein.

[9] Y. V. Subba Rao, D. E. De Vos, T. Bein, P. A. Jacobs, *Chem. Commun.* **1997**, 355–356.

[10] C. Zondervan, R. Hage, B. L. Feringa, *Chem. Commun.* **1997**, 419–420.

[11] C. Bolm, D. Kadereit, M. Valacchi, *Synlett* **1997**, 687–688.

[12] K. Kawasaki, S. Tsumura, T. Katsuki, *Synlett* **1995**, 1245–1246.

[13] W. A. Nugent, *J. Am. Chem. Soc.* **1992**, *114*, 2768–2769.

[14] W. A. Nugent, R. L. Harlow, *J. Am. Chem. Soc.* **1994**, *116*, 6142–6148.

[15] F. Di Furia, G. Licini, G. Modena, R. Motterle, *J. Org. Chem.* **1996**, *61*, 5175–5177.

[16] M. Bonchio, S. Calloni, F. Di Furia, G. Licini, G. Modena, S. Moro, W. A. Nugent, *J. Am. Chem. Soc.* **1997**, *119*, 6935–6936.

[17] Review: W. A. Nugent, G. Licini, M. Bonchio, O. Bortolini, M. G. Finn, B. W. McCleland, *Pure Appl. Chem.* **1998**, *70*, 1041–1046.

[18] W. A. Nugent, *J. Am. Chem. Soc.* **1998**, *120*, 7139–7140.

[19] P. J. Pérez, M. Brookhart, J. L. Templeton, *Organometallics* **1993**, *12*, 261–262.

[20] a) C. J. Tokar, P. B. Kettler, W. B. Tolman, *Organometallics* **1992**, *11*, 2737–2739; b) M. C. Keyes, B. M. Chamberlain, S. A. Caltagirone, J. A. Halfen, W. B. Tolman, *Organometallics* **1998**, *17*, 1984–1992.

[21] M. J. Burk, R. L. Harlow, *Angew. Chem.* **1990**, *102*, 1511–1513; *Angew. Chem. Int. Ed. Engl.* **1990**, *29*, 1467.

[22] M. J. Burk, J. E. Feaster, R. L. Harlow, *Tetrahedron: Asymmetry* **1991**, *2*, 569–592.

Highly Enantioselective Catalytic Reduction of Ketones Paying Particular Attention to Aliphatic Derivatives

Renat Kadyrov and Rüdiger Selke

Institut für Organische Katalyseforschung an der Universität Rostock, Germany

Dedicated to Professor Dr. Klaus Peseke on the occasion of his 60th birthday

The research group of Zhang at Pennsylvania State University has climbed a mountain that had long defied all attempts – the complex-catalyzed, highly enantioselective (*ee* values of over 90 %) hydrogenation of purely aliphatic ketones [1]. Such high *ee* values have up till now remained the exclusive domain of enzymatic methods [2].

It has long been possible to hydrogenate a wide range of functionalized ketones asymmetrically achieving selectivities of over 90 % *ee*. Thus, pharmaceutically important 1,2–amino alcohol derivatives have been available since 1979 with over 95 % *ee* through the hydrogenation of amino ketones. This was achieved by Hayashi *et al.* [3] using planar chiral ferrocenebis(phosphanyl)rhodium(I) catalysts and above all by application of the *control concept* developed by Achiwa *et al.* [4] with BCPM rhodium(I) catalysts (BCPM = (2*S*,4*S*)-*N*-(*tert*-butoxycarbonyl)-4-(dicyclohexylphosphanyl)-2-(diphenylphosphanylmethyl)pyrrolidine) [4b]. Very high turnover rates (100 000) can thus be achieved. Likewise, α- and *β*-ketocarboxylic acid derivatives have been reduced (› 90 % *ee*) to the hydroxy acids or corresponding derivatives with high selectiv-

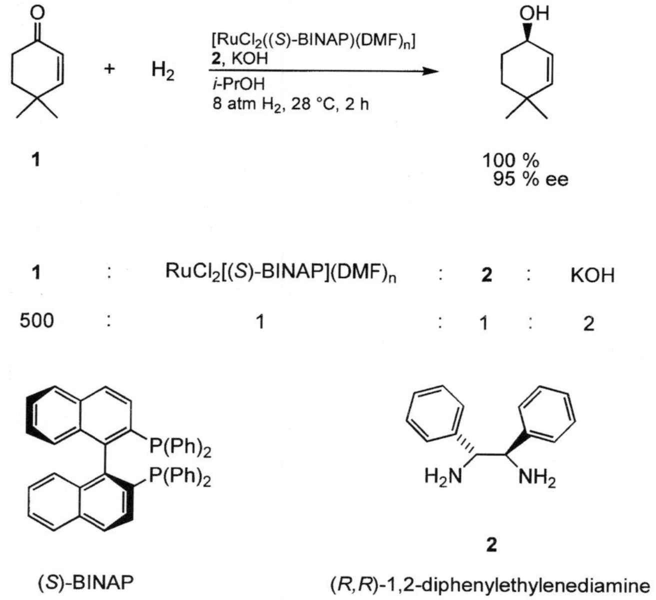

1 + H_2 → [$RuCl_2$((*S*)-BINAP)$(DMF)_n$], **2**, KOH; *i*-PrOH; 8 atm H_2, 28 °C, 2 h → 100 %, 95 % ee

1	:	$RuCl_2$[(*S*)-BINAP]$(DMF)_n$	:	2	:	KOH
500	:	1	:	1	:	2

$P(Ph)_2$ $P(Ph)_2$ — (*S*)-BINAP

H_2N NH_2 — **2** — (*R,R*)-1,2-diphenylethylenediamine

Scheme 1. Regioselective hydrogenation of unsaturated ketones.

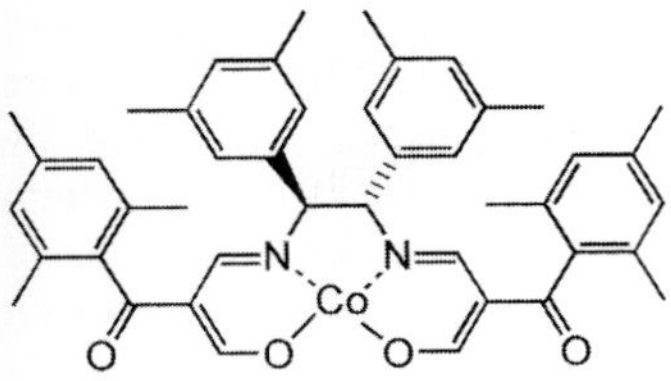

3 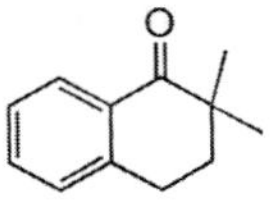 **4**

ity [4a, 5, 19a]; the most impressive results were obtained with chiral, modified heterogeneous catalysts [6]. Since the pioneering studies of Noyori *et al.* with BINAP-ruthenium(II) complexes [7] (BINAP = 2,2'-bis(diphenylphosphanyl)-1,1'-binaphthyl, see Scheme 1), the range of ketones which can be homogeneously hydrogenated with *ee* values of well over 90 % has been expanded to encompass among others γ-keto esters, hydroxy ketones, α- and β-diketones, and even phenyl thioketones [7f]. Also the important regioselective hydrogenation of unsaturated, and above all cyclic, ketones such as **1** was possible with 95 % *ee* (Scheme 1) [7d, e]. In this case, the otherwise preferred hydrogenation of the olefinic double bond [7g] is completely suppressed by the addition of a chiral diamine such as (*R*,*R*)-1,2–diphenylethylenediamine (**2**), which serves as a selectivity-promoting modifier (Scheme 1). Moreover, Noyori *et al.* have recently shown that the selective power of the enantiomerically pure diamine **2** alone is sufficient for a successful reaction, as the enantioselectivity only falls from 96 to 95 % if racemic TolBINAP[7h] is used instead of (*R*)-TolBINAP (TolBINAP is a BINAP derivative in which the four phenyl groups attached to the phosphorus atoms are replaced by four *p*-tolyl groups).

The transformation of unfunctionalized alkyl aryl ketones such as acetophenone and its analogs to the corresponding chiral alcohols was for a long time possible only by asymmetric hydrosilylation [8] or hydroboration [9], giving enantioselectivities of around 95 %. However, in 1995 this reaction was accomplished by activation of BINAP-ruthenium catalysts with chiral diamines and KOH [10]. In its optimized form the reaction needs less than 10^{-4} mol % of catalyst [10b]. Similarly high selectivities were also reported for cyclic alkyl aryl ketones with BINAP-iridium catalysts [11]. Use of sodium borohydride as the reductant and the cobalt catalyst **3** with tetradentate ligands derived from Schiff bases afforded reduction of ketone **4** with 94 % *ee*. However, the corresponding reaction with acetophenone was unsatisfactory, leading only to 68 % *ee* [12].

A successful approach in the field of transfer hydrogenation has been developed by Noyori [13a] using formic acid or isopropyl alcohol as the hydrogen source and ruthenium catalysts containing chiral N⌒P chelating ligands such as **5** or monosulfonated bisamines such as **6**. This breakthrough is particularly significant in the case of alkyl aryl ketones or ketones with triple bonds in the α position [13b]. The excess hydrogen pressure, normally necessary for ketone hydrogenations, is avoided here by the high concentration of the hydrogen donor. Enantioselectivities of between 95 and 99 % are afforded in most cases for aryl alkanols; such values can otherwise be obtained only with microbiological processes [14].

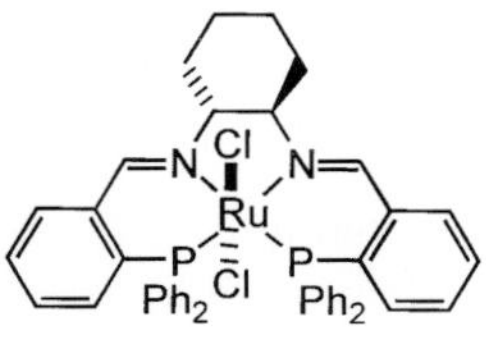

5 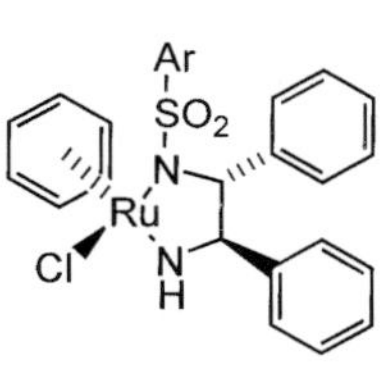 **6**

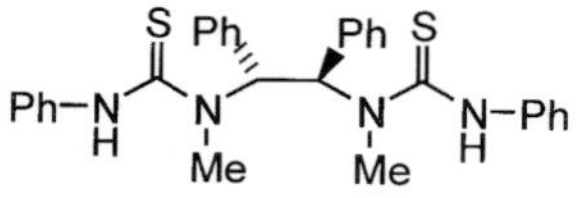

7

The number of publications describing new ligands that allow the transfer hydrogenation of aromatic ketones with over 90 % *ee* has grown in leaps and bounds since 1996 [15]. In these reactions the use of ruthenium [15a–f] and iridium [15g] as the catalytically active metals has recently been augmented by the use of phosphorus-free ligands such as chiral diamines, amino alcohols, and bisthioureas such as **7** [15a,e–g]. A ruthenium-catalyzed transfer hydrogenation with 92 % *ee* has even been reported for the aliphatic ketone pinacolone (*tert*-butyl methyl ketone) [16].

One problem that remains unsolved is the complex-catalyzed enantioselective hydrogenation of dialkyl ketones with dihydrogen gas, as these substrates lack the aryl substituent or second functional group necessary for effective enantiofacial selective binding to the catalyst. Corey *et al.* nevertheless succeeded in reducing pinacolone, in which the steric requirements of the methyl and *tert*-butyl substituents differ quite distinctly, by the hydroboration method and obtained the product alcohol in 97 % *ee* [9]. By contrast, the straight-chain octan-2–one gave an *ee* value of only 72 % [17], compared to 94 % *ee* with an alcohol dehydrogenase [2].

A particularly successful approach to the catalytic hydrogenation of dialkyl ketones with hydrogen has been the use of the heterogeneous contact catalyst system – Raney nickel chirally modified with tartaric acid [18]. Here too, selectivity is enhanced by branching of the alkyl substituent in the alkyl methyl ketones (e.g., 85 % *ee* for the hydrogenation of isopropyl methyl ketone). With

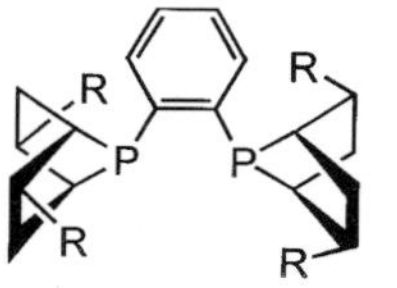

PennPhos **8a** R = Me
8b R = *i*Pr

DuPhos **9**
(R = Me, Et, *n*Pr, *i*Pr)

straight-chain ketones the highest *ee* values were achieved for hexan-2-one and octan-2-one (80 %). Catalyst selectivity was optimized by the use of pivalic acid as a modifier, which through association clearly blocks one side of the chirality-inducing tartaric acid molecules on the surface of the nickel. Owing to the sterically demanding *tert*-butyl group of the pivalic acid, the range of options for coordination of the ketone to the tartaric acid in the transition state is limited. It is, of course, perfectly conceivable that this differentiation effect is particularly pronounced for alkyl methyl ketones and decreases upon replacement of the methyl group with a larger alkyl moiety.

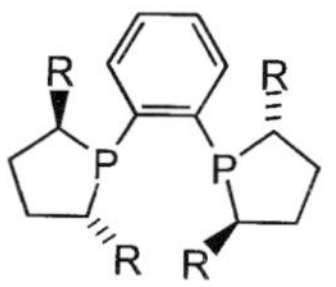

Ligands of type **8**, known as PennPhos, have given new impetus to the rhodium-catalyzed homogeneous hydrogenation of prochiral ketones [1]. In **8** the favorable properties of the strongly basic Duphos ligands **9**, developed by Burk *et al.* [19]. are cleverly enhanced by the incorporation of additional ring-forming bridges into the molecule, in accordance with the established principle of increasing the rigidity in the backbone. This at the same time increases the spatial requirements of the ligands. For the hydrogenations with **8** the enantiomeric excesses obtained are likewise optimized by the use of non-chiral modifiers such as bromides and weak bases. A remarkable feature of this reaction is that bases such as 2–methylimidazole and 2,6-lutidine, if added in substoichiometric amounts to the rhodium catalyst, achieve very similar increases in selectivity, whereas if equimolar quantities are used (base/rhodium catalyst: 1/1) the results obtained differ dramatically [1 % *ee* versus 95 % *ee*, (*S*)-alcohol]. Use of triethylamine as the added base even yields the *R* enantiomer. The turnover rate also shows optima that are dependent on the ratio of base to the rhodium catalyst. This is viewed by the authors as an opportunity to improve the still unsatisfactory length of time required for hydrogenations with the PennPhos rhodium catalyst, which in some cases is as long as several days at 30 atm hydrogen pressure and room temperature.

The influence of the alkyl substituents in the substrate on the enantioselectivity of the hydrogenation (Table 1) corresponds to experience from experiments with modified Raney nickel [18].

Table 1. Enantioselectivities for the hydrogenation of simple ketones RCOR'. [a]

R	R'	*ee* (*S*) [%]
*t*Bu	Me	94
C_6H_{11}	Me	92
*i*Pr	Me	84
*n*Bu	Me	75
Ph	Me	95 (without KBr)
Ph	Et	93
Ph	*i*Pr	72

[a] With 0.5 mol % $[Rh(COD)Cl]_2$, **8a**, lutidine, and KBr under 30 atm H_2 [1].

According to these, for purely aliphatic ketones the highest enantioselectivities are achieved for methyl ketones with a second, branched-chained alkyl substituent (84–94 % *ee*). The value of 75 % *ee* obtained with the straight-chain hexan-2–one is, to the best of our knowledge, in any case better than anything achieved to date with nonenzymatic systems and homogeneous catalysis. Higher selectivities have been reported for reductions with stoichiometric amounts of chiral borohydrides (e.g. 80 % *ee* for the reduction of octan-2-one) [20].

For alkyl aryl ketones, the use of the novel PennPhos rhodium catalyst afforded enantioselectivities of over 90 %, such as are routinely obtained by transfer hydrogenation [13], but, only for alkyl groups up to C_2. This is in contrast

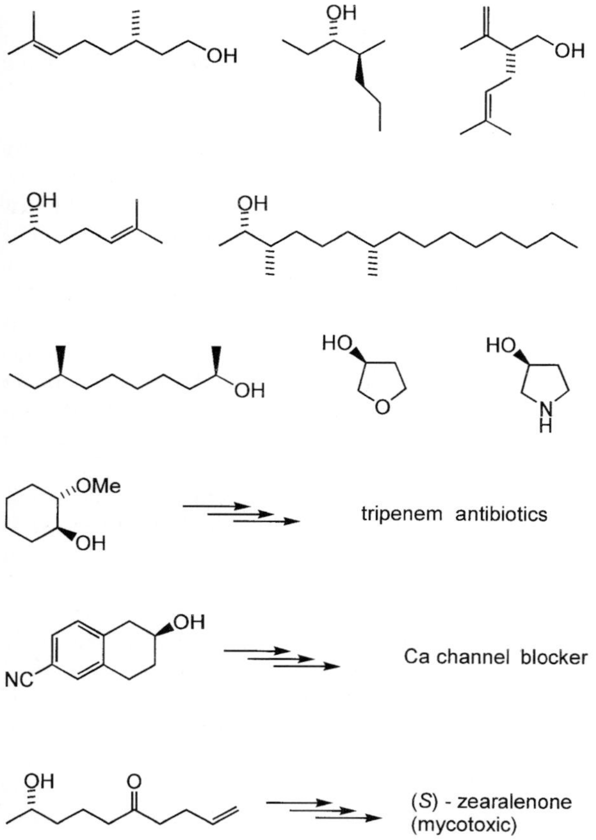

OH OH OH OH OH HO HO OMe OH OH OH O

tripenem antibiotics

NC

Ca channel blocker

(S) - zearalenone (mycotoxic)

Scheme 2. Optically active alcohols that can be used to synthesize biologically important chiral compounds [7c, 22].

to microbial reductions, where the selectivity rises with the number of carbon atoms in the alkyl group [14]. A recent report describes a most spectacular inversion of selectivity for a novel ruthenium-catalyzed transfer hydrogenation of alkyl phenyl ketones with the use of the bis(thiourea) ligand **7** derived from (*R*,*R*)-1,2–diphenylethylenediamine [21]. In this reaction the following enantioselectivities were obtained as a function of the alkyl group in the substrate molecule: Me: 89 %, (*S*); Et: 91 %, (*S*); *i*Pr: 94 %, (*S*); *t*Bu: 85 %, (*R*). By ordering the substrates according to the magnitude of the selectivities obtained, other studies have likewise demonstrated a marked dependence on the structure of the catalyst [10].

The significance for industrial processes is clear. Scheme 2 shows several examples from a survey recently published by Noyori [7c] of optically active alcohols that can be used to synthesize biologically important chiral compounds below along with some of relevance to the pharmaceutical industry [22]. Pheromones constitute a particularly diverse family of chiral aliphatic alcohols [23].

Finally, the bis(trialkylphosphane) ligands **10** developed by Imamoto *et al.* may, on account of their high basicity, also be suitable for the hydrogenation of ketones. Use of these ligands, which contain chiral phosphorus centers, affords enantiomeric excesses of over 99 % for the rhodium(I)-catalyzed hydrogenation of *N*-acyl-dehydroamino acids [24]. The tetrasubstituted substrates, which are generally difficult to hydrogenate enantioselectively, are of particular interest. The partially reversed orders of selectivity for the five ligands described leads us to suspect that high enantiomeric excesses may also be possible for the reduction of ketones. Further advances are conceivable by combining the methods developed by the research groups of Noyori and Zhang, with the use of bases as modifiers.

10

	R
a	*t*Bu
b	Et_3C
c	1-adamantanyl
d	c-C_5H_9
e	c-C_6H_{11}

The Progress of Asymmetric Reduction of Ketones since 1998

The actuality of the subject matter is underlined by the fact that since January 1998 up to the middle of May 1999 more than 140 papers were published in the field of enantioselective hydrogenation of ketones, which we will briefly discuss here only if the enantioselectivity surpasses 90 % *ee*. Among them, the particularly relevant manuscripts dealing with the hydrogenation of purely aliphatic ketones are in the minority. Four rapidly developing main lines can be recognized: enzymatic reduction, oxazaborolidine-catalyzed reduction by boron hydride, transition metal complex-catalyzed hydrogenation with hydrogen or transfer hydrogenation mainly with isopropanol [25].

Enzymatic Reductions

Exclusively *enzymatic reductions* give satisfying selectivities in the region of 90 to over 99 % *ee* for substrates possessing only the carbonyl as functional group [26]. Enantioselective reduction by a crude alcohol dehydrogenase from *Geotrichium candidum* is effective even for alkyl-trifluoromethyl ketones (96–99 % *ee*) [27]. The enantiopure synthesis of 3-hydroxypiperidine-2-one **12** is possible according to Scheme 3 by chemoenzymatic synthesis from 5-nitro-2-oxo-pentanoate **11**, with enzymatic reduction of the carbonyl group as the key step. Nearly enantiopure α-hydroxy ketones or α-hydroxy esters are obtained using baker's yeast reductase by reduction of the α-keto groups [28]. With D-hydroxyisocaproate dehydrogenase from genetically engineered H205Q mutants some *N*-protected (*S*)-4-amino-2-hydroxy esters could be synthesized in better yield than with the wild-type lactate dehydrogenases [29].

In addition to this, that an interesting novel emulsion membrane reactor concept overcomes the difficulties of the large solvent volume otherwise required for the reduction of poorly soluble ketones [30]. 2–Octanone was reduced by a carbonyl reductase from *Candida parapsilosis* to (*S*)-2-octanol with >99.5 % *ee* and total turnover number of 124 – the 9-fold value of that obtained in a classical enzyme reactor.

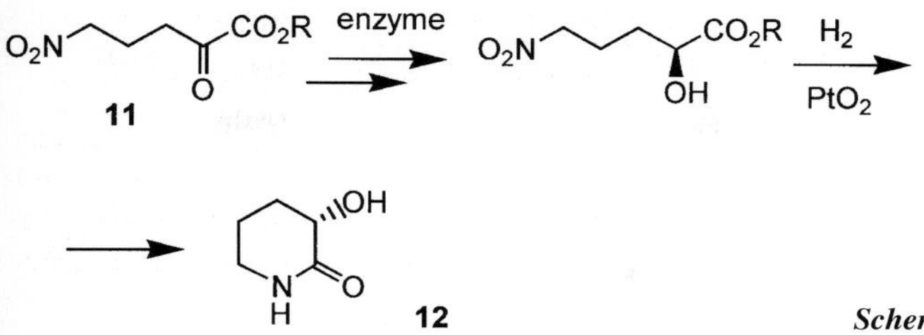

Scheme 3

The main disadvantage for all enzymatic hydrogenations is the indispensability of a coenzyme such as NADH, whose regeneration needs an additional dehydrogenase system and a hydrogen source such as formic acid. This can be overcome by the use of complete cells. Choosing the appropriate microorganism, the extremely difficult enantioselective reduction of ketone functions located in the middle of carbon chains (as in 4-chloro-5-nonanone) succeeded recently with $>98\,\%$ *ee* [31]. A first application of baker's yeast in an industrial process for the production of trimegestone by selective monoreduction of a triketone using glucose as hydrogen source was a new highlight [32]. Success with baker's yeast cells immobilized on chrysotile fibers is reported for large-scale preparation of chiral alcohols [33].

Oxazaborolidine-Catalyzed Reduction with Boron Hydride

For *aryl* ketones the Corey-Bakshi-Shibata (CBS) reduction using *oxazaborolidines* as catalysts for the boron hydride mediated hydrogenation is particularly useful, with maximum selectivities up to 99 % *ee* (see Scheme 4) [34]. The excellent review by Corey *et al.* [35] also shows clearly the power for chemo- and enantioselective reduction of purely aliphatic α,β-enones and -ynones only on the carbonyl group. In the reduction of enones such as **13**, a bulky R group leads to an enhanced yield and selectivity: R = $SiMe_3$ (94 %, 90 % *ee*), R = H (30 %, 76 % *ee*) [36]. **13** is a key precursor for the synthesis of atractyligenin, a naturally occurring adenosine diphosphate transport inhibitor. Recently, this method found application also for a key step in the synthesis of zaragozic acids investigated for the treatment of hypercholesterolemia [37]. Sulfur-containing phenylthioalkenones are proven to be excellent substrates for highly regio- and enantioselective reduction (89 – 98 % *ee*) [38].

For saturated methyl ketones with a branched alkyl group the enantioselectivities can exceed

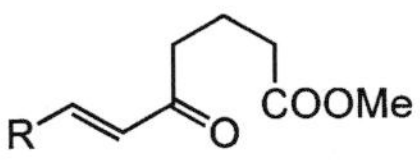

13

90 % *ee* (see Scheme 4) [39]. This convenient method serves even for the preparation of chiral ligand precursors **15** by reduction of 1,1'–diacylmetallocenes **14** with over 98 % *ee* (Scheme 5) [40]. The modified catalysts **16** have been applied recently in enantioselective hydrogenation of β-ketoesters (91–98 % *ee*) and for β-diketones, also giving excellent diastereoselectivities [41].

It seems important to note that unsaturated diketones **17** may be reduced selectively on

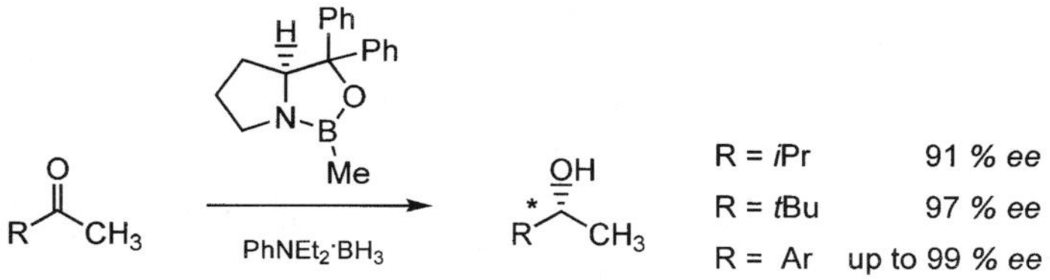

R = *i*Pr	91 % *ee*
R = *t*Bu	97 % *ee*
R = Ar	up to 99 % *ee*

Scheme 4

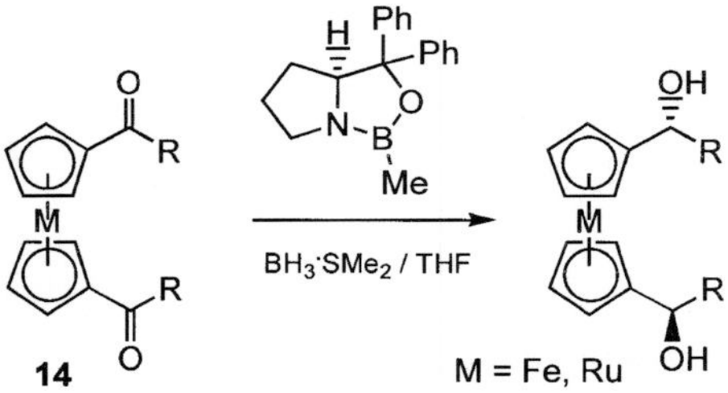

BH$_3$·SMe$_2$ / THF

14 M = Fe, Ru 15

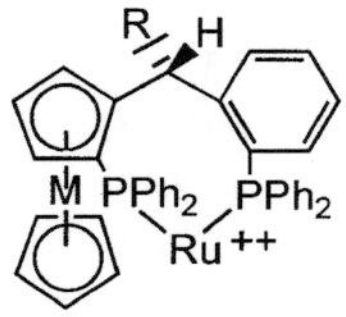

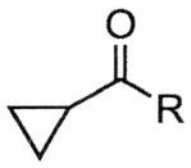

Scheme 5 16

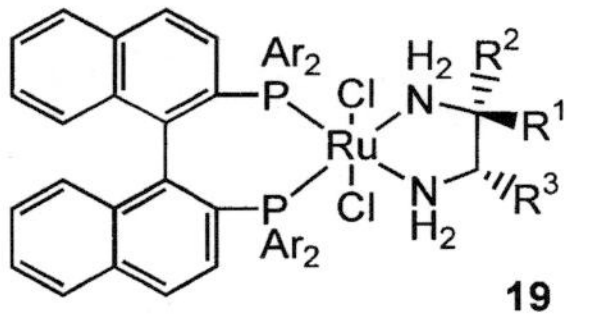

17 18

Scheme 6

Ar = 3,5-(CH$_3$)$_2$C$_6$H$_3$
R^1 = R^2 = 4-CH$_3$OC$_6$H$_4$
R^3 = *i*Pr

19 20

both carbonyl groups to unsaturated diols **18** (*meso* : dl = 13 : 87; the latter with 99 % *ee* (see Scheme 6).

The alternative use of LiH/BF$_3$OEt$_2$ as a reducing agent for the in situ generation of borane is used to save costs [42]. Polymer-enlarged oxazaborolidines have been applied with success in a membrane reactor [43].

Hydrogenation Catalyzed by Transition Metal Complexes

Impressive progress in the particularly important asymmetric hydrogenation of ketones by hydrogen using transition metal complex catalysts was demonstrated by Noyori et al. [44]. The introduction of XylBINAP with P-xylyl groups in the ruthenium chelate **19** allowed an increase in the enantioselectivity of the reduction of acetophenone to 99 % *ee* compared to 87 % *ee* with BINAP or TolBINAP. The fine tuning of the diamine substituents R^1, R^2 and R^3 enables **19** to reach 95 to 99 % *ee* for the fully selective hydrogenation of a broad spectrum of olefinic or cyclopropyl ketones such as **20** without affecting the double bond or the three-membered ring. Ru-BINAP or similar complexes derived from atropisomeric diphosphane ligands also catalyze the enantioselective reduction of 3-oxobutanoate or β-ketoesters bearing functional groups in 95 to 99 % *ee*, and the method could be widened to β-ketophosphonates and sulfur-containing ketones [45, 46]. The work of Genêt et al. [46] is worthy of notice regarding some interesting applications for the synthesis of multifunctional bioactive compounds e.g. 2-amino-3-hydroxy acids, some heterocycles such as (−)-swainsonine or some channel blockers such as Dilthiazem® or the side chain for Taxotere®. Ruthenium catalysts with the new ligand **21,** given the abbreviation BICP, were introduced by the Zhang group and are particularly effective for the hydrogenation of thiophene containing ketones such as **22** [47]. Rhodium chelates of BICP also allow the synthesis of β-aminoalcohols in 90–99 % *ee* by hydrogenation of *E,Z*-mixtures of protected β-hydroxy enamides **23** (see Scheme 7) [48]. In principle, this is a selective reductive transformation of the carbonyl group of α-hydroxyacetophenones via the oxime to an amino group.

Respectable results for the ruthenium chelate-catalyzed reduction of β-ketoesters were also obtained by Pye [49] with 2.2-PHANEPHOS (**24**) and Imamoto [50] with **10a** as the ligand.

An interesting stepwise reduction of 2,4-dioxovalerate (**25**) with BINAP-analogous ruthenium

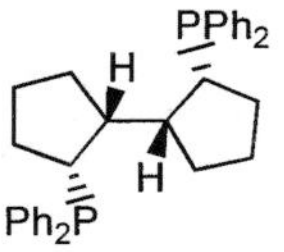

BICP 21 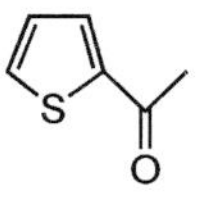 22

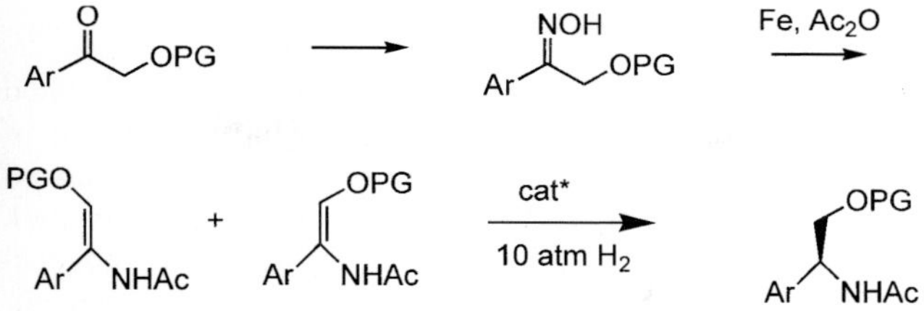

Scheme 7

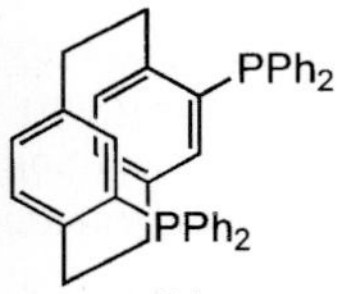

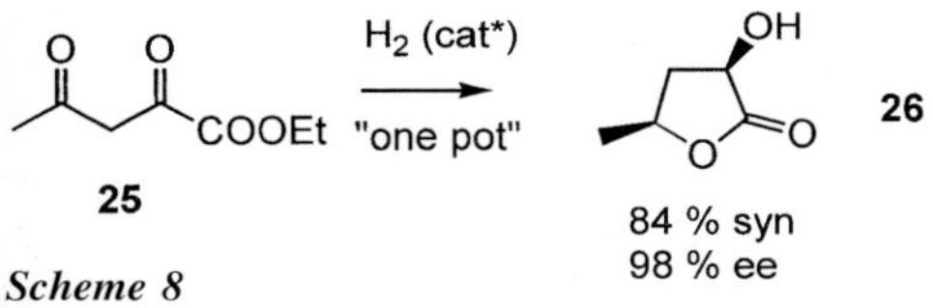

Scheme 8

chelates followed by cyclization gives 84 % of the *syn*-α-hydroxy-γ-valerolactone **26** in 98 % *ee* (see Scheme 8) [51]. For the hydrogenation of α-keto-esters, -amides or -amines, a further number of highly enantioselective rhodium(I) catalysts have found application (maximum 97 – 99 % *ee*) [52, 53]. Burk realized the synthesis and reduction of α-acyloxy-acrylates **27** in over 99 % *ee* by using a rhodium (*S,S*)-Et-DuPHOS (**9**) catalyst [54]. This method competes successfully with the reduction of α-keto esters and is remarkable because the high selectivities are obtained using the *E/Z*-mixtures of the substrates **27**.

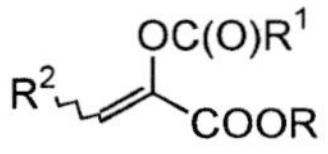

The very new highly enantioselective hydrogenation of enol acetates by a rhodium PennPhos chelate should be mentioned here [55]. Cyclic enol acetates are especially suitable for this hydrogenation, yielding the products in 99 % *ee.*

Exciting results using an iron gelatin complex supported on zeolite for heterogeneous asymmetric hydrogenation of purely aliphatic ketones under normal pressure at 20 °C have been reported recently by Ying-Yan Jiang et al. (Chinese Academy of Sciences, Peking) [66]. This procedure yields the corresponding chiral alcohols with highter than 95 % *ee* (polarimetric measurements).

Transfer Hydrogenantion Catalyzed by Complexes

Some progress has been made, particularly with the ruthenium complex-catalyzed transfer hydrogenation [56] For *tert*-butyl methyl ketone as the first purely aliphatic ketone the bench mark of 90 % *ee* has been crossed by application of the oxazolinylferrocenylphosphine **28**; the transfer hydrogenation by isopropanol under reflux in the presence of sodium hydroxide resulted in 93 % *ee* (*S*)-3,3–dimethyl-butan-2-ol [57]. For alkyl aryl ketones 92–95 % *ee* is obtained with

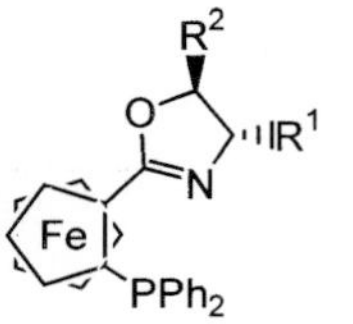

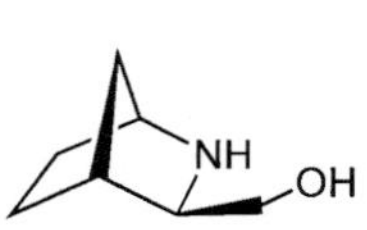

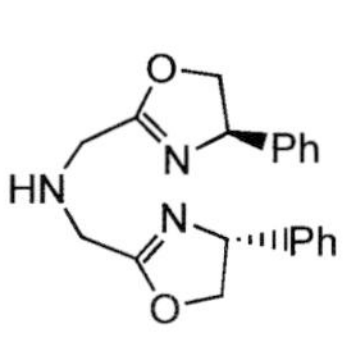

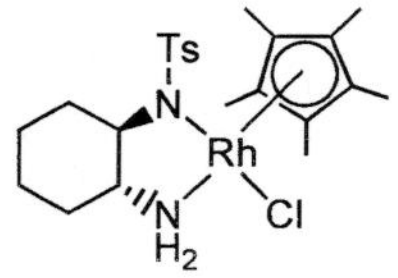

31

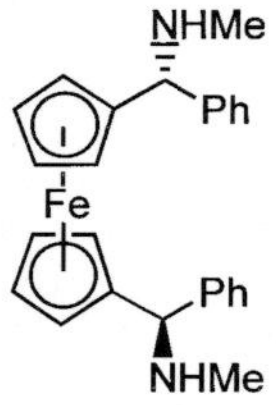

32

(1*S*,3*R*,4*R*)-2-azanorbornylmethanol **29** [58] and up to 98 % *ee* with Ph-Ambox **30** [59] as ligands free of phosphorus for ruthenium. Murata et al. worked on further optimization of the power of ruthenium chiral diamino-monotosylates **6** [60]. Other metals and amines were investigated. It seems that rhodium chelates are somewhat less active for the hydrogenation of acetophenone but more selective (up to 97 % *ee* for catalyst **31** with (1*R*,2*R*)-*N*-(*p*-toluenesulfonyl)-1,2-cyclohexanediamine). With iridium as the central metal 96 % *ee* is still possible, but the activity strongly decreases. Also analogous polymer-bound catalysts of ruthenium [61] and iridium [62] give high enantioselectivities. But, the selectivity decreases when the immobilized catalyst is reused (with ruthenium: first run >99 % *ee*, second run 96 % *ee* and third run 91 % *ee* in methylene chloride).

With ferrocenyl diamines such as **32**, the transfer hydrogenation of 1'-acetonaphthone reached 90 % *ee* at –30 °C with 2-propanol as the hydrogen source [63]. Even amino acids have been used as ligands for ruthenium [64], but, more than 90 % *ee* results only when tetralone is the substrate.

In the transfer hydrogenation of acetylacetates ephedrinium chelates of ruthenium led to 94 % *ee* [65].

Conclusion

As before, the enzymatic reduction is the method of choice for the enantioselective reduction of purely aliphatic ketones and only in the case of *tert*-butyl methyl ketone could the bench mark of 90 % *ee* be crossed by the transfer hydrogenation and both other catalytic hydrogenation methods. However, substantial success in the hydrogenation of aromatic ketones by transition metal complexes with respect to the enantioselectivity and the activity (TON) strengthens the confidence that further progress is possible, enabling us to use some advantages of these nonenzymatic processes for extended application in the near future, for example in the facilitation of product isolation.

References

[1] Q. Jiang, Y. Jiang, D. Xiao, P. Cao, X. Zhang, *Angew. Chem.* **1998**, *110*, 1203–207; *Angew. Chem. Int. Ed.* **1998**, *37*, 1100–1103.

[2] In the case of purely aliphatic ketones such as nonan-3-one, use of the alcohol dehydrogenase extracted from *Gluconobacter oxidans* (ATCC 621) results in enantioselectivities of up to 95 %: P. Adlercreuz, *Enzyme Mikrob. Technol.* **1991**, *13*, 9–14; see also review: R. Csuk, B. Glänzer, *Chem. Rev.* **1991**, *91*, 49–97.

[3] T. Hayashi, A. Katsumura, M. Konishi, M. Kumada, *TetrahedronLett.* **1979**, 425–428.

[4] a) T. Morimoto, H. Takayashi, K. Fujii, M. Chiba, K. Achiwa, *Chem.Lett.* **1986**, 2061–2064; b) H. Takeda, T. Tachinami, M. Aburatami, H. Takahashi, T. Morimoto, K. Achiwa, *Tetrahedron Lett.* **1989**, *30*, 363–366; c) K. Inoguchi, S. Sakuraba, K. Achiwa, *Synlett* **1992**, 169–178.

[5] A. Roucoux, M. Devocelle, J.-F. Carpentier, F. Agbossou, A. Mortreux, *Synlett* **1995**, 358–360.

[6] a) A. Tai, T. Kikukawa, T. Sugimura, Y. Inoue, S. Abe, T. Osawa, T. Harada, *Bull. Chem. Soc. Jpn.* **1994**, *67*, 2473–2477; b) H. U. Blaser, H. P. Jalett, J. Wiehl, *J. Mol. Catal.* **1991**, *68*, 215–222; c) H. U. Blaser, H. P. Jalett, F. Spindler, *J. Mol. Catal. A.* **1996**, *107*, 85–94; d) H. U. Blaser, H. P. Jalett, M. Müller, M. Studer, *Catal. Today* **1997**, *37*, 465–480.

[7] a) R. Noyori, T. Ohkuma, M. Kitamura, H. Takaya, N. Sayo, H. Kumobayashi, S. Akutagawa, *J. Am. Chem. Soc.* **1987**, *109*, 5856–5858; b) M. Kitamura, T. Ohkuma, S. Inoue, N. Sayo, H. Kumobayashi, S. Akutagawa, T. Ohta, H. Takaya, R. Noyori, *ibid.* **1988**, *110*, 629–631; for a review of subsequent work, see c) R. Noyori, *Asymmetric-*

Catalysisin Organic Synthesis, Wiley, New York, **1994**, 61–82; d) T. Ohkuma, H. Ooka, T. Ikariya, R. Noyori, *J. Am. Chem. Soc.* **1995**, *117*, 10417–10418; e) T. Ohkuma, H. Ikehira, T. Ikariya, R. Noyori, *Synlett* **1997**, 467–468; f) J.-P. Tranchier, V. Ratovelomanana-Vidal, J.-P. Genet, S. Tony, T. Cohen, *Tetrahedron Lett.* **1997**, *38*, 2951–2954; g) T. Ohta, T. Miyake, N. Seido, H. Kumobayashi, H. Takaya, *J. Org. Chem.* **1995**, *60*, 357–363; h) T. Ohkuma, H. Doucet, T. Pham, K. Mikami, T. Korenaga, M. Terada, R. Noyori, *J. Am. Chem. Soc.* **1998**, *120*, 1086–1087.

[8] a) H. Brunner, *Organometallics* **1984**, *3*, 1354–1359; b) H. Nishiyama, M. Kondo, T. Nakamura, K. Itoh, *ibid.* **1991**, *11*, 500–508. c) M. B. Carter, B. Schiott, A. Gutiérrez, S. L. Buchwald, *J. Am. Chem. Soc.* **1994**, *116*, 11667–11669.

[9] E. J. Corey, R. K. Bakshi, S. Shibata, C.-P. Chen, V. K. Singh, *J. Am. Chem. Soc.* **1987**, *109*, 7925–7926.

[10] a) T. Ohkuma, H. Ooka, S. Hashiguchi, T. Ikariya, R. Noyori, *J. Am. Chem. Soc.* **1995**, *117*, 2675–2676: b) H. Doucet, T. Ohkuma, K. Murata, T. Yokozawa, M. Kozawa, E. Katayama, A. F. England, T. Ikariya, R. Noyori, *Angew. Chem.* **1998**, *110*, 1792–1796, *Angew. Chem. Int. Ed.* **1998**, *37*, 1703–1707.

[11] X. Zhang, T. Taketomi, T. Yoshizumi, H. Kumobayashi, S. Akutagawa, K. Mashima, H. Takaya, *J. Am. Chem. Soc.* **1993**, *115*, 3318–3319.

[12] T. Nagata, K. Yorozu, T. Yamada, T. Mukaiyama, *Angew. Chem.* **1995**, *107*, 2309–2311; *Angew. Chem. Int. Ed. Engl.* **1995**, *34*, 2145–2147. For a recent review on the asymmetric reduction of carbonyl groups with hydrides, see J. Seyden-Penne, *Reductions by the Alumino- and Borohydrides in Organic Synthesis*, 2nd edn., Wiley, New York, **1997**, 55–84.

[13] Review: a) R. Noyori, S. Hashiguchi, *Acc. Chem. Res.* **1997**, *30*, 97–112; b) K. Matsumura, S. Hashiguchi, T. Ikariya, R. Noyori, *J. Am. Chem. Soc.* **1997**, *119*, 8738–8739.

[14] K. Nakamura, T. Matsuda, A. Ohno, *Tetrahedron:Asymmetry* **1996**, *7*, 3021–3024.

[15] a) J. Takehara, S. Hashiguchi, A. Fujii, S. Inoue, T. Ikariya, R. Noyori, *Chem. Commun.* **1996**, 233–234; b) T. Langer, G. Helmchen, *Tetrahedron Lett.* **1996**, *37*, 1381–1384; c) K. Püntener, L. Schwink, P. Knochel, *ibid.* **1996**, *37*, 8165–8168; d) T. Sammakia, E. L. Stangeland, *J. Org. Chem.* **1997**, *62*, 6104–6105; e) M. Palmer, T. Walsgrove, M. Wills, *ibid.* **1997**, *62*, 5226–5228; f) F. Touchard, P. Gamez, F. Fache, M. Lemaire, *TetrahedronLett.* **1997**, *38*, 2275–2278; g) S. Inoue, K. Nomura, S. Hashiguchi, R. Noyori, Y. Izawa, *Chem.Lett.* **1997**, 957–958.

[16] Y. Jiang, Q. Jiang, G. Zhu, X. Zhang, *Tetrahedron-Lett.* **1997**, *38*, 215–218.

[17] R. Berengeuer, J. Garcia, J. Villarasa, *Tetrahedron:Asymmetry* **1994**, *5*, 165–168.

[18] a) T. Osawa, T. Harada, A. Tai, *J. Mol. Catal.* **1994**, *87*, 333–342; b) T. Osawa, T. Harada, A. Tai, *Catal. Today* **1997**, *37*, 465–480.

[19] a) M. J. Burk, M. F. Gross, T. G. P. Harper, C. S. Kalberg, J. R. Lee, J. P. Martinez, *Pure Appl. Chem.* **1996**, *68*, 37–44; b) J. Albrecht, U. Nagel, *Angew. Chem.* **1996**, *108*, 444–446; *Angew. Chem. Int. Ed. Engl.* **1996**, *35*, 404–407.

[20] T. Imai, T. Tamura, A. Yamamuro, *J. Am. Chem. Soc.* **1986**, *108*, 7402–7404.

[21] F. Touchard, F. Fache, M. Lemaire, *Tetrahedron:Asymmetry* **1997**, *8*, 3319–3326.

[22] R. Stürmer (BASF AG), personal communication.

[23] Review: K. Mori, *Chem. Commun.* **1997**, *13*, 1153–1158.

[24] T. Imamoto, J. Watanabe, Y. Wada, H. Masuda, H. Yamada, H. Tsuruta, S. Matsukawa, K. Yamaguchi, *J. Am. Chem. Soc.* **1998**, *120*, 1635–1636.

[25] Earlier results see also the relevant chapters in: M. Beller, C. Bolm (Eds.), *Transition Metals for Organic Synthesis,* Wiley VCH, Weinheim, **1998**.

[26] N. Itoh, N. Mizuguchi, M. Mabuchi, *J. Mol. Catal. B: Enzym.* **1999**, *6*, 41–50.

[27] a) K. Nakamura, *J. Mol. Catal. B: Enzym.* **1998**, *5*, 129–132; K. Nakamura, T. Matsuda, M. Shimizu, T. Fujisawa, *Tetrahedron* **1998**, *54*, 8393–8402, c) K. Nakamura, T. Matsuda, *J. Org. Chem.* **1998**, *63*, 8957–8964.

[28] Y. Kawai, K. Hida, M. Tsujimoto, S. Kondo, K. Kitano, K. Nakamura, A. Ohno, *Bull. Chem. Soc. Jap.* **1999**, *72*, 99–102.

[29] A. Sutherland, C. L. Willis, *J. Org. Chem.* **1998**, *63*, 7764–7769.

[30] A. Liese, T. Zelinski, M.-R. Kula, H. Kierkels, M. Karutz, U. Kragl, C. Wandrey, *J. Mol. Catal. B: Enzym.* **1998**, *4*, 91–99.

[31] P. Besse, T. Sokoltchik, H. Veschambre, *Tetrahedron: Asymmetry* **1998**, *9*, 4441–4457.

[32] V. Crocq, C. Masson, J. Winter, C. Richard, G. Lemaitre, L. Lenay, M. Vivat, J. Buendia, D. Prat, *Org. Process Res. Dev.* **1997**, *1*, 2–13.

[33] R. Wendhausen, P. J. S. Moran, I. Joekes, J. A. R. Rodrigues, *J. Mol. Catal. B: Enzym.* **1998**, *5*, 69–73.

[34] a) Z. X. Shen, W. Huang, J. W. Feng, Y. W. Zhang, *Tetrahedron: Asymmetry* **1998**, *9*, 1091–1095; b) K. Manju, S. Trehan, *Tetrahedron: Asymmetry* **1998**, *9*, 3365–3369; c) R. Hett, C. H. Senanayake, S. A. Wald, *Tetrahedron Lett.* **1998**, *39*, 1705–1708; d) M. Shimizu, K. Tsukamoto, T. Matsutani, T. Fujisawa, *Tetrahedron* **1998**, *54*, 10265–10274; e) M.P. Sibi, G. R. Cook, P. R. Liu, *Tetrahedron Lett.* **1999**, *40*, 2477–2480.

[35] E. J. Corey, C. J. Helal, *Angew. Chem. Int. Ed.* **1998**, *37*, 1987–2012.

[36] E. J. Corey, A. Guzmann-Perez, S. E. Lazerwith, *J. Am. Chem. Soc.* **1997**, *119*, 11769–11776.
[37] J. Bach, M. Galobardes, J. Garcia, P. Romea, C. Tey, F. Urpi, J. Vilarrasa, *Tetrahedron Lett.* **1998**, *39*, 6765–6768.
[38] a) R. Berenguer, M. Cavero, J. Garcia, M. Munoz, *Tetrahedron Lett.* **1998**, *39*, 2183–2186; b) T. K. Yang, D. S. Lee, *Tetrahedron: Asymmetry* **1999**, *10*, 405–409.
[39] A. M. Salunkhe, E. R. Burkhardt, *Tetrahedron Lett.* **1997**, *38*, 1523–1526.
[40] L. Schwink, P. Knochel, *Chem. Eur. J.* **1998**, *4*, 950–968.
[41] T. Ireland, G. Großheimann, C. Wieser-Jeunesse, P. Knochel, *Angew. Chem. Int. Ed.* **1999**, *38*, 3212–3215.
[42] A. Ford, S. Woodward, *Synth. Commun.* **1999**, *29*, 189–192.
[43] G. Giffels, J. Beliczey, M. Felder, U. Kragl, *Tetrahedron: Asymmetry* **1998**, *9*, 691–696.
[44] T. Ohkuma, M. Koizumi, H. Doucet, T. Pham, M. Kozawa, K. Murata, E. Katayama, T. Yokozawa, T. Ikariya, R. Noyori, *J. Am. Chem. Soc.* **1998**, *120*, 13529–13530.
[45] U. Matteoli, V. Beghetto, A. Scrivanti, *J. Mol. Catal. A: Chem.* **1999**, *140*, 131–137.
[46] Review: V. Ratovelomanana-Vidal, J.-P. Genêt, *J. Organomet. Chem.* **1998** *567*, 163–171.
[47] P. Cao, X. Zhang, *J. Org. Chem.* **1998**, *64*, 2127–2129.
[48] G. Zhu, A. Casalnuovo, X. Zhang, *J. Org. Chem.* **1998**, *64*, 8100–8111.
[49] P. J. Pye, K. Rossen, R. A. Reamer, R. P. Volante, P. J. Reider, *Tetrahedron Lett.* **1998**, *39*, 4441–4444.
[50] T. Yamano, N. Taya, H. Kawada, T. Huang, T. Imamoto, *Tetrahedron Lett.* **1999**, *40*, 2577–2580.
[51] V. Blandin, J. F. Carpentier, A. Mortreux, *Tetrahedron: Asymmetry* **1998**, *9*, 2765–2768.
[52] C. Pasquier, J. Eilers, I. Reiners, J. Martens, A. Mortreux, F. Agbossou, *Synlett* **1998**, 1162.
[53] C. Pasquier, S. Naili, L. Pelinski, J. Brocard, A. Mortreux, F. Agbossou, *Tetrahedron: Asymmetry* **1998**, *9*, 193–196.
[54] M. J. Burk, C. S. Kalberg, A. Pizzano, *J. Am. Chem. Soc.* **1998**, *120*, 4345–4353.
[55] Q. Jiang, D. Xiao, Z. Zhang, P. Cao, X. Zhang, *Angew. Chem. Int. Ed.* **1999**, *38*, 516–518.
[56] a) M. J. Palmer, M. Wills, Review: Asymmetric Transfer Hydrogenation of C = O and C = N Bonds, *Tetrahedron: Asymmetry* **1999**, *10*, 2045–2061; b) J. A. Kenny, M. J. Palmer, A. R. C. Smith, T. Walsgrove, M. Wills, *Synlett* **1999**, 1615–1617.
[57] Y. Arikawa, M. Ueoka, K. Matoba, Y. Nishibayashi, M. Hidai, S. Uemura, *J. Organomet. Chem.* **1999**, *572*, 163–168.
[58] D. A. Alonso, D. Guijarro, P. Pinho, O. Temme, P. G. Andersson, *J. Org. Chem.* **1998**, *63*, 2749–2751.
[59] Y. Jiang, Q. Jiang, X. Zhang, *J. Am. Chem. Soc.* **1998**, *120*, 3817–3818.
[60] K. Murata, T. Ikariya, *J. Org. Chem.* **1999**, *64*, 2186–2187.
[61] D. J. Bayston, C. B. Travers, M. E. C. Polywka, *Tetrahedron: Asymmetry* **1998**, *9*, 2015–2018.
[62] R. ter Halle, E. Schulz, M. Lemaire, *Synlett* **1997**, 1257–1258.
[63] L. Schwink, T. Ireland, K. Puntener, P. Knochel, *Tetrahedron: Asymmetry* **1998**, *9*, 1143–1163.
[64] T. Ohta, S. Nakahara, Y. Shigemura, K. Hattori, I. Furukawa, *Chem. Lett.* **1998**, 491–492.
[65] K. Everare, J. F. Carpentier, A. Mortreux, M. Bulliard, *Tetrahedron: Asymmetry* **1998**, *9*, 2971–2974.
[66] X. Zhang, Y. Geng, M. Yin, M. Hang, Y. Jiang, *J. Mol. Sci.* **1999**, *15*, 243–244. Poster on the IXth International Symposium on Fine Chemistry and Functional Polymers in Haikou, P. R. China (December 1999).

Part II. Applications

A. Total Synthesis of Natural Products

Total Synthesis of Ikarugamycin

Oliver Schwarz and Hans-Günther Schmalz

Institut für Organische Chemie, Universität Köln, Germany

Acyclic η^4-butadiene $Fe(CO)_3$ complexes have repeatedly demonstrated their enormous value for organic synthesis in the last few years [1]. In this context, both the changed reactivity of the ligand and the steric effect(s) of the $Fe(CO)_3$ fragment have been exploited for the stereocontrolled generation of new chirality centers in the neighborhood of the butadiene-$Fe(CO)_3$ unit. It is important to note that unsymmetrically substituted complexes (e.g. of type **A** with $R^1 \neq R^2$) are chiral.

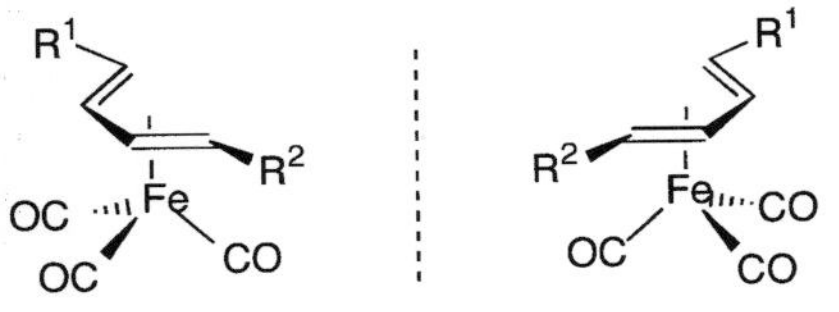

A *ent*-**A**

In order to utilize such a chiral substructure as a source of absolute stereochemical information in a synthesis, it is necessary to employ the complexes **A** and *ent*-**A** in non-racemic form. While the enantioselective preparation of such chiral complexes was achieved in the past more or less exclusively via resolution of racemic mixtures, the diastereoselective complexation of chirally modified ligands was shown more recently to be a practical alternative [2]. Another possibility, the enantioselective conversion of prochiral metal complexes by means of chiral reagents, has been achieved by W. R. Roush [3]. In a remarkable (formal) total synthesis of the antibiotic (+)-ikarugamycin (**1**), Roush et al. apply their method and demonstrate in a highly convincing fashion the synthetic usefulness of acyclic butadiene-$Fe(CO)_3$ complexes [4]. This synthesis will be discussed in some detail in the following.

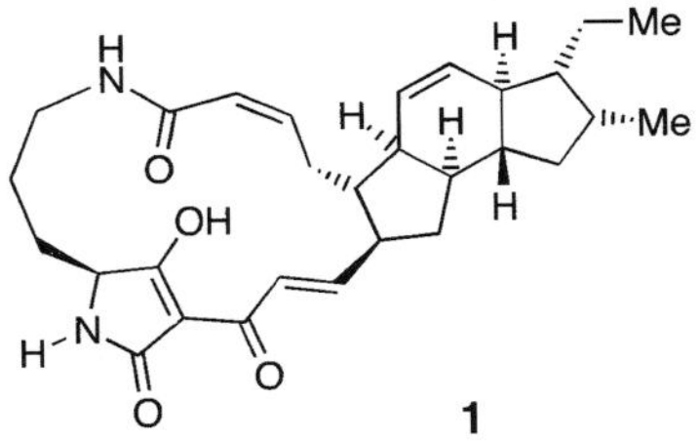

1

The Target Molecule:

In 1972 Ito and Hirata reported on the isolation and structure elucidation of (+)-ikarugamycin as the first representative of a new class of macrolactam antibiotics [5]. Besides its remarkable biological activity, ikarugamycin is of interest because of its unusual architecture and it represents an attractive target structure for organic synthesis [6, 7, 8].

According to R. K. Boeckmann [6], the retrosynthetic analysis of ikarugamycin leads (by disconnection of the double bonds within the macrocyclic ring) to the tricyclic building block **2** which, because of its eight subsequent stereocenters, is still of considerable structural complexity (Scheme 1). In the synthesis of Roush discussed here [4], the *cis-anti-cis* configured decahydro-*as*-indacene derivative **2** is formed by cyclization from the bicyclic dialdehyde **3**, which in turn is the product of an intramolecular Diels-Alder reaction of the acyclic dialdehyde **4**. The special feature of the Roush synthesis is the way the acyclic intermediate **5** is prepared from the iron complex **6**.

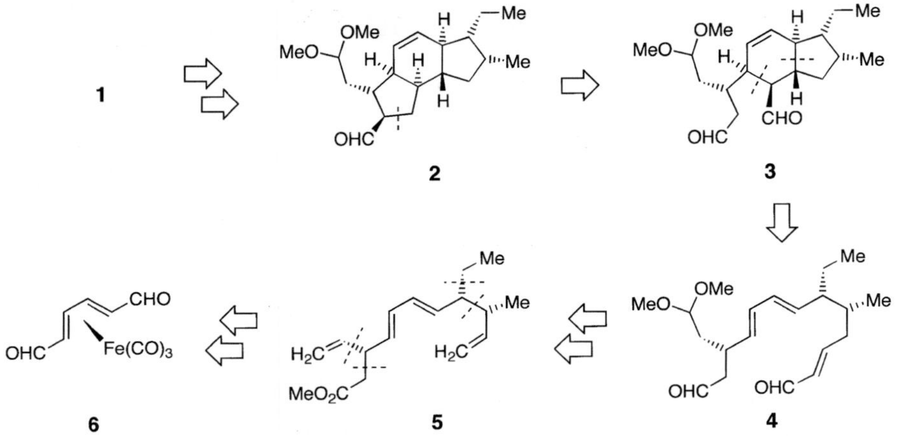

Scheme 1. Retrosynthesis of ikarugamycin according to W. R. Roush and R. K. Boeckmann.

Stereoselective Synthesis of the Acyclic Intermediate 5

The synthesis (Scheme 2) starts with an impressive transformation, i.e. the conversion of the prochiral complex **6**, which is easily prepared from diol **7** by complexation and oxidation, with the chiral crotylboronic ester **8**. After work-up, the chiral complex **9** is isolated in excellent yield as a virtually pure diastereomer in high enantiomeric purity. In this remarkable (reagent-controlled) reaction, two new stereocenters as

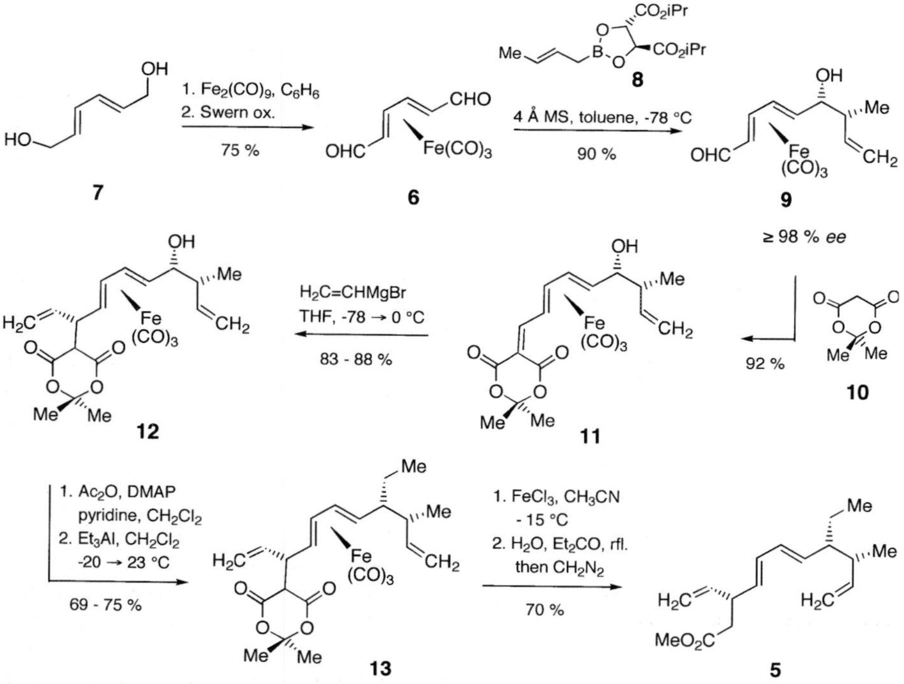

Scheme 2. Enantioselective synthesis of the acyclic intermediate **5** according to W. R. Roush.

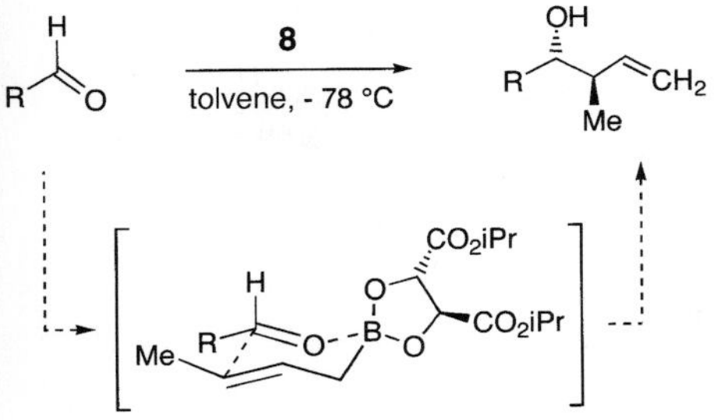

Figure 1

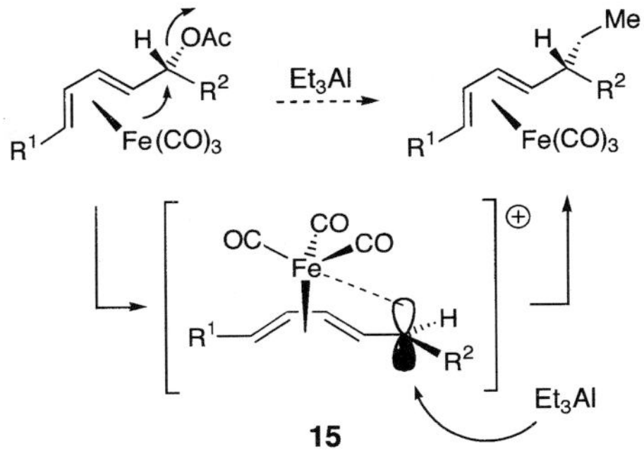

well as the chiral metal-complex substructure are generated with extremely high stereoselectivity. Practically, only one of eight possible diastereomers is formed.

The formation of the two chirality centers, with the relative and absolute configuration as shown in structure **9**, can be rationalized by assuming a chair-like transition state of type **B** (Fig. 1) [7]. More difficult to explain is how the reagent distinguishes between the two enantiotropic aldehyde functions in substrate **6** [3].

Condensation of complex **9** with Meldrums acid (**10**) gives complex **11** in good yield, which is then reacted with an excess of vinyl-MgBr to afford complex **12**. The complete diastereoselectivity of the latter reaction can be explained by assuming that the substrate prefers conformation **14** shown in Fig. 2 and that the nucleophile attacks the ligand exclusively from the less hindered π-face opposite to the $Fe(CO)_3$.

Next, the ethyl substituent is introduced via stepwise conversion of **12** with acetic anhydride and triethyl aluminum. This results in the completely diastereoselective formation of complex **13**. The fact that the substitution of the oxygen functionality occurs with complete retention of configuration can be explained as follows (Fig. 3). Supported by the iron fragment as a neighboring group, the acetoxy group leaves the molecule in an *exo* fashion. The resulting iron-stabilized cation **15** is then attacked by the nucleophile (Et_3Al) from the *exo* side, i.e. the side from which the acetoxy group left the molecule.

Figure 2

Figure 3

At this point of the synthesis the $Fe(CO)_3$ fragment has done its job and is oxidatively removed. Hydrolysis, decarboxylation and re-esterification then afford the desired intermediate **5**. With the aid of the small iron fragment, the synthesis of the acyclic intermediate **5** has been achieved in a short, highly selective sequence. Thus, the problem of the stereocontrolled generation of distant stereocenters (1,6 asymmetric induction) was solved in an elegant manner.

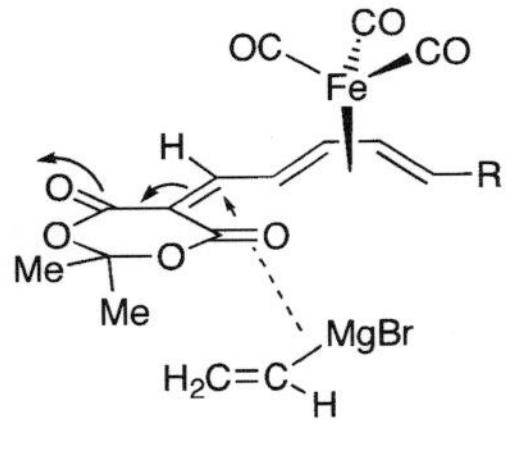

The Completion of the Ikarugamycin Synthesis

The transformation of the acyclic intermediate **5** into the tricyclic ikarugamycin precursor **2** was achieved by Roush following the route shown in Scheme 3.

At first, both terminal double bonds of **5** are hydroborated and the diol, obtained after oxidative work-up, is treated with mild acid (PPTs) resulting in the selective protection of one OH function through lactone formation. Now, the other OH group can be selectively oxidized to give the aldehyde **16**, which is then converted in a Wittig reaction to the α,β-unsaturated ester **17**. The transformation of **17** to the dialdehyde

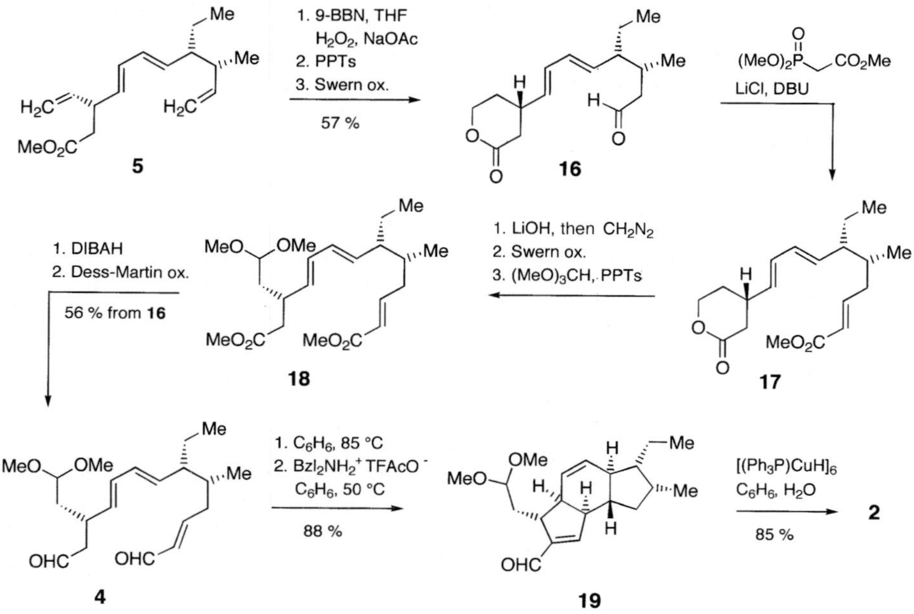

Scheme 3. Preparation of the tricyclic ikarugamycin building block **2** according to W. R. Roush.

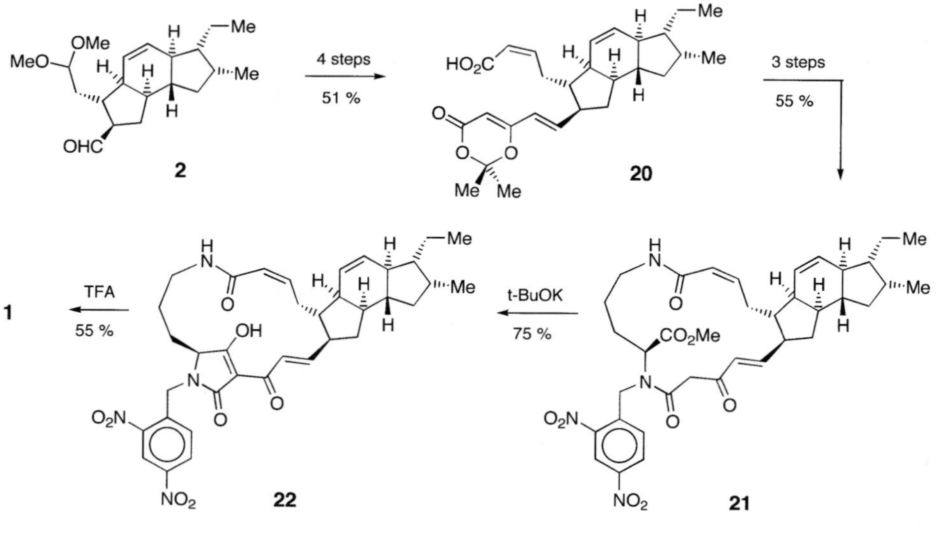

Scheme 4. Conversion of intermediale **7** into ikarugamycin (**1**) according to R. K. Boeckmann

4 is achieved in a 5-step sequence via the ester **18**. The intramolecular Diels-Alder reaction of **4** (→**3**) is directly followed by an aldol ring closure to give **19**, which is converted to the ikarugamycin precursor **2** by 1,4-reduction.

At this point, the synthesis of W. R. Roush [4] ends. It can, however, be considered as a formal total synthesis of ikarugamycin because the conversion of **2** into the target molecule **1** had been achieved by Boeckmann et al. (Scheme 4) [6a].

Conclusion

In comparison to the other syntheses of ikarugamycin [6, 8] or tricyclic precursors [9], the (formal) total synthesis according to Roush [4] and Boeckmann [6a] is particularly convincing because of its relatively small number of steps, its extremely high selectivity and its significant overall yield (>1 % over 28 steps).

The key of this success lies in the opening of the synthesis in which Roush and Wada utilize the various possibilities offered by the butadiene-$Fe(CO)_3$ chemistry for acyclic stereocontrol. They thus impressively demonstrate that the use of transition metal π-complexes as synthetic building blocks opens new and powerful strategies for the synthesis of complex target molecules.

References

[1] Reviews: a) A. J. Pearson, *Iron Compounds in Organic Synthesis*, Academic Press, London, **1994**, chapter 4; b) L. S. Hegedus, *Transition Metals in the Synthesis of Complex Organic Molecules*, University Science Books, Mill Valley, CA, **1994**, chapter 7.3; c) R. Grée, *Synthesis* **1989**, 341; d) M. Franck-Neumann, in: *Organometallics in Organic Synthesis*; A. de Meijere, H. tom Dieck (eds.) Springer-Verlag, Berlin, **1987**, p. 247.

[2] a) A. J. Pearson, K. Chang, D. B. McConville, W. J. Youngs, *Organometallics* **1994**, *13*, 4; b) H.-G. Schmalz, E. Heßler, J. W. Bats, G. Dürner, *Tetrahedron Lett.*, **1994**, *35*, 4543.

[3] W. R. Roush, J. C. Park, *Tetrahedron Lett.* **1990**, *31*, 4707.

[4] W. R. Roush, C. K. Wada, *J. Am. Chem. Soc.* **1994**, *116*, 2151.

[5] a) S. Ito, Y. Hirata, *Tetrahedron Lett.* **1972**, 1181; 1185; 2557; b) S. Ito, Y. Hirata, *Bull. Chem. Soc. Japn.* **1977**, *50*, 227; 1813.

[6] First total synthesis: a) R. K. Boeckmann, Jr., C. H. Weidner, R. B. Perni, J. J. Napier, *J. Am. Chem. Soc.* **1989**, *111*, 8036; b) R. K. Boeckmann, Jr., J. J. Napier, E. W. Thomas, *J. Org. Chem.* **1983**, *48*, 4152 [GSE]

[7] W. R. Roush, K. Ando, D. P. Powers, A. D. Palkowitz, R. L. Haltermann, *J. Am. Chem. Soc.* **1990**, *112*, 6339.

[8] Second total synthesis: a) L. A. Paquette, D. Macdonald, L. G. Anderson, J. Wright, *J. Am. Chem. Soc.* **1989**, *111*, 8037; b) L. A. Paquette, J. L. Romie, H.-S. Lin, J. Wright, *J. Am. Chem. Soc.* **1990**, *112*, 9284; c) L. A. Paquette, D. Macdonald, L. G. Anderson, *J. Am. Chem. Soc.* **1990**, *112*, 9292.

[9] For the synthesis of ikarugamycin precursors, see: a) M. J. Kurth, D. H. Burns, M. J. OBrien, *J. Org. Chem.* **1984**, *49*, 733; b) J. K. Whitesell, M. A. Minton, *J. Am. Chem. Soc.* **1987**, *109*, 6403; c) R. C. F. Jones, R. F. Jones, *Tetrahedron Lett.* **1990**, *31*, 3363; d) R. C. F. Jones, R. F. Jones, *Tetrahedron Lett.* **1990**, *31*, 3367.

Palladium-Catalyzed Synthesis of Vitamin D-Active Compounds

Sandra Krause and Hans-Günther Schmalz

Institut für Organische Chemie, Universität Köln, Germany

Vitamin D_3 (**1**) is an essential factor for the life of animals and man. It is formed in the skin under the influence of UV light from provitamin D_3 (**2**) and is one of the most important regulators of calcium metabolism. For instance, children lacking vitamin D develop rickets and adults suffer from osteoporosis.

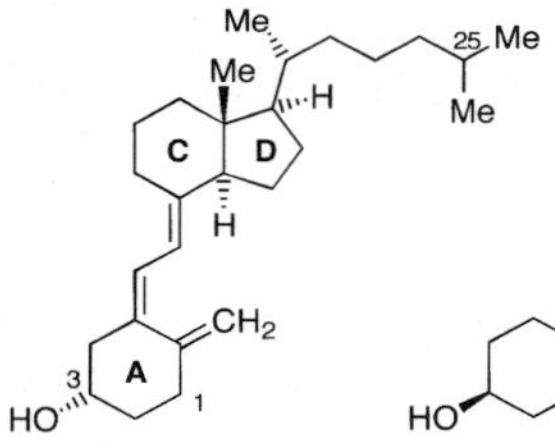

The discovery that vitamin D_3 (**1**), also called calciol [1], is actually a pro-hormone and not a vitamin as previously assumed has induced intense worldwide research activities within the last 20 years. Nowadays, it is known that the prohormone is transformed in liver and kidney into physiologically much more active metabolites by hydroxylation. In particular, the 1α,25-dihydroxylated derivative, calcitriol (**3**), performs a key function in the regulation of different physiological events [2]. Some hydroxylated vitamin D derivatives and structural analogs are currently being clinically tested as drugs for the treatment of a range of human diseases such as cancer, psoriasis or immune defects.

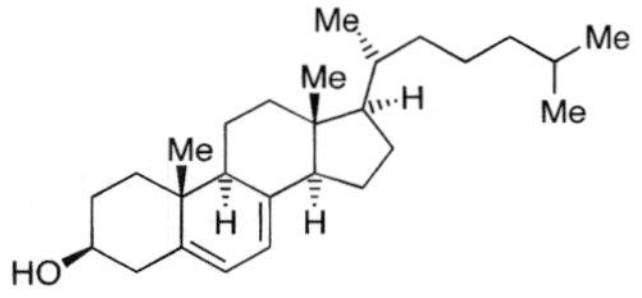

In order to cover the increasing demand for vitamin D-active substances and to make available labeled or structurally modified derivatives for research purposes, chemists are challenged to develop efficient and flexible synthetic routes to such compounds.

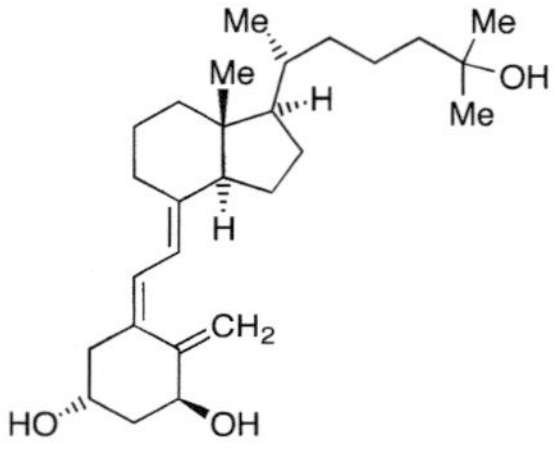

Since the completion of the first total synthesis of calciol (**1**) by H. H. Inhoffen, H. Burkhardt and G. Quinkert in 1959 [3], several research groups have focussed their activities on the synthesis of vitamin D derivatives [4]. In the course of these investigations, two particularly powerful strategies have been identified, which still today [5] form the basis of many syntheses of 1α-hydroxylated vitamin D derivatives (Scheme 1).

The biomimetic approach (strategy A) starts from the intact steroid skeleton and follows the construction principle ABCD → ACD. Irradiation of suitable functionalized dehydrocholesterols (**4**) leads to the formation of (6*Z*)-tacalciol derivatives (**5**) by electrocyclic ring opening. These intermediates are subsequently converted into the desired products (**6**) by thermal isomerization (sigmatropic 1,7-H-shift). The main disadvantage of this method is the circumstance that the functionalized starting materials have to be prepared in long linear sequences, which usually results in poor overall yields.

The second strategy is based on the modular construction principle A + CD → ACD and ac-

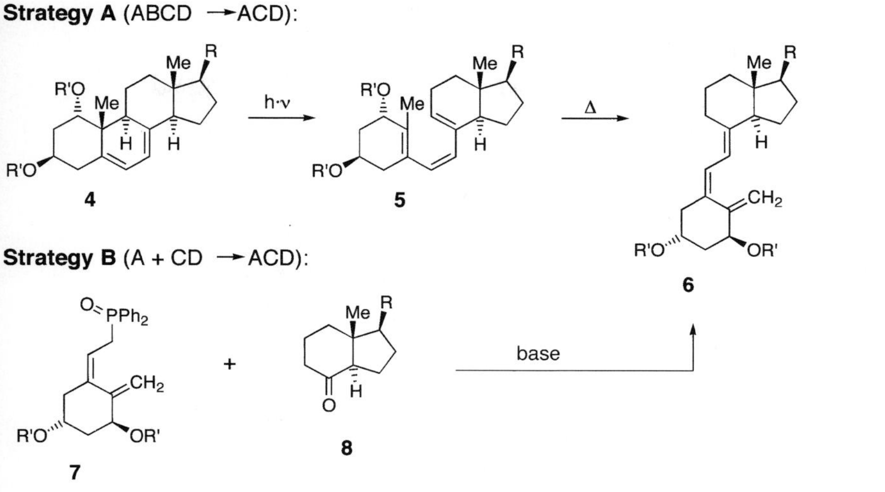

Strategy A (ABCD → ACD):

Strategy B (A + CD → ACD):

Scheme 1

cordingly offers high flexibility. Following a protocol developed by B. Lythgoe [6], ring A building blocks of type **7** are coupled with CD building blocks of type **8** by Wittig olefination. Whereas useful total synthetic methods have been established for the preparation of ring A building blocks, no really competitive approaches are yet available for the total synthesis of CD building blocks [7]. Therefore, the latter are usually prepared from Grundmann's ketone (**9**) or the Inhoffen-Lythgoe diol (**10**), which are obtained by partial synthesis (degradation) of vitamin D_3 or vitamin D_2, respectively.

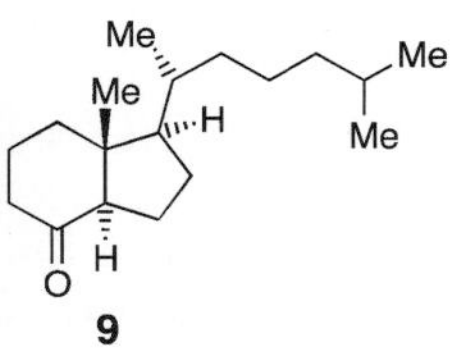

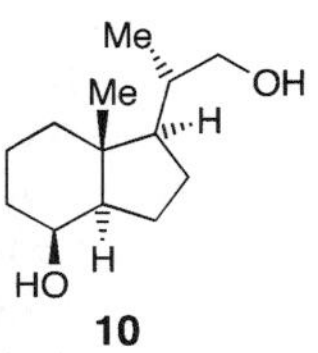

In recent years, the development of powerful methods for palladium-catalyzed C–C bond formation [8] has led to the development of novel and highly efficient synthetic avenues to vitamin D-active compounds, which, at least in many respects, are superior to the "classical" routes mentioned above.

One of these new strategies was developed in the laboratories of A. Mouriño [9] and W. H. Okamura [10] and has been successfully applied in the synthesis of a variety of vitamin D derivatives [9, 10]. Following this strategy (Scheme 2), the target molecules **11** are derived from ynedienes of type **12**, which in turn represent products of a palladium-catalyzed coupling reaction [8] between the building blocks **13** and **14**. Enol triflates of type **13** are easily accessible from Grundmann's ketone (**15**) or related ketones, while ring A building blocks of type **14** can be prepared, for instance, from (*S*)-carvone (**16**).

As an example, a remarkable synthesis of calcitriol (**3**) will be discussed here (Scheme 3), which was described by A. Mouriño [9g]. This synthesis starts with the Inhoffen-Lythgoe diol (**10**) which is first converted to the iodide **17**. From this intermediate, the CD building block **18** is obtained by coupling with acrylic acid methyl ester in aqueous ethanol under sonochemical conditions. Using Pd-catalyzed coupling, triflate **18** is then coupled to the ring A building block **19**, which is accessible in seven steps (36 % yield) from (*S*)-carvone (**16**) [11]. The coupling product **20** is transformed by hydrogenation and isomerization in excellent yield to **21**, from which calcitriol **3** is finally obtained by treatment with methyl lithium and tetrabutyl ammonium fluoride. All in all, this synthesis impresses because of its low number of steps and high overall yield. In addition, since the tertiary OH group in the side chain is generated in the last step of the synthesis, there is only a minimum need for protecting groups.

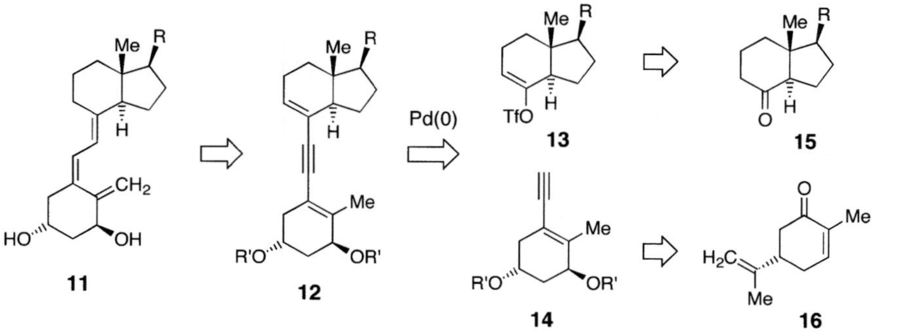

Scheme 2

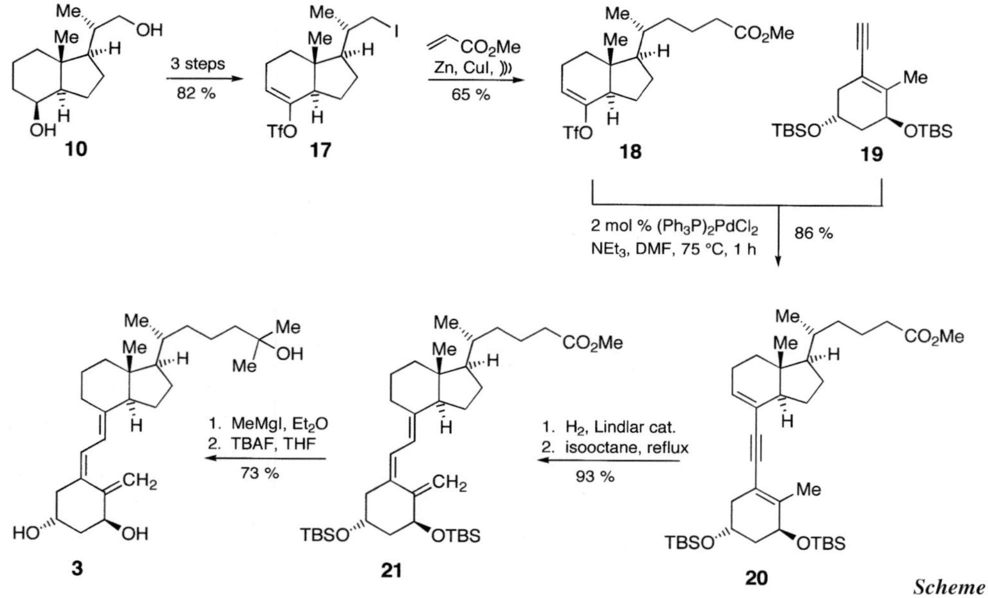

Scheme 3

Another very elegant approach towards the vitamin D skeleton has been developed by B. M. Trost and his co-workers [12]. In contrast to all other strategies, the A ring need not be preformed in this case. According to Trost, the target molecules (**11**) are retrosynthetically disconnected to form CD building blocks of type **22** and enynes of type **23**.

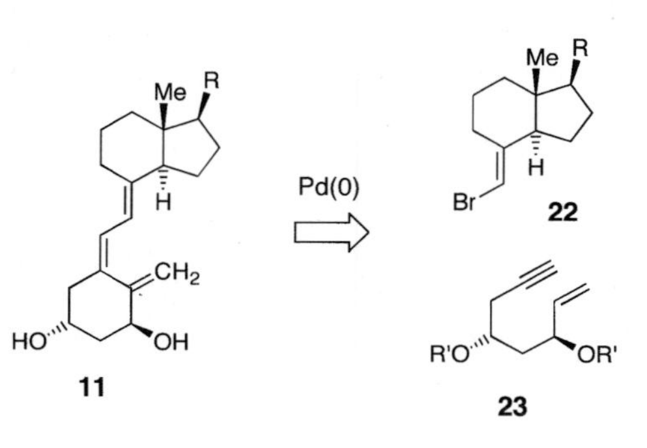

By Pd-catalyzed coupling of the components **22** and **23**, the construction of ring A and its attachment to the CD building block (under stereoselective establishment of the complete triene unit) is achieved in a single synthetic operation. Mechanistically, this magic-looking transformation can be rationalized as follows (Scheme 4). In the first step a Pd(0) species reacts with the alkenyl bromide **22** to an intermediate of type **24** (oxidative addition). A subsequent *syn*-inser-

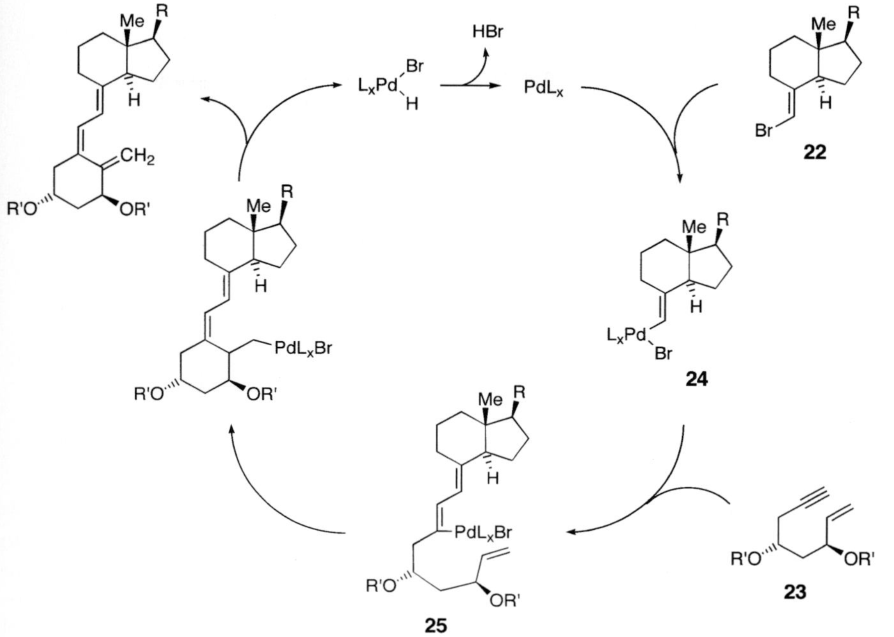

Scheme 4

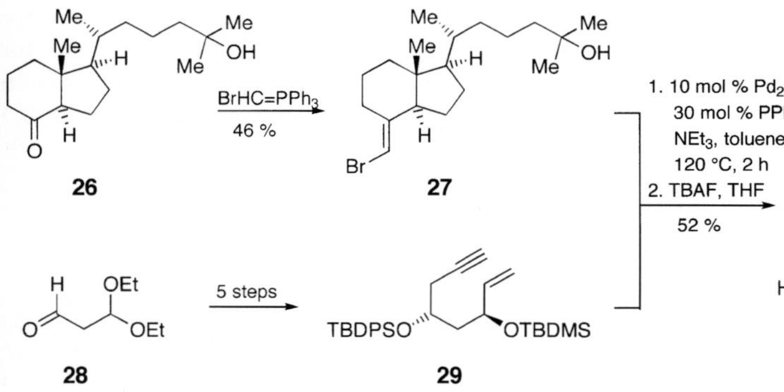

Scheme 5

tion of the alkyne **23** then leads to a complex of type **25**, which is finally transformed into the desired product in a Heck-type reaction, i.e. by olefin insertion (cycloisomerization) and β-H-elimination.

The Trost synthesis of calcitriol (**3**) is shown in Scheme 5. In a Wittig reaction, the hydroxylated Grundmann's ketone **26** (obtained by partial synthesis) is transformed into the alkenyl bromide **27** with astonishingly high diastereoselectivity ($E/Z \geq 50:1$). The chiral enyne **29**, which is needed as the second building block for the coupling step, is prepared from the aldehyde **28** in only a few steps. The enantioselectivity is achieved by kinetic resolution (applying Sharpless' method). The crucial Pd-catalyzed coupling step then proceeds smoothly (despite the presence

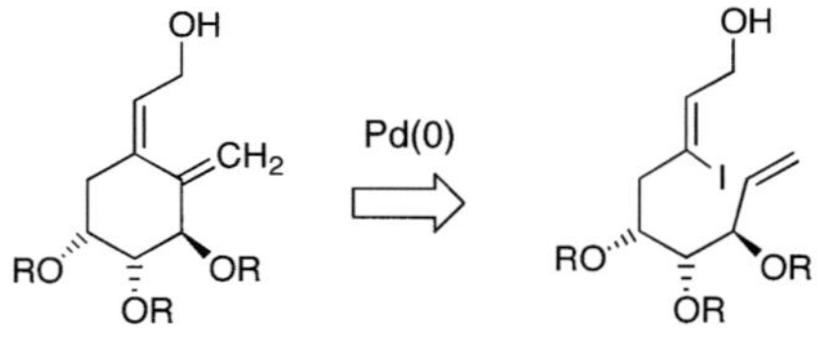

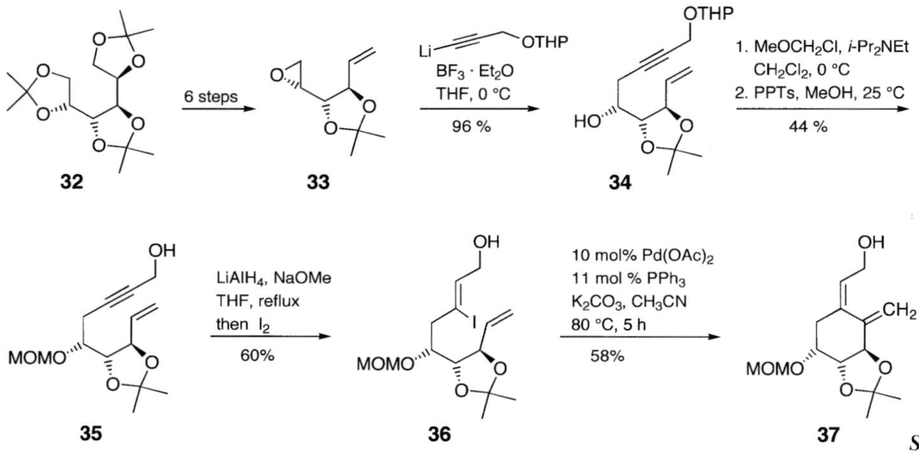

32 → 6 steps → 33

$Li-C\equiv C-CH_2-OTHP$, $BF_3 \cdot Et_2O$, THF, 0 °C, 96 % → 34

1. $MeOCH_2Cl$, i-Pr_2NEt, CH_2Cl_2, 0 °C
2. PPTs, MeOH, 25 °C

44 % → 35

$LiAlH_4$, NaOMe, THF, reflux, then I_2, 60% → 36

10 mol% $Pd(OAc)_2$, 11 mol % PPh_3, K_2CO_3, CH_3CN, 80 °C, 5 h, 58% → 37

Scheme 6

of the unprotected hydroxyl group). After fluoride-induced cleavage of the silyl protecting groups, isomerically pure calcitriol (**3**) is obtained in excellent overall yield.

An approach to trihydroxylated ring A building blocks of type **30**, which is also based on an intramolecular Heck reaction (of substrates of type **31**), was described by T. Takahashi [13].

In this synthesis (Scheme 6), the C_2-symmetrical triacetonide of *D*-mannitol (**32**) is converted via the epoxide **33** and its nucleophilic addition product **34** to the propargylic alcohol derivative **35**. From this intermediate, the *Z*-configured vinyl iodide **36** is stereoselectively obtained by hydroalumination/iodination. The Pd-catalyzed Heck cyclization then affords the isomerically pure product **37**, which represents a potential building block for the synthesis of $1\alpha,2\beta,25$-trihydroxy-vitamin D_3 following the classical Wittig strategy of Lythgoe.

Of course, one could also think about opening the epoxide **33** with lithium acetylide to the enyne **38**, which could eventually be further reacted by Trost's method (e.g. with the vinyl bromide **27**).

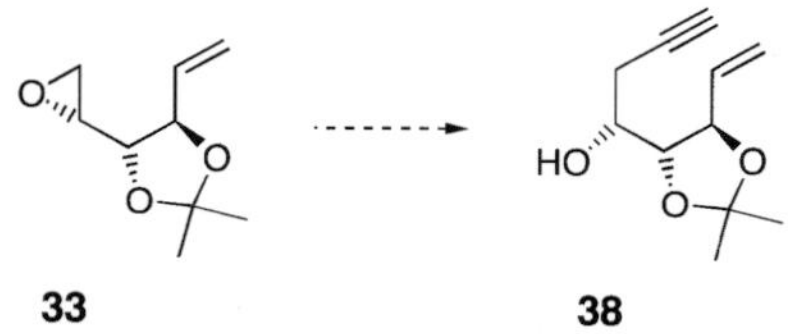

33 → 38

The examples discussed in this paper demonstrate that Pd-catalyzed coupling reactions can be successfully utilized in the convergent (modular) synthesis of vitamin D-active compounds. The new synthetic routes open an efficient and highly selective access to a variety of vitamin D analogs. While reliable methods exist today for the construction of the triene system and for the synthesis of the ring A precursors, the search for efficient total synthetic approaches to the CD building blocks still remain a challenging task for the future.

References

[1] For the nomenclature of vitamin D and related compounds, see: *Pure Appl. Chem.* **1982**, *54*, 1511; *ibid.* **1989**, *61*, 1783.

[2] Reviews: a) H. F. DeLuca, J. Burmester, H. Darwish, J. Krisinger, *Comprehensive Medicinal Chemistry*, Pergamon, New York, **1990**, Vol. 3, 1129; b) A. W. Norman, R. Bouillon, M. Thomasset (eds.) *Vitamin D: Gene Regulation, Structure Function Analysis and Clinical Application*, de Gruyter, Berlin, **1991**; c) A. W. Norman, R. Bouillon, M. Thomasset (eds.) *Vitamin D: Chemistry, Biology and Clinical Application of the Steroid Hormone*, Vitamin D workshop, Inc: Riverside, CA, **1997**.

[3] H. H. Inhoffen, H. Burkardt, G. Quinkert, *Chem. Ber.* **1959**, *92*, 1564.

[4] a) B. Lythgoe, *Chem. Soc. Rev.* **1980**, 449 and refs. cited therein; b) G.-D. Zhu, W. H. Okamura, *Chem. Rev.* **1995**, *95*, 1877; c) For a concise collection of total- and partial syntheses of vitamin D active compounds, see: G. Quinkert (Ed.), *Synform* **1985**, *3*, 41; *ibid.* **1986**, *4*, 131; *ibid.* **1987**, *5*, 1; d) H. Dai, G. H. Posner, *Synthesis* **1994**, 1383, e) see also ref. 5c and refs. cited therein.

[5] Selected more recent publications: a) K. Yamamoto, J. Takahashi, K. Hamano, S. Yamada, *J. Org. Chem.* **1993**, *58*, 2530; b) G. H. Posner, H. Dai, K. Afarinkia, N. N. Murthy, K. Z. Guyton, T. W. Kensler, *J. Org. Chem.* **1993**, *58*, 7209; c) M. de los Angeles Rey, J. A. Martínez Pérez, A. Fernandez-Gacio, K. Halkes, Y. Fall, J. R. Granja, A. Mouriño, *J. Org. Chem.* **1999**, 64, 3196.

[6] B. Lythgoe, T. A. Moran, M. E. N. Nambudiry, J. Tideswell, P. W. Wright, *J. Chem. Soc. Perkin Trans.* I **1978**, 590.

[7] P. Jankowski, S. Marczak, J. Wicha, *Tetrahedron*, **1998**, 12071.

[8] See, for instance: a) R. F. Heck, *Palladium Reagents in Organic Synthesis*, Academic Press, London, **1985**, b) F. Diederich, P. J. Stang (eds.) *Metal-catalyzed Cross-coupling Reactions*, Wiley-VCH, Weinheim, **1998**.

[9] a) L. Castedo, A. Mouriño, L. A. Sarandeses, *Tetrahedron Lett.* **1986**, *27*, 1523; b) L. Castedo, J. L. Mascareñas, A. Mouriño, *Tetrahedron Lett.* **1987**, *28*, 2099; c) L. Castedo, J. L. Mascareñas, A. Mouriño, L. A. Sarandeses, *Tetrahedron Lett.* **1988**, *29*, 1203; d) J. L. Mascareñas, L. A. Sarandeses, L. Castedo, A. Mouriño, *Tetrahedron* **1991**, *47*, 3485; e) M. Torneiro, Y. Fall, L. Castedo, A. Mouriño, *Tetrahedron Lett.* **1992**, *33*, 105; f) L. A. Sarandeses, M. J. Vallés, L. Castedo, A. Mouriño, *Tetrahedron* **1993**, *49*, 731; g) J. P. Sestelo, J. L. Mascareñas, L. Castedo, A. Mouriño, *J. Org. Chem.* **1993**, *58*, 118; h) J. R. Granja, L. Castedo, A. Mouriño, *J. Org. Chem.* **1993**, *58*, 124.

[10] a) S. A. Barrack, R. A. Gibbs, W. H. Okamura, *J. Org. Chem.* **1988**, *53*, 1790; b) M. L. Curtin, W. H. Okamura, *J. Am. Chem. Soc.* **1991**, *113*, 6958; c) A. S. Lee, A. W. Norman, W. H. Okamura, *J. Org. Chem.* **1992**, *57*, 3846; d) A. S. Craig, A. W. Norman, W. H. Okamura, *J. Org. Chem.* **1992**, *57*, 4374; e) W. H. Okamura, H. Y. Elnagar, M. Ruther, S. Dobreff, *J. Org. Chem.* **1993**, *58*, 600; f) K. R. Muralidharan, A. R. de Lera, S. D. Isaeff, A. W. Norman, W. H. Okamura, *J. Org. Chem.* **1993**, *58*, 1895.

[11] W. H. Okamura, J. M. Aurrecoechea, R. A. Gibbs, A. W. Norman, *J. Org. Chem.* **1989**, *54*, 4072.

[12] a) B. M. Trost, J. Dumas, *J. Am. Chem. Soc.* **1992**, *114*, 1924; b) B. M. Trost, J. Dumas, M. Villa, *J. Am. Chem. Soc.* **1992**, *114*, 9836.

[13] T. Takahashi, M. Nakazawa, *Synlett* **1993**, 37.

Syntheses of Oligo(thiazoline) Natural Products

Sabine Hoppen and Ulrich Koert

Institut für Chemie, Humboldt-Universität, Berlin, Germany

Oligo(thiazolines) are a class of natural products with interesting structural features and promising biological activities (Scheme 1). Typical members are mirabazole C (**1**) [1], mirabazole B (**2**) [1], tantazole B (**3**) [2], and thiangazole (**4**) [3, 4].

The mirabazoles were isolated from the blue-green alga *Scytonemea mirabile* [1]. The tantazoles were named after the place of their first isolation, Mount Tantalus on Hawaii. The biological source of thiangazole is strain PI 3007 of the bacterium *Polyangium spec.* [3]. The mirabazoles and tantazoles offer selective cytotoxicity profiles [1, 2]. Thiangazole is a potent HIV-1 inhibitor [3]. The biosynthesis of the oligo(thiazolines) is considered to involve 2-methylcysteine, an amino acid of rare occurrence in nature. Stimulated by these interesting new lead structures, several research groups focussed on the total synthesis of oligo(thiazoline) natural products. Here we discuss the synthetic work of Fukuyama et al. [5] (tantazole B, 1993), Ehrler et al. [6] (thiangazole, 1994), Pattenden et al. [7] (thiangazole, 1994), Heathcock et al. [8] (mirabazole C, 1994), Wipf et al. [9] (thiangazole, 1995) and Kiso et al. [10] (mirabazole C, 1996). Special emphasis is given to

- the preparation of the enantiomerically pure (*R*)- and (*S*)-2-methylcysteine,
- the closure of the thiazoline rings
- the synthetic strategy (sequential or multiple formation of thiazoline rings).

The groups mentioned above have used their synthetic expertise to prepare other oligo(thiazoline) natural products as well [11]. This work is methodologically related to the total syntheses discussed below.

Enantiomerically Pure (*R*)- and (*S*)-2-Methylcysteine Building Blocks

Access to enantiomerically pure (*R*)- and (*S*)-2-methylcysteine is necessary to assemble these oligomeric heterocyclic compounds. The decisive

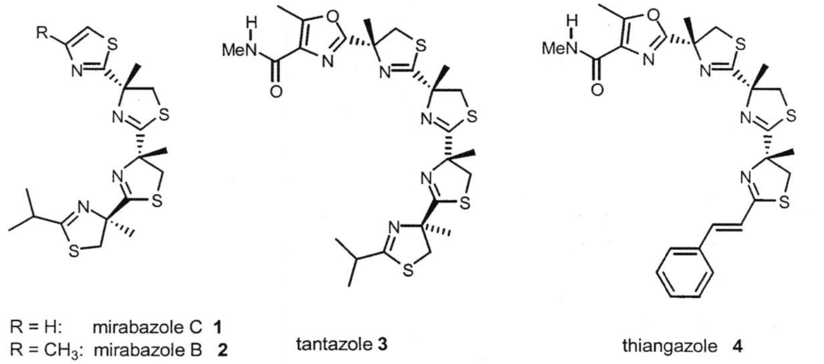

R = H: mirabazole C **1**
R = CH_3: mirabazole B **2**

tantazole **3**

thiangazole **4**

Scheme 1. Oligo(thiazoline) natural products.

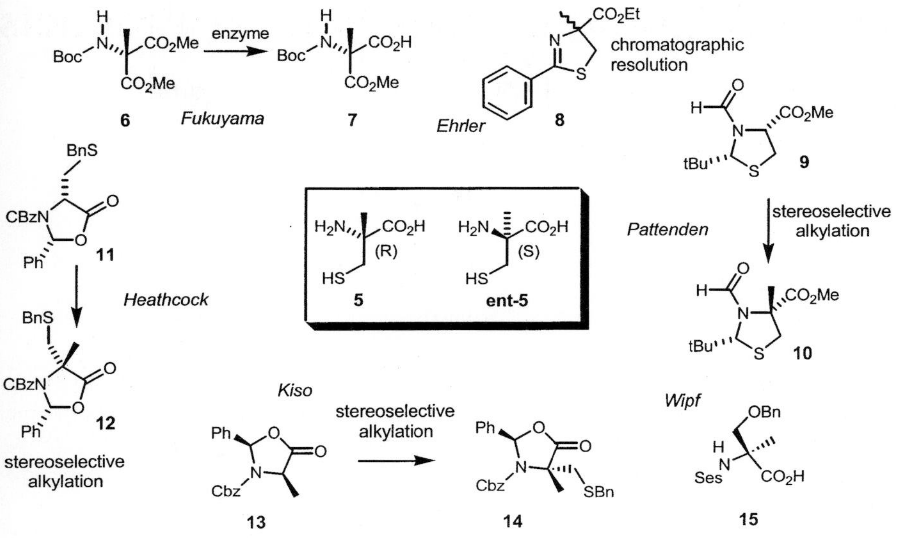

Scheme 2. Preparation of the enantiomerically pure (*R*)- and (*S*)-2-methylcysteine building blocks.

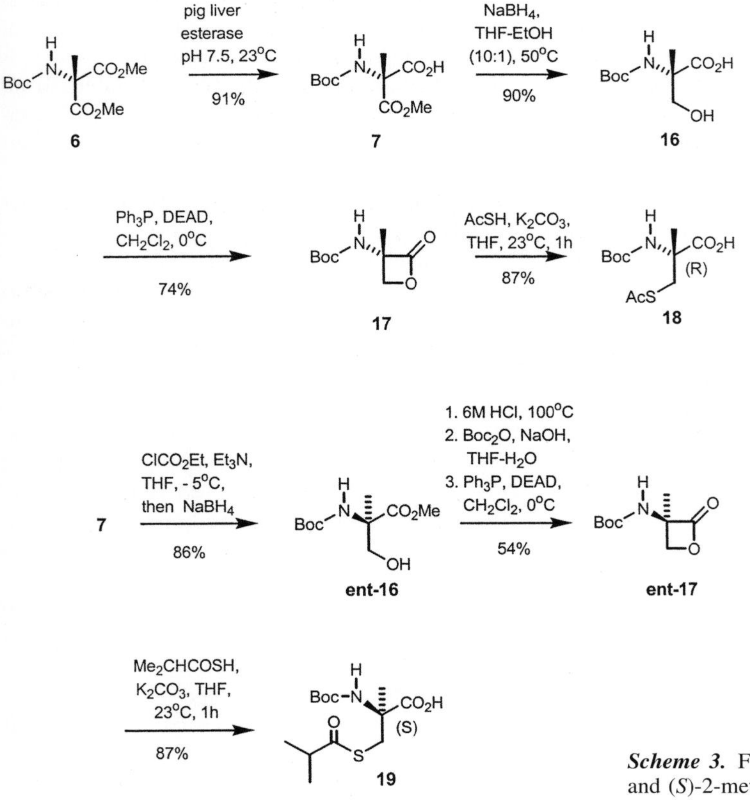

Scheme 3. Fukuyama's synthesis of the (*R*)- and (*S*)-2-methylcysteine building blocks.

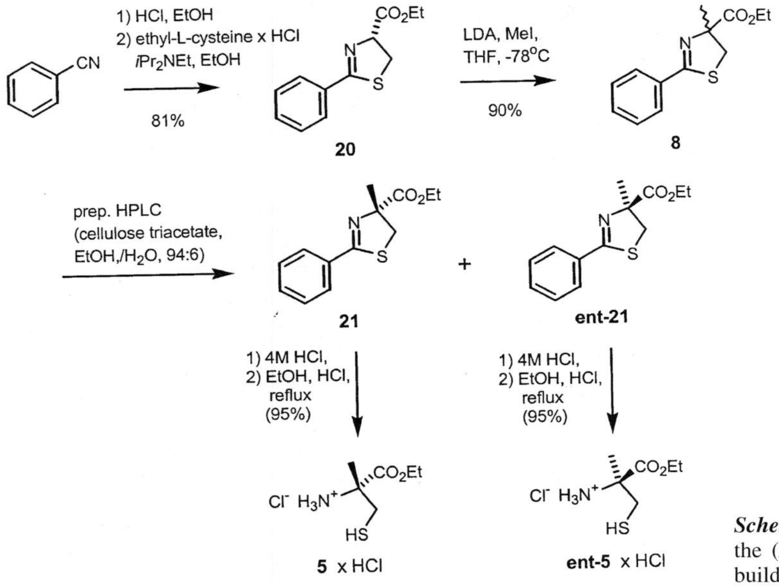

Scheme 4. Ehrler's synthesis of the (*R*)- and (*S*)-2-methylcysteine building blocks.

steps for the construction of the chiral center of 2-methylcysteine are shown in Scheme 2.

Fukuyama [5] obtains the building blocks by an enzymatic ester hydrolysis (**6** → **7**). As both enantiomers are needed in the synthesis, Ehrler [6] uses the separation of a racemate by chromatography of **8**. A stereoselective alkylation is the key step in the work of Pattenden [7] (**9** → **10**), Heathcock [8] (**11** → **12**) and Kiso [10] (**13** → **14**). Wipf [9] starts from (*S*)-methylserine **15** (Scheme 3).

The preparation of protected (*R*)-2-methylcysteine by Fukuyama starts with the enantioselective discrimination of the prochiral ester groups in **6** with pig liver esterase (Scheme 3) [5]. The ester function of the resulting product **7** is selectively reduced (**7** → **16**). Cyclization to the β-lactone gives compound **17**. Attack of the thioacetate at the β-lactone methylene carbon atom provides the (*R*)-compound **18**. Selective reduction of the carboxylic acid function in **7** gives the (*S*)-compound **19** in an analogous fashion.

Ehrler's synthesis was carried out in the Ciba-Geigy laboratories. There exists not only plenty of expertise on chromatography on chiral columns, but also the equipment for chromatographic resolution on a preparative scale. Therefore Ehrler's key step for the synthesis of (*R*)- and (*S*)-2-methylcysteine (Scheme 4) is the separation of thiazoline (**8**) on cellulose triacetate.

Thiazoline (**8**) is readily available from cysteine ethyl ester and benzonitrile. The separated enantiomers **18** and *ent-18* can be converted to (*R*)- and (*S*)-2-methylcysteine ethyl ester hydrochloride **5** × HCl and *ent-5* × HCl under standard conditions.

Pattenden's synthesis of (*R*)- and (*S*)-2-methylcysteine [12] is based on Seebach's self regeneration of chirality [13]. Scheme 5 shows the synthesis of the (*R*)-isomer **5b**.

From the (*R*)-cysteine methyl ester **22** the thiazolidine **9** is obtained, which can be deprotonated with LDA. The attack of the methyl iodide on the enolate takes place from the side opposite to the bulky *tert*-butyl group (**23** → **10**). The auxiliary chiral center is removed under acidic conditions and (*R*)-2-methylcysteine methyl ester hydrochloride **5b** × HCl is obtained. (*S*)-2-Methylcysteine methyl ester hydrochloride **ent-5b** × HCl can be prepared in the same way from (*S*)-cysteine.

The synthesis of the (*R*)- and (*S*)-2-methylcysteine by Heathcock (Scheme 6) uses a similar stereoselective alkylation for the introduction of the methyl group [8].

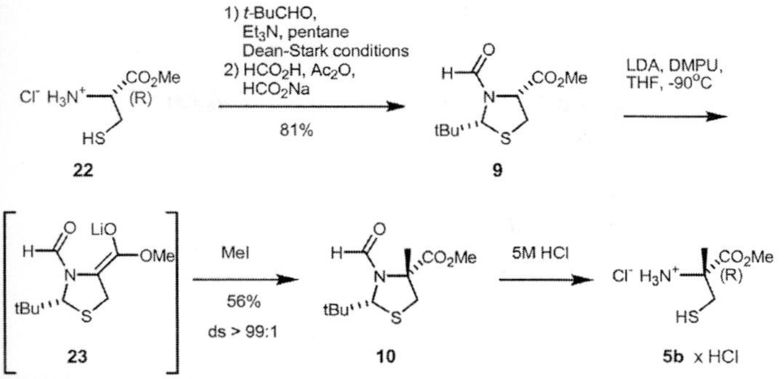

Scheme 5. Pattenden's synthesis of the (*R*)- and (*S*)-2-methylcysteine building blocks.

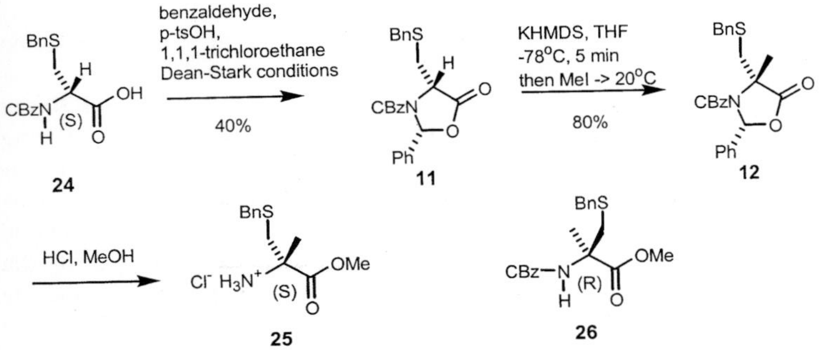

Scheme 6. Heathcock's synthesis of the (*R*)- and (*S*)-2-methylcysteine building blocks.

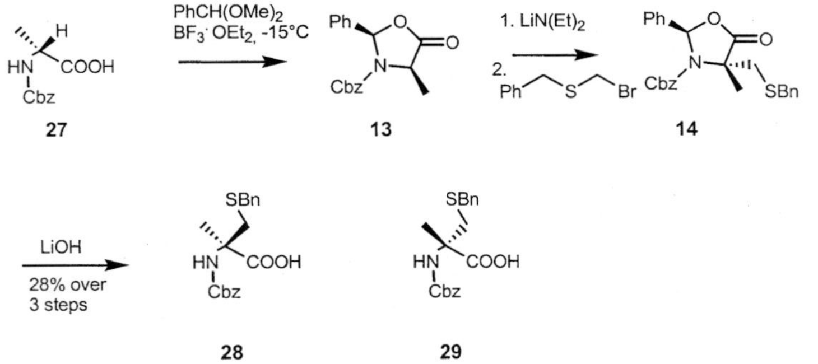

Scheme 7. Kiso's synthesis of the (*R*)- and (*S*)-2-methylcysteine building blocks.

First, the diprotected cysteine **24** is converted into the oxazolidinone **11**. Then a stereoselective alkylation takes place introducing the methyl group from the side opposite to the phenyl group (**11** → **12**). After removal of the auxiliary chiral center, the (*S*)-2-methylcysteine compound **25** is obtained. The (*R*)-2-methylcysteine compound **26** is accessible along the same route. The use of oxazolidinones of the type **11** for the enantioretentive alkylation of acyclic amino acids originates from the work of Karady et al. [14]. A modification of the Karady method is used by Kiso [10] (Scheme 7).

$BF_3 \cdot OEt_2$-mediated condensation of Cbz-protected D-alanine with benzaldehyde dimethyl acetal provides the oxazolidinone **13**. Stereoselective alkylation of **13** with bromomethyl ben-

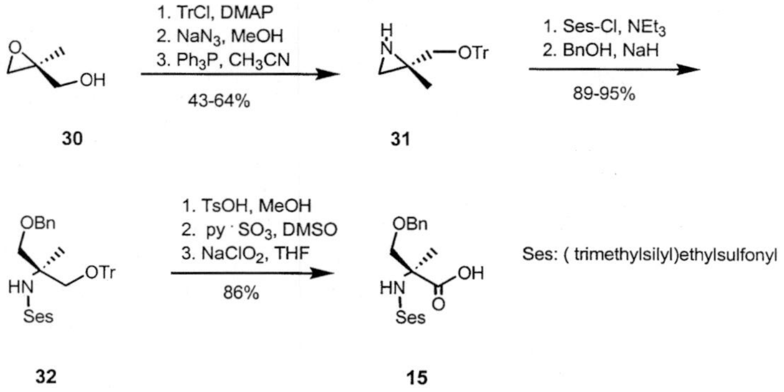

1. TrCl, DMAP
2. NaN_3, MeOH
3. Ph_3P, CH_3CN
43-64%

1. Ses-Cl, NEt_3
2. BnOH, NaH
89-95%

1. TsOH, MeOH
2. py · SO_3, DMSO
3. $NaClO_2$, THF
86%

Ses: (trimethylsilyl)ethylsulfonyl

Scheme 8. Wipf's synthesis of diprotected (*S*)-2-methylserine.

zylsulfide leads to the oxazolidinone **14**. Again the electrophile is introduced from the less hindered side. The choice of base was crucial. Only with lithium diethylamide was a reasonable yield achieved. Subsequent hydrolysis of **14** provides the (*R*)-2-methylcysteine derivative **28**. The related (*S*)-enantiomer **29** is available on the same way, starting with Cbz-protected L-alanine.

To simplify the handling of intermediates, Wipf introduces the sulfur atoms very late in the course of the synthesis [9]. Here 2-methylserine **15** serves as masked 2-methylcysteine (Scheme 8).

Starting from the (*S*)-2-methylglycidol **30** the aziridine **31** is obtained via O-tritylation, ring opening with sodium azide and subsequent reductive cyclisation. After N-activation of the aziridine **31**, treatment with sodium benzyloxide leads to a regioselective ring opening of the aziridine. The product **32** is detritylated and the resulting primary alcohol is oxidized to the 2-methylserine derivative **15**.

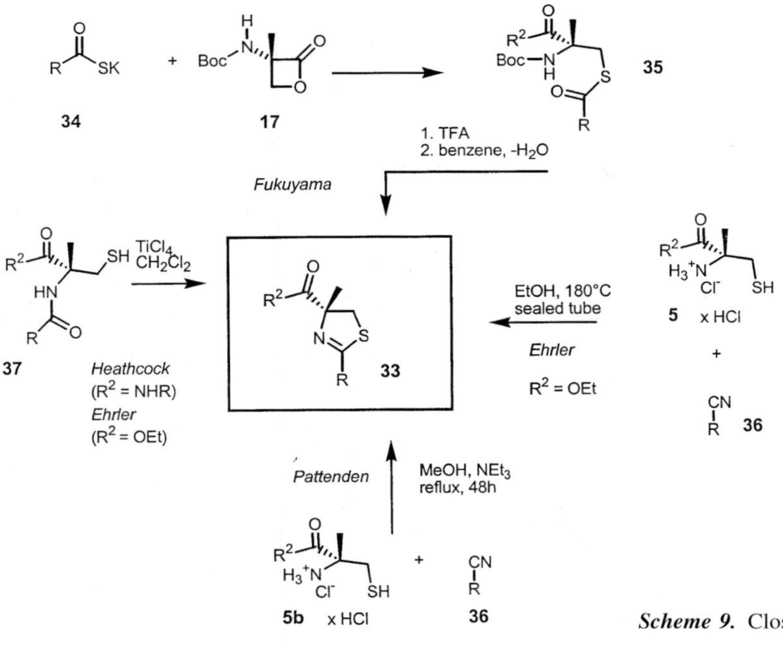

1. TFA
2. benzene, $-H_2O$

Fukuyama

$TiCl_4$, CH_2Cl_2

Heathcock (R^2 = NHR)
Ehrler (R^2 = OEt)

EtOH, 180°C
sealed tube
Ehrler
R^2 = OEt

Pattenden
MeOH, NEt_3
reflux, 48h

Scheme 9. Closure of the thiazoline rings.

Closure of the Thiazoline Rings

After the preparation of the 2-methylcysteine compounds, the ring closure to the thiazoline **33** is the next step in most of the synthetic strategies. The key reactions used are summarized in Scheme 9.

Fukuyama [5] obtained the thioester **35** from the reaction of the thiocarboxylic acid anion **34** with the *β*-lactone compound **17**. After acidic deprotection of the Boc group, the thiazoline ring is successfully closed in benzene under Dean-Stark conditions (**35** → **33**) in yields between 60 and 80 %. Ehrler [6] and Pattenden [7] choose a more classical way. They close the thiazoline ring by condensation of the aminothiol hydrochloride **5** × HCl with a nitrile. The yields are generally modest (45 and 55 %) except in one case [6]. Installation of the necessary nitrile function hinders its repetitive use, as this procedure causes a considerable loss of material. Heathcock [8] and Ehrler [6] choose the thiolamide **37** as a key compound. Titanium tetrachloride in dichloromethane turned out to be the best choice for the thiazoline ring closure. Thus, Heathcock successfully closes all four rings of mirabazole C at once in a remarkable 45 % yield. In one synthetic step all four thiazoline rings are set up! Ehrler uses titanium tetrachloride for the consecutive construction of single thiazoline rings. The yield for one closure is between 30 and 35 %. Wipf [9] and Kiso [10) use a multiple ring closure with titanium tetrachloride according to Heathcock's procedure.

Synthesis of Tantazole B by Fukuyama

Fukuyama starts with the synthesis of three of the four thiazoline rings of tantazole B in a linear fashion (Scheme 10).

Closure of the (*S*)-2-methylcysteine compound **19** (see Scheme 3) provides the thiazoline carboxylic acid **38**. For the addition of the next thiazoline ring, the carboxylic acid **38** has to be converted to a thiocarboxylic acid. Fukuyama

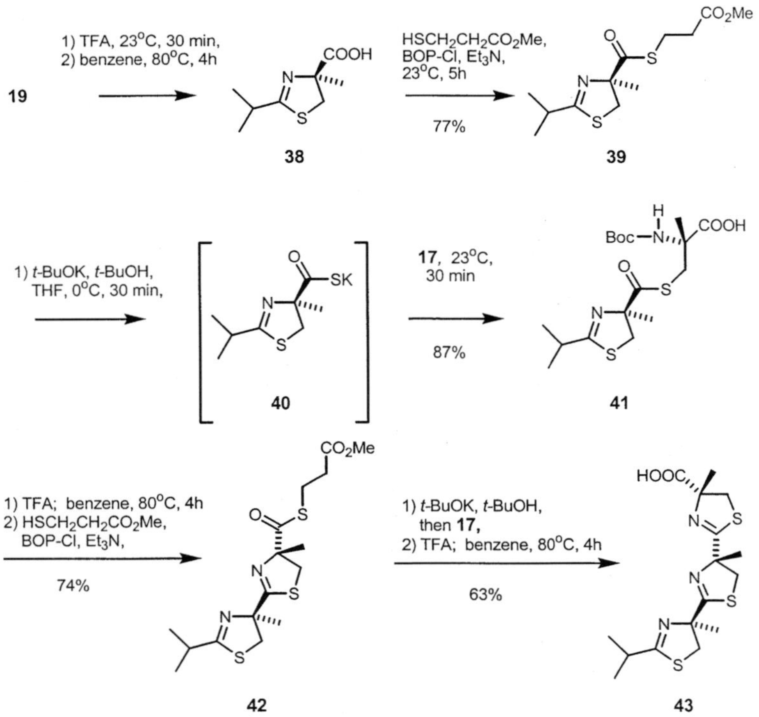

Scheme 10. Fukuyama's synthesis of tantazole B (part 1).

introduces the stable thioester **39** to circumvent the very unstable free thioacid. After a β-elimination of **39** under basic conditions, the desired thiocarboxylate **40** is generated *in situ*. Addition of the β-lactone **17** gives the thioester **41**. Thiazoline ring closure leads to the corresponding bis(thiazoline) carboxylic acid derivative **42**. The tris(thiazoline) carboxylic acid **43** is accessible by addition of another thiazoline unit.

The target molecule tantazole B contains an oxazole ring as well as the four thiazoline rings. Because of the less nucleophilic oxygen, oxazoles or oxazolines have to be prepared under more drastic conditions than those for sulfur counterparts. Under these conditions, the thiazolines could be damaged. Because of that, Fukuyama prepares the oxazole first. Then he couples the oxazole with the tris(thiazoline) carboxylic acid **43** (Scheme 11).

The synthesis of the oxazole compound **45** starts with the coupling of the N-protected (*R*)-methylcysteine compound **18** with threonine *tert*-butyl ester using bis(2-oxo-3-oxazolidinyl)phosphinyl chloride (BOP-Cl) [15] as a coupling reagent. Jones oxidation of the threonine hydroxy group leads to the ketoamide **44**. The desired oxazole ring is closed by treatment with thionylchloride/pyridine. After deprotection, the oxazole, compound **45** is obtained. In the next step the oxazole compound **45** is coupled with the tris(thiazoline) compound **43** to yield the thioester **46**. Now Fukuyama closes the fourth and last thiazoline ring (**46** → **47**). After conversion of the carboxylic acid function into a methyl-

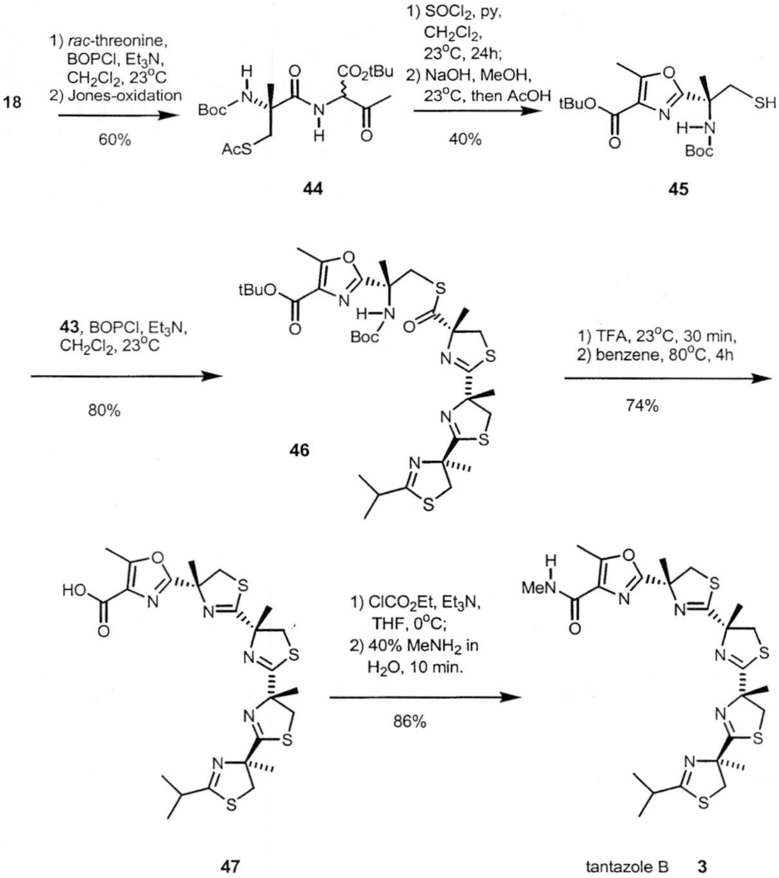

18 → 1) *rac*-threonine, BOPCl, Et_3N, CH_2Cl_2, 23°C 2) Jones-oxidation; 60% → **44** → 1) $SOCl_2$, py, CH_2Cl_2, 23°C, 24h; 2) NaOH, MeOH, 23°C, then AcOH; 40% → **45**

45 → **43**, BOPCl, Et_3N, CH_2Cl_2, 23°C; 80% → **46** → 1) TFA, 23°C, 30 min, 2) benzene, 80°C, 4h; 74% → **47**

47 → 1) $ClCO_2Et$, Et_3N, THF, 0°C; 2) 40% $MeNH_2$ in H_2O, 10 min.; 86% → tantazole B **3**

Scheme 11. Fukuyama's synthesis of tantazole B (part 2).

amide, the target molecule tantazole B (**3**) is reached.

Ehrler's Synthesis of Thiangazole

Thiangazole contains an oxazole ring as well as three thiazoline rings and a styrene group. Ehrler [6] decides to introduce the styrene group by a Wittig-type reaction. He prepares the necessary diphenylphosphinoxide function right at the beginning and carries it successfully through the whole synthesis. So he has the option to install either the styrene function or a related group at every synthetic step. This synthetic strategy is flexible enough to obtain several analogs on the way to the lead structure. The biological activity of all these compounds can be evaluated.

The synthesis starts with the condensation of nitrile **48** with the (*R*)-2-methylcysteine compound **5** × HCl (Scheme 12).

The resulting ethyl ester **49** is hydrolyzed and coupled with another (*R*)-2-methylcysteine (**5**) to the amide **50**. Using the method of Heathcock, the thiazoline ring closure with titanium tetrachloride leads to the bis(thiazoline) **51**. Another coupling with **5** and subsequent thiazoline ring closure provides the tris(thiazoline) **52**. The further progress of the synthesis is shown in Scheme 13.

First the styrene function is generated (**52** → **53**). Now only the construction of the oxazole remains to be accomplished. In contrast to the ideas of Fukuyama, Ehrler decides to prepare the oxazole in the presence of the other thiazoline rings. The conversion of the ester **53** into the amide **54** succeeds with 76 % yield well enough. But the yields drop in the following cyclocondensation with the bromoketoester **55**. After conversion of the ester to the methylamide (**56** → **4**), thiangazole is isolated in only 20 % yield. Nevertheless, Ehrler successfully reached the target.

Pattenden's Synthesis of Thiangazole

The synthetic strategy of the thiangazole synthesis of Pattenden [7] is similar to the strategy of the tantazole B synthesis of Fukuyama. The oligo(thiazoline) compound and the oxazole compound are synthesized separately. At the end, the two compounds are coupled to give the last

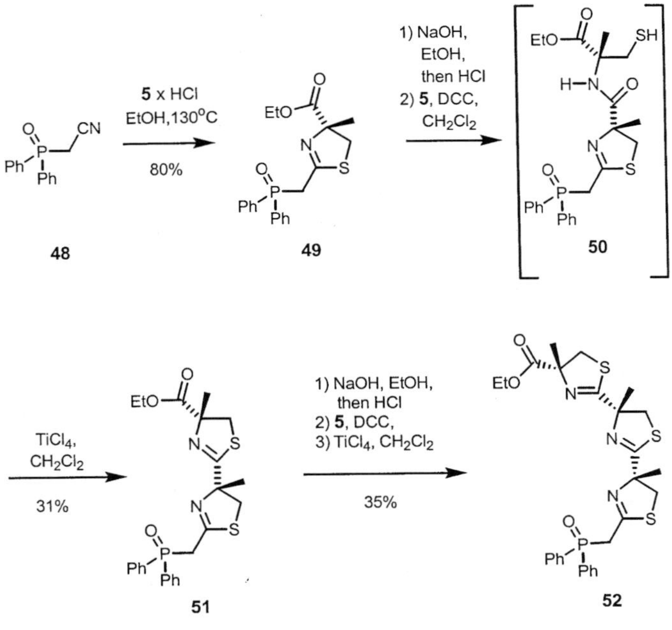

48 → (5 x HCl, EtOH, 130°C, 80%) → 49 → (1) NaOH, EtOH, then HCl; 2) 5, DCC, CH_2Cl_2) → [50] → ($TiCl_4$, CH_2Cl_2, 31%) → 51 → (1) NaOH, EtOH, then HCl; 2) 5, DCC; 3) $TiCl_4$, CH_2Cl_2; 35%) → 52

Scheme 12. Ehrler's synthesis of thiangazole (part 1).

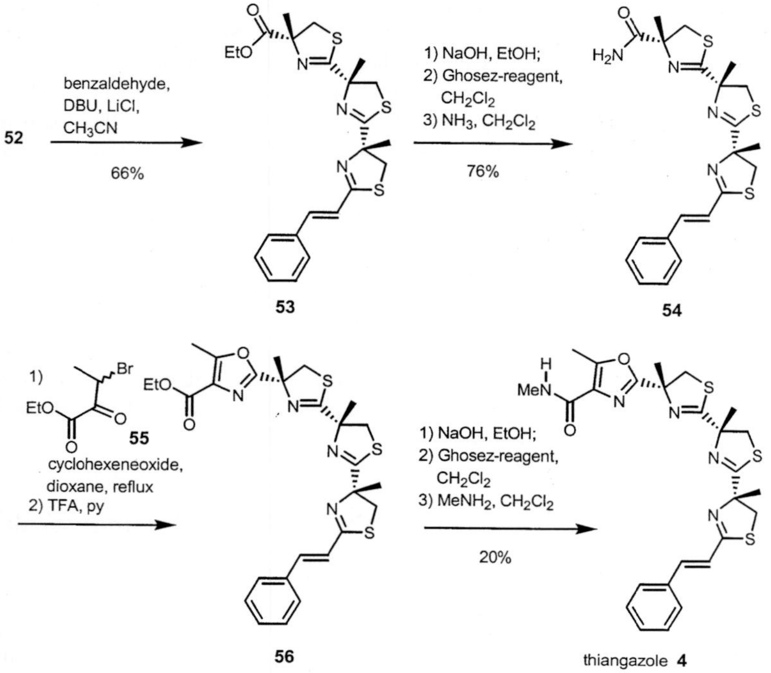

Scheme 13. Ehrler's synthesis of thiangazole (part 2).

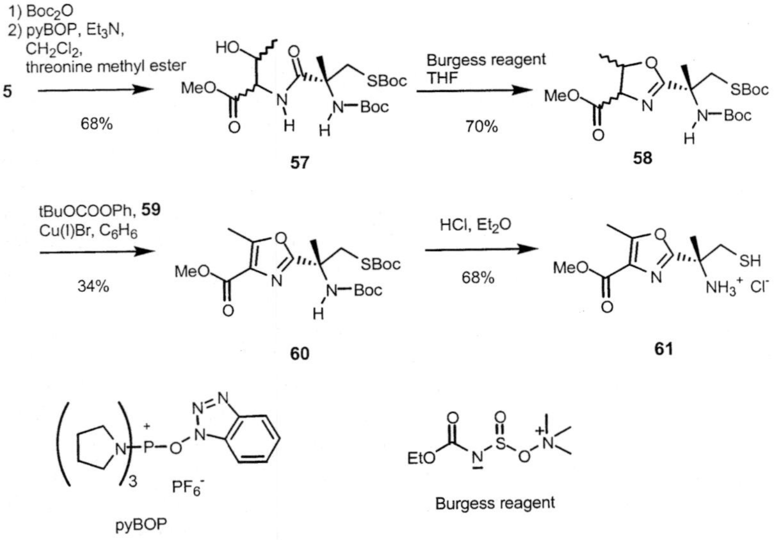

Scheme 14. Pattenden's synthesis of thiangazole (part 1).

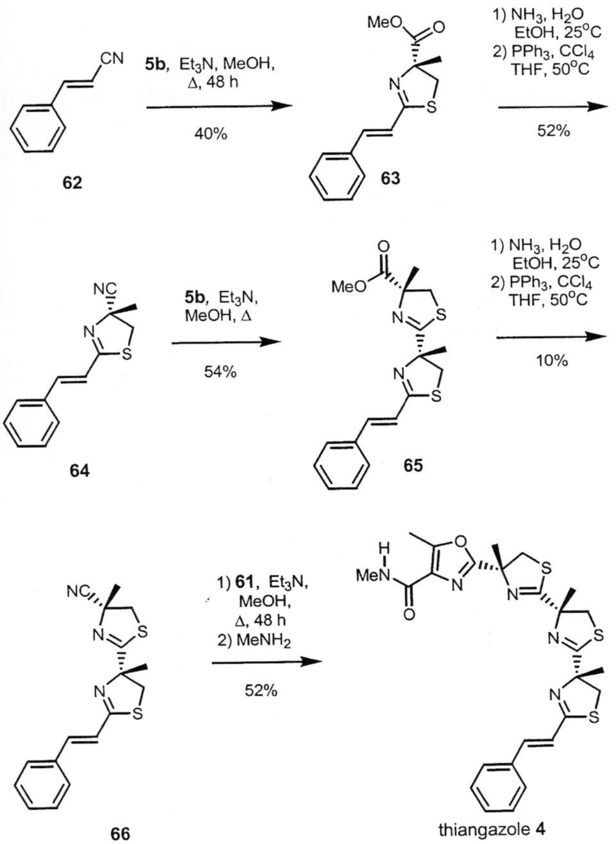

Scheme 15. Pattenden's synthesis of thiangazole (part 2).

thiazoline ring. The synthesis of the oxazole compound **61** is shown in Scheme 14.

First, the (*R*)-2-methylcysteine compound **5** is N- and S-protected and coupled with the threonine methyl ester to the hydroxyamide **57** with benzotriazole-1-yloxy-tripyrrolidino-phosphonium hexafluorophosphate (pyBOP) [16]. Burgess reagent [17] turns out to be the best choice for the conversion of **57** into **58**. The use of Burgess reagent for the synthesis of oxazoline was examined extensively by Wipf et al. [18]. For other methods of synthesizing oxazoles, which were developed in connection to the synthesis of calyculin A, see [19].

The oxidation of oxazoline **58** to oxazole **60** turns out to be problematical in Pattendens synthesis. Even with special oxidation techniques (*tert*-butyl peroxybenzoate **59**/copper (I) bromide), only the moderate yield of 34 % is achieved. The acidic deprotection of the Boc group finally leads to the oxazole compound **61**. The further route of Pattenden's thiangazole synthesis is shown in Scheme 15.

Cyclocondensation of the nitrile **62** with the (*R*)-2-methylcysteine building block **5b** provides the thiazoline ester **63**. After conversion of the ester to the nitrile **64**, cyclocondensation with **5b** leads to the bis(thiazoline) **65**. The subsequent conversion of the ester to the nitrile function (**65** $\rightarrow$ **66**) nearly failed, with 10 % yield. In the last synthetic sequence the bis(thiazoline) compound **66** is coupled with the oxazole compound **61** to provide the third and last thiazoline ring. Formation of the methylamide is the final step of the thiangazole synthesis of Pattenden.

Heathcock's Synthesis of Mirabazole C

The strategy of this synthesis is different from that for the examples described above. Heathcock assembles a peptide chain of the S-protected 2-methylcysteines and closes all four thiazoline rings simultaneously at the end. The synthesis of the peptide (Scheme 16) starts with the coupling of the 2-methylcysteine compounds ***ent*-25** and **26**.

Bromo-tris-pyrolidinophosphonium hexafluorophosphate (PyBrOP) is used as a benzotriazole-free coupling reagent [20]. After deprotection of the Boc group, the dipeptide **67** is obtained in 90 % yield. The same coupling is repeated with the (*R*)-2-methylcysteine compound ***ent*-26** to yield the tripeptide **68**. After deprotection of the Boc group, an isobutyryl amide **69** is installed. The peptide precursor **70** for the multiple ring closure is prepared from **69** in 93 % yield. The multiple ring closure for the construction of the tetra(thiazoline) backbone is shown in Scheme 17.

After reductive removal of the benzyl groups of **70**, the tetrathiol **71** is obtained. **71** is cyclisized to the tetra(thiazoline) **72** using titanium tetrachloride (45 % yield!). Final oxidation of the terminal thiazoline to the thiazole with nickel dioxide provided mirabazole C (**1**).

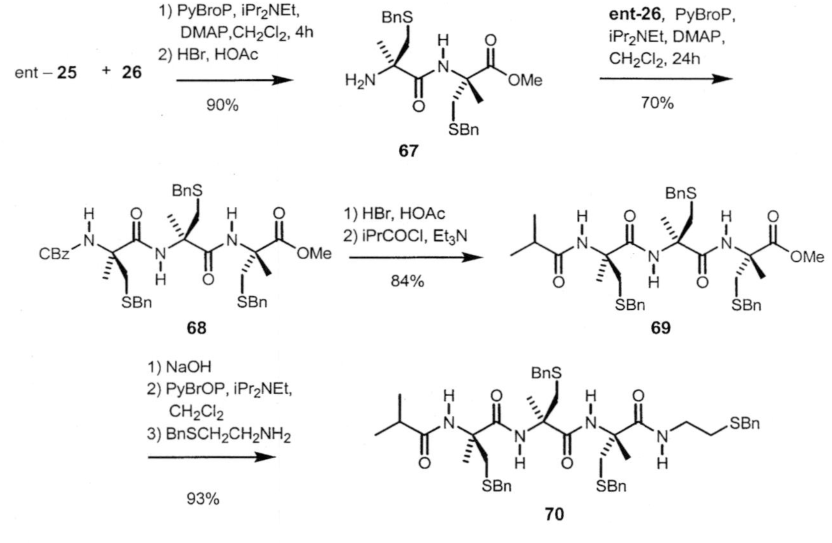

Scheme 16. Heathcock's synthesis of mirabazole C (part 1).

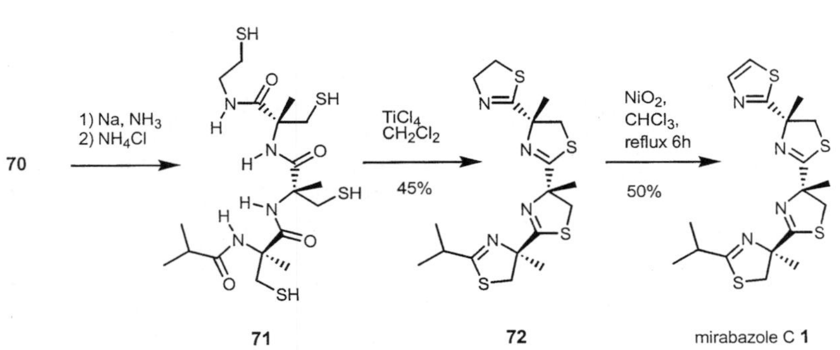

Scheme 17. Heathcock's synthesis of mirabazole C (part 2).

Synthesis of Thiangazole by Wipf

The synthetic strategy of Wipf [9] is quite different. He decides to prepare oligo(oxazolines) instead of oligo(thiazolines). Then a new multiple oxazoline → thiazoline conversion is used. Wipf chooses the oxazoline route for two reasons. The first reason is a synthetic one: oxazolines are easier to prepare than thiazolines. The second reason is a pharmacological one: the oxazolines obtained along this route may be interesting drug candidates.

Wipfs enantiomerically pure building block is not (*S*)-2-methylcysteine, but the (*S*)-2-methylserine compound **15** (Scheme 18).

Coupling of **15** with D-threonine methyl ester in the presence of PyBroP and DMAP gives the dipeptide **73**. The oxazole **74** is formed after Dess-Martin oxidation and cyclodehydration with triphenylphosphine/iodine. After formation of the methylamide function, an iterative sequence of deprotection with TBAF and coupling with **15** leads to the dipeptide **75**, then to the tripeptide **76** and finally to the tetrapeptide **77**. The coupling reagent is PyBroP in each case. The overall yield (**74** → **77**) is 21 %. After catalytic hydrogenation of **77** with $Pd(OH)_2$, a subsequent multiple ring formation with Burgess's reagent [18] leads to the tris(oxazoline) **78** with 60 % yield (Scheme 19).

Now an oxazoline → thiazoline conversion is achieved by nucleophilic opening of the tris(oxazolines) to the 2-methylcysteine peptide **79** with thioacetic acid. For the ring closure Wipf uses titanium tetrachloride according to Heathcocks protocol. The side chain of tris(thiazoline) **80** is oxidized with benzeneselenic acid to provide the desired natural product thiangazole **4**.

Kiso's Synthesis of Mirabazole C

Kiso's synthetic strategy follows the work of Heathcock [8]. He also uses titanium tetrachloride as the reagent for the multiple ring closure. His work focusses on an effective preparation of the peptide precursor for the multiple ring closure. For this purpose he introduces a new coupling reagent: 2-chloro-1,3-dimethylimidazolidium hexafluorophosphate (CIP) **81** (Scheme 20).

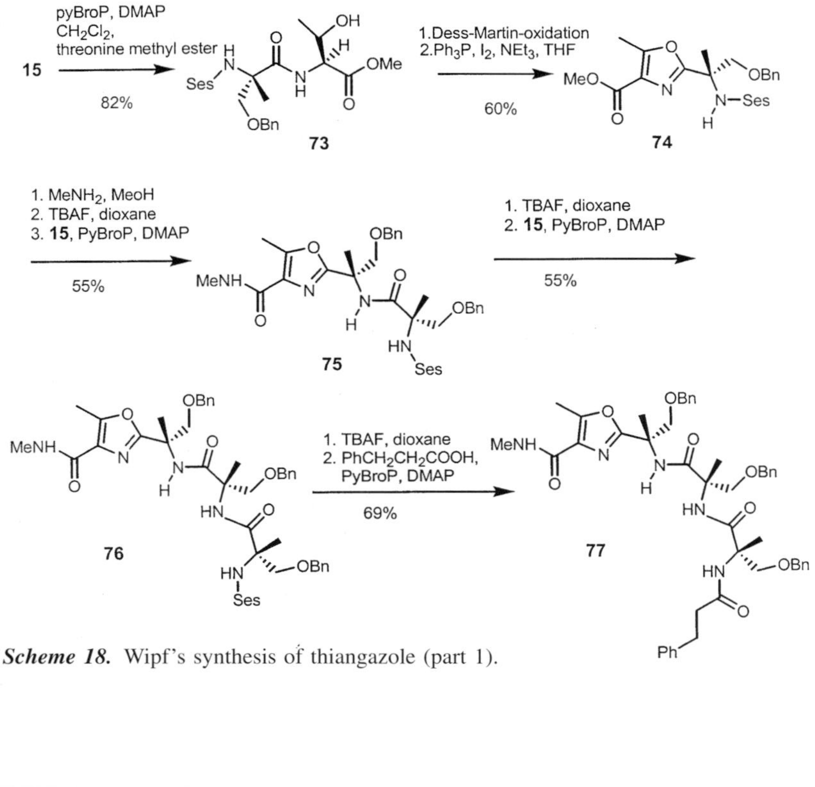

15 → (pyBroP, DMAP, CH_2Cl_2, threonine methyl ester; 82%) → 73 → (1. Dess-Martin-oxidation; 2. Ph_3P, I_2, NEt_3, THF; 60%) → 74 → (1. $MeNH_2$, MeoH; 2. TBAF, dioxane; 3. **15**, PyBroP, DMAP; 55%) → 75 → (1. TBAF, dioxane; 2. **15**, PyBroP, DMAP; 55%) → 76 → (1. TBAF, dioxane; 2. $PhCH_2CH_2COOH$, PyBroP, DMAP; 69%) → 77

Scheme 18. Wipf's synthesis of thiangazole (part 1).

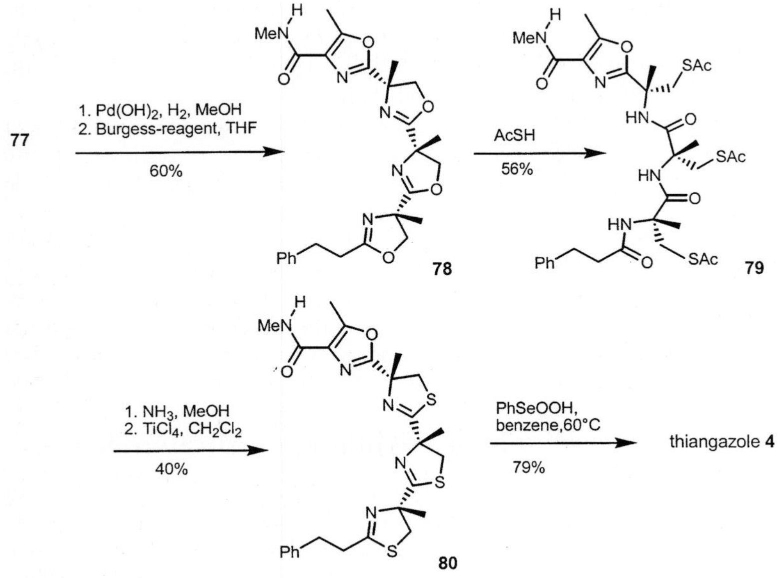

77 → (1. Pd(OH)$_2$, H$_2$, MeOH; 2. Burgess-reagent, THF; 60%) → 78 → (AcSH; 56%) → 79

79 → (1. NH$_3$, MeOH; 2. TiCl$_4$, CH$_2$Cl$_2$; 40%) → 80 → (PhSeOOH, benzene, 60°C; 79%) → thiangazole **4**

Scheme 19. Wipf's synthesis of thiangazole (part 2).

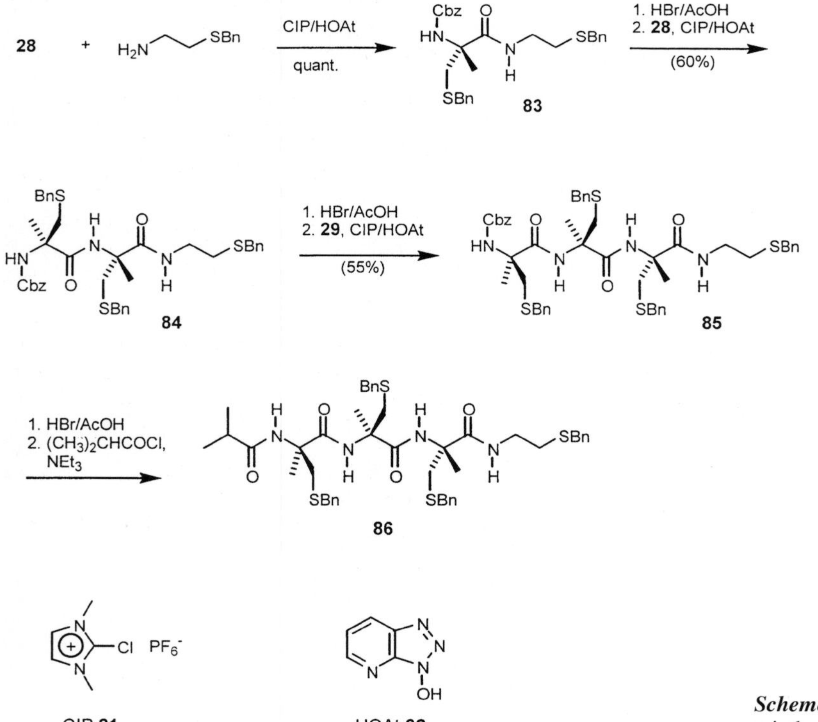

28 + H$_2$N–CH$_2$CH$_2$–SBn → (ClP/HOAt; quant.) → 83 → (1. HBr/AcOH; 2. **28**, ClP/HOAt; (60%)) → 84

84 → (1. HBr/AcOH; 2. **29**, ClP/HOAt; (55%)) → 85

85 → (1. HBr/AcOH; 2. (CH$_3$)$_2$CHCOCl, NEt$_3$) → 86

ClP **81**   HOAt **82**

Scheme 20. Kiso's synthesis of mirabazole C

The (*R*)-2-methylcysteine derivative **28** is coupled with (*S*)-benzyl-2-aminoethanethiol using CIP/HOAt [21] (**81/82**) to provide **83** quantitatively. After N-deprotection with HBr/AcOH, coupling with **28** using CIP/HOAt gives the dipeptide **84**, which, after deprotection, is coupled again using CIP/HOAt with the (*S*)-2-methylcysteine compound **29** to provide the tripeptide **85**. The yields are 60 % and 55 %, respectively. **85** is N-deprotected with HBr/AcOH, and the resulting amine is acylated with isobutyryl chloride to obtain the precursor **86** for the multiple ring closure. From this stage onwards, the synthetic route follows that one of Heathcock (Scheme 17).

In summary, these examples show an impressive insight into modern heterocyclic chemistry. Basically, two strategies have been adopted for the total syntheses of this class of bioactive compounds. One is the sequential formation of the thiazoline ring and the other is the multiple ring closure of a peptide precursor. Heathcock's strategy of $TiCl_4$-mediated simultaneous formation of several thiazoline rings proved to be the most efficient. Wipf's approach via oxazolines is ideally suited for the construction of oligo(thiazoline) analogs, with potential applications in biology and medicine.

References

[1] S. Carmeli, R. E. Moore, G. M. L. Patterson, *Tetrahedron Lett.* **1991**, *32*, 2593.

[2] S. Carmeli, R. E. Moore, G. M. L. Patterson, T. Corbett, F. A. Valeriote, *J. Am. Chem. Soc.* **1990**, *112*, 8195.

[3] R. Jansen, B. Kunze, H. Reichenbach, E. Jurkiewicz, G. Hunsmann, G. Höfle, *Liebigs Ann. Chem.* **1992**, 357.

[4] R. Jansen, D. Schomburg, G. Höfle, *Liebigs Ann. Chem.* **1993**, 701.

[5] T. Fukuyama, L. Xu, *J. Am. Chem. Soc.* **1993**, *115*, 8449.

[6] J. Ehrler, S. Farooq, *Synlett*, **1994**, 702.

[7] R. J. Boyce, G. C. Mulqueen, G. Pattenden, *Tetrahedron Lett.* **1994**, *35*, 5705.

[8] R. L. Parsons Jr., C. H. Heathcock, *Tetrahedron Lett.* **1994**, *35*, 1379.

[9] P. Wipf, S. Venkatraman, *J. Org. Chem.* **1995**, *60*, 7224; P. Wipf, S. Venkatraman, C. P. Miller *Tetrahedron Lett.* **1995**, *21*, 3639.

[10] K. Akaji, N. Kuriyama, Y.Kiso, *J. Org. Chem.* **1996**, *61*, 3350.

[11] Pattenden: G. Pattenden, S. M. Thom, *Synlett*, **1992**, 533 and G. Pattenden, S. M. Thom, *J. Chem. Soc. Perkin Trans. 1*, **1993**, 1629 (epimer of didehydromirabazole A); R. J. Boyce, G. Pattenden, *Synlett*, **1994**, 587 (didehydromirabazole A); R. J. Boyce, A. G. C. Mulqueen, G. Pattenden, *Tetrahedron* **1995**, *26*, 7321 (thiangazole). B. Heathcock: R. L. Parsons Jr., C. H. Heathcock, *Tetrahedron Lett.* **1994**, *35*, 1383 (mirabazole B); M. A. Walker, C. H. Heathcock, *J. Org. Chem.* **1992**, *57*, 5566 (epimer of mirabazole C); R. L. Parsons Jr., C. H. Heathcock, *J. Org. Chem* **1994**, *59*, 4733 (thiangazole); R. L. Parsons Jr., C. H. Heathcock, *Synlett* **1996**, 1168 (tantazole B). N. Kuriyama, K. Akaji, Y. Kiso, *Tetrahedron* **1997**, *25*, 8323 (mirabazole B). P. Wipf, S. Venkatraman, *Synlett* **1997**, 1 (thiangazole review).

[12] G. Pattenden, S. M. Thom, M. F. Jones, *Tetrahedron* **1994**, *49*, 2131.

[13] A. Jeanguenat, D. Seebach, *J. Chem. Soc. Perkin Trans. 1*, **1991**, 2291.

[14] S. Karady, J. S, Amato, L. M. Weinstock, *Tetrahedron Lett.* **1984**, *35*, 4337.

[15] J. Cabre, A. L. Palomo, *Synthesis* **1994**, 413.

[16] J. Coste, D. Le-Nguyen, D. Castro, *Tetrahedron Lett.* **1970**, *21*, 205.

[17] G. M. Atkins, E. M. Burgess, *J. Am. Chem. Soc.* **1968**, *90*, 4744.

[18] a) P. Wipf, P. C. Fritch, *Tetrahedron Lett.* **1994**, *35*, 5397; b) P. Wipf, C. P. Miller, *Tetrahedron Lett.* **1992**, *33*, 907; c) P. Wipf, C. P. Miller, *J. Org. Chem.* **1993**, *58*, 3604.

[19] a) D. A. Evans, J. R. Gage, J. L. Leighton, *J. Am. Chem. Soc.* **1992**, *114*, 9434; b) H. A. Vaccaro, D. E. Levy, A. Sawabe, T. Jaetsch, S. Masamune, *Tetrahedron Lett.* **1992**, B. *33*, 1937; c) B. A. Salvatore, A. B. Smith III, *Tetrahedron Lett.* **1994**, *35*, 1329.

[20] J. Coste, E. Frerot, P. Jouin, B. Castro, *Tetrahedron Lett.* **1991**, *32*, 1967.

[21] L. A. Carpino, *J. Am. Chem. Soc.* **1993**, *115*, 4397.

Camptothecin – Synthesis of an Antitumor Agent

Stefan Bäurle and Ulrich Koert

Institut für Chemie, Humboldt Universität, Berlin, Germany

Camptothecin (**1**) was first isolated in 1966 from the Chinese tree *camptotheca acuminata* [1]. As shown in Fig. 1, the five rings of the natural compound are named A, B, C, D and E. The stereogenic center in the E ring is *S*-configured. The antitumor activity of camptothecin quickly made the compound an interesting target for synthetic chemists. After a short time, successful total syntheses were reported. Especially three synthetic approaches should be mentioned here: those of Stork (1971) [2], Winterfeldt (1972) [3] and Corey (1975) [4].

However, the high clinical expectations for camptothecin turned to disappointment in the 1970s because in some clinical studies no effect was observed. So the compound disappeared from the scene. Later, the reason for that failure was discovered. Instead of compound **1**, which has a very poor water solubility, the water-soluble sodium salt **2** was used in these early clinical

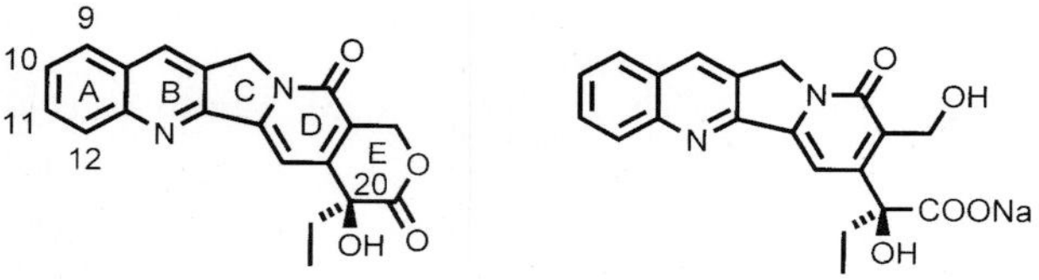

Camptothecin (**1**)

Camptothecin, sodium salt (**2**)

Figure 1

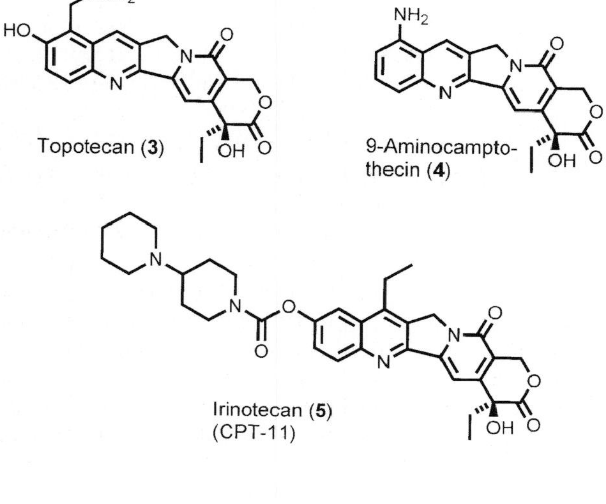

Topotecan (**3**)

9-Aminocamptothecin (**4**)

Irinotecan (**5**) (CPT-11)

Figure 2

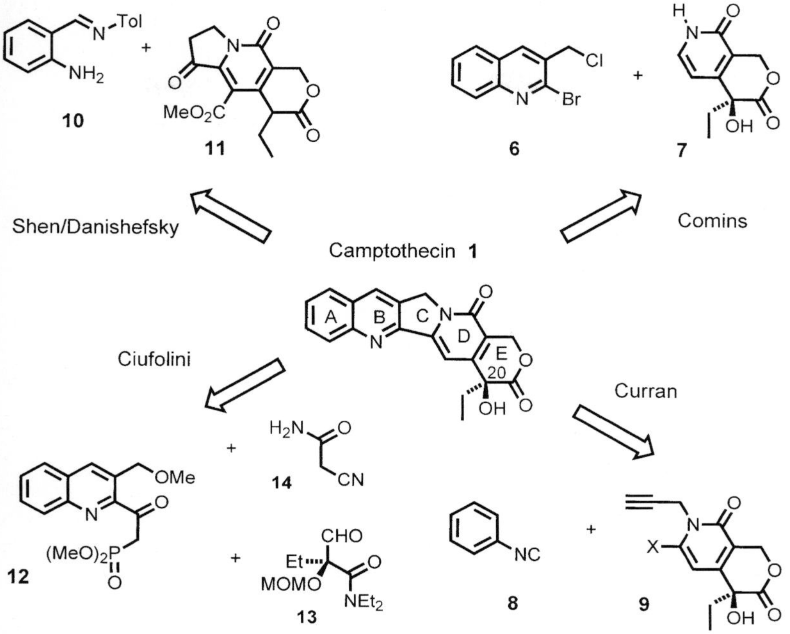

Scheme 1. Retrosynthetic analysis of camptothecin (**1**) according to Comins, Curran, Shen/Danishefsky and Ciufolini.

studies, and compound **2** with the open E ring proved to be pharmacologically ineffective. In trying to get good bioavailability, the pharmacophor had been destroyed by the change from **1** to **2**. Keeping that in mind, the solubility problem was solved by additional substituents in the A ring in positions 9 to 12 without loss of pharmacological effect. Then a second important discovery was made. The mode of action of camptothecin is based on the inhibition of topoisomerase 1-DNA complex [5]. The DNA double strand in the cell is coiled and knotted. For replication and cell division, this DNA tangle has to be unwound. And that is the exact function of the topoisomerases. In continually replicating cancer cells, topoisomerase 1 is overexpressed. Camptothecin is thought to interact at this point. It binds at the topoisomerase 1-DNA complex, inhibits the activity of the enzyme and stops cell division [6].

Today, several soluble camptothecin derivatives such as topotecan (**3**), 9-aminocamptothecin (**4**) and irinotecan (CPT-11) (**5**) are of clinical importance. Positive results were achieved, especially against intestinal, breast and ovarian cancer. After camptothecin returned like the phoenix from the ashes, it again became a target for the synthetic chemists. Recent developments are summarized in this review. It is interesting to note how the number of steps could be reduced from 15–20 in the 1970s to under 10 nowadays with the help of the modern synthetic methods of the 1990s.

Scheme 1 shows a retrosynthetic synopsis of the syntheses of Comins [7–9], Curran [10–14], Shen/Danishefsky [16] and Ciufolini [17].

Comins connects an A,B ring building block (**6**) with a D,E building block **7** and closes the C ring last – a classic example of a convergent synthetic strategy! Curran uses an A ring building block (**8**) and a D,E ring building block (**9**) in order to build up the rings B and C in one step. Shen and Danishefsky take the classical route. They connect an A ring building block (**10**) with a C,D,E ring building block (**11**), creating the B ring last. Ciufolini starts with an A,B quinoline (**12**) and closes the C ring in the final step.

The camptothecin synthesis of Comins (Scheme 2) commences with the *ortho*-directed

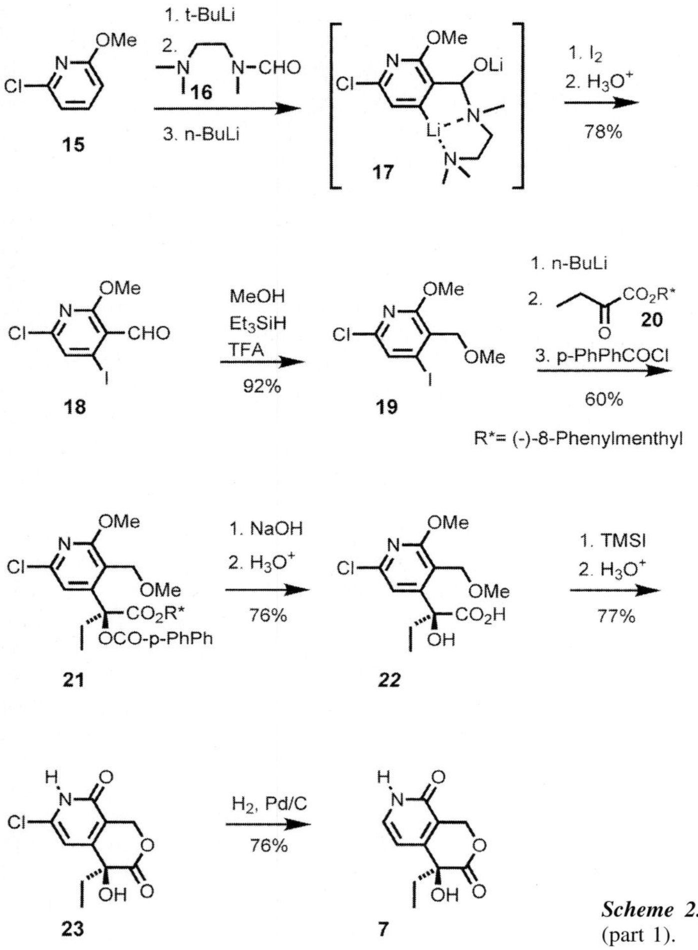

15 → [17] : 1. t-BuLi; 2. 16 (N,N,N'-trimethylethylenediamine formamide); 3. n-BuLi

17 → 18 : 1. I_2; 2. H_3O^+; 78%

18 → 19 : MeOH, Et_3SiH, TFA; 92%

19 → 21 : 1. n-BuLi; 2. 20 (CO_2R^*); 3. p-PhPhCOCl; 60%

R*= (-)-8-Phenylmenthyl

21 → 22 : 1. NaOH; 2. H_3O^+; 76%

22 → 23 : 1. TMSI; 2. H_3O^+; 77%

23 → 7 : H_2, Pd/C; 76%

Scheme 2. Comins' synthesis of camptothecin (part 1).

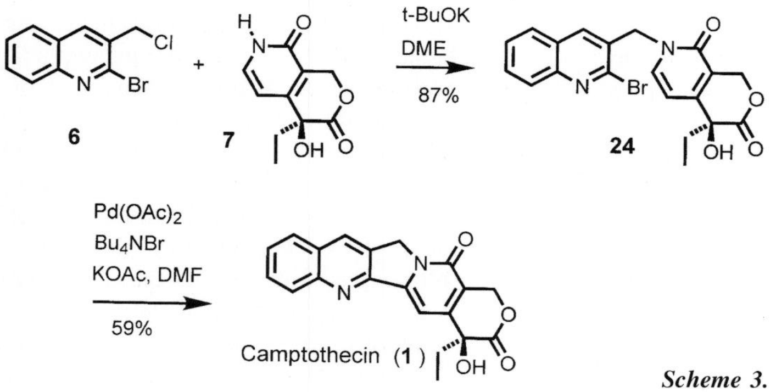

6 + 7 → 24 : t-BuOK, DME; 87%

24 → Camptothecin (1) : $Pd(OAc)_2$, Bu_4NBr, KOAc, DMF; 59%

Scheme 3. Comins' synthesis (part 2).

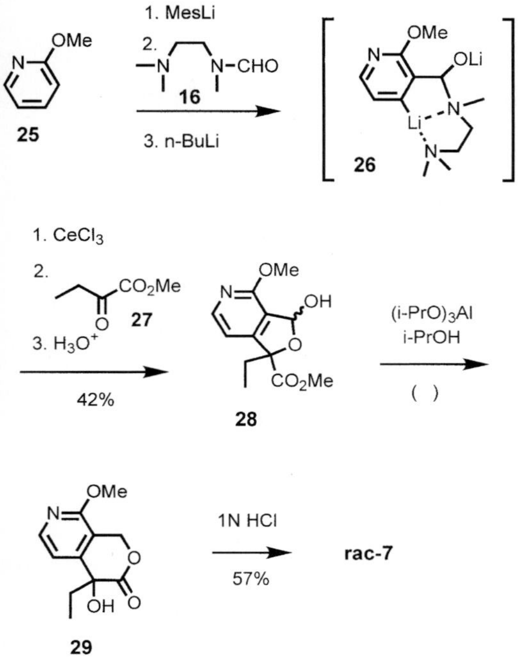

Scheme 4. Comins' synthesis of the D,E ring building block.

lithiation of pyridine (**15**). The aryllithium compound obtained is allowed to react with the formamide (**16**) to yield an α-amino alkoxide, which is converted to the intermediate **17** by directed *ortho*-lithiation. Reaction of **17** with iodine affords the aryl iodide **18**. Conversion of the aldehyde function of **18** to a methyl ether group provides compound **19**. Iodine-lithium exchange at **19** and subsequent reaction with α-keto ester **20** yields ester **21**. Comins uses (−)-8-phenylmenthol as chiral auxiliary. The conversion of **19** to **21** shows a stereoselectivity of 93 : 7. After recrystallisation, the product (**21**) is obtained as a pure stereoisomer in 60 % yield. The chiral auxiliary is cleaved off (**21** → **22**) and the E ring of camptothecin is closed (**22** → **23**). The chlorine substituent can be removed by catalytic hydrogenation.

Next, the D,E ring building block **7** is coupled with the A,B ring fragment **6** by N-alkylation to provide compound **24** (Scheme 3). The final step of Comins' synthesis of camptothecin is an intramolecular Heck reaction which closes the C ring.

Comins was able to shorten the (racemic) synthesis of the D,E ring building block to three steps by further optimization of the *ortho*-directed

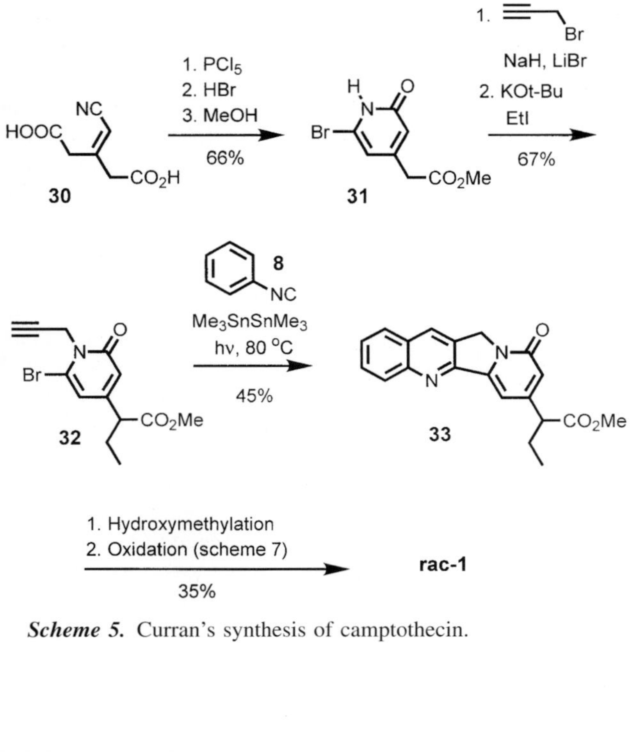

Scheme 5. Curran's synthesis of camptothecin.

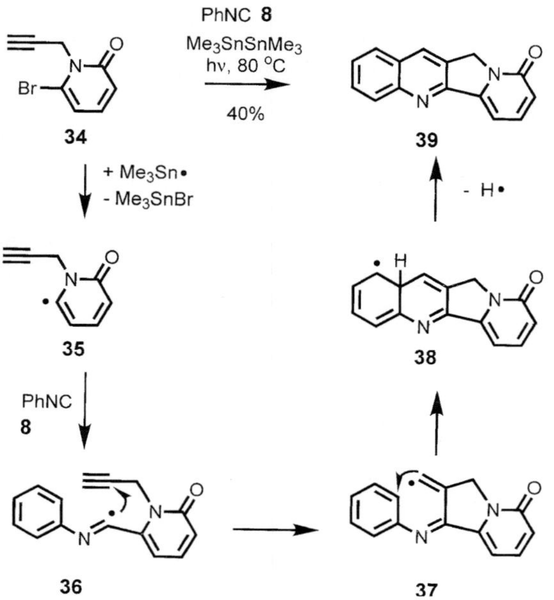

Scheme 6. Curran's synthesis: mechanism of the radical key step.

metalation (Scheme 4). The decisive improvement is based on the transmetalation of the aryllithium intermediate **26** to the corresponding arylcerium compound. The organocerium compound, as a weaker base, allows the addition to the α-keto ester **27** and the formation of the lactol **28**. Meerwein-Ponndorf-Verley reduction and acidic cleavage of the pyridone-*O*-methyl ester **29** affords the desired D,E ring building block **rac-7**.

Curran's synthesis (Scheme 5) is an instructive example of applied radical chemistry. His racemic synthesis [10, 11] of camptothecin starts with compound **30**. Via **31**, the D ring building block is reached by some standard steps. Treatment of **32** with phenyl isonitrile (**8**) provides compound **33** in one step. The final construction of the E ring follows Danishefsky's synthesis (see below).

The simultaneous construction of the B and C rings leading to compound **33** was accomplished by a radical cascade reaction. The mechanistic details of this cascade are summarized in Scheme 6, where the reaction of **34** with phenyl isonitrile (**8**) is shown [12]. First, a trimethylstannyl radical, derived from hexamethyldistannane, attacks the C-Br bond of **34**. The resulting pyridone radical **35** reacts intermolecularly with the isonitrile **8** to yield the radical intermediate **36**. An intramolecular attack of the radical center in **36** on the alkyne functionality leads to radical **37**. Finally, the A ring is attacked by the radical center of **37** leading to **38**, which rearomatizes to the desired compound **39**.

The second generation of Curran's camptothecin synthesis [13, 14] is based on the same radical key step, but an enantiomerically pure D,E ring fragment **7** is used. His asymmetric approach to compound **7** follows Comins' strategy. A research team at Glaxo [15] elaborated this synthesis in a similar way. A Sharpless dihydroxylation (AD reaction) is applied to introduce the stereogenic center at C20. Scheme 7 shows the enantioselective synthesis of **7** according to the Glaxo group.

An *ortho*-directed lithiation allows the conversion of **25** to aryl iodide **40**. Reductive ether formation of aldehyde **40** with crotyl alcohol yields compound **41**. Intramolecular Heck reaction of **41** affords a mixture of the olefins **42** and **43**. The undesired alkene **42** can be isomerized quantitatively to the desired enol ether **43** with Wilkinson's catalyst. Sharpless dihydroxylation (*ee* 94 %) of the enol ether **43** provides lactol **44**, which is oxidized directly to lactone **45**. Finally, the pyridone-*O*-methyl ester is cleaved under acid conditions (**45** $\rightarrow$ **7**).

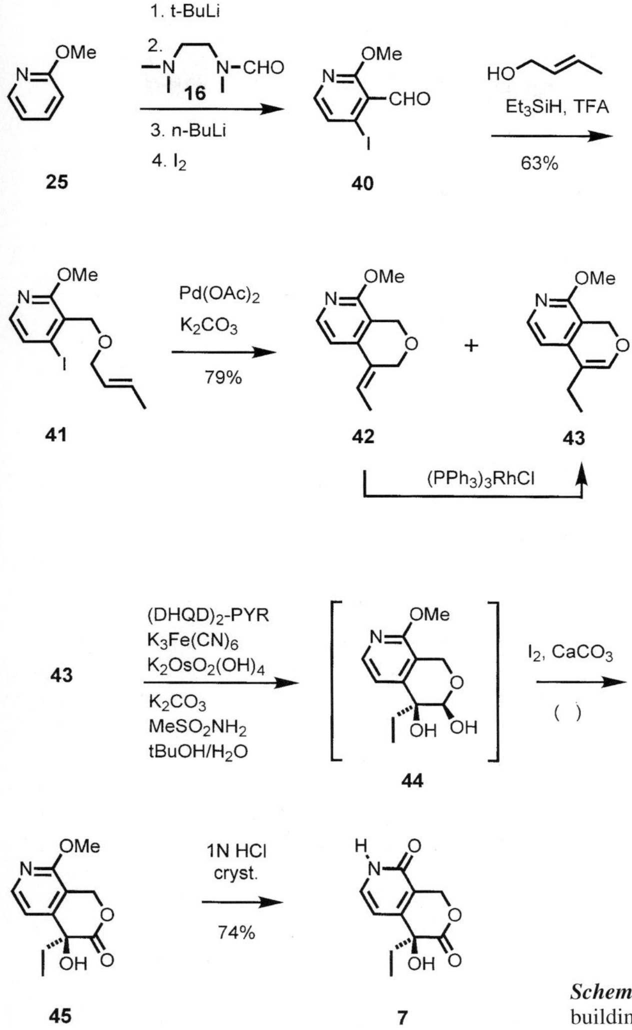

Scheme 7. Asymmetric synthesis of the D,E ring building block.

The camptothecin synthesis of Shen/Danishefsky [16] (Scheme 8) starts with the construction of the D ring: reaction of **46** with **47** affords **48**. The next step (**48** → **49**) introduces the C20 ethyl group. Hydroxymethylation of **49** delivers lactone **50**. The B ring of the camptothecin precursor **52** should be accomplished by a Friedländer condensation of **10** with **11**. For this purpose, a keto group at C2 of the C,D,E ring building block had to be installed. Shen and Danishefsky solved this problem by combination of an aldol condensation (**50** → **51**) and a subsequent ozonolysis (**51** → **11**). Important for the aldol condensation is the methyl ester functionality. It allows the regioselective deprotonation at the benzylic position. After the methyl ester has done its job it is removed by heating with HBr to afford compound **53**. The final step of Shen/Danishefsky's camptothecin synthesis is the oxidative introduction of the alcohol function at C20. Unfortunately, this hydroxylation leads only to the racemic target compound.

Ciufolini [17] reaches camptothecin in 5 steps, connecting the building blocks **12**, **13** and **14** (Scheme 9). Condensation of the quinoline phosphonate **12** with aldehyde **13** afforded enone **54**.

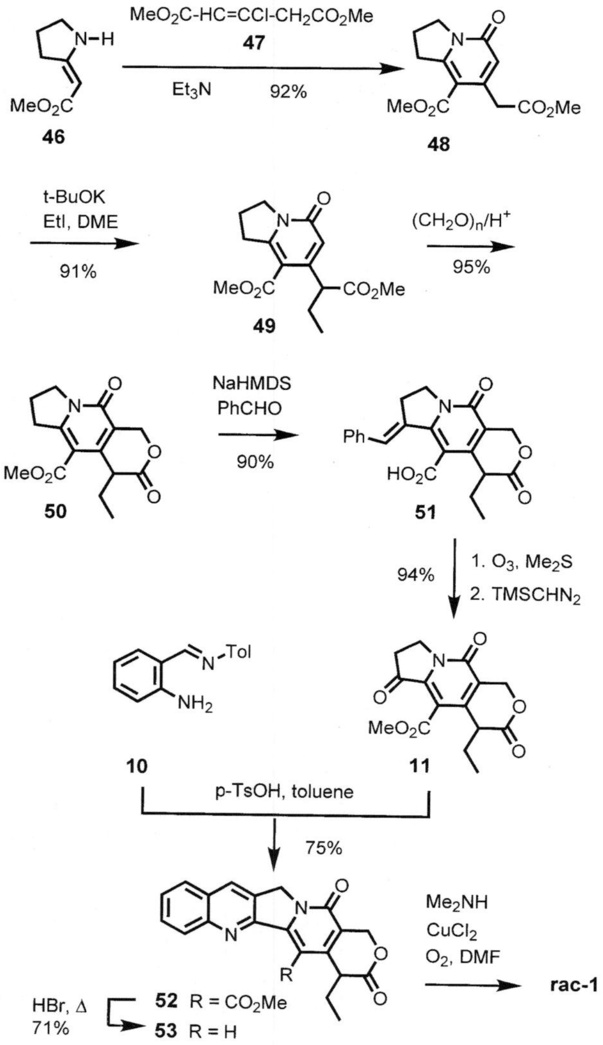

Scheme 8. Danishefsky's synthesis of camptothecin.

Michael addition of the potassium enolate of cyanacetamide (**14**) provided the precursor **55** of the pyridone **56**. This intermediate was obtained by oxidation of **55** with *t*-BuOOH in the presence of 20 mol % SeO_2 on SiO_2. The addition of 10 % H_2SO_4 to the reaction mixture directly delivered the lactone **56**. Lactones like **56** are not easy to reduce, but treatment with $NaBH_4/CeCl_3$ provided the diol **57** in 95 % yield. **57** could be easily converted to camptothecin by heating it to 115 °C in 60 % H_2SO_4 in EtOH.

The preparation of compound **12** started with the quinoline derivative **58**, which can be carbomethoxylated in the presence of $[Pd(dppp)_2Cl_2]$. Radical bromination, methanolysis and reaction with $(MeO)_2P(O)CH_2Li$ delivers **12** (Scheme 10).

The stereogenic center at C20 is introduced by enantioselective enzymatic hydrolysis of MOM-protected malonic acid dimethyl ester derivative **60** (Scheme 10) with pig liver esterase (PLE). The asymmetric compound **61** is obtained in 90 % yield and 98 % *ee*. Amide formation with Mu-

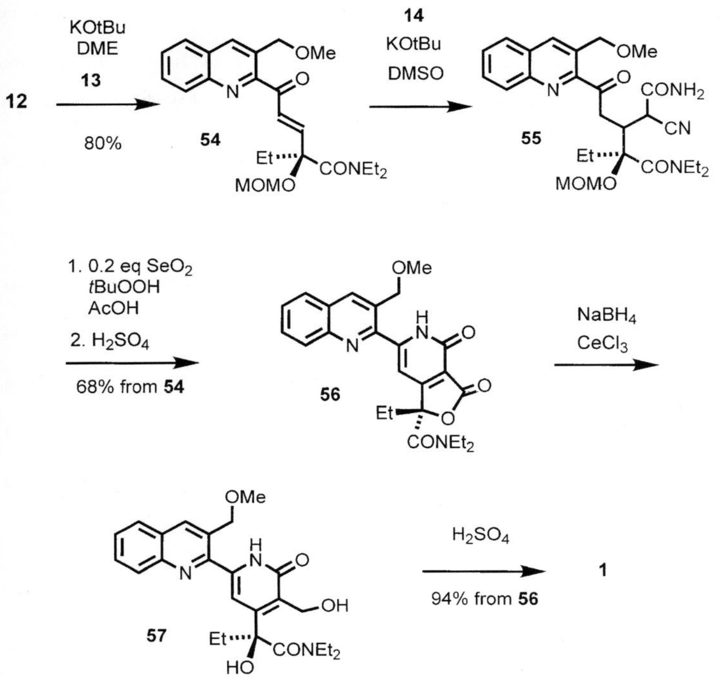

Scheme 9. Ciufolini's synthesis of camptothecin (part 1).

kaiyama's reagent and DIBAH reduction provides aldehyde **13**.

Ciufolini's strategy allows the synthesis of camptothecin in 10 steps and 30 % overall yield, starting with dimethyl malonate acid dimethyl ester. Like the other synthetic strategies discussed, this synthesis is open to variations in the A ring and thus allows the preparation of pharmacologically interesting camptothecin derivatives. Shen/Danishefsky's approach is efficient but has to be improved to an asymmetric level. The stereoselective 2nd generation of Curran's synthesis constructs the B and C ring in one elegant step. New variations even allow the regioselective introduction of different substituents in positions 9–11 [14].

In summary, the development in the field of total syntheses of the antitumor agent camptothecin nicely illustrates the progress of modern synthesis over the last decade. Radical reactions and organometallic coupling reactions, for example, have reached such a level of maturity that they now belong to the standard repertoire of key steps in the construction of complex molecules.

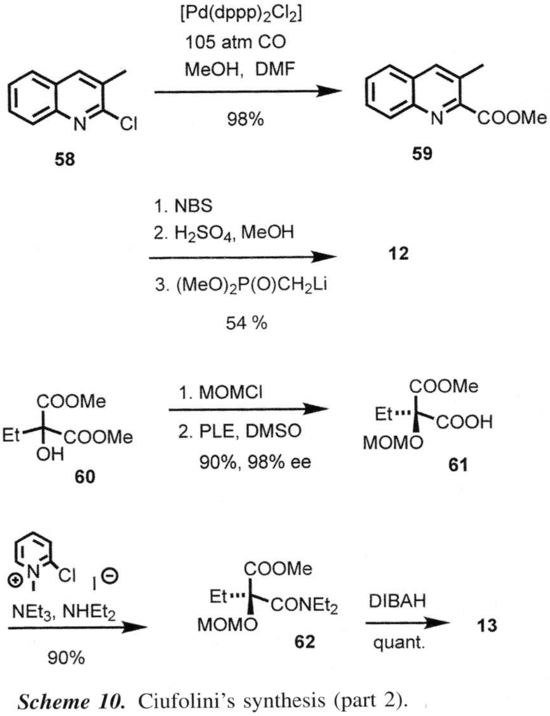

Scheme 10. Ciufolini's synthesis (part 2).

References

[1] M. E. Wall, M. C. Wani, C. E. Cook, K. H. Palmer, A. T. McPhail, G. A. Sim, *J. Am. Chem. Soc.* **1966**, *88*, 3888.
[2] G. Stork, A. G. Schultz, *J. Am. Chem. Soc.* **1971**, *93*, 4074.
[3] a) M. Boch, T. Korth, J. M. Nelke, D. Pike, H. Radunz, E. Winterfeldt, *Chem. Ber.* **1972**, *105*, 2126; b) K. Krohn, E. Winterfeldt, *Chem. Ber.* **1975**, *108*, 3030.
[4] E. J. Corey, D. N. Crouse, J. E. Anderson, *J. Org. Chem.* **1995**, *40*, 2140.
[5] a) Y. H. Hsiang, R. P. Hertzberg, S. M. Hecht, L. F. Liu, *J. Biol. Chem.*, **1985**, *260*, 14873; b) C. D. Lima, J. C. Wang, A. Mondragon, *Nature* **1994**, *367*, 138.
[6] W. J. Slichenmeyer, E. K. Rowinsky, R. C. Donehower, S. H. Kaufmann, *J. Nat. Canc. Inst.* **1993**, *85*, 271.
[7] D. L. Comins, M. F. Baevsky, H. Hong, *J. Am. Chem. Soc.* **1992**, *114*, 10971.
[8] D. L. Comins, H. Hong, J. K. Saha, G. Jianhua, *J. Org. Chem.* **1994**, *59*, 5120.
[9] D. L. Comins, H. Hong, G. Jianhua, *Tetrahedron Lett.* **1994**, *35*, 5331.
[10] D. P. Curran, H. Liu, *J. Am. Chem. Soc.* **1992**, *114*, 5863.
[11] D. P. Curran, S.-B. Ko, *J. Org. Chem.* **1994**, *59*, 6139.
[12] U. Koert in J. Mulzer, H. Waldmann (eds.), *Organic Synthesis Highlights III*, Wiley-VCH, Weinheim **1998**, 235.
[13] D. P. Curran, S.-B. Ko, H. Josien, *Angew. Chem.* **1995**, *107*, 2948.
[14] H. Josien, S.-B. Ko, D. Bom, D.P. Curran, *Chem. Eur. J.* **1998**, *4*, 67.
[15] F. G. Fang, S. Xie, M. W. Lowery, *J. Org. Chem.* **1994**, *59*, 6142.
[16] W. Shen, C. A. Coburn, W. G. Bornmann, S. J. Danishefsky, *J. Org. Chem.* **1993**, *58*, 611.
[17] M. A. Ciufolini, F. Roschangar, *Angew. Chem.* **1996**, *108*, 1789.

Polycyclic Guanidines From Nature's Shaped Cations to Abiotic Anion Hosts

Hans-Dieter Arndt and Ulrich Koert

Institut für Chemie, Humboldt Universität, Berlin, Germany

Marine rather than terrestrial organisms produce a large variety of structurally intriguing metabolites endowed with guanidine functions [1]. The most well known probably are saxitoxin (**1**), an extremely potent Na^+ channel blocker, and tetrodotoxin (**2**), the powerful poison of the puffer fish family (Scheme 1).

Recently, several new alkaloids containing guanidines in complex polycyclic frameworks have been isolated from sponges. Their novel structural features stimulated considerable synthetic efforts, supported by interesting bioactivity profiles. An early account was the isolation of ptilocaulin (**3**) by the Rinehart group, which displayed a broad range of activities (Scheme 2) [2]. In 1989, ptilomycalin A (**4**) was isolated from the Caribbean sponge *Ptilocaulis spiculifer* and from the Red Sea sponge *Hemimycale sp* [3]. Its unique pentacyclic guanidinium core is linked to an ω-hydroxy fatty acid side chain terminated by a N^2-spermidine amide. The closely related crambescidines 800, 816 and 844 (**5–7**) from *Crambe crambe* [4], celeromycalin (**8**) and fromiamycalin from *Celerina heffernani* and *Fromia monilis* respectively were reported there after [5].

Because of to pronounced differences in their antiviral and cytotoxic properties, the major role of the pentacyclic guanidine in the strong biological activity was soon identified. Using ptilomycalin A and simplified model compounds, a host-like behavior towards anions became evident experimentally [6]. The precise mode of action has yet to be elucidated, but the competitive interaction with ATP at its binding site in $Na^+/K^+/Ca^{2+}$-ATPase suggests anion receptor behavior and highlights these compounds in resembling artificial anion hosts (see below) [7].

The remarkable hexacyclic alkaloid palauamine (**10**) from *Stylotella agminata* was reported in 1993 (Scheme 3) [8]. Its high immunosuppresive activity – whilst being reasonably nontoxic – has provoked preclinical studies [8c]. Interestingly, the isomer styloguanidine (**11**) is a powerful chitinase inhibitor [8b].

Targeting the treatment of AIDS, a group at Smith Kline Beecham established a screening method to identify bioactive substances which inhibit the HIV binding to human $CD4^+$ receptor.

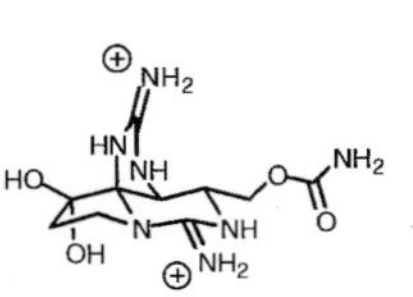

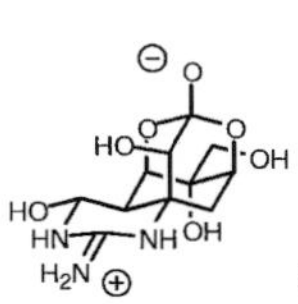

Scheme 1.

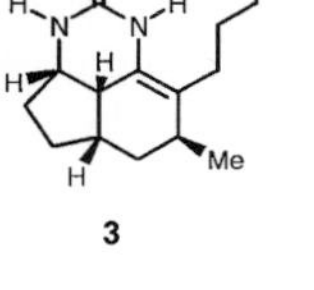

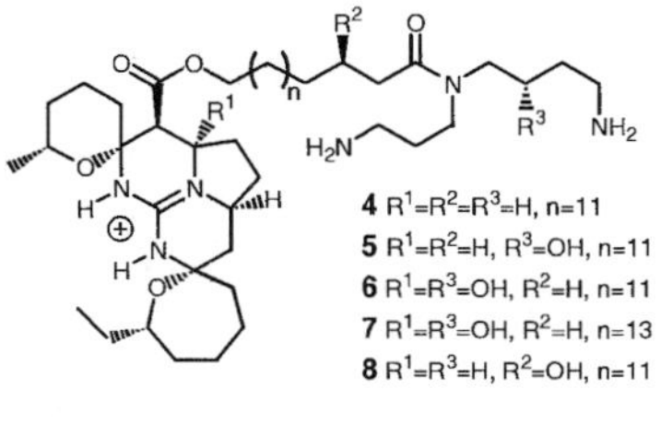

4 $R^1=R^2=R^3=H$, n=11
5 $R^1=R^2=H$, $R^3=OH$, n=11
6 $R^1=R^3=OH$, $R^2=H$, n=11
7 $R^1=R^3=OH$, $R^2=H$, n=13
8 $R^1=R^3=H$, $R^2=OH$, n=11

Scheme 2.

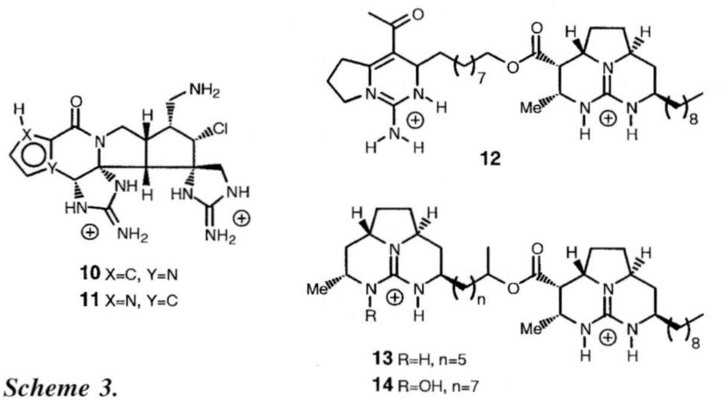

Scheme 3.

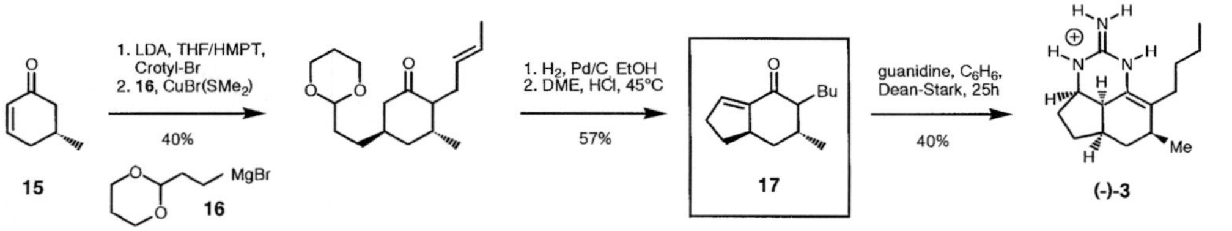

Scheme 4. Biomimetic synthesis of (−)-ptilocaulin by Snider et al.

On scanning numerous natural sources only the extracts from the Caribbean sponge *Batzella sp.* proved to be active [9]. This led to the isolation of the batzelladines A (**12**) and B as the first low-molecular-weight compounds able to inhibit this interaction. Subsequently, batzelladines F (**13**) and G (**14**) were shown to induce dissociation of thyrosin kinase p56^{lck} and CD4 cells highly specific in an immunosuppressivity-based assay [10].

Several of the compounds mentioned show similarities, and in fact some of them are produced together from the same species. A biosynthetic rationale for their synthesis was suggested early by Rinehart [2a] for (+)-ptilocaulin, and the findings of Snider [11a] and Roush [11b] strongly supported him. Condensation and Michael addition of guanidine to an enone could deliver these systems, and their thermodynamic equilibration is feasible. The brevity of this pioneering synthesis remains challenging (Scheme 4) [11a]. The kinetic enolate of enone **15**, accessible from natural (+)-pulegone, was crotylated, and a 3,5-*trans* specific addition of a Grignard-derived cuprate, followed by hydrogenation and acid-catalyzed ring closure, furnished enone **17** in 23 % overall yield. Treatment of **17** with excess guanidine under dehydrating conditions delivered unnatural (−)-ptilocaulin [(−)-**3**] (40 %, as its nitrate salt).

As (−)-pulegone is not readily available, subsequent work addressed the natural enantiomer by different methods [12], but all completed their synthesis via Snider's enone (**17**). Asaoka utilised a TMS group as temporary directing volume (Scheme 5) [12a]. Cossy applied a photoreductive cyclopropane ring opening, which was obtained by a diastereoselective Simmons-Smith reaction [12b]. Schmalz finally used a planar-chiral arene complex to synthesise enone **17** with interesting $Cr(CO)_3$ chemistry [12c].

The pentacyclic ptilomycalin A was originally targeted by conceptually related biomimetic strategies. A retrosynthetic analysis comprising the Snider [13] and Murphy [14] approaches is outlined in Scheme 6. After disconnection of the aminal functions (A and E rings), the tricyclic precursor could principally be formed by wrapping a bis-enone around guanidine via consecutive Michael addition and condensation reactions. The apparent simplicity of this approach is opposed by difficulties in stereocontrol, as a directing influence of the pendant C13 and C19 stereocenters remains speculative at this stage.

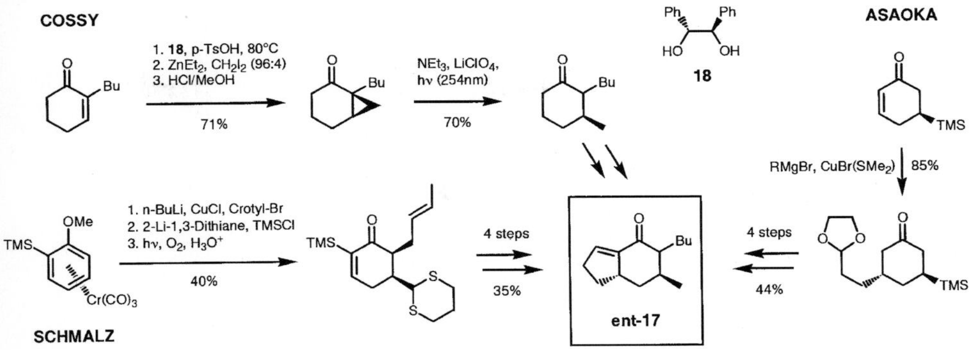

Scheme 5. Methods approaching natural (+)-ptilocaulin (**3**.)

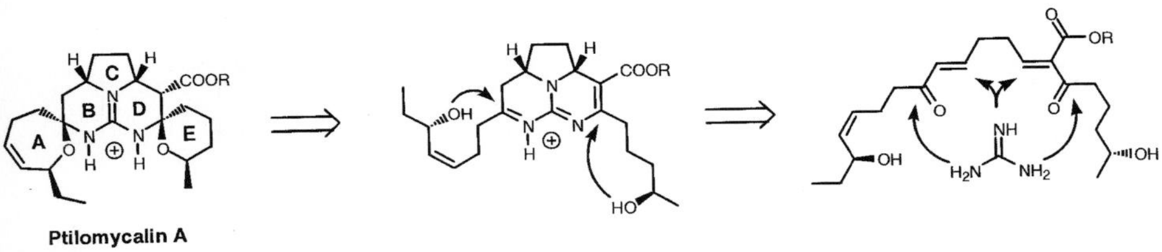

Scheme 6. Apparent retrosynthetic disconnection of ptilomycalin A (**4**).

In their successful synthesis of the ptilomycalin A core structure, Snider et al. had to divide the crucial tricyclization into two steps. The attempted multiple ring closure failed to yield the product sought (Scheme 7) [13]. After Knoevenagel condensation of β-keto ester **19** with aldehyde **20**, bis-enone **21** was condensed with *O*-methylisourea to close the C and D rings to give **22**. Treatment with buffered ammonia then furnished the B ring, albeit as a 1 : 1 mixture of diastereomers **23** and **24**. After separation and silyl ether cleavage, the A ring closed spontaneously (**25**), whereas the E Ring then needed forcing assistance by NEt_3. Again, stereoselection was negligible (**26/27** = 4 : 3), and the procedure required tedious purification.

The outlined "zipper" concept does indeed work, on condition that the C14 ester is omitted (Scheme 8) [14]. Treating a such simplified precursor as **28** with guanidine and then HCl gave racemic pentacyclic guanidine **29**, in one step and in 20 % yield, as its tetrafluoroborate salt, which could be characterized by X-ray analysis.

Being more concerned about specific stereocontrol, Overman et al. decided to install the C ring stereocenters first before elaborating the polycyclic framework [15]. In their successful total synthesis of ptilomycalin A (Scheme 9), the central C ring building block evolves from a reliable β-keto ester reduction of **30**. Mitsunobu displacement with azide, double reduction and urea formation gave after ozonolysis the ureido aminal **31** in 42 % total yield (98 % *ee*). A tethered Biginelli condensation [16] with β-keto ester **32** (available from methyl acetoacetate) as the key step led to *cis*-fused dihydropyrimidine **33** (7.5 : 1), which was deprotected and cyclized to the C,D,E ring fragment **34** in 59 % combined yield. Unfortunately the less strained C-14 α-epimer was formed exclusively, a significant drawback.

Preparing to install the A and B rings, **34** was converted to isourea **35**, remarkably without erosion of the C7 stereochemistry. The missing part was introduced with Grignard compound **36**, and after oxidation and *O*-deprotection the two final rings were closed via guanidine formation with buffered ammonia (20 % overall) to yield pentacycle **37**.

Having established the spermidine amide **38**, the wrong stereochemistry at C14 had to be corrected. After three cycles of partial epimerization

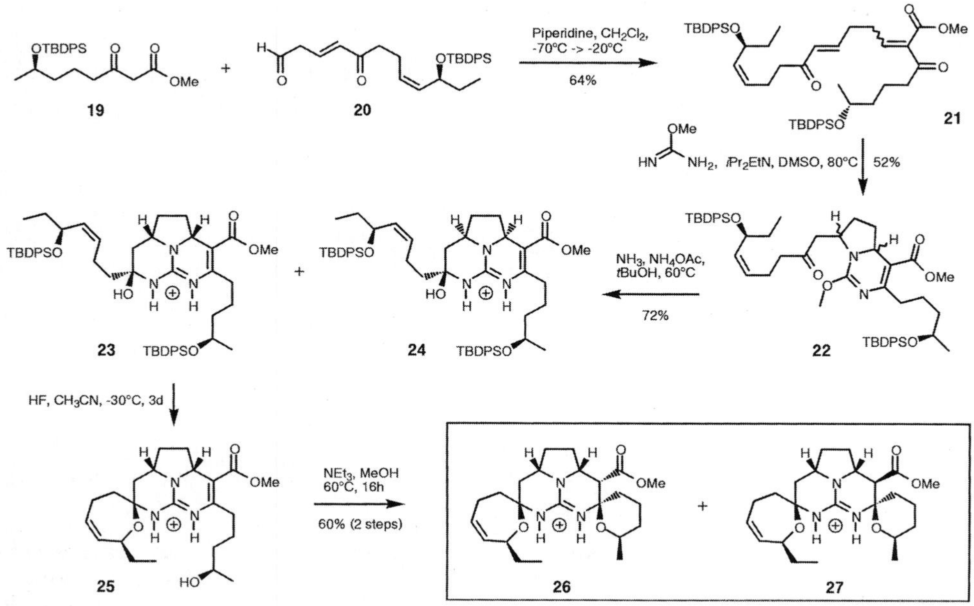

Scheme 7. Synthesis of the ptilomycalin A pentacyclic scaffold.

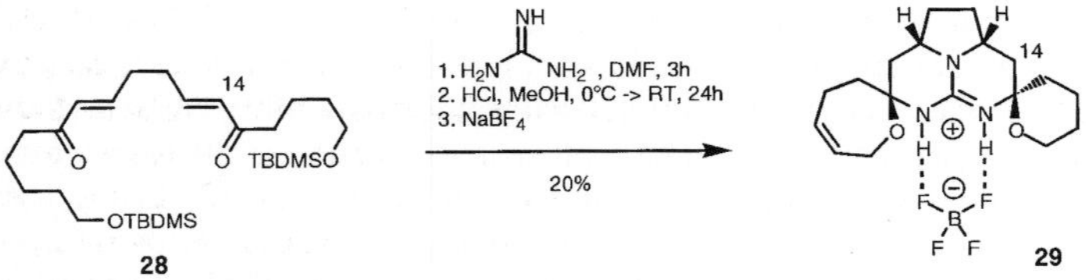

Scheme 8. "Zipper" condensation according to Murphy et al.

and separation, the natural isomer so obtained (50 % yield) was deprotected with neat formic acid to yield (−)-ptilomycalin A quantitatively (18 steps, 1 % total yield).

The isolation of the batzelladines attracted the attention of several groups, and some work has appeared in recent years concerned with partial syntheses, mainly in the racemic series [17, 18, 19]. The efforts of Snider culminated in a nine-step synthesis of racemic batzelladine E, following the biomimetic strategy [18b]. The only enantioselective synthesis reported to date established the absolute stereochemistry of batzelladine B (Scheme 10) [20a]. Diol **39** (obtained by β-keto ester chemistry) was transformed into guanidine **41** via the diazide. An improved Biginelli type condensation with methyl acetoacetate in trifluoroethanol gave the methanolysis product of batzelladine B (**42**) with 10 : 1 selectivity in excellent 94 % yield. Further experiments proved fruitful [20b]. *N*-sulfonylguanidine aldehyde (**43**) could be transformed selectively to the respective *cis*- (7 : 1) and *trans*-products (20 : 1) just by changing the reaction conditions. This approach seems ideally suited for delivering the batzelladines and related molecules using an acid-labile sulfonamide as the protecting and modulating moiety.

The synthesis of palau'amine (**10**) was addressed by Overman [21]. Its dense functionalization, with two quaternary carbons being most prominent, presents a formidable challenge to synthesis. Taking advantage of the molecules folded shape, a retrosynthesis reduces the structure's

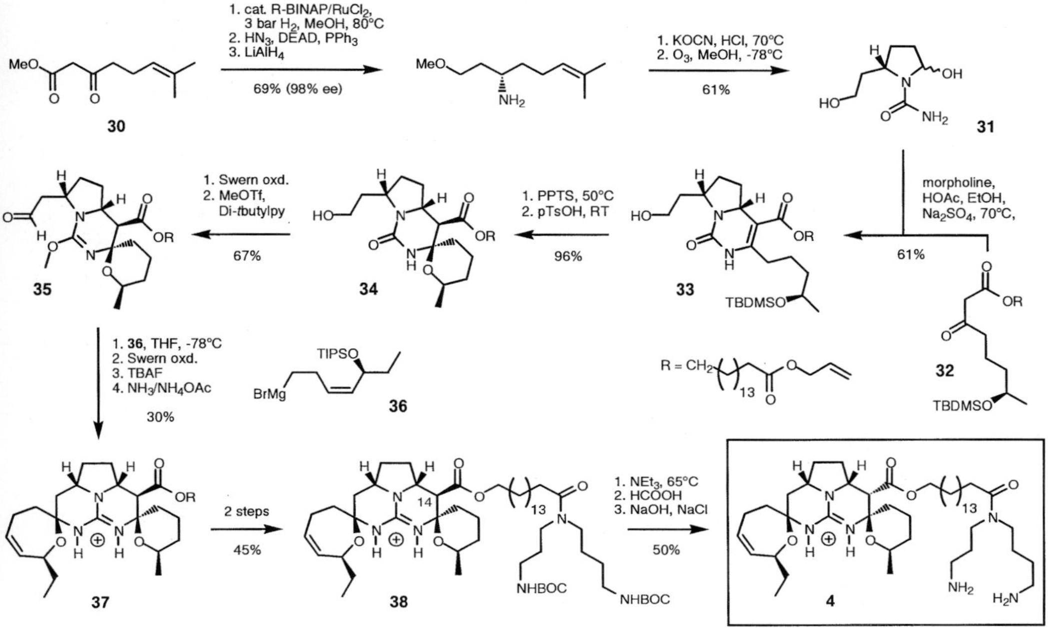

Scheme 9. Total synthesis of ptilomycalin A (**4**) by Overman et al.

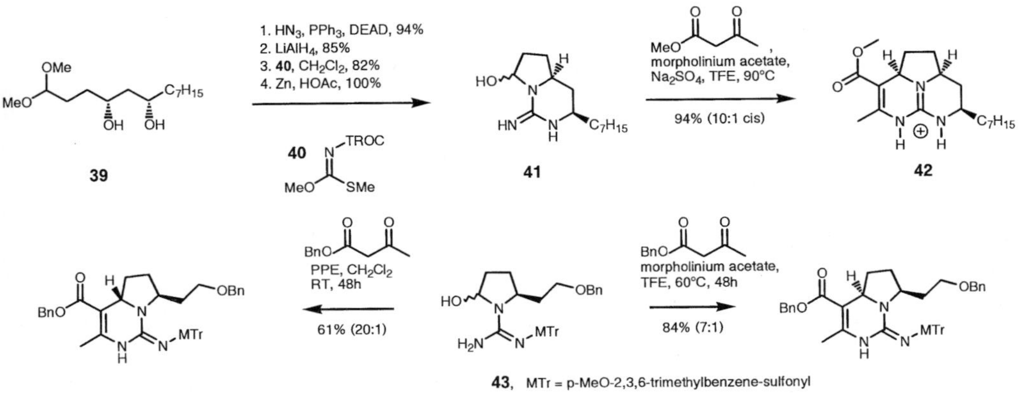

Scheme 10. Partial synthesis of batzelladine B and improved Biginelli condensation.

complexity (Scheme 11). A crucial azomethine ylide cycloaddition reveals a dihydropyrrole bearing an α-keto ester side chain. In their elegant model study (Scheme 11), α-keto ester **44** was condensed with thiosemicarbazide to deliver tetracycle **45** in excellent yield. When $OP(NCS)_3$ was applied in excess to form the isothiocyanate, not only ring closure but also reductive cleavage of the strained N–N bond was observed. After guanidinylation, protected bis(spiroguanidine) **47** was reached in 55 % overall yield.

Bearing their biological activity in mind, the total synthesis of the compounds mentioned will be mastered sooner or later. Up to now the

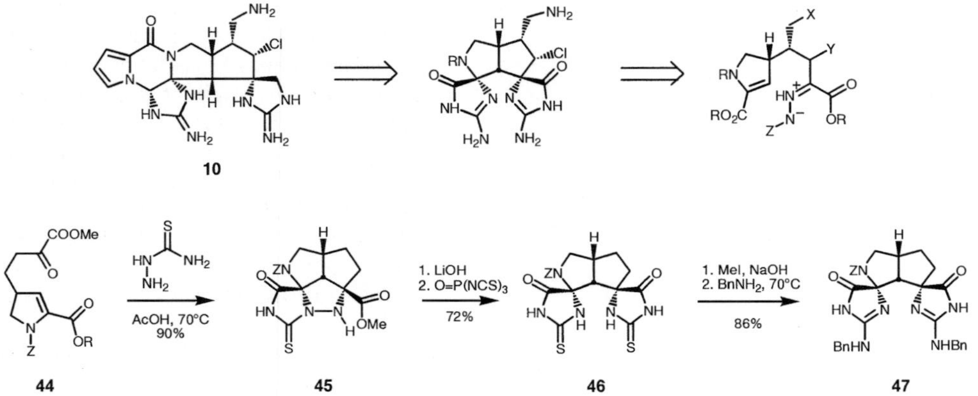

Scheme 11. Disconnection and partial synthesis of palau'amine (**10**).

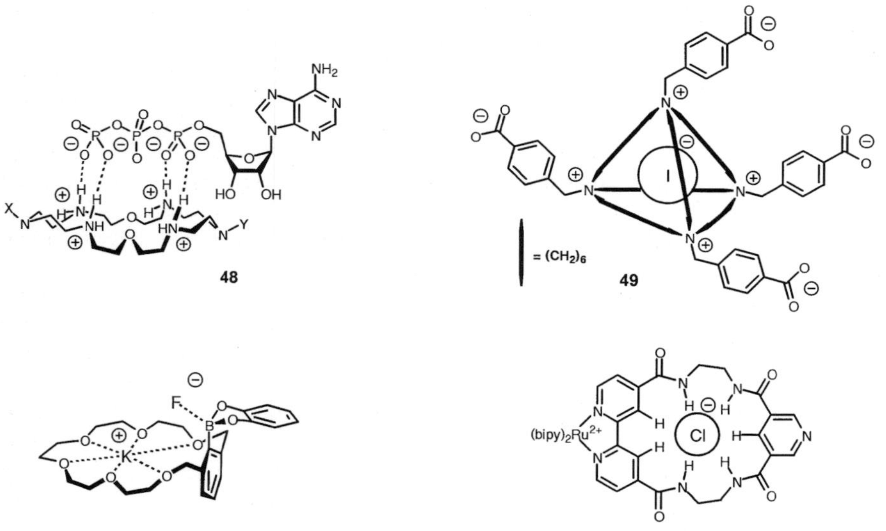

Scheme 12. Representative anion receptors.

so-called biomimetic approaches suffer from quite drastic reaction conditions and unpredictable thermodynamic control, often leading to product mixtures. This severely limits their applicability. Lacking the cell's enzymatic armament, chemistry has to develop new methods and strategic improvements. Thus these nitrogen-rich, oxidized synthetic targets apparently remain a challenge.

But, apart from synthetic endeavors, which factors make these compounds biologically active? Guanidine functions are fully protonated in aqueous media (pK_a ca. 13–14), and thus these compounds behave as large, lipophilic cations of defined shape. Their competitive interaction with (de-)phosphorylating enzymes or with sulfated recognition motives and so modulating signal transduction seem to be two likely modes of action. In other terms, they may behave as anion receptors.

Studying the molecular recognition of anions has been a growing part of supramolecular chem-

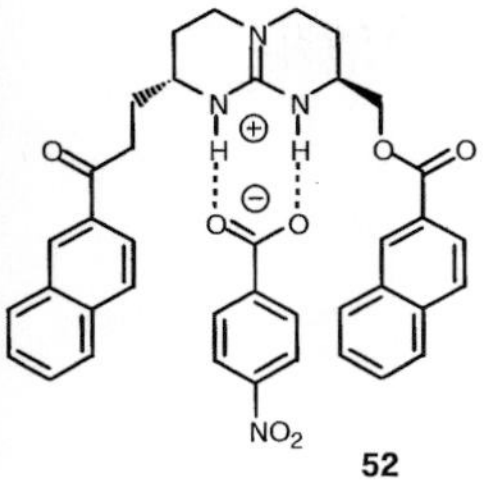

Scheme 13. Functional guanidinium-based receptors.

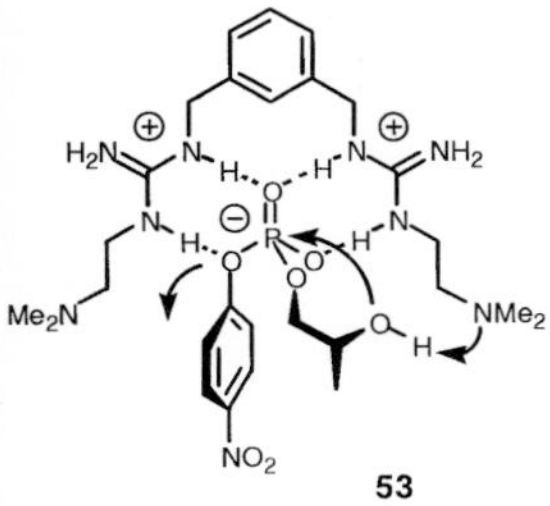

istry for a long time [22]. Its biochemical importance is clear: any kinase or phosphatase will primarily have to complex phosphate(s). Sulfated sugars and sugar acids – e.g. neuraminic acid – are often involved in recognition processes on the cell surface. And a hereditary mutation of merely one residue on the chloride ion channel protein induces its malfunction, leading to cystic fibrosis.

So recently more and more groups have become involved in the design of synthetic anion receptors. These topics are covered by several reviews [23], and a representative selection may illustrate the field (Scheme 12). Lehn and Hosseini have devoted a lot of effort to aza crown complexes with ATP **48**, showing them to be true yet weak ATPase mimics [24]. Schmidtchen reported tetrahedral zwitterions **49** to discriminate between spherical anions [25]. Reetz was able to show that ditopic receptor **50**, which includes both Lewis-acidic and basic centers will complex a whole ion pair [26]. Incorporation of transition metals into anion receptors such as **51** renders neutral species positively charged and susceptible to spectroscopic and electrochemical manipulations [27].

Guanidines have been implemented early as recognition elements, guided by the apparent function of arginine in protein structures. The C_2-symmetric, chiral anion receptor **52** was introduced by Lehn, Schmidtchen and de Mendoza consecutively and studied in various modifications (Scheme 13) [23c]. For example, an elaborate system based on **52** provided reasonable enantioselective recognition of amino acids [23c, 28]. Furthermore, bis(guanidinium) compounds catalyze RNA hydrolysis in the presence of external base via phosphodiester complexation [29]. These functional elements were joined in receptor **53** to yield a functional transesterification catalyst [30].

An elegant modular synthesis of the **52** scaffold originates from methionine, which is transformed to isothiocyanate **54** by standard operations (Scheme 14) [31]. Condensation with an appropriate 1,3-diamine, here **55** derived from aspartate, is followed by exhaustive S-methylation inducing a smooth double ring closure. After deprotection, guanidine **58** was isolated in gratifying 85 % yield.

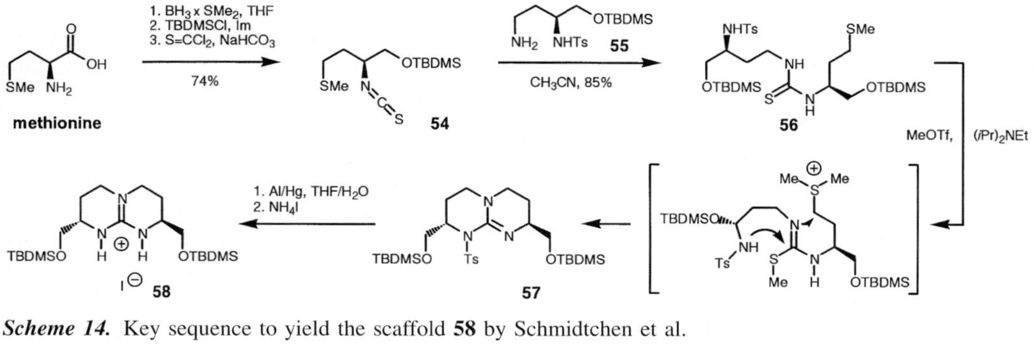

Scheme 14. Key sequence to yield the scaffold **58** by Schmidtchen et al.

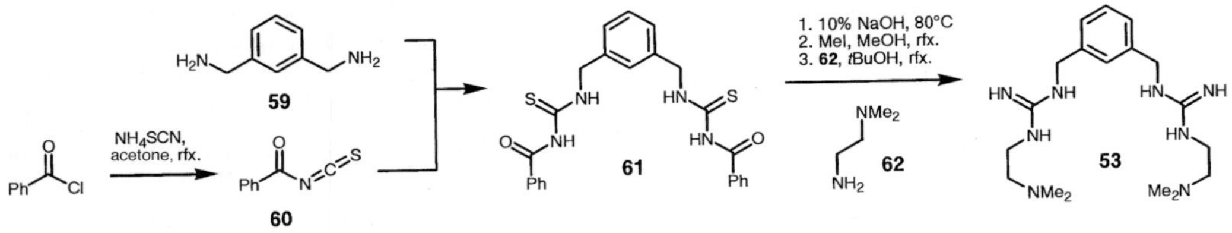

Scheme 15. Synthesis of the RNAse mimic **53**.

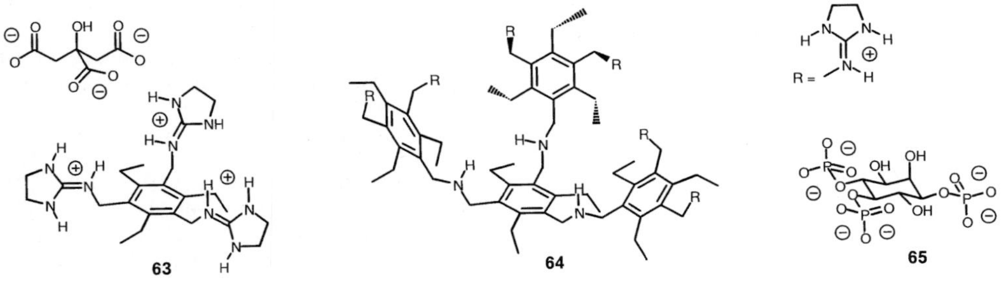

Scheme 16. Guanidinum hosts as polyanion sensors.

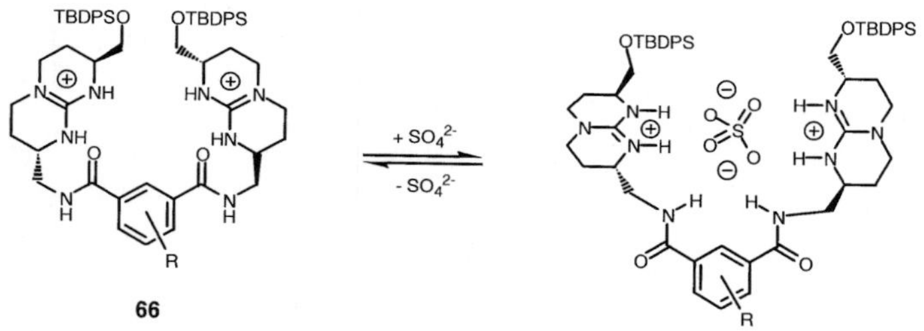

Scheme 17. Entropy-driven SO_4^{2-} complexation found for **66**.

The RNAse model compound of Hamilton was constructed in a few steps (Scheme 15) [30]. Benzoyl isothiocyanate was allowed to react with diamine **59** to yield bis(thiourea) **61** after deprotection. S-methylation activated the urea moieties for guanidine formation with *N,N*-dimethyl-ethylenediamine **62** to reach the desired compound **53**.

Receptors showing useful selectivity for polyanions have been invented by the Anslyn group [31]. Tris(guanidinium) compound **63** displays a binding constant of 6900 for citrate in water with remarkable selectivity (Scheme 16) [31a]. This was utilized to set up a competition assay for the direct determination of citrate in the mM range [31b]. Steric gearing forces the recognition elements to one face of the hexa-substituted benzene ring, which was further used for the hexacationic receptor **64**. Applying **64** to the competition assay, the inositol-1,4,5-triphosphate **65** concentration – an important second messenger in cell signalling – could be measured down to 1 μM [31c].

The strong Coulomb interactions present in the guanidine-anion complexes may be largely balanced out by solvation energies in a polar medium. By the means of isothermal titration calorimetry, Schmidtchen et al. studied the exceptionally high binding constants of receptor type **66** for SO_4^{2-} (10^6-10^7 in MeOH) [32]. A large binding entropy gain overrules an unfavorable ΔH_a in this case, indicating a solvophobic collapse (Scheme 16). These findings coincide with modern models about protein-ligand interactions [33] and should lead to new design concepts, overcoming the "rigid cavity"-type receptors.

In summary, all of the examples shown outline the significance of the guanidine function for anion recognition in general, whether in bioactive natural products or man-made tools. Joining the concepts and methods should seed ideas in the respective fields and lead to a better understanding of supramolecular interaction in the future.

References

[1] Berlinck, R.G.S.; *Nat. Prod. Rep.* **1996**, *13*, 377 and references cited therein.

[2] a) Harbour, G.C.; Tymiak, A.A.; Rinehart, K.L.; Shaw, P.D.; Hughes Jr., R.G.; Mizsak, S.A.; Coats, J.H.; Zurenko, G.E.; Li, L.H.; Kuentzel, S.L.; *J. Am. Chem. Soc.* **1981**, *103*, 5604. On related metabolites see b) Berlinck, B.G.S.; Braekman, J.C.; Daloze, D.; Hallenga, K.; Ottinger, R.; Bruno, I.; Riccio, R.; *Tetrahedron Lett.* **1990**, *31*, 6531; c) Tavares, R.; Daloze, D.; Braekman, J.C.; Hajdu, E.; van Soest, R.W.M.; *J. Nat. Prod.* **1995**, *58*, 1139; d) Barrow, R.A.; Murray, L.M.; Lim, T.K.; Capon, R.J.; *Aust. J. Chem.* **1996**, *49*, 767; e) Patil, A.D.; Freyer, A.J.; Offen, P.; Bean, M.F.; Johnson, R.K.; *J. Nat. Prod.* **1997**, *60*, 704.

[3] Kashman, Y.; Hirsh, S.: McConnell, O.J.; Ohtani, I.; Kusumi, T.; Kakisawa, H.; *J. Am. Chem. Soc.* **1989**, *111*, 8925. The assignment of the species as *Ptilocaulis* has been questioned [2c, 9].

[4] Jares-Eriyman, E.A.; Sakai, R.; Rinehart, K.L.; *J. Org. Chem.* **1991**, *56*, 5712; Jares-Eriyman, E.A.; Ingrum, A.L.; Carney, J.R.; Rinehart, K.L.; Sakai, R.; *J. Org. Chem.* **1993**, *58*, 4805; Berlinck, R.G.S.; Braekman, J.C.; Daloze, D.; Bruno, I.; Riccio, R.; Ferri, S.; Spampinato, S.; Speroni, E.; *J. Nat. Prod.* **1993**, *56*, 1007.

[5] Palagiano, E.; de Martino, S.; Minale, L.; Riccio, R.; Zollo, F.; Iorizzi, M.; Carre, J.B.; Debitus, C.; Lucarain; L.; Provost, J.; *Tetrahedron* **1995**, *51*, 3675.

[6] a) Ohtani, I.; Kusumi, T.; Kakisawa, H.; Kashman, Y., Hirsh, S.; *J. Am. Chem. Soc.* **1992**, *114*, 8472. b) Murphy, P.J.; Williams, H.L.; Hibbs, D.E.; Hursthouse, M.B.; Abdul Malik, K.M.; *Chem. Commun.* **1996**, 445.

[7] Ohizumi, Y.; Sasaki, S.; Kusumi, T.; Ohtani, I.; *Eur. J. Pharmacol.* **1996**, *310*, 95.

[8] a) Kinnel, R.B.; Gehrken, H.-P.; Scheuer, P.J.; *J. Am. Chem. Soc.* **1993**, *115*, 3376; b) Kato, T.; Shizuri, Y.; Izumida, H.; Yokoyama, A.; Endo, M.; *Tetrahedron Lett.* **1995**, *36*, 2133. Several brominated analogues: c) Kinnel, R.B.; Gehrken, H.-P.; Swali, R.; Skoropowski, G.; Scheuer, P.J.; *J. Org. Chem.* **1998**, *63*, 3281.

[9] Patil, A.P.; Kumar, N.V.; Kokke, W.C.; Bean, M.F.; Freyer, A.J.; de Brosse, C.; Mai, S.; Truneh, A.; Faulkner, D.J.; Carte, B.; Breen, A.L.; Hertzberg, R.P.; Johnson, R.K.; Westley, J.W.; Potts, B.C.M.; *J. Org. Chem.* **1995**, *60*, 1182. The structures of Batzelladine A, D and E have been revised by partial synthesis [17a, b].

[10] Patil, A.P.; Freyer, A.J.; Taylor, P.B.; Carté, B.; Zuber, G.; Johnson, R.K.; Faulkner, D.J.; *J. Org. Chem.* **1997**, *62*, 1814. The relative stereochemistry of Batzelladine F is in doubt [18, 19a].

[11] a) Snider, B.B.; Faith, W.C.; *J. Am. Chem. Soc.* **1984**, *106*, 1443; b) Roush, W.R.; Walts, A.E.; *J. Am. Chem. Soc.* **1984**, *106*, 721.

[12] a) Asaoka, M.; Sakurai, M.; Takei, H.; *Tetrahedron Lett.* **1990**, *31*, 4159; b) Cossy, J.; BouzBouz, S.; *Tetrahedron Lett.* **1996**, *37*, 5091; c) Schellhaas, K.; Schmalz, H.-G.; Bats, J.W.; *Chem. Eur. J.* **1998**, *4*. For stimulating racemic work see d) Hassner, A.; Keshava Murthy, K.S.; *Tetrahedron Lett.* **1986**, *27*, 1407; e) Uyehara, T.; Furuta, T.; Kabawawa, Y.; Yamada, J.; Kato, T.; Yamamoto, Y.; *J. Org. Chem.* **1988**, *53*, 3669.

[13] Snider, B.B.; Shi, Z.; *J. Am. Chem. Soc.* **1994**, *116*, 549.

[14] a) Murphy, P.J.; Williams, H.L.; Hibbs, D.E.; Hursthouse, M.B.; Abdul Malik, K.M; *Tetrahedron* **1996**, *52*, 8315; b) Howard-Jones, A.; Murphy, P.J.; Thomas, D.A.; *J. Org. Chem.* **1999**, *64*, 1039.

[15] Overman, L.E.; Rabinowitz, M.H.; Renhowe, P.A.; *J. Am. Chem. Soc.* **1995**, *117*, 2657.

[16] a) Biginelli, P.; *Gazz. Chim. Ital.* **1893**, *23*, 360; review: b) Kappe, C.O.; *Tetrahedron* **1993**, *49*, 6937; for recent improvements see c) Kappe, C.O.; Falsone, S.F.; *Synlett* **1998**, *718*.

[17] a) Snider, B.B.; Chen, J.; Patil, A.D.; Freyer, A.J.; *Tetrahedron Lett.* **1996**, *37*, 6977; b) Snider, B.B.; Chen, J.; *Tetrahedron Lett.* **1998**, *39*, 5697.

[18] Black, G.P.; Murphy, P.J.; Walshe, N.D.A.; *Tetrahedron* **1998**, *54*, 9481.

[19] a) Rama Rao, A.V.; Gurjar, M.K.; Vasudevan, J.; *J. Chem. Soc. Chem. Commun.* **1995**, *20*, 1369; b) Gurjar, M.K.; Lalitha, S.V.S.; *Pure & Appl. Chem.* **1998**, *70*, 303.

[20] a) Franklin, A.S.; Ly, S.K.; Mackin, G.H.; Overman, L.E.; Shaka, A.J.; *J. Org. Chem.* **1999**, *64*, 1512; b) McDonald, A.I.; Overman, L.E.; *J. Org. Chem.* **1999**, *64*, 1520.

[21] Overman, L.E.; Rogers, B.N.; Tellew, J.E.; Trenkle, W.C.; *J. Am. Chem. Soc.* **1997**, *119*, 7159.

[22] Lehn, J.-M.; *Supramolecular Chemistry*, VCH, Weinheim **1995**.
[23] Reviews: a) Dietrich, B.; *Pure Appl. Chem.* **1993**, *65*, 1457; b) Scheerder, J.; Engbersen, J.F.J.; Reinhoudt, D.N.; *Recl. Trav. Chim. Pays-Bas* **1996**, *115*, 307; c) Schmidtchen, F.P.; Berger, M.; *Chem. Rev.* **1997**, *97*, 1609; d) Antonisse, M.M.G.; Reinhoudt, D.N.; *Chem. Commun.* **1998**, 443.
[24] Hosseini, M.W.; Lehn, J.-M.; Jones, K.C.; Plute, K.E.; Mertes, K.B.; Mertes, M.P.; *J. Am. Chem. Soc.* **1989**, *111*, 6330.
[25] Worm, K.; Schmidtchen, F.P.; *Angew. Chem.* **1995**, *107*, 71; *Angew. Chem. Int. Ed.* **1995**, *34*, 65.
[26] Reetz, M.T.; Johnson, B.M.; Harms, K.; *Tetrahedron Lett.* **1994**, *35*, 2525.
[27] Beer, P.D.; *Acc. Chem. Res.* **1998**, *31*, 71.
[28] For receptors discriminating amino acid derivatives see: Davies, A.P.; Lawless, L.J.; *Chem. Commun.* **1999**, 9 and references cited therein.
[29] Oivanen, M.; Kuusela, S.; Lönnberg, H.; *Chem. Rev.* **1998**, *98*, 961.
[30] Juiban, V.; Veronese, A.; Dixon, R.P.; Hamilton, A.D.; *Angew. Chem.* **1995**, *107*, 1343; *Angew. Chem. Int. Ed.* **1995**, *34*, 1237. A more elaborate Zn^{2+}-based catalyst: Molenveld, P.; Kapsabelis, S.; Engbersen, J.F.J.; Reinhoudt, D.N.; *J. Am. Chem. Soc.* **1997**, *119*, 2948–49.
[31] a) Metzger, A.; Lynch, V.M.; Anslyn, E.V.; *Angew. Chem.* **1997**, *109*, 911; *Angew. Chem. Int. Ed.* **1997**, *36*, 862; b) Metzger, A.; Anslyn, E.V.; *Angew. Chem.* **1998**, *110*, 682; *Angew. Chem. Int. Ed.* **1998**, *37*, 649; c) Niikura, K.; Metzger, A.; Anslyn, E.V.; *J. Am. Chem. Soc.* **1998**, *120*, 8533.
[32] Berger, M.; Schmidtchen, F.P.; *Angew. Chem.* **1998**, *110*, 2840; *Angew. Chem. Int. Ed.* **1998**, *37*, 2694.
[33] a) Williams, D.H.; Westwell, M.S.; *Chem. Soc. Rev.* **1998**, *27*, 57; b) Davis, A.M., Teague, S.J.; *Angew. Chem.* **1999**, *111*, 778; *Angew. Chem. Int. Ed. Engl.* **1999**, *38*, 736.

Synthetic Access to Epothilones – Natural Products with Extraordinary Anticancer Activity

Ludger A. Wessjohann and Günther Scheid

Bio-organic Chemistry, Vrije Universiteit, Amsterdam, The Netherlands

General Aspects and Biology

Epothilone B (**2b**) and derivatives (**2a**, **2c–e**) are among the most exciting drug candidates of the new millennium. They have an extraordinary potential to become the most important antitumor agents of their time, either directly as active drug components or through the enhanced possibility of finally obtaining enough data from quantitative structure activity relationships to understand antimitotic action. Although structurally unrelated to paclitaxel (Taxol®), discodermolide and eleutherobine, they exhibit the same if not a better effect, inducing the polymerization of tubulin to stabilized microtubules. The consequent massive complication during mitosis [1], among possible other actions, finally triggers apoptosis (programmed cell death).

This type of tubulin activity has so far been exclusively found in the four above-mentioned natural products and some derivatives, although far more then 140 000 synthetic compounds and extracts have been tested. Of these four compounds, epothilones appear to be the best candidates. They are equally or even more active, e.g. up to 35 000 times better then Taxol in resistant cell lines [2]. They also have better cytotoxic potential connected to the tubulin activity, as not all microtubule stabilizers lead to sufficient cell death, and they allow extensive derivatization much faster then Taxol or discodermolide [3, 4]. Also, improvements in the applicability to patients compared to the sparingly soluble Taxol are expected, eliminating some of the severe side effects connected to the latter drug. Since the binding sites of Taxol and epothilones overlap, epitope comparisons and models of binding have been proposed, as well as synthetic combinations of the two structures synthesized or suggested [5, 6].

Epothilones were isolated from the myxobacterium *Sorangium cellulosum* by the groups of Höfle and Reichenbach in Germany [7, 8]. The constitution of the epothilones suggests that they are derived from polyketide metabolism, like similarly built macrolides. The synthesis apparently commences with *N*-acetylcysteine, which later becomes part of the thiazole. Apart from this ring, the epoxide and the geminal dimethyl group are among the less common features. Under basic conditions, epothilones can undergo retro-aldol cleavage of the C3–C4 bond when the C3 hydroxy group is deprotonated.

Retrosynthetic Analysis

So far, the groups of Danishefsky [9–12], Nicolaou [13–20], Schinzer [21, 22], Mulzer [23], White [24] and Grieco [25] have published total syntheses. Many other groups submitted advanced identical intermediates by different routes, i.e. delivered formal total syntheses. Overall, a multitude of permutations of existing building blocks and strategies were published, pushed to the extreme by the numerous permutating contributions of Nicolaou et al. Thus, single start-to-finish syntheses will not be discussed here, but the most important concepts and building blocks, arranged as far as possible according to the following retrosynthetic analysis. Epothilone numbering is used in all cases.

The first retrosynthetic transformation on epothilones A and B (**2a**,**b**) is the removal of

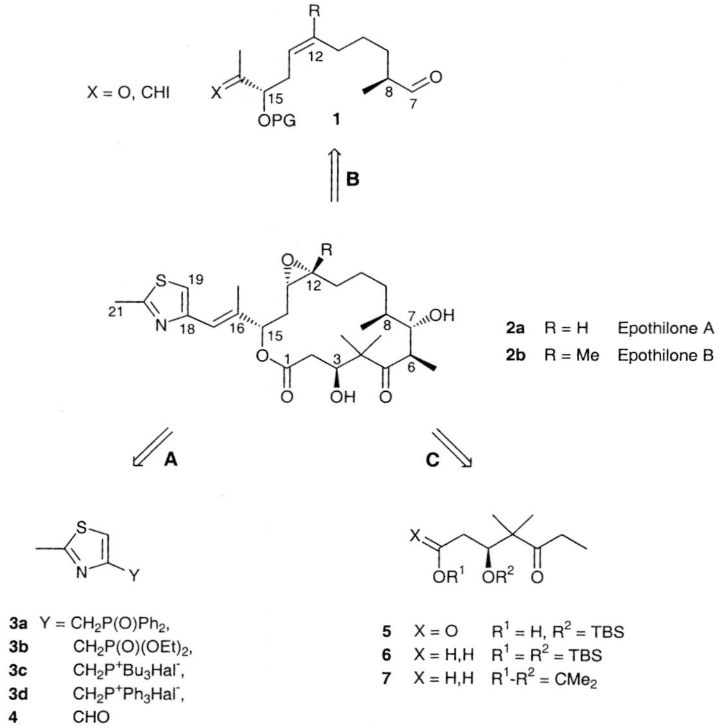

X = O, CHI

1

2a R = H Epothilone A
2b R = Me Epothilone B

3a Y = $CH_2P(O)Ph_2$,
3b $CH_2P(O)(OEt)_2$,
3c $CH_2P^+Bu_3Hal^-$,
3d $CH_2P^+Ph_3Hal^-$,
4 CHO

5 X = O R^1 = H, R^2 = TBS
6 X = H,H R^1 = R^2 = TBS
7 X = H,H R^1-R^2 = CMe_2

Scheme 1. Epothilone A and B and the retrosynthetic analysis followed in this highlight. (PG = Protecting Group, TBS = *tert*-butyldimethylsilyl, Hal = Cl, Br, I).

the epoxide to give the C12–C13 double-bonded desoxyepothilones (= epothilones C and D – see Scheme 3: **2c** and **2d**, resp.). The thus simplified molecules allow many reasonable disconnections of macrocycle and side chain of which almost all possibilities were followed. However, some fragmentations have been more successful (Scheme 1). These also separate regions of different functionalization within the molecule as they are often found in macrocyclic polyketides in general: a heteroaromatic side chain (*cf.* **A**), a lipophilic macrocycle half (*cf.* **B**), and a more hydrophilic aldol-type macrocycle half (*cf.* **C**).

All strategies, with the exception of Danishefskys first-generation approach [9], placed disconnections at the C1 ester bond and the C6–C7 aldol bond to leave behind the aldol fragment **C** (*cf.* **5–6** and Chapter 6). The thiazole fragment **A** was usually split off by breaking the C16–C17 double bond, but modern palladium methods alternatively render the C17–C18 bond strategic (*cf.* Chapter 4).

The remaining "northern half" **B** (C7–C16/C17, *cf.* Chapter 5) superficially appears to offer fewer strategic disconnections for a convergent approach, with only a C-15-hydroxy-group, and a double bond in the middle part at C12-C13. The latter being the obvious disconnection, it has been the most frequent one too, with the additional possibility of attaching the corresponding subfragments of this disconnection to A and C respectively and using the connection C12–C13 for the macrocyclization. However, almost all other disconnections conceivable have been used in part **B** (*cf.* **1**), although they are often not immediately obvious. This makes a unified description of synthetic approaches to this part of the molecule difficult.

Of all disconnections, only three have been used for the macrocyclization, the starting point for the following discussion.

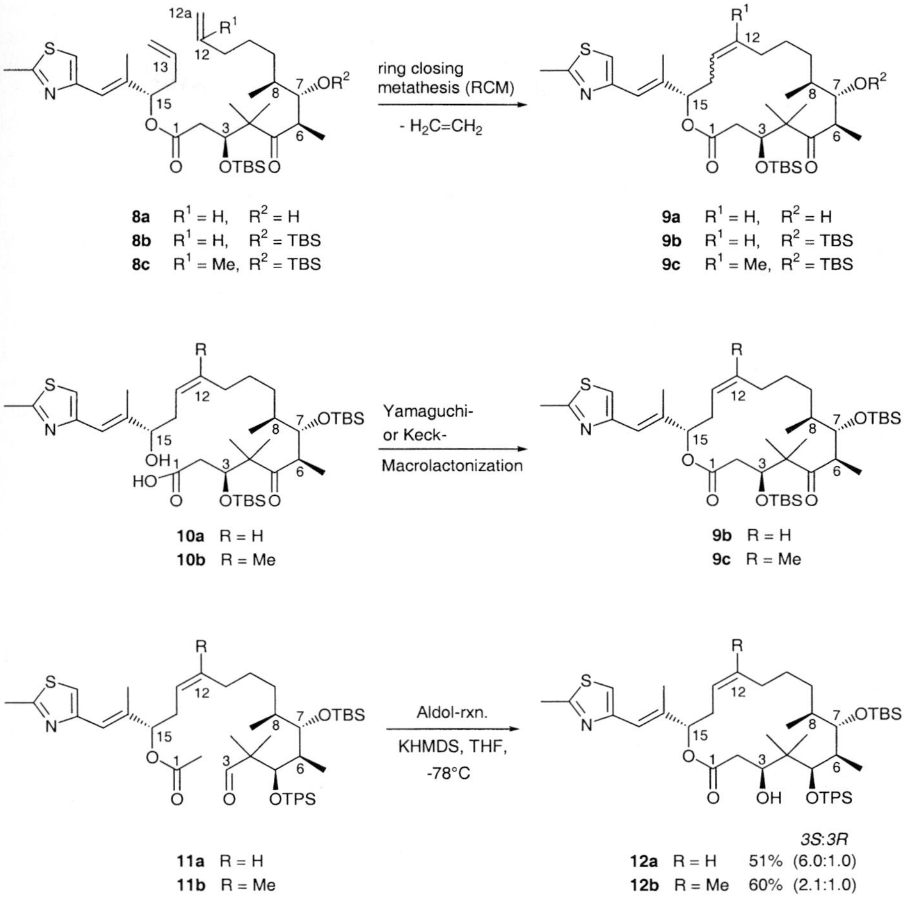

ring closing metathesis (RCM)

- $H_2C{=}CH_2$

8a R^1 = H, R^2 = H
8b R^1 = H, R^2 = TBS
8c R^1 = Me, R^2 = TBS

9a R^1 = H, R^2 = H
9b R^1 = H, R^2 = TBS
9c R^1 = Me, R^2 = TBS

Yamaguchi- or Keck- Macrolactonization

10a R = H
10b R = Me

9b R = H
9c R = Me

Aldol-rxn.
KHMDS, THF, -78°C

11a R = H
11b R = Me

			3*S*:3*R*
12a	R = H	51%	(6.0:1.0)
12b	R = Me	60%	(2.1:1.0)

Scheme 2. Macrocyclization reactions applied in epothilone total syntheses. From top to bottom: ring-closing metathesis (RCM); macrolactonization, and C2-C3 macroaldolization. (TBS = *tert*-butyldimethylsilyl, TPS = triphenylsilyl, HMDS = hexamethyldisilazide).

Macrocyclization and Epoxidation

The final steps, macrocyclization and epoxidation, will be discussed first, in order to later have a better understanding of the positioning and requirements of the building blocks leading to the open chain precursors.

For the cyclization to the 16–membered macrolactone structure of epothilones C and D (= desoxyepothilones A and B, resp. [26]), three different strategies have been used successfully so far: (1) Ring-closing olefin metathesis (RCM) between C12 and C13. RCM is a comparably new method in total synthesis and underwent enormous improvements in recent years, especially because defined stable catalysts have been developed. Because of the complexity of the open chain bisolefin this approach was challenging initially but had the appeal of novelty. Certainly the epothilone results contributed considerably to the current success of this method in total syntheses; (2) macrolactonization – probably the safest and most promising method with many successful variations known; and (3) the macroaldolization between C2 and C3 as the least conventional approach.

Ring-Closing Metathesis (RCM) Reactions

The RCM of **8a–c** shown in Scheme 2 (top) allows the synthesis of the macrocycles **9a–c**. Nicolaou et al. were the first to publish a successful RCM of **8a** with 15 mol % Grubb's catalyst [$RuCl_2$(= CHPh)$(Pcy_3)_2$] under high dilution conditions (0.006 M in CH_2Cl_2) [20]. They obtained a mixture of cyclized olefins in 85 % yield with a *Z:E* ratio of 1.2 : 1.0. Generally, this ratio could not be improved much towards the desired Z-isomer, something which also holds true for RCMs of other precursors. Danishefsky et al. studied a variety of derivatives of **8** (e.g. other or no protective groups or other stereoisomers at C3 and C7 or a C5–hydroxygroup etc.) with 50 mol% catalyst (boiling benzene, 0.006 M). In general, the yields of cyclized products were greater than 80 %, but the *Z:E* ratio changed with small variations in the bisolefin structure [11, 27]. Their cyclization of **8b** to **9b** (86 % yield) with a *Z:E*-ratio of just 1.7 : 1.0 is still the best result reported so far. In CH_2Cl_2, Schinzer could improve the yield to an excellent 94 %, but the *Z:E* ratio dropped to 1.0 : 1.0 [21].

The cyclization of **8c** to **9c** created a trisubstituted double bond, thus preparing new ground for the application of RCM in complex systems. Actually the substrate failed to cyclize with the ruthenium-based Grubb's catalyst, but 20 mol% of a molybdenum-based catalyst described by Schrock led to the cyclized product **9c** in 86 % yield (benzene, 55 °C), unfortunately again with a 1.0 : 1.0 ratio of *Z* to *E* isomers [11].

In a solid phase bound synthesis of epothilone A and derivatives, Nicolaou et al., in an early stage of the synthesis, connected the vinylic carbon C12a (*cf.* **8a**) via a linker chain to Merrifield resin. After building up the C12a-attached **8a**-precursor, RCM not only formed the macrocycle **9a**, but at the same time released it from the resin, leaving C12a=C13a behind [20].

Overall, the new possibilities of RCM were impressively demonstrated on the epothilones. Unfortunately it is too early to address substantial problems with RCM in a satisfactory way. The relatively large amounts of catalyst (15–50 mol%) and the low diastereoselectivity make the syntheses very expensive, because half of the advanced material lost in one of the last steps of the syntheses in form of the wrong double-bond isomer. With the current efforts and the excitement about the truely unique possibilities of metathesis, it is predictable that these drawbacks will be overcome in the future, especially regarding catalyst quantity.

Macrolactonization

The 15-hydroxy acids **10a** and **10b** (Scheme 2) were cyclized with the reliable Yamaguchi (2,4,6-trichlorobenzoylchloride, TEA, DMAP, toluene) [10, 11, 13, 16] and Keck methods (DCC, DMAP, DMAP×HCl, $CHCl_3$) [23, 28]. Both delivered the macrolactones in good to excellent yields, the first (90 % **9b** and 78 % **9c**) being superior to the latter.

Macroaldolization

In the first syntheses of epothilone A [9] and B [12] by Danishefsky et al., the acetates **11a** and **11b** were treated with the base K-HMDS in THF at low temperature to generate the C1, C2

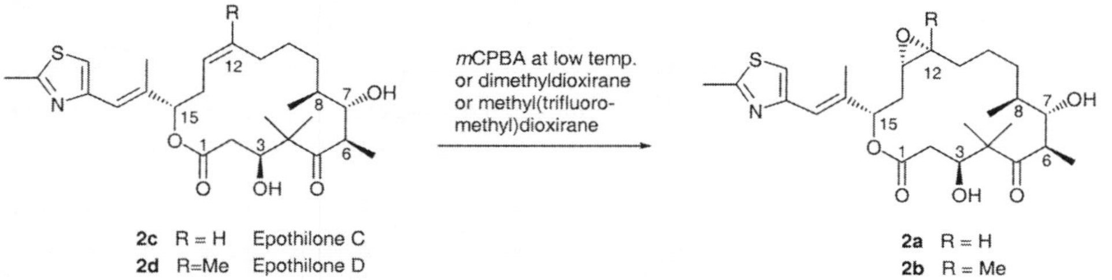

*m*CPBA at low temp. or dimethyldioxirane or methyl(trifluoromethyl)dioxirane

2c R = H Epothilone C
2d R=Me Epothilone D

2a R = H
2b R = Me

Scheme 3. Chemo-, regio- and stereoselective oxidation of epothilones C and D (**2c** and **2d**) to epothilones A and B, respectively (*m*CPBA = *meta*-chloroperbenzoic acid).

enolates, which reacted intramolecularly with the C3 aldehyde moiety to obtain the cyclized products **12a** and **12b**. The yields of 51 % and 60 % are comparable, but the considerable change in diastereoselectivity at C3 demonstrates the somewhat unpredictable influence of small changes in the structure. Thus, the selectivity drops from 6 : 1 to only ca. 2 : 1 with just one additional methyl group at C12 in the linear precursor, i.e. a small steric change 8 atoms away from the actually formed stereocenter.

It should be noted that this macroaldolization approach requires the protection of the C5 keto group, in this case as triphenylsilyloxy group, in order to avoid the otherwise facile retroaldolization along the C3–C4 bond (see below).

Completion of the Syntheses, Epoxidation

The syntheses of epothilone A and B are usually completed by the epoxidation of the C12-C13 double bond in epothilone C (desoxyepothilone A, **2c**) and epothilone D (desoxyepothilone B, **2d**), obtained from the deprotection of the precursors **9a–c**, or, after an additional oxidation at C5, of **12a–b**.

To epothilones C (**2c**) and D (**2d**), four epoxidation reagents of two classes were applied, betting on the regio- and stereoselective formation of epothilones A (**2a**) and B (**2b**) respectively. The electron-rich C12-C13 double bond is indeed attacked preferentially by all reagents applied and the dominant diastereoselectivity is in agreement with the natural structures too (Scheme 3). Epoxidations to epothilone B are superior to those to epothilone A with respect to both yield and diastereoselectivity.

It turned out that mCPBA is the least selective reagent (*dr*s from 2.8 : 1 [20] to 4 : 1 [23]), also giving rise to many side products e.g. by oxidation of the side chain double bond and of the aromatic system, depending on the conditions [26]. Nicolaou et al. used the alternative H_2O_2/MeCN-system with $KHCO_3$ in methanol to epoxidize desoxyepothilone E (**2e**, Scheme 13) derivatives with good yields and selectivities [18].

The best reagents so far are dimethyldioxirane and methyl(trifluoromethyl)dioxirane, the first allowing the epoxidation of **2c** to **2a** in 49 % yield and $dr > 16:1$ and of **2d** to **2b** in 97 % yield and $dr > 20:1$ [11]. The fluorinated dioxirane was superior in yield in the formation of **2a** but in all other aspects was inferior to dimethyldioxirane [13].

Synthesis of the Thiazole Fragment A

We begin our discussion of the retrosynthetic fragments with the thiazole moiety A (*cf.* Scheme 1), i.e. the least problematic building block. Indeed, they appear to be so easily accessible that some authors do not mention their synthesis or references, and sometimes even "forget" to count the necessary steps in their final conclusion.

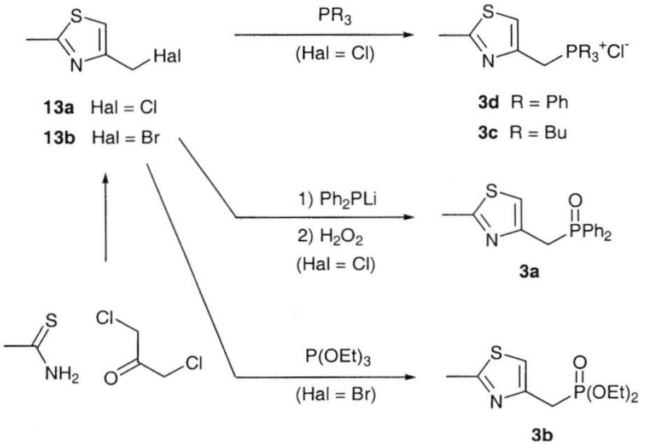

Scheme 4. Syntheses of 2-methyl thiazole-4-methyl phosphorus reagents (fragment A building blocks).

Two different strategies for the incorporation of the thiazole moiety were used so far: either Stille coupling reactions of 4-stannylthiazoles by one group [18, 19] or Wittig-type olefination reactions by all others. The latter approach requires either 2-methyl-thiazole-4-carbaldehyde (**4**, Scheme 1) available from the corresponding ester by reduction [13, 20, 29, 30] or suitable 4-phosphorusmethyl derivatives (**3a–d**, Schemes 1 and 4) for the more common inverse approach.

The phosphorus derivatives **3a**, **3c** and **3d** were synthesized from **13a**, readily available from 1,3-dichloroacetone and thioacetamide followed by dehydration [31]. Mulzer reported the synthesis of Wittig salt **3d** in 85 % overall yield (no reaction conditions given) [32]. Schlosser salt **3c** can be obtained in 87 % yield by heating **13a** and PBu_3 without solvent at 70 °C [23, 33]. Phosphinoxide **3a** was used by Danishefsky; no yields were given [9, 34]. In a different approach, bromide **13b** synthesized from cysteine methyl ester hydrochloride and acetaldehyde [35], gave phosphonate **3b** in 89 % yield with $P(OEt)_3$ in an Arbuzov reaction [36].

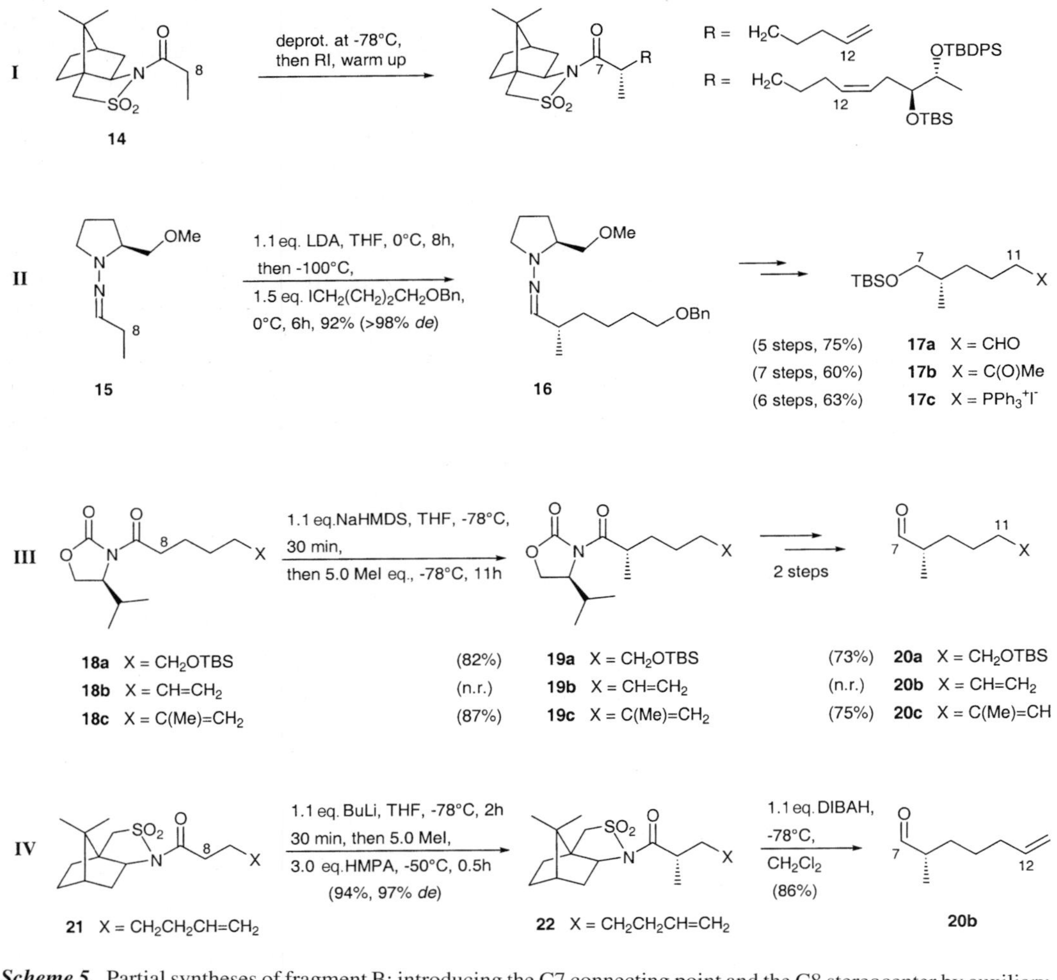

Scheme 5. Partial syntheses of fragment B: introducing the C7 connecting point and the C8 stereocenter by auxiliary induced alkylation (n.r. = no yield reported).

Synthesis of the Northern Half of epothilones (Fragment B) and the Introduction of the Thiazole Side Chain (Fragment A)

The "deoxy-northern half" C7–C15 is barely functionalized apart from the marginal connection points, i.e. the central C8 to C14 area contains only a double bond with imperative *cis* configuration and one stereocenter at C8 (instead of three in the final target). Possible disconnections are much less predetermined than in the C1-C7 "aldol-region". Therefore all six C-C-bonds within the seven-carbon chain from C8 to C14 were used as strategic bonds.

Synthesis of Building Blocks C7-C12 and C13-C21 (Including Side Chain)

The control of the C8 stereocenter was achieved by alkylation directed by an auxiliary at C7 involving Oppolzer sultams, Enders hydrazones and Evans oxazolidinones (Scheme 5: I–III, resp.). Either the corresponding propionate [14, 37] or propionaldehyde [13, 16] equivalents were alkylated with an alkyl iodide representing the principal part of the northern half of epothilones (Scheme 5: I, II), or C-8 of a sui-table chain was methylated diastereoselectively (Scheme 5: III, IV).

While the sultame and the oxazolidinone auxiliaries represent carboxylate equivalents, which have to be reduced (and sometimes re-oxidized) to the required aldehyde function at C7, the strength of Enders SAMP and RAMP auxiliaries is their direct use as aldehyde and ketone equivalents. However, Nicolaou et al. [13, 16] for the synthesis of the protected building blocks **17a–c** had to give up the correct oxidation state in order to allow necessary later manipulations. Considering the necessity of reduction, the cheaper and recoverable Evans-oxazolidinones **18** appear to be the auxiliaries of choice, as demonstrated by Schinzer et al. [21, 22, 36]. A similar methylation is described in an early publication of De Brabander et al. [38] where sultame **21** was methylated and reduced to the α-methylaldehyde **20b** in only two steps in good yield and enantiomeric excess.

The thiazole-substituted homoallylic alcohol **25** (Scheme 6) is a key intermediate, not only for RCM strategies, but also for other routes. Thiazole aldehyde **4** (Chapter 3) after homologation to enal **24** (90 % yield) [11, 20] was subjected to asymmetric allylation with allylboron and tin reagents. Interestingly **25** with identical absolute stereochemistry was synthesized by Nicolaou et al. with (+)-Ipc_2B(allyl) in 96 % yield and > 97 % *ee* [13, 20], and by Danishefsky et al. with the enantiomeric (−)-Ipc_2B(allyl) in 83 % yield and > 95 % *ee* [11], i.e. in one case an er-

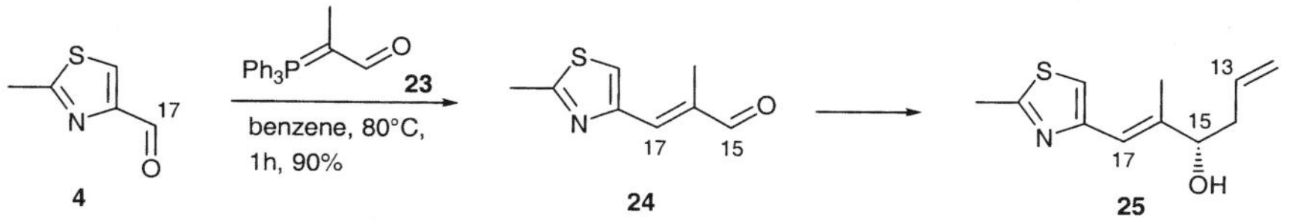

4 —(Ph_3P=C(Me)CHO **23**, benzene, 80°C, 1h, 90%)→ **24** → **25**

Scheme 6. Synthesis of a C13-C21 building block (a partial A-B fragment).

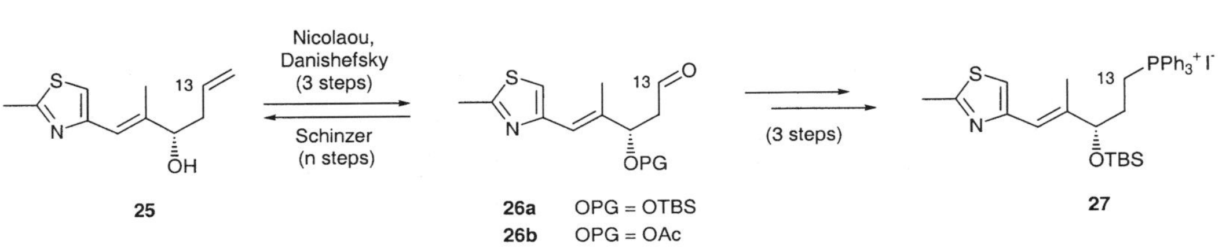

25 —(Nicolaou, Danishefsky (3 steps))→ **26a** OPG = OTBS, **26b** OPG = OAc —(3 steps)→ **27**; **26** —(Schinzer (n steps))→ **25**

Scheme 7. Switching the C13-functionalization of C13-C21 building blocks.

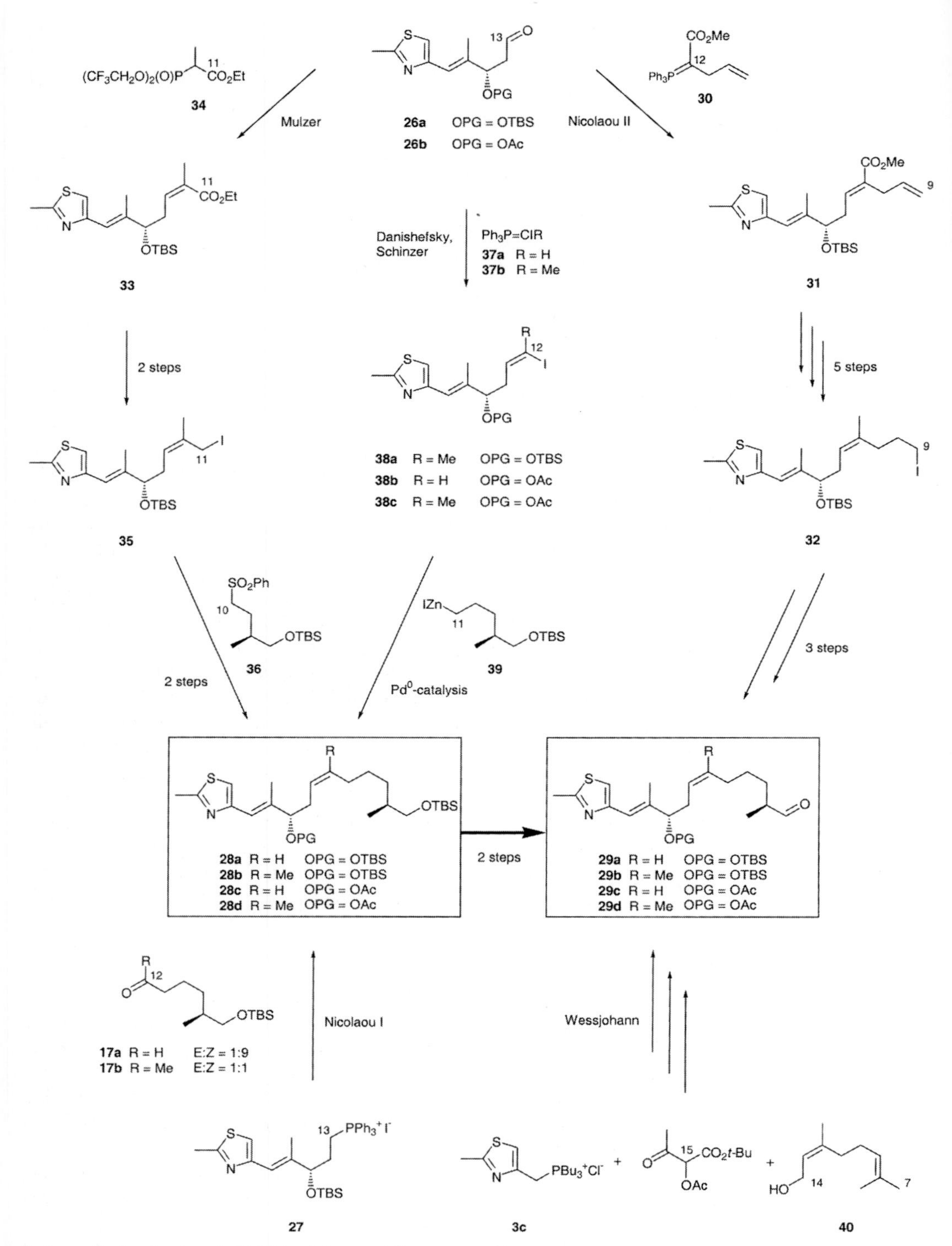

Scheme 8. Syntheses of C7-C21 of the epothilones according to Schinzer (& Danishefsky), Nicolaou II, Wessjohann, Nicolaou I and Mulzer (clockwise from 12 : 00).

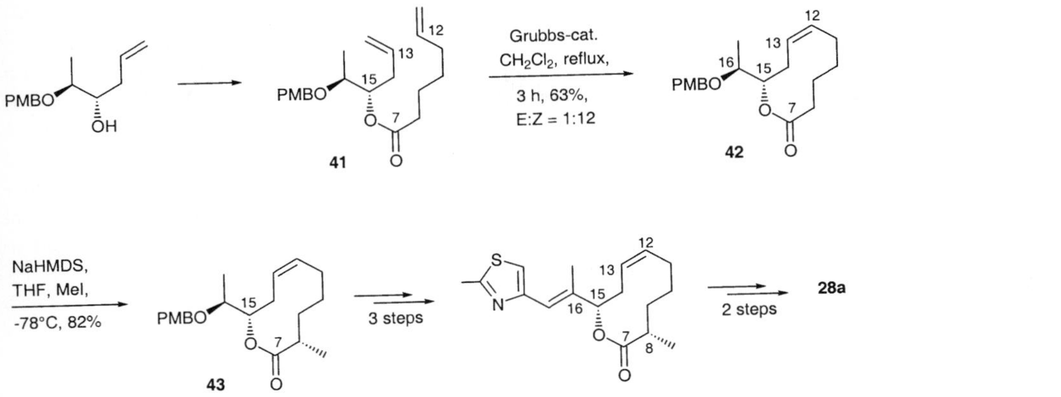

Scheme 9. Diastereoselective ring-closing metathesis and C8 methylation according to Kalesse et al. to synthesize C7-C21 (fragment A-B) (PMB = *p*-methoxybenzyl).

roneous configuration was published. The catalytic asymmetric allylation with tri-*n*-butylallylstannane was achieved in the presence of 10 mol% of a Lewis acid catalyst formed in situ from $Ti(O\textit{i}Pr)_4$ and (*S*)-BINOL in CH_2Cl_2 at −20 °C, a method originally described by Keck. This procedure was independently used by Taylor et al. (62 %, > 86 % *ee*) [30] and Danishefsky et al. (> 95 % *ee*) [11].

Compound **25**, if not used in the RCM strategy, can be converted to aldehydes **26** by protection of the alcohol group and oxidative double bond cleavage (Scheme 7) [11, 13]. It is interesting to note that Schinzer et al. go the opposite way and synthesize alkene **25** from this aldehyde **26a** obtained via a different route [olefination of protected (2*S*)-2-hydroxybutyrolactone with $ArCH_2P(O)(OEt)_2$ (**3b**) [21, 22] or $ArCH_2PBu_3^+Cl^-$ (**3c**) [23]]. Nicolaou then transformed aldehyde **26b** to the phosphonium salt **27** in three steps.

Syntheses of Complete A-B Fragments (Northern Half and Side Chain)

An overview of the most important routes to complete A-B fragments from the intermediates discussed in the previous chapter is given in Scheme 8.

Nicolaou (I, Scheme 8: lower left corner) coupled the phosphoniumiodide **27** with the carbonyl compounds **17a** and **17b**, bringing the synthesis of northern half building blocks **28a** and **28b** to completion after cleavage of the C7-TBS group and oxidation to the aldehydes **29a** and **29b** [13]. Unfortunately, while the double bond in **28a** is formed in high *Z* selectivity, **28b** is formed without diastereoselection. The inverse combination of phosphonium iodide **17c** (*cf.* Scheme 5) and aldehyde **26a** lead to **28a** with almost identical results. The problem of *E*:*Z* selectivity in **28b** was overcome when aldehyde **26a** was olefinated with the stabilized ylide **30**, giving **31** in 95 % yield exclusively with the desired configuration (Scheme 8: upper right) [13]. However, now the methyl ester had to be transformed to the C12 methylgroup with an additional three steps. Further transformations afforded iodide **32** which can be used for asymmetric alkylation with a C_3 building block according to Scheme 5 I–II [13].

Mulzer (Scheme 8: upper left) obtained the α,β-unsaturated ester **33** with *Z* configuration from aldehyde **26a** via a Still-Gennari olefination with phosphonate ester **34**. Reduction of the ester with DIBAH and application of I_2-imidazole-PPh_3 gives allylic iodide **35**. This acts as electrophile on the α-anion of sulfone **36**. After reductive removal of the phenylsulfone, group **28b** is obtained [23].

Danishefsky et al. (Scheme 8: upper center) were able to extend aldehyde **26b** using ylides $Ph_3P=CHI$ (**37a**) and $Ph_3P=C(Me)I$ (**37b**) and provide vinylic iodides **38b** and **38c**, which were then connected to a C3-C11 borane (not shown) via Pd^0-catalyzed Suzuki coupling to give precursors of **11a** and **11b** (*cf.* Scheme 2) [11, 12]. The same reaction sequenze was used later by Schinzer et al., synthesizing iodoolefin **38a** with phosphorane **37b** followed by Pd^0-coupling with zinc organyl **39** to building block **28b** [22].

Several other publications offered alternative solutions for northern half building blocks. Selected examples are briefly discussed. Wessjohann et al. (Scheme 8: lower right) [39–41] use the cheap C_{10} alcohol nerol (**40**) as a C7-C14 synthon with the correct branching and correctly configured double bond already in place, thus avoiding C12-C13 *E/Z* problems and non-functionalized carbon-carbon bond formations altogether. In another creative strategy, in order to circumvent the C12-C13 *E/Z* diastereoselectivity problem, Kalesse et al. (Scheme 9) [42, 43], use the RCM of **41** to obtain a strained ten-membered ring, ensuring the excess formation of *Z* olefin **42** under thermodynamic conditions. It was shown, that the reaction conditions also allow the isomerisation of pure *E* olefin to the thermodynamic ratio of E:Z = 1 : 12. The methylation of lactone **42** at C8 is noteworthy also, because it luckily gives 82 % **43** with perfect control of the stereochemistry. Overall, the new chiral centers at C8 and C15 are introduced via the starting material (*S*)-ethyllactate without the need of additional chiral auxiliaries. Further transformations led to the known intermediate **28a**.

Synthesis of the Aldol Fragments C and Aldol Reactions between C6 and C7

The syntheses of the C1–C6 aldol fragment (C) and similar building blocks with defined stereochemistry at C3 reveal several general problems. On one hand, it is difficult to transfer an acetate unit to an aldehyde with good β-induction, this usually is referred to as the "acetate-problem" [44]. On the other hand the C1-C5

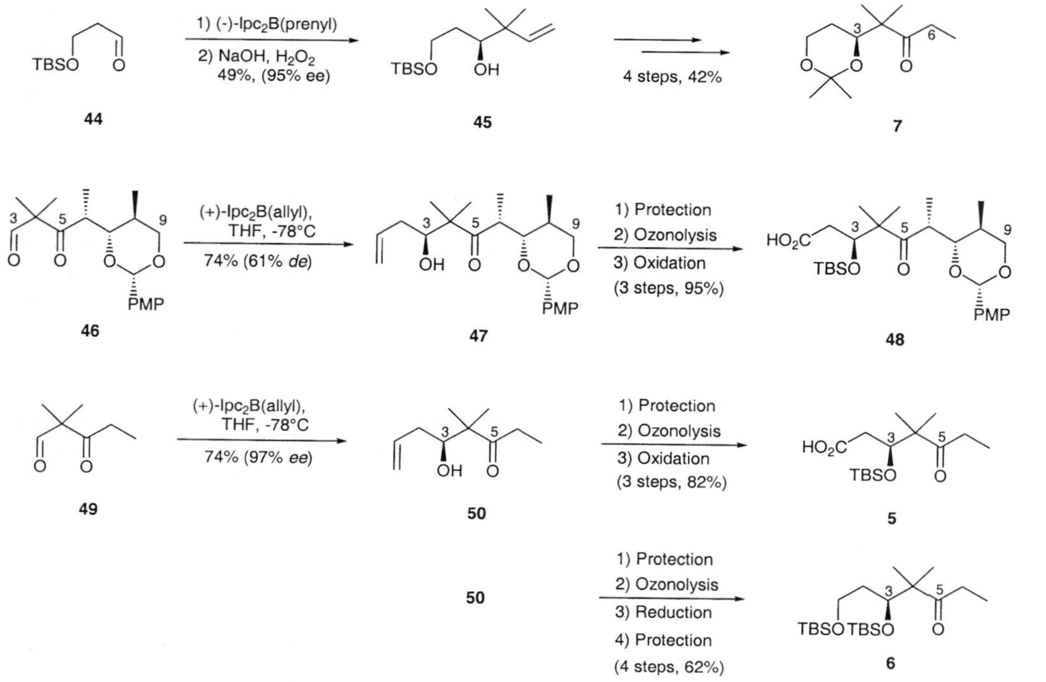

Scheme 10. The allylation/C=C-degradation strategy (Ipc = isopinocampheyl).

part of epothilones is prone to retroaldolization, especially between C3 and C4 enhanced by the geminal C4 dimethyl group, rendering the bond very labile under basic conditions. A somewhat minor problem is the chemoselective addition of carbon nucleophiles to one of two carbonyl groups, e.g. in β-ketoaldehyde **49**. In view of these facts it is not astonishing, that the first syntheses avoided the obvious aldol approach to build up the C3 stereocenter, but instead opted for indirect multistep reaction sequences.

The first solutions to this problem were devised by Schinzer et al. [36] and Mulzer et al. [45] with allylic isopinocamphenylborane reagents; later Nicolaou et al. [13, 20, 46] also followed this strategy (Scheme 10). In Schinzers approach, (−)-Ipc_2B(prenyl) as C4–C5 substitute is added to aldehyde **44**, obtained in two steps from 1,3-

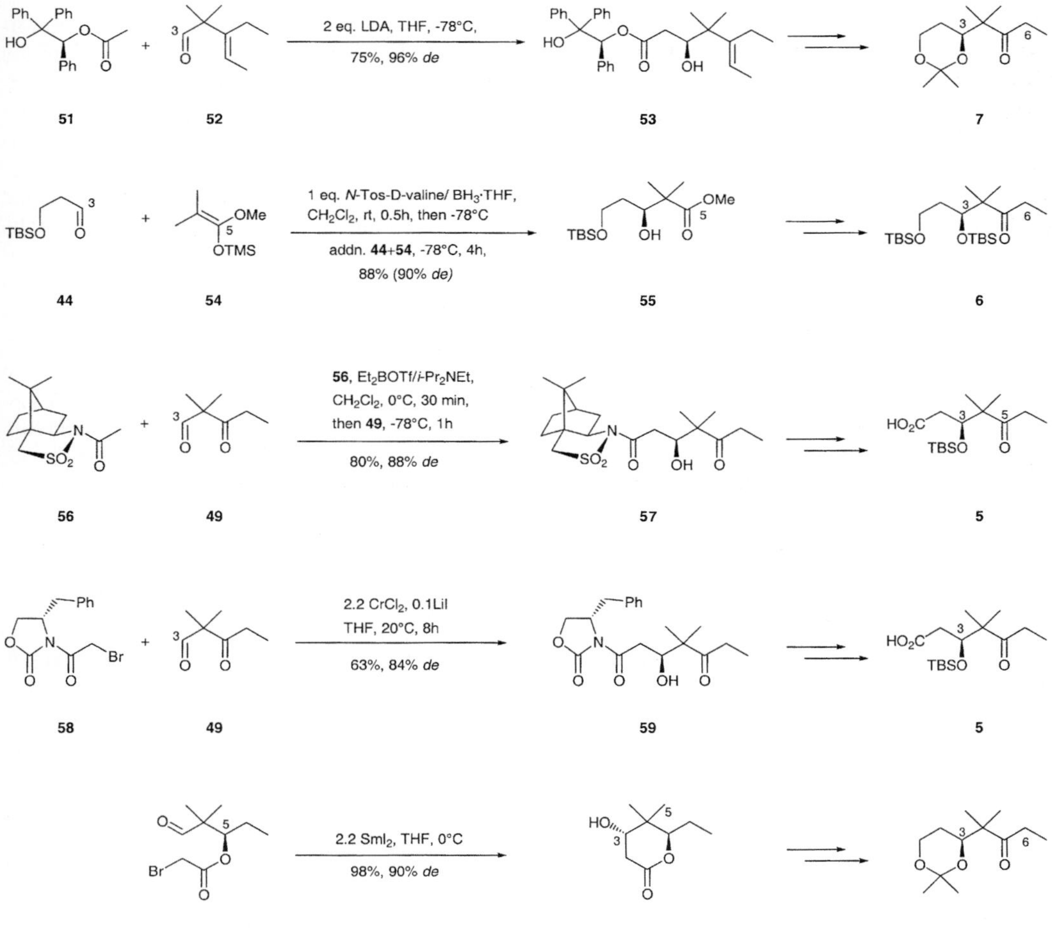

Scheme 11. Aldol-type strategies to the C1-C6 fragment C.

propandiol in 74 % yield. Further manipulations, including protection, oxidative cleavage of the double bond to the aldehyde, Grignard addition of EtMgBr and TPAP-oxidation of the resulting alcohol to the ketone, lead to the protected 1,3-diol **7** in 42 % (15 % overall) yield.

The other authors used (+)-Ipc_2B(allyl) as a C1-C2 building block. Mulzer applied it on the advanced C3-C9 building block **46** in 74 % yield, but with only 61 % *de*, whereas addition to *β*-ketoaldehyde **49** gave Nicolaou homoallyl alcohol **50** with 97 % *ee* and the same yield, and further on ketoacid **5** (Scheme 10). Mulzers approach obviously had to face negative double stereodifferentiation. Because Schinzers building block **7** showed better diastereoselectivities in the later C6-C7 connection, Nicolaou synthesized the similar C1-hydroxy derivative **6** from **50** in four steps instead of acid **5** [13]. Unfortunately, all allylation approaches required lengthy redox and protection procedures.

In a more recent approach (Scheme 11), Schinzer solved the problem of the C4–C5 retro-aldol reaction with Braun's (*S*)-HYTRA (**51**) [44] by replacing the keto group in *β*-ketoaldehyde **49** with a C=C double bond (*cf.* **52**, derived in four steps from ethyl-2-bromo-*iso*-butyrate and 3-pentanone in 13 % overall yield). The thus formed intermediate **53** is later deprotected and cleaved oxidatively to give the desired C5 ketone **7** in 52 % yield and 96 % *ee* from aldehyde **52** [22].

Also Mulzer applied a new approach, a stereoselective Mukaiyama-type aldol reaction of methyl trimethylsilyl dimethylketene acetal **54** to the known aldehyde **44** mediated by a chiral borane reagent formed in situ from *N*-tosyl-D-valine and $BH_3 \times THF$ [23]. Noteworthy is the fact that the conversion of the ester intermediate **55** to ethylketone **6** was not performed by following the common Weinreb procedure, but made use of the Peterson reagent $TMSCH_2Li$ as a sterically protected methyllithium equivalent to avoid double addition to the ester. The TMS group was then "substituted" by a methylgroup to yield the ethylketone (*cf.* **6**).

Direct aldol reactions on building block **49** appear to be much more straightforward, but are problematic because of the retroaldol problem. Nevertheless, two successful non-basic methods were discovered (Scheme 11). De Brabander et al. [38] used the boron enolate from Oppolzer sultam **56**. Wessjohann et al. [41] favored the readily available 2-bromoacylated Evans oxazolidinones (in two ways) in their newly developed chromium(II)-mediated Reformatsky reaction, and could e.g. add an acetate-enolate generated from **58** to aldehyde **49** in a one-pot reaction (Scheme 11). The intermediates **57** and **59** can be converted to carboxylic acid **5** in almost identical overall yield and *ee*.

An intramolecular diastereoselective Reformatsky-type aldol approach was demonstrated by Taylor et al. [47] with an Sm(II)-mediated cyclization of the chiral bromoacetate **60**, resulting in lactone **61**, also an intermediate in the synthesis of Schinzer's building block **7**. The alcohol oxidation state at C5 in **61** avoided retro-reaction and at the same time was used for induction, with the absolute stereochemistry originating from enzymatic resolution (Scheme 11). Direct resolution of racemic C3 alcohol was also tried with an esterase adapted by directed evolution [48]. In other, somewhat more lengthy routes to C1-C6 building blocks, Shibasaki et al. used a catalytic asymmetric aldol reaction with heterobimetallic asymmetric catalysts [49], and Kalesse et al. used a Sharpless asymmetric epoxidation [50].

Putting together Fragments (A-)B and C: The C6-C7-Aldol Reaction

With C1-C6 aldol fragments C and fragments B (C7-Cn) accessible, the strategic aldol reaction between C6 and C7 (Scheme 12), with the demand to form two stereocenters in the right way, is the final major task prior to the already discussed macrocyclization. Stereocenters C6 and C7 are in *syn* relationship (Please note that in Scheme 12 the main chain is twisted between these atoms, thereby pointing *syn* substituents in different directions). Therefore a *Z*-enolate of the ethylketone is required, and on the electrophile component an *anti*-Cram approach regarding the influence of C8 on C7. While control of enolate geometry is often possible, little can affect the adjustment of the relative stereochemistry between C7–C8 apart from trying different substrates.

The first C6-C7-aldol reactions were reported by Nicolaou's group with the dianion of keto-carboxylic acid **5** (Scheme 12). Aldehydes **29a**, **29b** and **20b** give the desired aldols **62**, **63** and **66** in "high yields" (?) at $-78\,^{\circ}\mathrm{C}$. As expected, excellent control of the enolate configuration had taken place to result in a perfect C6-C7 *syn* relationship. Nevertheless, there was no or minimal *anti*-Cram selectivity, up to a mere 2 : 1 maximum [13, 16, 20].

The main improvement was achieved by Schinzer et al. with acetonide **7**. The desired aldol products **65**, **67** and **68** were obtained in high yield and perfect formation of the 6*R*,7*S* stereochemistry, with ratios generally better than 9 : 1 for the correct C7-C8 stereochemistry by means

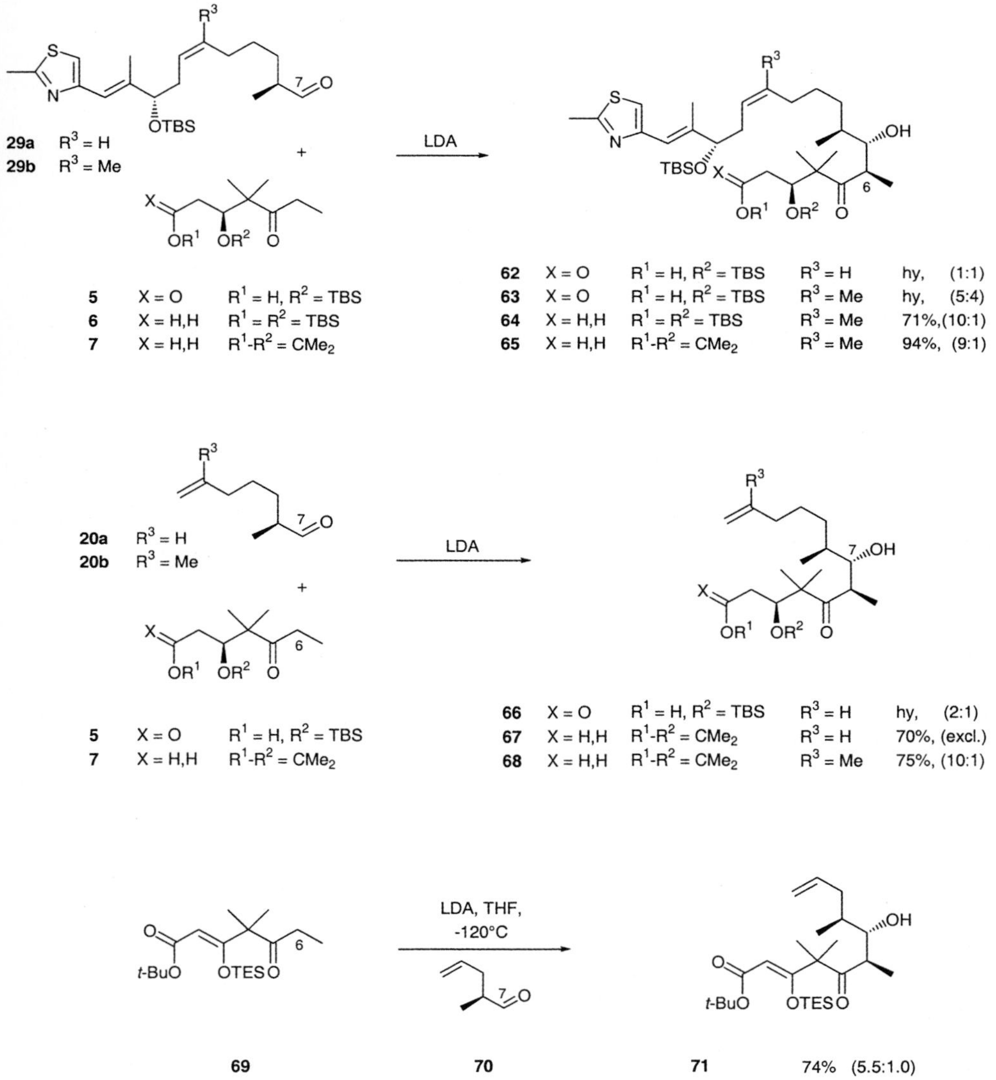

29a R^3 = H
29b R^3 = Me

LDA

5	X = O	R^1 = H, R^2 = TBS	
6	X = H,H	R^1 = R^2 = TBS	
7	X = H,H	R^1-R^2 = CMe_2	

62	X = O	R^1 = H, R^2 = TBS	R^3 = H	hy, (1:1)
63	X = O	R^1 = H, R^2 = TBS	R^3 = Me	hy, (5:4)
64	X = H,H	R^1 = R^2 = TBS	R^3 = Me	71%,(10:1)
65	X = H,H	R^1-R^2 = CMe_2	R^3 = Me	94%, (9:1)

20a R^3 = H
20b R^3 = Me

LDA

5	X = O	R^1 = H, R^2 = TBS
7	X = H,H	R^1-R^2 = CMe_2

66	X = O	R^1 = H, R^2 = TBS	R^3 = H	hy, (2:1)
67	X = H,H	R^1-R^2 = CMe_2	R^3 = H	70%, (excl.)
68	X = H,H	R^1-R^2 = CMe_2	R^3 = Me	75%, (10:1)

LDA, THF, -120°C

69 **70** **71** 74% (5.5:1.0)

Scheme 12. The C6-C7 aldol reaction has to proceed with double stereodifferentiation. The ratios refer to the *anti-syn* diastereoselectivity at the C7–C8 bond, of which the main component, the natural *6-syn, 7-anti* compound is shown (TES = triethylsilyl, hy = high yield).

of double stereodifferentiation [21, 22]. Following these convincing results, Nicolaou et al. also used doubly protected dialcohol **6** and thus improved the *anti*-Cram selectivity with aldehyde **29b**, first to a ratio of 3 : 1 and and later to 10 : 1 with optimized reaction conditions [13, 51]. Mulzer used the same disilylated diol **6** to obtain aldol **64** in 69 % yield with 4 : 1 selectivity [23].

An unusual enolate of the 3-triethylsilyl-protected 1,3,5-tricarbonyl compound **69** was applied to aldehyde **70** by Danishefsky et al., forming aldol **71** in 74 % yield and with a 5.5 : 1.0 ratio – remarkable considering that in this case no double stereodifferentiation improves the induction [10, 52]. A systematic study with different aldehydes revealed that an interaction between the double bond and the carbonyl group of the aldehyde is superior to minimization of steric hindrance in the transition state, thus leading to the desired C7-C8 *anti* relationship [53]. Later in the synthesis of epothilone B, in Danishefsky's approach, the triethylsilyl group was removed and the C3 ketone converted to the desired C3 alcohol by enantioselective catalytic Noyori reduction [10].

Compounds **62**–**65** by selective deprotection could be processed to macrolactonization precursors **10**; whereas compounds **66**–**68** needed attachment of (protected) **25** either by metathesis reaction (C12-C13) or esterification (C1-O-C15), prior to macrolactonization or RCM respectively (*cf.* Scheme 2).

Derivatives of Epothilones: Syntheses and Structure-Activity Relationship

With the established synthetic routes in hand, many of the more obvious derivatives have been synthesized in order to obtain a quantitative structure-activity relationship (QSAR) and to maybe enhance the activitiy and pharmacological profile of epothilones. These derivatives also include epothilone E (**2e**, Scheme 13), which is emphasized because it also is an active natural derivative, hydroxylated at C21.

Synthesis of Aromatic Derivatives of Epothilones, Including Epothilone E

The group of G. Höfle, the discoverer of the epothilones, was able to isolate sufficient epothilones from fermentations to examine classic derivatization. They examined the products of "Oxidative and reductive transformations of epothilone A" [54], "Derivatizations of the C12–C13 functional groups of epothilones A, B and C" [55] as well as "Substitutions at the thiazole moiety of epothilone" [56]. They were among the first to discover that the epoxide can be formed with predominantly correct diastereoselectivity in a final transformation from (protected) epothilones C and D. Recently a more thorough investigation of the epoxidation with *m*CPBA to the epothilones A and B was finally

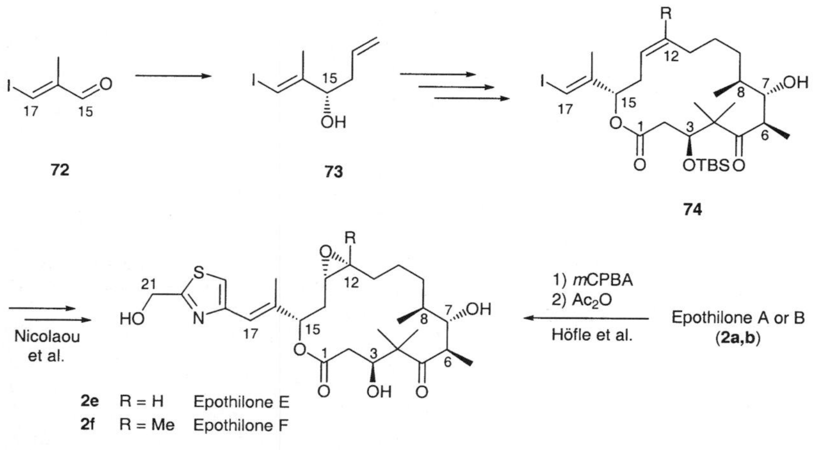

Scheme 13. Total and partial synthesis routes to epothilone E (**2e**).

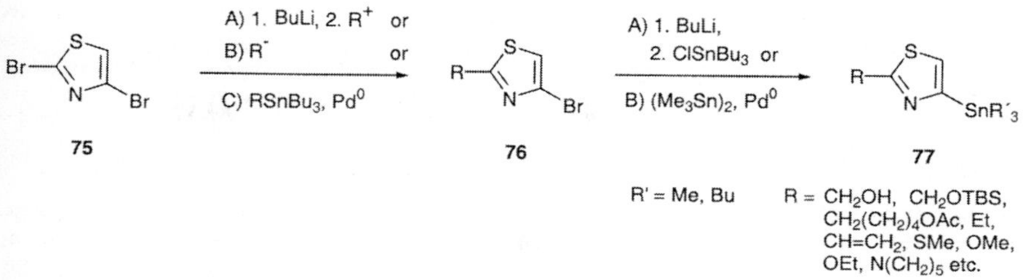

A) 1. BuLi, 2. R^+ or
B) R^- or
C) $RSnBu_3$, Pd^0

A) 1. BuLi,
2. $ClSnBu_3$ or
B) $(Me_3Sn)_2$, Pd^0

75 76 77

R' = Me, Bu R = CH_2OH, CH_2OTBS, $CH_2(CH_2)_4OAc$, Et, $CH{=}CH_2$, SMe, OMe, OEt, $N(CH_2)_5$ etc.

Scheme 14. Synthesis of substituted thiazole building blocks for QSAR studies.

published [26]. It was shown that at low temperature, the desired epothilones A and B were obtained (see Chapter 3), while at room temperature oxidation of the thiazole nitrogen to the corresponding *N*-oxide also occurs, a reaction not reported in the early total syntheses using this reagent. The epothilone A and B *N*-oxides could be rearranged to epothilone E (**2e**) and F with activated carboxylic acids such as acetic anhydride (Scheme 13).

In the total synthesis of epothilone E and other C20-derivatives, substituted thiazole (and other aromatics) have been coupled according to Stille's method to the macrocyclic vinyl iodide **74** obtained with selective RCM, leaving the vinyl iodide unaffected (Scheme 13) [18, 19]. The 2-substituted 4-trialkylstannyl-thiazoles **77** were synthesized from 2,4-dibromothiazole **75** in five steps, as shown in Scheme 14.

Structure-Activity Relationships

Structure-activity relationships of published epothilone derivatives have been included in the review article and publications of Nicolaou et al. [57, 58] and Danishefsky et al. [2, 59, 60].

The most active compound published (!) so far remains epothilone B. All derivatives show decreased activity, but areas of high, medium or low impact upon modification can be assigned to the molecule. These interestingly follow somewhat the common retrosynthetic separation, being C1-C8, C9-C15, and the side chain respectively [59]. The only variation accepted well in the aldol region is a C2-C3 *trans* double bond. Removing or adding methylene groups between C9 and C11 in order to decrease or increase the ring size also results in significant loss of activity [59, 61]. The C12-C13 *E*-configured desoxy-compounds have decreased biological activity, as have C15 stereo inverted ones. The side chain requires an olefinic spacer between macrocyclic and aromatic ring. As at C12, also at C16 a methyl group is optimal. An oxazole ring is tolerated, as well as substitutions at C20 (*cf.* epothilone E), if they are sterically not too demanding.

With the data published, it should be kept in mind that comparability of results from different sources may greatly deviate depending on the test used. Simple microtubule assays do not necessarily give relevant data for selective cytotoxicity or induction of apoptosis, as these two properties are not always coupled sufficiently. Also the significance of tests in vitro, even on cell lines, sometimes bears little significance for in vivo application. Thus it was reported that epothilone D (**2d**) despite a lower activity, is a much better drug candidate because of its greatly reduced toxic side effects, which therefore have to be attributed to the oxirane-ring [60]. Also, while studying publications at this stage of development, one should keep in mind that the most interesting derivatives are probably not yet published, even if known, owing to the looming commercial potential for a billion dollar drug.

References

[1] D. M. Bollag, P. A. McQueney, J. Yhu, O. Hensens, L. Koupal, J. Liesch, M. Goetz, E. Lazarides, C. M. Woods, *Cancer Res.* **1995**, *55*, 2325–2333.
[2] T. C. Chou, X. G. Zhang, C. R. Harris, S. D. Kuduk, A. Balog, K. A. Savin, J. R. Bertino, S. J. Danishefsky, *Proc. Natl. Acad. Sci. USA* **1998**, *95*, 15798–15802.

[3] L. Wessjohann, *Angew. Chem.* **1994**, *106,* 1011–1013; *Angew. Chem. Int. Ed. Engl.* **1994**, *33,* 959–961.
[4] L. Wessjohann, *Angew. Chem.* **1997**, *109,* 739–742; *Angew. Chem. Int. Ed. Engl.* **1997**, *36,* 738–742.
[5] I. Ojima, S. Chakravarty, T. Inoue, S. Lin, L. He, S. B. Horwitz, S. D. Kuduk, S. J. Danishefsky, *Proc. Natl. Acad. Sci. USA* **1999**, *96,* 4256–4261.
[6] M. Wang, X. Xia, Y. Kim, D. Hwang, J. M. Jansen, M. Botta, D. C. Liotta, J. P. Snyder, *Org. Lett.* **1999**, *1,* 43–46.
[7] G. Höfle, N. Bedorf, H. Steinmetz, D. Schomburg, K. Gerth, H. Reichenbach, *Angew. Chem.* **1996**, *108,* 1971–1673; *Angew. Chem. Int. Ed. Engl.* **1996**, *35,* 1567–1569.
[8] K. Gerth, N. Bedorf, G. Höfle, H. Irschik, H. Reichenbach, *J. Antibiot.* **1996**, *49,* 560–563.
[9] A. Balog, D. Meng, T. Kamenecka, P. Bertinato, D.-S. Su, E. J. Sorensen, S. J. Danishefsky, *Angew. Chem.* **1996**, *108,* 2976–2978; *Angew. Chem. Int. Ed. Engl.* **1996**, *35,* 2801–2803.
[10] A. Balog, C. Harris, K. Savin, X. G. Zhang, T. C. Chou, S. J. Danishefsky, *Angew. Chem.* **1998**, *110,* 2821–2824; *Angew. Chem. Int. Ed. Engl.* **1998**, *37,* 2675–2678.
[11] D. Meng, P. Bertinato, A. Balog, D.-S. Su, T. Kamenecka, E. J. Sorensen, S. J. Danishefsky, *J. Am. Chem. Soc.* **1997**, *119,* 10073–10092.
[12] D.-S. Su, D. Meng, P. Bertinato, A. Balog, E. J. Sorensen, S. J. Danishefsky, Y.-H. Zheng, T.-C. Chou, L. He, S. B. Horwitz, *Angew. Chem.* **1997**, *109,* 775–777; *Angew. Chem. Int. Ed. Engl.* **1997**, *36,* 757–759.
[13] K. C. Nicolaou, S. Ninkovic, F. Sarabia, D. Vourloumis, Y. He, H. Vallberg, M. R. V. Finlay, Y. Yang, *J. Am. Chem. Soc.* **1997**, *119,* 7974–7991.
[14] K. C. Nicolaou, Y. He, D. Vourloumis, H. Vallberg, F. Roschangar, F. Sarabia, S. Ninkovic, Z. Yang, J. I. Trujillo, *J. Am. Chem. Soc.* **1997**, *119,* 7960–7973.
[15] K. C. Nicolaou, N. Winssinger, J. Pastor, S. Ninkovic, F. Sarabia, Y. He, D. Vourloumis, Z. Yang, T. Li, P. Giannakakou, E. Hamel, *Nature (London)* **1997**, *387,* 268–272.
[16] K. C. Nicolaou, F. Sarabia, S. Ninkovic, Z. Yang, *Angew. Chem.* **1997**, *109,* 539–540; *Angew. Chem. Int. Ed. Engl.* **1997**, *36,* 525–527.
[17] K. C. Nicolaou, N. Winssinger, J. Pastor, S. Ninkovic, F. Sarabia, Y. He, D. Vourloumis, Z. Yang, T. Li, P. Giannakakou, E. Hamel, *Nature (London)* **1997**, *390,* 100.
[18] K. C. Nicolaou, Y. He, F. Roschangar, N. P. King, D. Vourloumis, T. Li, *Angew. Chem.* **1998**, *110,* 89–92; *Angew. Chem. Int. Ed. Engl.* **1998**, *37,* 84–87.
[19] K. C. Nicolaou, N. P. King, M. R. V. Finlay, Y. He, F. Roschangar, D. Vourloumis, H. Vallberg, F. Sarabia, S. Ninkovic, D. Hepworth, *Bioorg. Med. Chem.* **1999**, *7,* 665–697.
[20] Z. Yang, Y. He, D. Vourloumis, H. Vallberg, K. C. Nicolaou, *Angew. Chem.* **1997**, *109,* 170–172; *Angew. Chem. Int. Ed. Engl.* **1997**, *36,* 166–168.
[21] D. Schinzer, A. Limberg, A. Bauer, O. M. Böhm, M. Cordes, *Angew. Chem.* **1997**, *109,* 543–544; *Angew. Chem. Int. Ed. Engl.* **1997**, *36,* 523–524.
[22] D. Schinzer, A. Bauer, J. Schieber, *Synlett* **1998**, 861–863.
[23] J. Mulzer, A. Mantoulidis, E. Öhler, *Tetrahedron Lett.* **1998**, *39,* 8633–8636.
[24] J. D. White, R. G. Carter, S. K.F., *J. Org. Chem.* **1999**, *64,* 684–685.
[25] S. A. May, P. A. Grieco, *J. Chem. Soc., Chem. Commun.* **1998**, 1597–1598.
[26] G. Höfle, N. Glaser, M. Kiffe, H.-J. Hecht, F. Sasse, H. Reichenbach, *Angew. Chem.* **1999**, *111,* 2090–2093.
[27] D. Meng, D.-S. Su, A. Balog, P. Bertinato, E. J. Sorensen, S. J. Danishefsky, Y.-H. Zheng, T.-C. Chou, L. He, S. B. Horwitz, *J. Am. Chem. Soc.* **1997**, *119,* 2733–2734.
[28] A. Balog, P. Bertinato, D.-S. Su, D. Meng, E. J. Sorensen, S. J. Danishefsky, Y.-H. Zheng, T.-C. Chou, L. He, S. B. Horwitz, *Tetrahedron Lett.* **1997**, *38,* 4529–4532.
[29] M. W. Bredenkamp, C. W. Holzapfel, W. J. van Zyl, *Synth. Commun.* **1990**, *20,* 2235–2249.
[30] R. E. Taylor, J. D. Haley, *Tetrahedron Lett.* **1997**, *38,* 2061–2064.
[31] G. Marzoni, *J. Heterocyclic. Chem.* **1986**, *23,* 577–580.
[32] J. Mulzer, A. Mantoulidis, E. Öhler, *Tetrahedron Lett.* **1997**, *38,* 7725–7728.
[33] Z.-Y. Liu, C.-Z. Yu, R.-F. Wang, G. Li, *Tetrahedron Lett.* **1998**, *39,* 5261–5264.
[34] D. Meng, E. J. Sorensen, P. Bertinato, S. J. Danishefsky, *J. Org. Chem.* **1996**, *61,* 7998–7999.
[35] S. Mensching, M. Kalesse, *J. Prakt. Chem.* **1997**, *339,* 96–97.
[36] D. Schinzer, A. Limberg, O. M. Böhm, *Chem. Eur. J.* **1996**, *2,* 1477–1482.
[37] P. Bijoy, M. A. Avery, *Tetrahedron Lett.* **1998**, *39,* 209–212.
[38] J. De Brabander, S. Sosset, G. Bernardinelli, *Synlett* **1997**, 824–826; *ibid.* **1998**, 328; *ibid.* **1998**, 692.
[39] G. Scheid, L. A. Wessjohann, U. Bornscheuer, *Manuscript in preparation* **2000**.
[40] L. A. Wessjohann, M. Kalesse (Anmelder: Wessjohann, L.), Deutsche Offenlegungsschrift **1998**, *DE 19713970.1.*
[41] T. Gabriel, L. A. Wessjohann, *Tetrahedron Lett.* **1997**, *38,* 1363–1366.

[42] K. Gerlach, M. Quitschalle, M. Kalesse, *Tetrahedron Lett.* **1999**, *40,* 3553–3556.
[43] K. Gerlach, M. Quitschalle, M. Kalesse, *Synlett* **1998**, 1108–1109.
[44] M. Braun, S. Gräf, *Org. Synth.* **1993**, *72,* 38–47.
[45] J. Mulzer, A. Mantoulidis, *Tetrahedron Lett.* **1996**, *37,* 9179–9182.
[46] K. C. Nicolaou, H. Vallberg, N. P. King, F. Roschangar, Y. He, D. Vourloumis, C. G. Nicolaou, *Chem. Eur. J.* **1997**, *3,* 1957–1970.
[47] R. E. Taylor, M. G. Galvin, K. A. Hilfiker, Y. Chen, *J. Org. Chem.* **1998**, *63,* 9580–9583.
[48] U. T. Bornscheuer, A. J., H. H. Meyer, *Biotechnology & Bioengineering* **1998**, *58,* 554–559.
[49] N. Yoshikawa, Y. M. A. Yamada, J. Das, H. Sasai, M. Shibasaki, *J. Am. Chem. Soc.* **1999**, *121,* 4168–4178.
[50] E. Claus, A. Pahl, P. G. Jones, H. M. Meyer, M. Kalesse, *Tetrahedron Lett.* **1997**, *38,* 1359–1362.
[51] K. C. Nicolaou, D. Hepworth, M. R. V. Finlay, N. P. King, B. Werschkun, A. Bigot, *J. Chem. Soc., Chem. Commun.* **1999**, 519–520.
[52] C. R. Harris, S. D. Kuduk, K. Savin, A. Balog, S. J. Danishefsky, *Tetrahedron Lett.* **1999**, *40,* 2263–2266.
[53] C. R. Harris, S. D. Kuduk, A. Balog, K. A. Savin, S. J. Danishefsky, *Tetrahedron Lett.* **1999**, *40,* 2267–2270.
[54] M. Sefkow, M. Kiffe, D. Schummer, G. Höfle, *Bioorg. Med. Chem. Lett.* **1998**, *8,* 3025–3030.
[55] M. Sefkow, M. Kiffe, G. Höfle, *Bioorg. Med. Chem. Lett.* **1998**, *8,* 3031–3036.
[56] G. Höfle, M. Sefkow, *Heterocycles* **1998**, *48,* 2485–2488.
[57] K. C. Nicolaou, F. Roschangar, D. Vourloumis, *Angew. Chem.* **1998**, *110,* 2120–2153; *Angew. Chem. Int. Ed. Engl.* **1998**, *37,* 2015–2045.
[58] K. C. Nicolaou, D. Vourloumis, T. Li, J. Pastor, N. Winssinger, Y. He, S. Ninkovic, F. Sarabia, H. Vallberg, F. Roschangar, N. P. King, M. R. V. Finlay, P. Giannakakou, P. Verdier-Pinard, E. Hamel, *Angew. Chem.* **1997**, *109,* 2181–2187; *Angew. Chem. Int. Ed. Engl.* **1997**, *36,* 2097–2103.
[59] D.-S. Su, A. Balog, D. Meng, P. Bertinato, S. J. Danishefsky, Y.-H. Zheng, T.-C. Chou, L. He, S. B. Horwitz, *Angew. Chem.* **1997**, *109,* 2178–2187; *Angew. Chem. Int. Ed. Engl.* **1997**, *36,* 2093–2096.
[60] T. C. Chou, X. G. Yhang, A. Balog, D. S. Su, D. F. Meng, K. Savin, J. R. Bertino, S. J. Danishefsky, *Proc. Natl. Acad. Sci. USA* **1998**, *95,* 9642–9647.
[61] K. C. Nicolaou, F. Sarabia, S. Ninkovic, M. R. V. Finlay, C. N. C. Boddy, *Angew. Chem.* **1998**, *110,* 85–89; *Angew. Chem. Int. Ed. Engl.* **1998**, *37,* 81–84.

Total Syntheses of the Marine Natural Product Eleutherobin

Thomas Lindel

Pharmazeutisch-chemisches Institut, Universität Heidelberg, Germany

In 1994, the soft coral *Eleutherobia* sp. was discovered as the source of a marine natural product with outstanding biological activity. In a bioassay-guided fractionation of its extract, Fenical et al. (Scripps Institution of Oceanography, San Diego, USA) were able to attribute the cytotoxic activity to the glycosylated diterpenoid eleutherobin (**1**, Fig. 1). The structure of **1** was elucidated by extensive 2D NMR spectroscopy and mass spectrometry [1]. The eunicellane carbon skeleton of eleutherobin (**1**) is exclusive to natural products from gorgonians and alcyonaceans, and was observed for the first time in eunicellin (**2**) from *Eunicella stricta* in 1968 [2]. Compounds **1** and **2** differ with respect to the position of their oxygen bridges between C2 and C9 in **2**, and C4 and C7 in **1**. Other members of the small group of natural products with a 4,7-oxaeunicellane skeleton include the eleuthosides [3] and the non-glycosylated sarcodictyins [4] and valdivones [5] from related soft corals.

Eleutherobin (**1**) competes with paclitaxel for its binding at the microtubuli, inhibits their depolymerization, and thereby prevents the division of

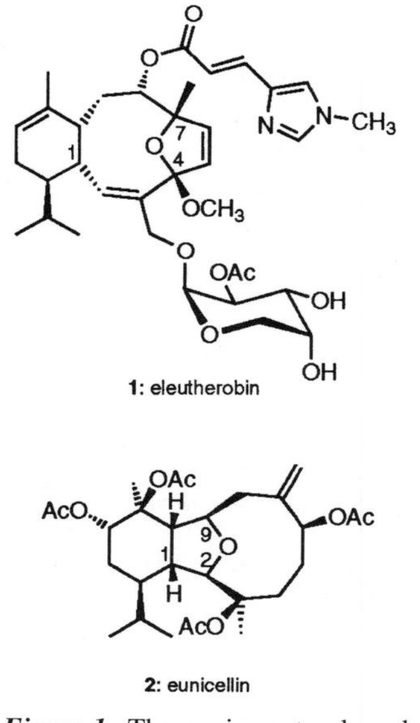

1: eleutherobin

2: eunicellin

Figure 1. The marine natural products eleutherobin (**1**) and eunicellin (**2**).

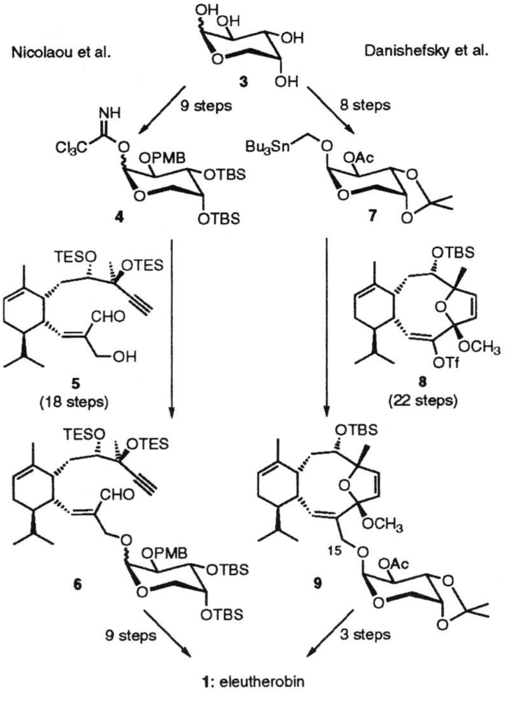

Nicolaou et al. Danishefsky et al.

3 → 9 steps → **4**; **3** → 8 steps → **7**

5 (18 steps) → **6**; **8** (22 steps) → **9**

6 → 9 steps → **1**: eleutherobin; **9** → 3 steps → **1**: eleutherobin

Scheme 1. The synthetic strategies towards eleutherobin (**1**) of Nicolaou et al. (*left*) and Danishefsky et al. (*right*).

cancer cells [6]. Until 1994, paclitaxel (Taxol) had remained the only compound showing that mode of action [7], despite intense efforts. Since then, additional natural products, the epothilones [8], discodermolide [9], and laulimalide [10], have shown very similar effects. Eleutherobin (**1**) showed an in vitro cytotoxicity of about 10–15 nM (IC50) against a diverse panel of tumor tissue cell lines with an approximate 100-fold increased potency toward selected breast, renal, ovarian, and lung cancer cell lines (National Cancer Institute). It exhibited a similar tumor-type selectivity

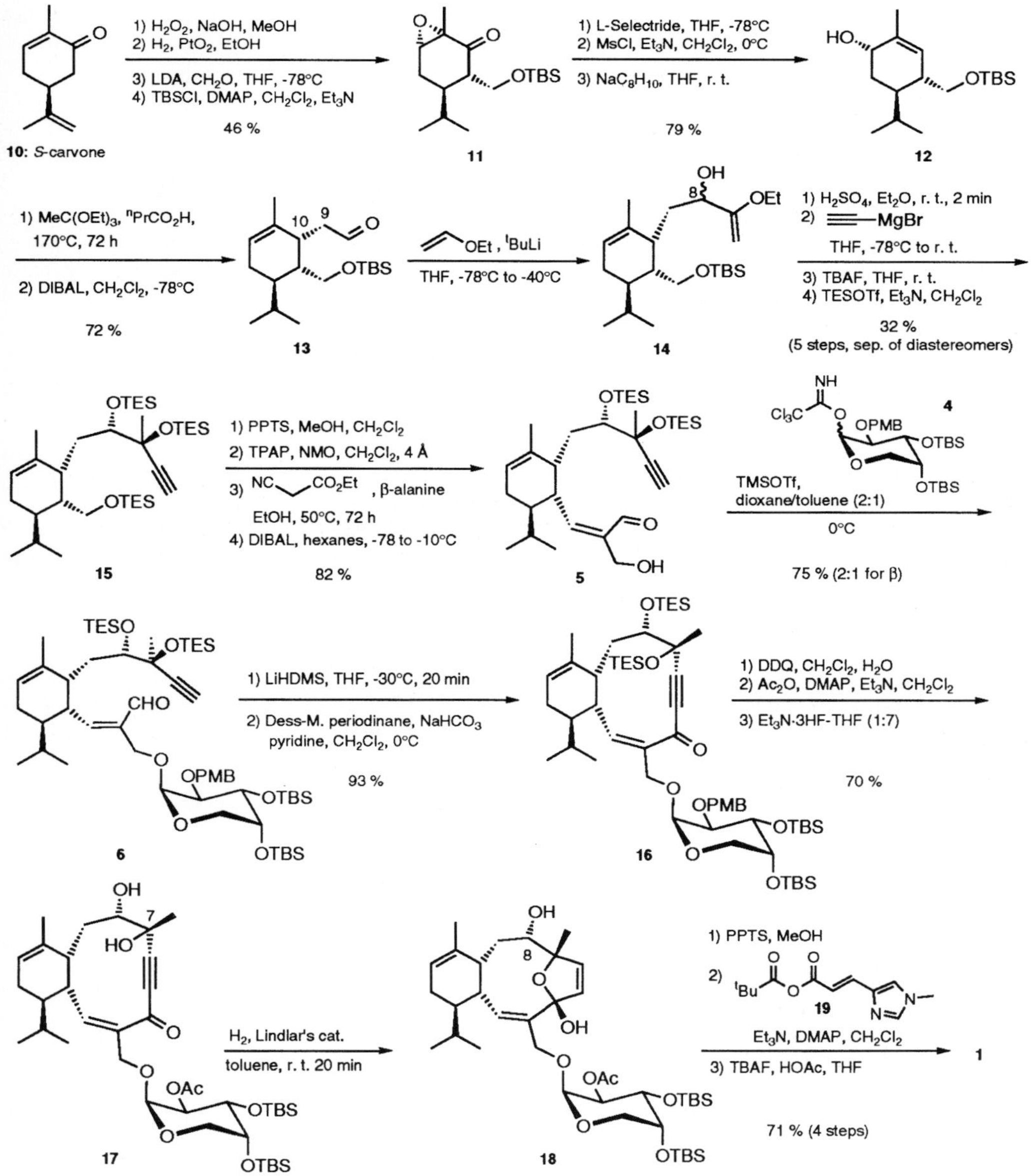

Scheme 2. Complete total synthesis of eleutherobin (**1**) by Nicolaou et al.

as paclitaxel (correlation coefficient 84 %, COMPARE protocol [11]). Recently, the microtubule-stabilizing natural products were compared with respect to their biological effects, confirming the potency of eleutherobin (**1**) [12].

From natural sources, eleutherobin (**1**) and the eleuthosides are available only in very small amounts (0.01 – 0.02 % of the dry weight of the rare alcyonacean *Eleutherobia* sp.). The lack of material for comprehensive biological testing was overcome by two total syntheses of eleutherobin (**1**) by Nicolaou et al. [13] and by Danishefsky et al. [14]. Full details are given in Schemes 2 and 3, respectively.

Both research groups take advantage of the chiral pool and start from D-arabinose to synthesize their glycosylation building blocks **4** and **7**, respectively (Scheme 1). The two approaches

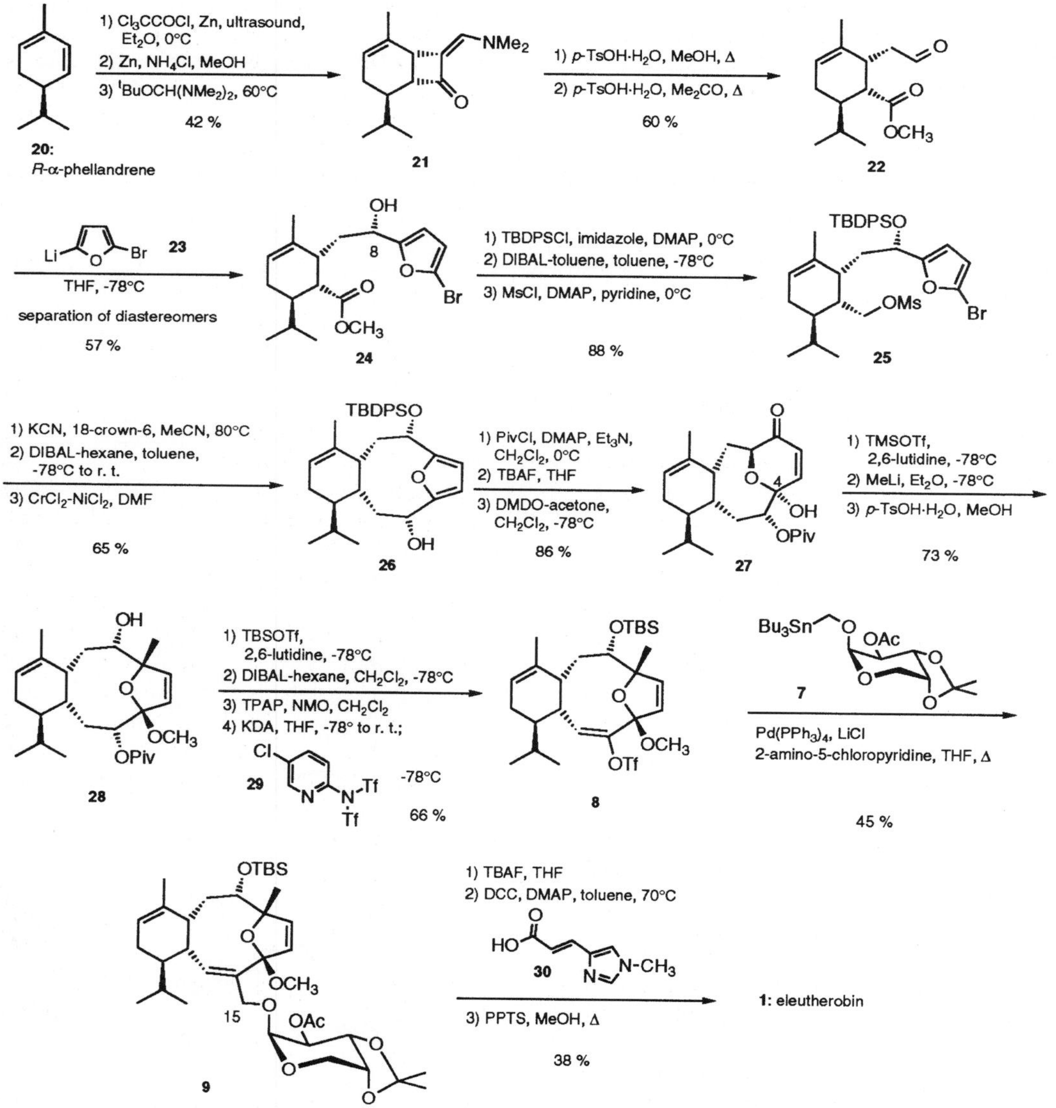

Scheme 3. Complete total synthesis of eleutherobin (**1**) by Danishefsky et al.

strategically differ with regard to the order of the glycosylation step and the contruction of the oxygen-bridged, ten-membered ring. While Nicolaou et al. make use of Schmidt's method, employing a trichloroacetimidate to glycosylate the monocyclic allylic alcohol **5**, [15] Danishefsky et al. first generate the norterpenoid tricycle **8**, which is then subjected to a modified Stille coupling, simultanously introducing the carbon atom C15 and the arabinose moiety [16].

In both approaches, the required diastereomeric purity of the glycosylated intermediates **6** and **9**, respectively, is reached in a laborious way. Nicolaou et al. obtain **6** as a mixture of anomers from which the desired β-form is separated by column chromatography (α: 28 %, β: 54 %). Danishefsky et al. stereospecifically couple the anomerically pure arabinosyloxymethyl stannane **7** to the vinyl triflate **8**, but they achieve a yield of only 40–50 %. In addition, **7** has to be separated from its α-anomer prior to use.

The monoterpenes *S*(+)-carvone (**10**; Scheme 2, Nicolaou et al.) and *R*(−)-α-phellandrene (**20**; Scheme 3, Danishefsky et al.) are chosen as the starting materials for the syntheses of the diterpenoid skeleton of eleutherobin (**1**). Again, both routes require the separation of diastereomers by chromatography. The addition of 1-ethoxyvinyllithium to the TBS-protected aldehyde **13** (Scheme 2) leads to a mixture of diastereomeric alcohols (5 : 4 ratio in favor of the desired configuration at C8), which is separated by Nicolaou et al. after the alkynylation with ethynyl magnesium bromide in a later step. **13** was synthesized in close analogy to Trost et al. with bond formation between C9 and C10 via Claisen rearrangement (Scheme 2) [17].

Danishefsky et al. have to separate a mixture of diastereomeric alcohols after addition of 2-bromo-5-lithiofuran to their aldehyde **22** (Scheme 3), obtaining the desired diastereomer in a yield of 57 %. **22** was synthesized in an interesting pathway via the ring opening of the cyclobutanone **21**.

Key step of Nicolaou's synthesis is the stereoselective hydrogenation of the unsaturated cyclododecanone **17** which was obtained via intramolecular acetylide-aldehyde condensation of **6**. The intermediate (*Z*)-olefin immediately rearranges to the tricyclic dihydrofuran **18** (Scheme 2). Solely the hydroxyl group at the quaternary carbon atom C7 takes part in the intramolecular formation of the hemiacetal.

Danishefsky et al. achieve the ring closure in a Nozaki-Kishi reaction, coupling the bromofuran ring to the aldehyde generated by homologization of **25**. Protection and oxidation of the furan **26** (Scheme 3) by dimethyldioxirane at −78 °C give the dihydropyranone **27**. In an improvement of their synthesis of eleutherobin (**1**), Danishefsky et al. silylate the hemiketal hydroxyl group at C4 of **27** prior to the subsequent nucleophilic methylation [14c]. In comparison to the earlier version, the overall sequence from **27** to **28** is shortened by one step and proceeds in a higher yield. The conversion of the intermediate silylated rearrangement product to the dihydrofuran **28** proceeds smoothly via acid-catalyzed methanolysis.

Both total syntheses of eleutherobin (**1**) are completed by the introduction of the (*E*)-*N*(6')-methylurocanic acid moiety via acylation of the free hydroxyl function at C8, followed by the removal of the TBS or isopropylidene protecting groups, respectively. The longest linear sequences cover 28 (Nicolaou et al., 2.4 % overall yield) and 26 steps (Danishefsky et al., 0.6 % overall yield), respectively.

The reactivity of the strained, tricyclic skeleton of the 4,7-oxaeunicellanes was studied by Pietra in 1988 [4]. Treatment of sarcodictyin A (**31**) with methanolic potassium hydroxide led to the formation of the butenolide **32** (Scheme 4). The Michael acceptor property of the carbon atom C2 favors this rearrangement, leading to a relaxation of the strained ring system. After methanolysis of the *N*-methylurocanate, the newly formed hydroxyl group attacks at C2 and the fragmentation of the carbon–carbon bond between C3 and C4 immediately follows.

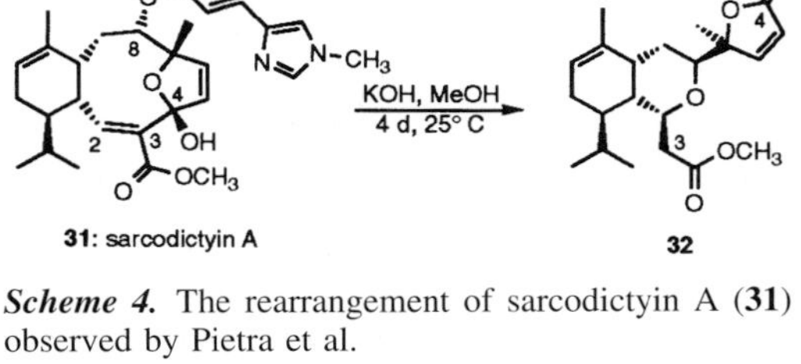

KOH, MeOH
4 d, 25° C
31: sarcodictyin A
32

Scheme 4. The rearrangement of sarcodictyin A (**31**) observed by Pietra et al.

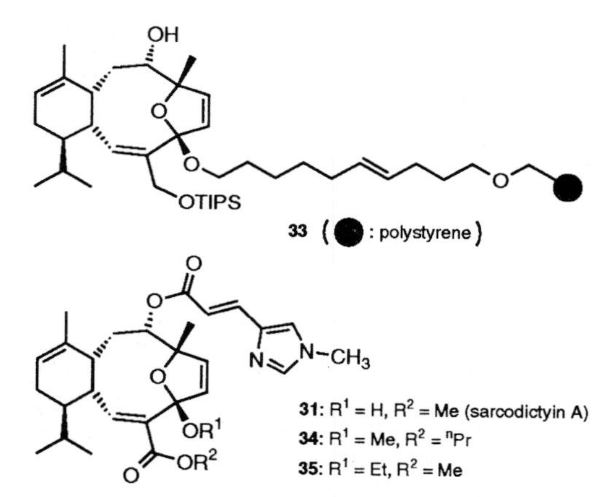

Figure 2. The sarcodictyin scaffold linked to a polystyrene resin (Nicolaou et al.). The derivatives **34** and **35** showed improved cytotoxicity and tubulin polymerization, respectively.

The absolute stereochemistry of eleutherobin (**1**) had not been determined when the compound was first isolated, because biological testing was given preference. However, results obtained by Pietra et al. strongly suggested the proposed absolute configuration of the diterpenoid 4,7-oxaeunicellane skeleton [4]. Nicolaou et al. and Danishefsky et al. unambiguously identified the sugar unit of **1** as D-arabinose. In addition to a comparison of optical rotations, Danishefsky et al. synthesized the L-arabinosyl diastereomer neo-eleutherobin.

Eleutherobin (**1**) and the non-glycosylated sarcodictyins continue to be subject of extensive biological testing. Although the sarcodictyins are about ten times less cytotoxic than eleutherobin (**1**), their simpler structure gave rise to the development of a solid-phase synthesis [18]. Nicolaou et al. loaded the sarcodictyin scaffold onto a polystyrene resin via a 10-membered linker chain (**33**, Fig. 2). The double bond was generated via a Wittig olefination, with the resin bearing the triphenylphosphorane part. The synthesized library of more than 60 sarcodictyin derivatives allowed some structure-activity relationships to be assigned. The presence of the unmodified urocanyl side chain appears to be required for both tubulin-polymerizing and cytotoxic activity, while different ketal substitution is tolerated. In the absence of the sugar side chain, the reduction of the ester function of the sarcodictyins results in loss of activity. The highest cytotoxicity was measured for the n-propyl ester **34** (IC50 3–5 nM against the ovarian cancer cell line 1A9 and the paclitaxel-resistant 1A9PTX10 and 1A9PTX22), while the highest degree of tubulin polymerization was observed for the ethyl ketal **35** (85 % [19]; sarcodictyin A (**31**): 67 %).

Overman et al. were the first to synthesize the eunicellane carbon skeleton [20]. In contrast to eleutherobin (**1**), (−)-7-deacetoxyalcyonin acetate (**45**) from the marine soft coral *Cladiella* sp. [21] shows the oxygen bridge between C2 and C9 instead of C4 and C7. Starting from *S*(+)-carvone, an elegant Prins-pinacol conden-

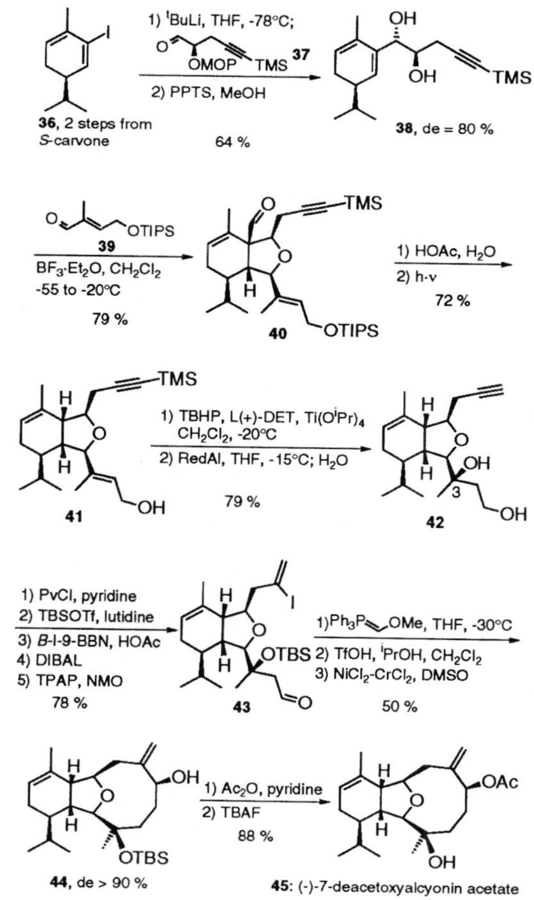

Scheme 5. Total synthesis of (−)-7−deacetoxyalcyonin acetate (**45**) by Overman et al.

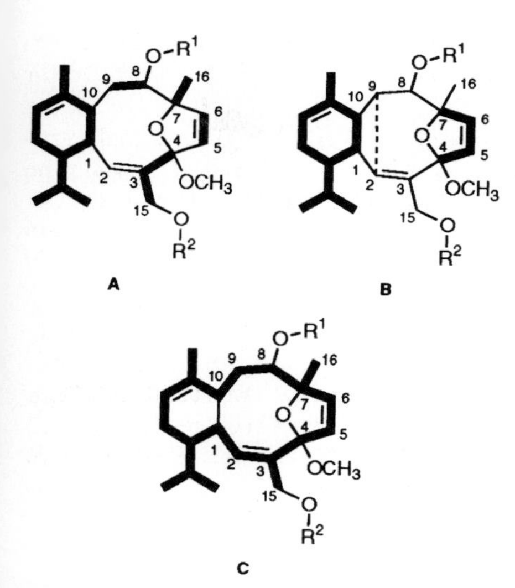

Figure 3. Carbon–carbon bonds introduced from synthetic building blocks by Nicolaou et al. (**A**) and Danishefsky et al. (**B**, C2–C9 broken after ketene cycloaddition). In the putative biosynthesis of the diterpenoid 4,7-oxaeunicellanes from cembranoid precursors, only one additional carbon–carbon bond (C1-C10) would have to be formed (**C**). (1: R1 = (E)-N(6')-methylurocanyl, R2 = beta-D-*O*(2)-acetylarabinopyranosyl; stereochemistry has been omitted for clarity).

sation-rearrangement was used to assemble the 2–oxabicyclo[4.3.0]non-4-ene **40** (Scheme 5) from the diol **38** and the aldehyde **39** in a fully stereoselective manner. Deformylation to **41** was subsequently achieved photochemically. Sharpless epoxidation followed by Red-Al reduction stereoselectively introduced the hydroxyl group at C3. After homologization the oxonane ring (**44**) was assembled, making use of a remarkably stereoselective Nozaki-Kishi reaction (*de* > 90 %). The synthesis of (–)-7-deacetoxyalcyonin acetate (**45**) proved the proposed structure and was achieved in 17 steps and a total yield of 9.8 %.

In conclusion, three independent syntheses of marine eunicellane diterpenoids have been developed, which will be the basis of the future development of this exciting natural product class. The biosynthesis of the eunicellanes probably involves the oxidative cyclization of a cembranoid precursor lacking solely the bond between C1 and C10 (C, Fig. 3). Considering the number of carbon–carbon bonds formed in the course of the stepwise syntheses by Nicolaou et al. (six), Danishefsky et al. (eight), and Overman et al. (four), it can be stated that the efficiency of nature still sets the standard for modern organic synthesis.

References

[1] a) W. Fenical, P. R. Jensen, T. Lindel (University of California), US Patent No 5,473,057, **1995** [*Chem. Abstr.* **1996**, 124, P194297z]; b) T. Lindel, P. R. Jensen, W. Fenical, B. H. Long, A. M. Casazza, J. Carboni, C. R. Fairchild, *J. Am. Chem. Soc.* **1997,** *119,* 8744–8745.

[2] O. Kennard, D. G. Watson, L. Riva di Sanseverino, B. Tursch, R. Bosmans, C. Djerassi, *Tetrahedron Lett.* **1968**, *9,* 2879–2884.

[3] S. Ketzinel, A. Rudi, M. Schleyer, Y. Benayahu, Y. Kashman, *J. Nat. Prod.* **1996**, *59,* 873–875.

[4] a) M. D'Ambrosio, A. Guerriero, F. Pietra, *Helv. Chim. Acta* **1987**, *70*, 2019–2027; b) *ibid.* **1988**, *71,* 964–976.

[5] Y. Lin, C. A. Bewley, D. J. Faulkner, *Tetrahedron* **1993**, *49,* 7977–7984.

[6] B. H. Long, J. M. Carboni, A. J. Wasserman, L. A. Cornell, A. M. Casazza, P. R. Jensen, T. Lindel, W. Fenical, C. R. Fairchild, *Cancer Res.* **1998**, *58,* 1111–1115.

[7] For the chemistry and biology of paclitaxel, see K. C. Nicolaou, W.-M. Dai, R. K. Guy, *Angew. Chem.* **1994**, *106*, 38–69; *Angew. Chem. Int. Ed. Engl.* **1994**, *33,* 15–44.

[8] a) G. Höfle, N. Bedorf, H. Steinmetz, D. Schomburg, K. Gerth, H. Reichenbach, *Angew. Chem.* **1996**, *108,* 1671–1673; *Angew. Chem. Int. Ed. Engl.* **1996**, *35,* 1567–1569; b) see also L. Wessjohann, *Angew. Chem.* **1997**, *109,* 739–742; *Angew. Chem. Int. Ed. Engl.* **1997**, *36,* 715–718.

[9] a) E. ter Haar, R. J. Kowalski, E. Hamel, C. M. Lin, R. E. Longley, S. P. Gunasekera, H. S. Rosenkranz, B. W. Day, *Biochemistry* **1996**, *35,* 243–250; b) R. Balachandran, E. ter Haar, M. J. Welsh, S. G. Grant, B. W. Day, *Anti-Cancer Drugs* **1998,** *9,* 67–76.

[10] S. L. Mooberry, G. Tien, A. H. Hernandez, A. Plubrukarn, B. S. Davidson, *Cancer Res.* **1999**, *59,* 653–660.

[11] Information with regard to the COMPARE protocol established at the NCI is available on the internet: http://dtp.nci.nih.gov/docs/compare/compare.html.

[12] E. Hamel, D. L. Sackett, D. Vourloumis, K. C. Nicolaou, *Biochemistry* **1999,** *38,* 5490–5498.

[13] a) K. C. Nicolaou, F. van Delft, T. Ohshima, D. Vourloumis, J. Xu, S. Hosokawa, J. Pfefferkorn, S. Kim, T. Li, *Angew. Chem.* **1997**, *109,* 2630–

2634; *Angew. Chem. Int. Ed. Engl.* **1997**, *36,* 2520–2524; b) K. C. Nicolaou, J.-Y. Xu, S. Kim, T. Ohshima, S. Hosokawa, J. Pfefferkorn, *J. Am. Chem. Soc.* **1997**, *119,* 11353–11354.

[14] a) X.-T. Chen, C. E. Gutteridge, S. K. Bhattacharya, B. Zhou, T. R. R. Pettus, T. Hascall, S. J. Danishefsky, *Angew. Chem.* **1998**, *110,* 195–197; *Angew. Chem. Int. Ed. Engl.* **1998**, *37,* 185–187; b) X.-T. Chen, B. Zhou, S. K. Bhattacharya, C. E. Gutteridge, T. R. R. Pettus, S. J. Danishefsky, *ibid.* **1998**, *110,* 835–838; *ibid.* **1998**, *37,* 789–792; c) S. K. Bhattacharya, X.-T. Chen, C. E. Gutteridge, S. J. Danishefsky, *Tetrahedron Lett.* **1999**, *40,* 3313–3316.

[15] R. R. Schmidt, K.-H. Jung in *Preparative Carbohydrate Chemistry* (Hrsg. S. Hanessian), Marcel Dekker, New York, **1997**, S. 283–312.

[16] J. K. Stille, *Angew. Chem.* **1986**, *98*, 504–519; *Angew. Chem. Int. Ed. Engl.* **1986**, *25,* 508–524.

[17] B. M. Trost, A. S. Tasker, G. Rüther, A. Brandes, *J. Am. Chem. Soc.* **1991**, *113,* 670–672.

[18] K. C. Nicolaou, N. Winssinger, D. Vourloumis, T. Ohshima, S. Kim, J. Pfefferkorn, J.-Y. Xu, T. Li, *J. Am. Chem. Soc.* **1998**, *120,* 10814–10826.

[19] Percentage of the value obtained on incubation of tubulin with 0.5 M GTP, 10 % glycerol in 100 mM MEM buffer. D. M. Bollag, P. A. McQueney, J. Zhu, O. Hensens, L. Koupal, J. Liesch, M. Goetz, E. Lazarides, C. M. Woods, *Cancer Res.* **1995**, *55,* 2325–2333.

[20] D. W. C. MacMillan, L. E. Overman, *J. Am. Chem. Soc.* **1995**, *117,* 10391–10392.

[21] Y. Uchio, M. Kodama, S. Usui, Tetrahedron *Lett.* **1992**, *33,* 1317–1320.

Selectin Inhibitors

Markus Rösch and Karola Rück-Braun

Institut für Organische Chemie, Universität Mainz, Germany

Adhesive cell-cell interactions determine several biological events, such as embryogeneses and immunological defence. On a molecular level, cell adhesion and migration are mediated by recognition and binding effects between cell specific adhesion molecules and their ligands of high affinity. The most important adhesion molecules on cell surfaces are cadherins, selectins (E-, P- and L- selectin), glycoproteins of the immunoglobulin superfamily and integrins [1, 2]. There is now general agreement that adhesion events are mediated by protein-protein interactions as well as by carbohydrate-protein interactions. Selectins belong to the family of carbohydrate-recognizing proteins (lectins) carrying calcium-dependent binding domains for carbohydrate structures. Sialyl Lewis^x tetrasaccharide was identified as a weak ligand of all three selectins by in vitro investigations (Scheme 1). It is a common segment of the natural ligand and involved in adhesion processes during inflammatory responses.

Inflammatory Processes

Carbohydrate selectin-mediated cell-cell recognition plays an important role in physiological processes such as inflammation reactions. The inflammation cascade begins with the stimulation of the endothelium by cytokines to express P-selectin followed by massive formation and presentation of E-selectin receptors. Neutrophile granulocytes and monocytes display glycoproteins such as polylactosamine sialomucine PSGL-1 [3] or sialoglycoproteins such as ESL-1 [4] on their surface. These are the natural ligands to E- and P-selectin-presenting sialyl Lewis^X and related carbohydrates. Carbohydrate-selectin interactions result in rolling of the circulating leucocytes on the endothelium. Other adhesion molecules such as integrins are simultaneously activated, mediating the binding of the leucocytes to the endothelium. This stronger protein-protein interaction leads to the migration of leucocytes through the endothelial layer to the site of the injury.

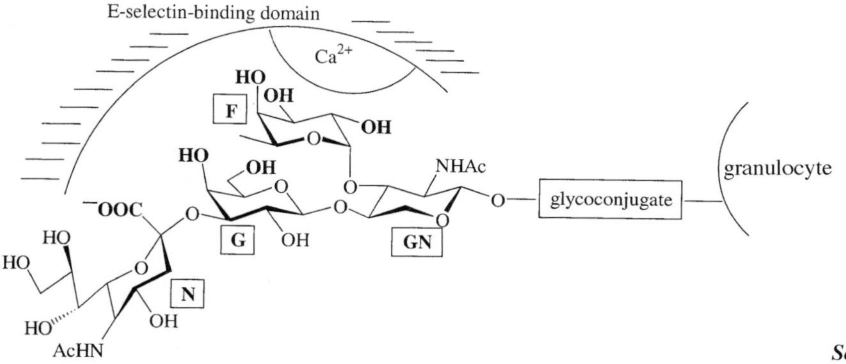

E-selectin-binding domain

Ca^{2+}

F, G, GN, N — glycoconjugate — granulocyte

Scheme 1

Also, transplant rejection is an inflammatory process characterized by lymphocyte and monocyte infiltration into the transplant tissue. On a molecular level, this process seems to depend largely on L-selectin/sialyl Lewis^X recognition [5]. Lymphocytes express constitutively L-selectin on their cell surface. The homing of lymphocytes is mediated by the L-selectin ligand GlyCAM 1 [6]. This glycoprotein is exclusively expressed on the endothelial cells of the postcapilar venoles in the lymph nodes and presents a high density of multiple oligosaccharide units with sialyl Lewis^X and sulfated sialyl Lewis^X structures. It has recently been shown that rejection of kidney transplants is initiated by a massive de novo expression of sialyl Lewis^X on the transplant's peritubular endothelial cells [5].

Selectin-dependent adhesion inhibitors offer a new therapeutical approach to the prevention of transplant rejection. They should also enable treatment and prophylaxis of various other pathological inflammation processes and diseases (i.e. rheumatic arthritis, dermatitis and bacterial meningitis) initiated by a massive invasion of leucocytes [1, 7].

First clinical trials of a tetrasaccharide sialyl Lewis^X analog as an inflammation blocker are already complete [7, 8]. But saccharidical sialyl Lewis^X analogs are inactive in oral application, and intravenous injection leads to a fast enzymatic decomposition: this is a challenge to clinical pharmacology. Furthermore, chemical synthesis of oligosaccharides is sophisticated, time-consuming and expensive. For a therapeutical approach, metabolic-stable selectin inhibitors have to be developed that are easily available on a large scale.

Inhibitor Design

A thorough understanding of structure-activity relationships is essential for the rational design of adhesion inhibitors based on partial structures of natural ligands [9–11]. The required amino acids for the binding between glycoprotein E-selectin and sialyl Lewis^X were identified by X-ray structure analysis and site-directed mutagenesis. On the other hand, binding studies of selectively modified sialyl Lewis^X analogs revealed the functional groups involved in the carbohydrate-selectin interactions. In the course of these studies, several structural analogs of the tetrasaccharide were synthesized, in which the carbohydrate building blocks were systematically modified or replaced. By biological examination in vitro and in vivo, respectively, six essential functional groups of the carbohydrate epitope for the binding to E-selectin were identified: all three OH groups of fucose, the 4- and 6-OH group of galactose and the carboxylic acid function of neuraminic acid. In the case of P-selectin binding, the 2- and 4-OH groups of fucose seem not to be critical. NMR studies of sialyl Lewis^X/selectin complexes, combined with molecular-mechanic calculations, lead to additional information about the bioactive conformations of the ligands [9–11]. Apparently, the conformation of sialyl Lewis^X in the L-selectin complex [9] corresponds to the preferential conformation in solution. The E-selectin complex, however, shows another bioactive conformation, with the carboxylate of neuraminic acid oriented differently (Scheme 2). Especially the flexibility of the neuraminic acid-galactose linkage (N1-3G) seems to be responsible for the different specificity of sialyl Lewis^X to E-, P- and L-selectin as shown in biological adhesion assays. The staggered, rather rigid alignment of the trisaccharide fragment galactose-fucose-glucosamine seems not to be involved in this specificity.

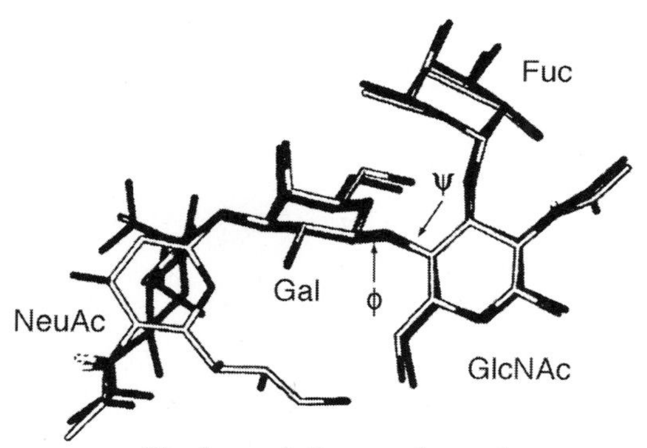

Scheme 2

Sialyl $Lewis^X$ Mimetics

A number of functional sialyl $Lewis^X$ mimetics have been synthesized. Their activities in vitro are equal or even better than those of the tetrasaccharide itself. To overcome synthetic problems, efficient stereoselective glycosylations as well as new chemoenzymatic methods for C–C bond formations had to be developed. The substitution of neuraminic acid by (*S*)-phenyl- and (*S*)-cyclohexyl lactic acid, as less flexible glycol acid residues, turned out to be very successful [10]. Also, a phosphate and a sulfate group, respectively, mimic neuraminic acid without loss of activity [11]. (*S*)-Cyclohexyl lactic acid-mimetic **2** shows a more than ten-fold efficacy compared with sialyl $Lewis^X$, whereas the corresponding (*R*)-isomer **3** is almost inactive [10]. The deviating orientation of the carboxylic acid functionality compared to the bioactive sialyl $Lewis^X$ conformation leads to the examined loss of activity. It was shown by transfer-NOE measurements of the corresponding E-selectin complexes that the coordinates of the bioactive conformation of sialyl $Lewis^X$ and of compound **2** are similar. Consequently structure **2** should bind to E-selectin in the same manner as that of sialyl $Lewis^X$ [10a, b].

The *N*-acetylglucosamine unit seems to act as a spacer, positioning the other sugar units. Glucose, diols and amino alcohols, e.g. (*R,R*)-cyclohexandiol (see Scheme 3, compounds **1–3**) or (*R,R*)-2-amino cyclohexanol, as well as amino acid-building blocks such as L-threonine (**5**, **6**), proved to be an effective replacement for *N*-acetylglucosamine [10–13].

Several mimetics carry dihydroxyamino acids, other diols and residues with carboxylic acid groups [13, 14], such as glutaminic acid, instead of galactose (see Scheme 3, compound **7**). Fucose is mainly substituted by glycerine, mannose **7** or galactose **8** [12–14]. Fucose compound **5** bears 2-amino-3,4-dihydroxy butanoic acid, which mimics the 4- and 6-hydroxyl group of galactose and a glutaric residue to replace the carboxylic acid functionality of neuraminic acid [13, 14]. The glycosylation of the tribenzyl fucosylphosphite **9** and the amino acid benzylester **10** with trifluoromethanesulfonic acid as catalyst was achieved in 84 % yield and 99 % α-selectivity using the phosphite method (Scheme 4) [13]. *C*-Fucopeptide **11b** as well as compound **7** bear a long-chain hydrophobic group which causes an activity eight times that of the unsubstituted compound **11a** [14]. In compound **8** L-fucose is

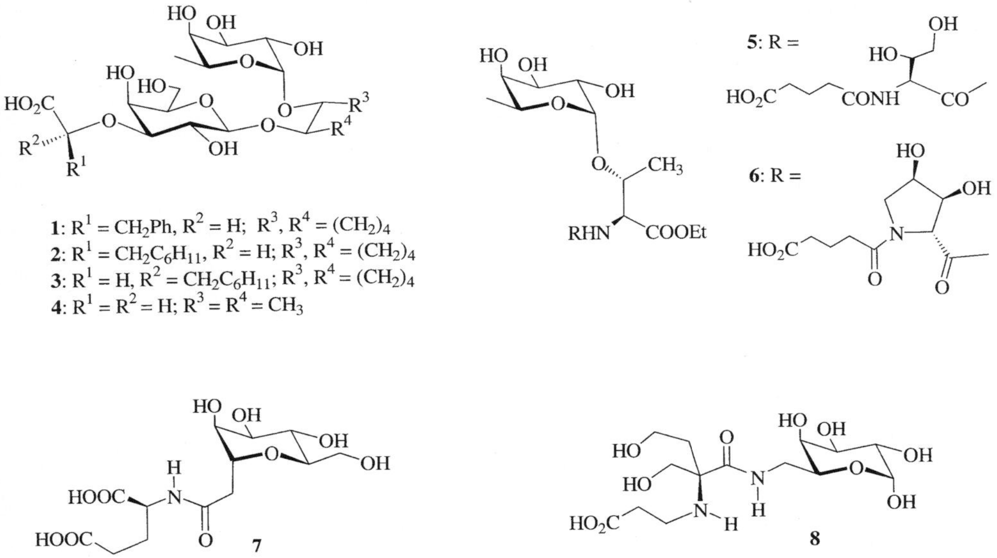

1: $R^1 = CH_2Ph$, $R^2 = H$; $R^3, R^4 = (CH_2)_4$
2: $R^1 = CH_2C_6H_{11}$, $R^2 = H$; $R^3, R^4 = (CH_2)_4$
3: $R^1 = H$, $R^2 = CH_2C_6H_{11}$; $R^3, R^4 = (CH_2)_4$
4: $R^1 = R^2 = H$; $R^3 = R^4 = CH_3$

5: R =

6: R =

7

8

Scheme 3

replaced by L-galactose. In this glycosylated peptide, once again dihydroxy-α-amino acid mimics galactose [15]. These amino acid building blocks are enzymatically synthesized utilizing L-threonine-aldolase starting from glycine (Scheme 4) [16]. Compound **7**, bearing two carboxylate residues, shows five times the activity of sialyl Lewis^X, although galactose is missing. Other modifications at the galactose and fucose residues, especially substitutions of essential hydroxyl groups, resulted in loss of activity. The most active sialyl Lewis^X mimetics are shown in Schemes 3 and 4. The efficacies of these compounds were examined by established selectin-adhesion assays, but different methods and standards complicate the direct comparison of the results obtained.

Concept of Multivalency/Polyvalent Mimetics

Tetrasaccharide sialyl Lewis^X represents only a part of the natural recognition region on selectin ligands, and a number of experimental data show that it is not exclusively responsible for the specific receptor binding. PSGL-1 [3], expressed on cell surfaces of human neutrophile granulocytes, is one of the natural ligands of the selectins. A sulfated peptide unit of this *O*-glycosidic polylactosamino sialomucin was identified as an element for recognition of P-selectin. In vitro there is only a weak interaction between an isolated sialyl Lewis^X molecule and its receptor, whereas in vivo a stronger binding between the natural ligands and selectins is based on several recognition events. For example, natural glycoproteins such as PSGL-1 or GlyCAM-1 bear a great number of oligosaccharide units with sialyl Lewis^X structure. Therefore, multivalent binding is expected to enhance the efficiency of binding. To obtain a deeper understanding of the nature of selectin-ligand recognition, artificial, multivalent sialyl Lewis^X conjugates were synthesized and examined for their efficacy in vitro [17–19]. The coupling of tetrasaccharide sialyl Lewis^X with polyfunctional templates such as nitromethane-trispropionic acid, cyclohexan-1,3-diol or cyclic and acyclic peptides leads to artificial, multivalent mimetics. The reaction of partially protected sialyl Lewis^X amine **12** (see Scheme 5) with the succinate of nitromethane-trispropionic acid gives, after a two-stage reductive and hydrogenolytic cleavage sequence, the tris-coupled product in 82 % overall yield [17a]. Fragment condensation of **12** with a cyclopeptide template was realized in 48 % yield to furnish sialyl Lewis^X-*N*-glycopeptide **13** [17b]. Inhibition analysis in vivo, however, does not show a superior inhibition of leucocyte adhesion compared to the biological efficacy using monovalent ligands. Poly-

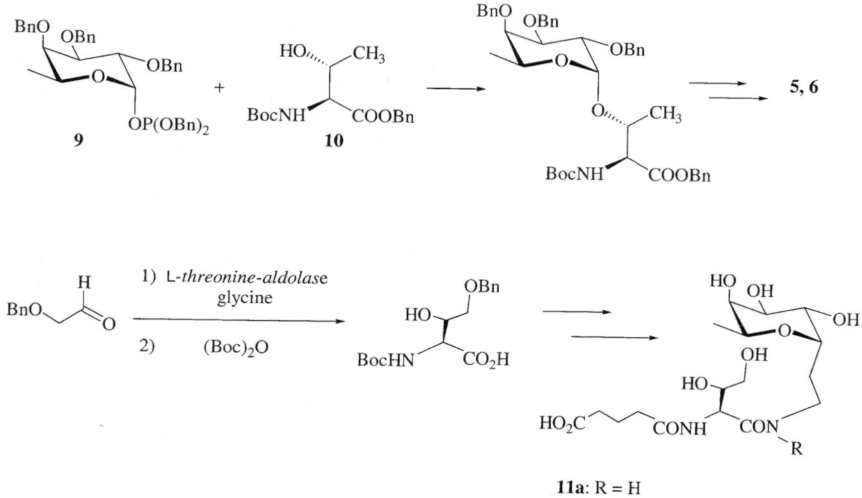

BnO OBn OBn O OP(OBn)$_2$ **9** + HO CH$_3$ BocNH COOBn **10** → BnO OBn OBn O O CH$_3$ BocNH COOBn → → **5, 6**

BnO H O — 1) *L-threonine-aldolase* glycine 2) (Boc)$_2$O → OBn HO BocHN CO$_2$H → → HO OH OH O OH HO HO$_2$C CONH CON R

11a: R = H
11b: R = CH$_2$CONH(CH$_2$)$_{13}$CH$_3$

Scheme 4

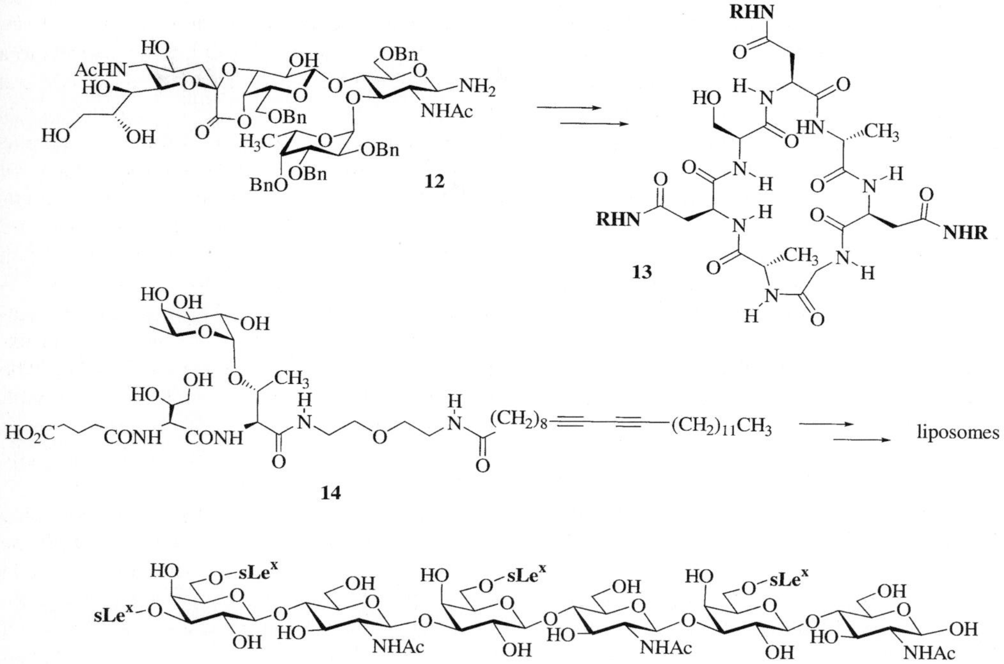

Scheme 5

valent presentation of sialyl Lewis^X mimetic **5**, realized by connecting monomer **14** with liposomes, resulted in only a poor increase of activity [18]. A ten-fold binding activity to L- and E-selectin was achieved with polymers made of sialyl Lewis^X acrylamide monomers [19]. Tetravalent compound **15** bearing sialyl Lewis^X moieties on an oligolactosamine backbone was identified as a very effective L-selectin inhibitor [20]. Initial investigations with oligo- and polyvalent sialyl Lewis^X derivatives and mimetics confirm the cluster model after all. The true functional binding situation, however, remains unclear. Further investigations identifying the size and orientation of sialyl Lewis^X/selectin cluster complexes are essential for a rational design of polyvalent adhesion inhibitors.

Conclusion

Based on the results obtained so far, the design of potent, non-saccharidical, orally available sialyl Lewis^X mimetics of low molecular weight should be possible in the near future. Besides indications already mentioned above, selectin antagonists could offer a new approach in cancer therapy to prevent metastasis. A number of tumor cells present tissue- and tumor-specific oligosaccharides with sialyl Lewis^X structures on their cell membrane. Cleavage or release during an operation leads to circulation of tumor cells in the blood stream that are able to bind to activated endothelium by a selectin-mediated mechanism [21]. Recent efforts show that this process is mainly responsible for migration of tumor cells into the tissue and the subsequent formation of secondary tumors. The design of anti-adhesive and anti-inflammatory drugs based on new biological binding assays would intensify the understanding of the structure of natural selectin ligands and their bioactive conformation. The complexity of the involved molecular structures and biological processes distinguish glycobiology [2] as an exciting and challenging field for physicians, biologists and chemists.

References

[1] A. Giannis, *Angew. Chem. Int. Ed. Engl.* **1994**, *106*, 178.
[2] R.A. Dwek, *Chem. Rev.* **1996**, *96*, 683.
[3] PSGL-1 = P-selectin-glycoprotein-ligand-1; a) D. Sako, K.M. Comess, *Cell* **1995**, *83*, 323; b) T. Pouyani, B. Seed, *ibid.* **1995**, *83*, 333.
[4] ESL-1 = E-selectin-ligand-1; M. Steegmaier, A. Levinowitz, S. Isenmann, E. Borges, M. Lenter, H. Kocher, B. Kleuser, D. Verstweber, *Nature* **1995**, *373*, 615.
[5] A. Seppo, J.P. Turunen, L. Penttila, A. Keane, O. Renkonen, R. Renkonen, *Glycobiology* **1996**, *6*, 65.
[6] GlyCAM-1 = glycosylated "Cell-adhesion-molecule-1"; a) D. Crommie, S. D. Rosen, *J. Biol. Chem.* **1995**, *270*, 22614; b) S. Komba, H. Ishida, M. Kiso, A. Hasegawa, *Bioorg. Med. Chem.* **1996**, *4*, 1833.
[7] J.C. McAuliffe, O. Hindsgaul, *Chem. & Ind.* **1997**, 170.
[8] T. Murohara, J. Margiotta, L.M. Phillips, J.C. Paulson, S. DeFrees, S. Zalipsky, L.S.S. Guo, A.M. Lefer, *Cardiovasc. Res.* **1995**, *30*, 965.
[9] L. Poppe, G.S. Brown, J.S. Philo, P.V. Nikrad, B.H. Shah, *J. Am. Chem. Soc.* **1997**, *119*, 1727.
[10] a) H.C. Kolb, B. Ernst, *Pure Appl. Chem.* **1997**, *69*, 1879; b) R. Bänteli, B. Ernst, *Tetrahedron Lett.* **1997**, *38*, 4059; c) W. Jahnke, H.C. Kolb, M.J.J. Blommers, J.L. Magnani, B. Ernst, *Angew. Chem. Int. Ed. Engl.* **1997**, *109*, 2603.
[11] G. J. McGarvey, C.-H. Wong, *Liebigs Ann./Recueil* **1997**, 1059.
[12] M. J. Bamford, M. Bird, P. M. Gore, D. S. Holmes, R. Priest, J. C. Prodger, V. Saez, *Bioorg. Med. Chem. Lett.* **1996**, *6*, 239.
[13] C.-C. Lin, M. Shimazaki, M.-P. Heck, S. Aoki, R. Wang, T. Kimura, H. Ritzen, S.Takayama, S.-H. Wu, G. Weitz-Schmidt, C.-H. Wong, *J. Am. Chem. Soc.* **1996**, *118*, 6826.
[14] a) C.-H. Wong, F. Moris-Varas, S.-C. Hung, T. G. Marron, C.-C. Lin, K. W. Gong, G. Weitz-Schmidt, *J. Am. Chem. Soc.* **1997**, *119*, 8152; b) S.-H. Wu, M. Shimazaki, C.-C. Lin, L. Qiao, W.J. Moree, G. Weitz-Schmidt, C.-H. Wong, *Angew. Chem. Int. Ed. Engl.* **1996**, *108*, 88.
[15] M. W. Cappi, W. J. Moree, L. Qiao, T. G. Marron, G. Weitz-Schmidt, C.-H. Wong, *Angew. Chem. Int. Ed. Engl.* **1996**, *108*, 2346.
[16] S. Takayama, G. J. McGarvey, C.-H. Wong, *Chem. Soc. Rev.* **1997**, *26*, 407.
[17] a) G. Kretzschmar, U. Sprengard, H. Kunz, E. Bartnik, W. Schmidt, A. Toepfer, B. Hörsch, M. Krause, D. Seiffge, *Tetrahedron* **1995**, *51*, 13015; b) U. Sprengard, M. Schudok, W. Schmidt, G. Kretzschmar, H. Kunz, *Angew. Chem. Int. Ed. Engl.* **1996**, *108*, 321.
[18] C.-C. Lin, T. Kimura, S.-H. Wu, G. Weitz-Schmidt, C.-H. Wong, *Bioorg. Med. Chem. Lett.* **1996**, *6*, 2755.
[19] O. Renkonen, S. Topipila, L. Penttila, H. Salminen, H. Maaheimo, C. E. Costello, J. P. Turunen, R. Renkonen, *Glycobiology* **1997**, *7*, 453.
[20] H. Miyauchi, M. Tanaka, H. Koike, N. Kawamura, M. Hayashi, *Bioorg. Med. Chem. Lett.* **1997**, *7*, 985.
[21] a) T. Nakashio, T. Narita, M. Sato, S. Akiyama, Y. Kasai, M. Fujiwara, K. Ito, H. Takagi, R. Kannagi, *Anticancer-Res.* **1997**, *17(1A)*, 293; b) R. Renkinen, P. Mattila, M.-L. Majuri, J. Rabina, J. P. Turunen, O. Renkonen, T. Paavonen, *Glycoconjugate J.* **1997**, *14*, 593; c) R. Kannagi, *ibid.* **1997**, *14*, 577.

Crossing the Finishing Line: Total Syntheses of the Vancomycin Aglycon

Holger Herzner and Karola Rück-Braun

Institut für Organische Chemie, Universität Mainz, Germany

The antibiotic vancomycin **1** (Scheme 1), discovered in 1956, has been used for the treatment of infections caused by gram – positive bacteria since 1958 [1]. Up to now, the compound has been commercially obtained from microorganisms. In clinics throughout the world, vancomycin is generally seen as the last line of defense for the treatment of infections caused e.g. by bacteria multi-resistant to the commonly used β-lactam antibiotics or in cases of penicillin hypersensitivity. The glycopeptide (**1**), structurally elucidated in the early 1980s [1], consists of an arylglycine-rich heptapeptide aglycon (**2**), the main characteristics of which are three atropisomeric moieties.

Caused by the high rotational barrier of the biaryl linkage, an element of axial chirality emerges. In addition, the rotation of the aryl residues of the amino acids 6 and 2 (Scheme 1) around their longitudinal axes is hindered, resulting in two elements of planar chirality. The unusual complexity of the aglycon's structure (**2**) is a special challenge for synthetic chemists. The construction of this aglycon unit and of some of its individual fragments has kept a number of research groups busy over the last years and has led to the development of a number of new synthetic methods. Applying an elaborate concept, Evans et al. recently reached the finishing line and presented the first total synthesis of the vancomycin aglycon [2, 3]. Only a few days later, Nicolaou and co-workers presented their total synthesis [4–6]. In the following, strategic considerations about the concepts of these syntheses as well as their key steps and key reactions will be presented.

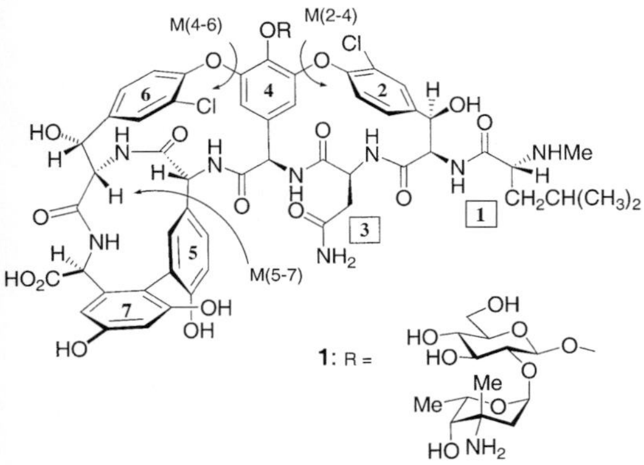

Scheme 1

Total Synthesis by Evans et al.

For several years Evans and co-workers focussed on synthetic strategies and methods for the construction of target structure **2** (Scheme 1). After the development of diastereoselective amino acid syntheses with oxazolidinones as chiral auxiliaries [2, 3], new methods for the construction of the biaryl- and biarylether-containing macrocyclic subunits were investigated [7, 8]. During this time, serious set-backs had to be dealt with. Despite earlier studies, the thallium(III)-mediated oxidative cyclization strategy for the construction of the macrocyclic biarylether units, published in 1989 by Yamamura et al. and Evans and co-workers [8], had to be abandoned. In the course of the synthesis of the bisdechloro vancomycin aglycon of orienticin C, unexpected difficulties were encountered [9]. However, to achieve the total synthesis of **2**, Evans and co-

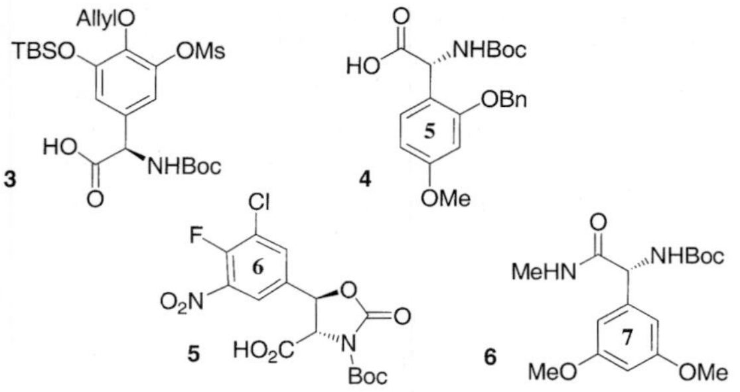

Scheme 2. TBS = *tert*-butyldimethylsilyl, Boc = *tert*-butyloxycarbonyl.

workers adjusted their strategy to the results obtained from studies of intramolecular S_NAr-cyclizations [9b, 10–13].

The synthetic route is based on the preparation of the M(4–6) and the M(2–4) subunit by intramolecular S_NAr cyclizations of suitable precursor peptides with *ortho*-nitro-substituted haloarene residues and a central phenolic unit (Schemes 2 and 4). This concept was originally developed by Zhu and co-workers and applied successfully to the synthesis of cyclic peptides of the vancomycin family, e. g. to the synthesis of K-13 [13].

Because of the versatility of the Sandmeyer reaction ($NO_2 \rightarrow H$ or $NO_2 \rightarrow Cl$), the new synthetic strategy permits the synthesis of the desired M(4–6)-chloro atropisomer independently of the stereochemical course of the macrocyclization (Scheme 3, **10** and **11**). According to the planned synthetic route, *ortho*-nitro-substituted halogen-containing β-hydroxyphenylalanine derivatives were prepared as synthons for the amino acids 6 and 2 (see, Scheme 2, compound **5**, and Scheme 4, compound **13**) by diastereoselective oxazolidinone methods, as well as a central phenolic arylglycine derivative for amino acid 4 (Scheme 2, compound **3**) [2, 3]. The fluorine-containing nitroarenes require accurate control of the reaction conditions to avoid undesired S_NAr reactions causing substitution of the fluorine [2]. The creation of the structure element of axial chirality has to be considered as a further key step within the total synthesis [2, 3, 7]. The *ortho*-alkoxy-substituted derivative of amino acid 5 (Scheme 2, compound **4**) has been designed in accordance with the synthetic route and was synthesized applying chiral oxazolidinone methods. The general synthetic strategy starts with the construction of the 12-membered (5–7)-tripeptide macrocycle from tripeptide **7** (Scheme 3).

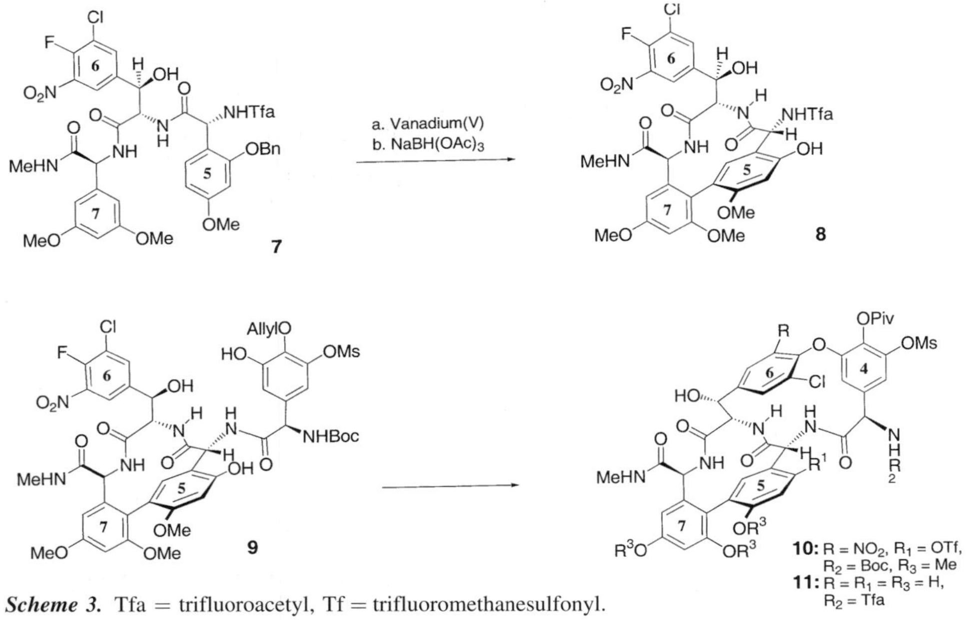

Scheme 3. Tfa = trifluoroacetyl, Tf = trifluoromethanesulfonyl.

A biomimetic oxidative cyclization procedure of these tripeptide precursors for the construction of the M(5–7) biaryl unit was described in the literature by Evans et al. in 1993 for the first time [7]. The vanadium(V)-mediated intramolecular coupling procedure is only successful in the presence of an alkoxy substituent in the *ortho*-position of the aryl moiety of amino acid 5. For the purpose of the total synthesis of the vancomycin aglycon **2**, the construction of the cyclic M(5–7)-tripeptide unit **8** (Scheme 3) was achieved by such an oxidative cyclization (VOF_3, $BF_3{\cdot}OEt_2$, $AgBF_4$, TFA/CH_2Cl_2) starting from the *N*-methylamide derivative **7**. By reductive work-up with $NaBH(OAc)_3$ with simultaneous fission of the amino acid 5 benzylether, the (*R*)-atropisomer **8** was isolated under preservation of the nitro group with high stereoselectivity (d.r. = 95 : 5, 65 % yield). In this case, an epimerization of the C-terminal stereocenter, previously seen with the corresponding amino acid-7-ester analog of **7**, was not observed. Thus, the *ortho*-alkoxy residue furnishes a high kinetic atropselectivity, though in favor of the unnatural (*R*)-isomer. Studies with the epimeric arylglycine synthon proved that A(1,3) strain and consequently the absolute configuration of the stereogenic center of amino acid 5 controls the stereochemical course of the construction of the M(5–7) biaryl unit [3, 7]. However, after building the rigid M(4–6) macrocycle (see Scheme 3, compound **11**), the desired (*S*)-biaryl configuration can be obtained (Scheme 4, compound **12**) by M(5–7) atropisomerization (MeOH, 24 h, 55 °C, d.r. = 95 : 5, 54 % yield). In the product, the 5–6 amide already exists in the desired *cis* configuration (see Scheme 1).

S_NAr-Macrocyclization

For the construction of the 16-membered M(4–6)-macrocycle, special attention had to be directed to the correct arrangement of the chloro-substituent at ring 6. The S_NAr macrocyclization gave the cyclization product diastereoselectively (d.r. = 5 : 1) in favor of the desired

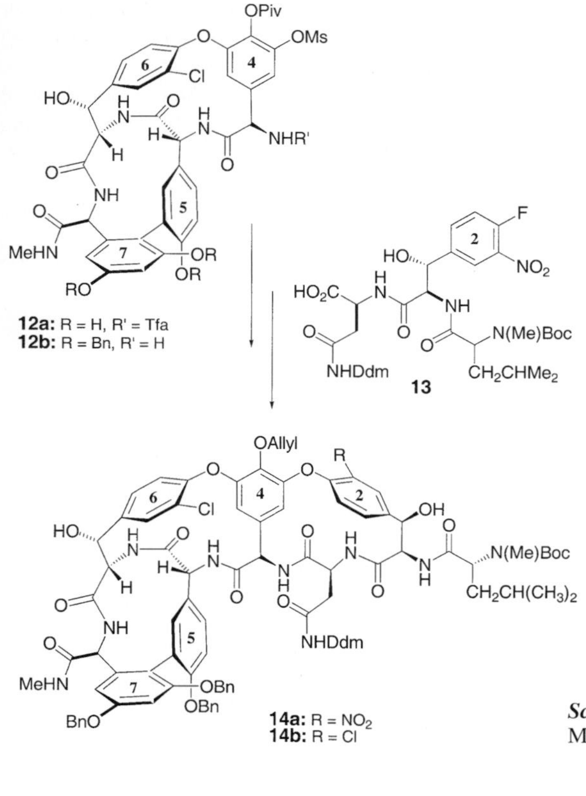

12a: R = H, R' = Tfa
12b: R = Bn, R' = H

13

14a: R = NO_2
14b: R = Cl

Scheme 4. Ddm = 4,4'-dimethoxydiphenylmethyl, Ms = methanesulfonyl atropisomer in 79 % yield (Na_2CO_3, DMSO, RT, 1.5 h, then TfNPh; see Scheme 3, compound **10**). The product **11** was obtained by reduction (Zn^0, HOAc, EtOH, 40 °C) and subsequent protecting group manipulations. After the M(5–7) atropisomerization to **12a**, described above, the coupling step with building block **13** was prepared. Thus, starting from **12b**, the corresponding heptapeptide was synthesized to afford the complete scaffold of the vancomycin aglycon by a second S_NAr cyclization (Scheme 4, **14**).

With CsF in DMSO at room temperature, the ring closure to the (*R*)-configured M(2–4) atropisomer **14a** was achieved in a yield of 95 % with a diastereoselectivity of 5 : 1. Detailed examinations of Evans et al. reveal that the diastereoselectivity of this M(2–4) cyclization is controlled by the natural conformation of the M(5–7) macrocycle [3]. The final functionalization to the chlorine-containing, protected vancomycin aglycon includes a chromatographic separation of the mixture of atropisomers by column chromatography.

In the course of the deprotection sequence yielding the vancomycin aglycon, the *N*-methyl amide at the carboxylic acid terminus was nitrosated highly regioselectively (N_2O_2, NaOAc, CH_2Cl_2, MeCN, 0 °C) and then cleaved with lithium hydroperoxide, furnishing the corresponding carboxylic acid in 68 % yield [2, 14].

Triazene Method According to Nicolaou et al.

Key reactions of the total synthesis of the vancomycin aglycon presented by Nicolaou and coworkers [4–6] are a Suzuki coupling [15] for the synthesis of the M(5–7) unit (Scheme 5) and the triazene method, described recently for the construction of macrocyclic biaryl ethers (Scheme 6) [16].

Starting material for the bicyclic peptide **22a** of desired configuration are the amino acid building blocks **15** and **16**, and the dipeptide **19**, shown in Scheme 5.

Compound **19** was synthesized starting from **17** and **18** by the Suzuki reaction ($[Pd(PPh_3)_4]$, Na_2CO_3) as a 2 : 1 mixture of atropisomers (84 % overall yield) in which the natural (*S*)-diastereomer predominated. By introduction of the azido function under inversion of the configuration followed by ester hydrolysis, compound **19** was finally obtained.

Key step of the synthesis of compound **15**, a derivative of amino acid 6, is an asymmetric Sharpless aminohydroxylation. The central building block **16** (amino acid 4), however, was built up from 4-aminobenzoic acid by an asymmetric dihydroxylation (AD) in 12 steps. Coupling of the biaryl fragment **19** with the corresponding amino acid derivatives (Scheme 6) gave tripeptide **20**. By treatment with CuBr · SMe_2, K_2CO_3 and pyridine in acetonitrile under re-

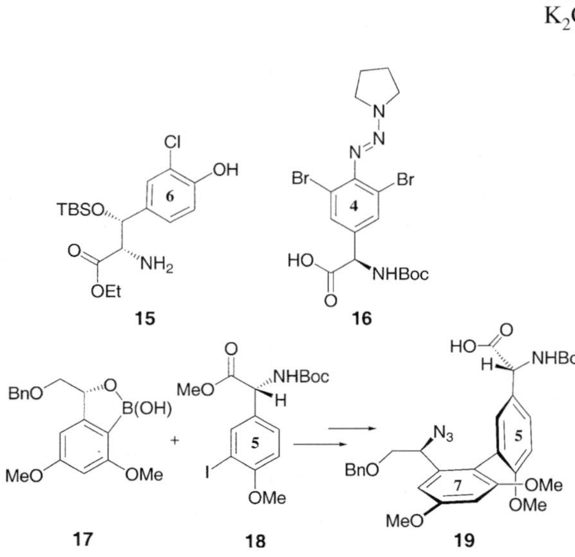

Scheme 5

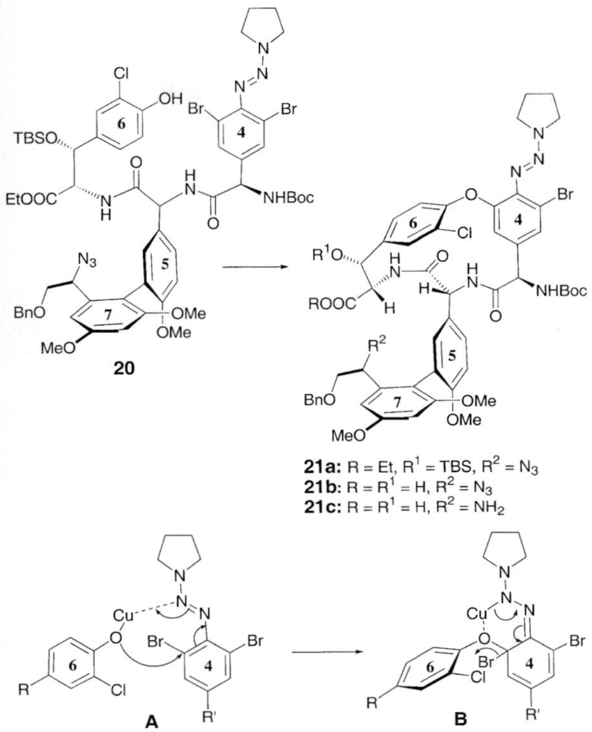

21a: R = Et, R^1 = TBS, R^2 = N_3
21b: R = R^1 = H, R^2 = N_3
21c: R = R^1 = H, R^2 = NH_2

Scheme 6

flux, the 16-membered M(4–6) macrocycle is obtained within 20 min as an inseparable mixture of atropisomers (d.r. = 1 : 1) in 60 % overall yield. In this reaction, the triazene unit functions as a coordinative anchoring group for metal ions, exerting a directing effect on the metalated nucleophile (Scheme 6, structures **A** and **B**).

In addition, it activates the arene and thus facilitates the substitution of the halogen atom in the *ortho*-position by its electronic influence. Beyond that, the triazene residue can be transformed into various other functional groups [16].

Following protecting group manipulations to prepare the next coupling step, the macrolactamization to the bicyclic building block **22a**, shown in Scheme 7, is carried out under the action of FDPP (pentafluorophenylphosphinate, 3.0 equiv. FDPP, 5.0 equiv. *i*-Pr_2NEt, DMF, 25 °C, 12 h, 71 % yield).

Subsequent coupling of the tripeptide **23** in the presence of EDC/HOAt ([(1-(3-dimethylaminopropyl)-3-ethylcarbodiimide-hydrochloride]/[7-aza-1-hydroxy-1*H*-benzotriazole]) yielded the corresponding heptapeptide in 81 % yield. The derivative of amino acid 2 (Scheme 7, compound **23**), required for the construction of the 16-membered M(2–4) macrocycle, was synthesized in four steps, including an enantioselective Sharpless dihydroxylation.

The final ring closure to the M(2–4)-macrocycle **24** (Scheme 7) produced an atropisomeric mixture in 72 % yield, favoring the unnatural (*S*)-atropisomer (d.r. = 3 : 1). However, after chromatographic separation, the latter can be converted thermally into a 1 : 1 mixture of atropisomers (1,2-dichlorobenzene, 140 °C, 4 h), thus opening up a stepwise route to the desired isomer **24**.

Last Hurdles

To complete the total synthesis, the unexpectedly difficult transformation of the triazene unit into the corresponding phenol had to be undertaken as a last hurdle [15, 6]. At first, the triazene unit was transformed into the aniline derivative **25** by re-

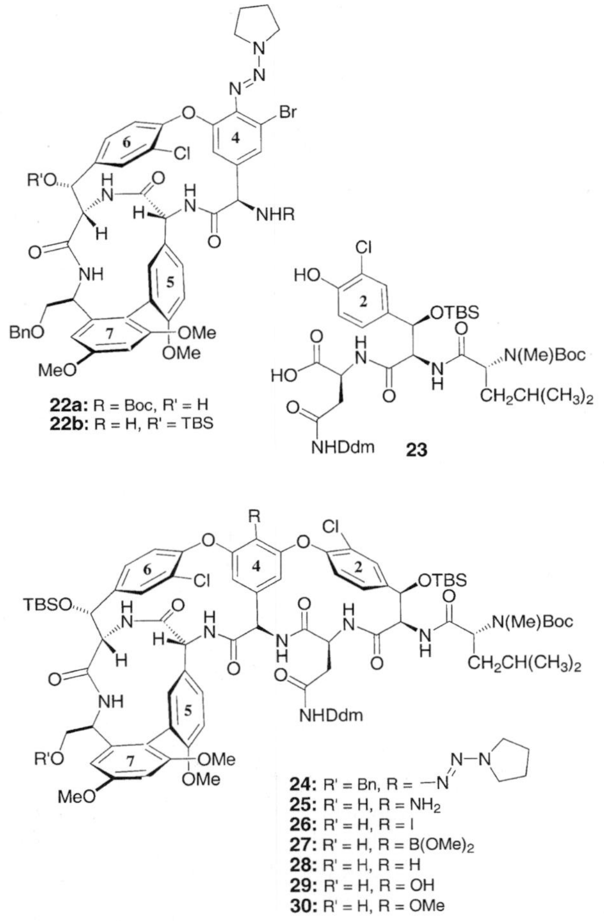

22a: R = Boc, R' = H
22b: R = H, R' = TBS

23

24: R' = Bn, R = —N=N—N(pyrrolidinyl)
25: R' = H, R = NH_2
26: R' = H, R = I
27: R' = H, R = $B(OMe)_2$
28: R' = H, R = H
29: R' = H, R = OH
30: R' = H, R = OMe

Scheme 7

duction (Scheme 7), resulting in simultaneous cleavage of the benzyl protecting group of amino acid 7. Diazotation (HBF_4, isoamyl nitrite) and subsequent treatment with potassium iodide yielded aryl iodide **26**. Halogen–metal exchange and simultaneous deprotonation of all NH groups and of the homobenzylic OH group with an excess of MeMgBr and *i*-PrMgBr, transformation to the boronic acid ester **27** and final treatment with alkaline hydrogen peroxide solution gave the desired product **29** with the phenolic hydroxyl group in 50 % overall yield, as well as the corresponding reduction product **28** (40 % yield). Prior to the cleavage of all protecting groups, starting from **30**, the homobenzylic hydroxyl group was oxidized to the carboxyl group of the carboxy terminus (Dess-Martin periodinan, CH_2Cl_2, then $KMnO_4$, *t*-BuOH, 5 % aqueous Na_2HPO_4 solution) and subsequently transformed to the methyl ester with diazomethane (90 % yield).

Only recently, the synthesis of the complete vancomycin framework was described by Nicolaou et al. [17]. Prior to the glycosylation of the aglycon **2** at the phenolic hydroxy group of amino acid 4, all hydroxyl groups were silylated with TBSOTf/lutidine. In addition, the C-terminal carboxylic acid was protected with diazomethane and the amino terminus with the Cbz group. Subsequently, the TBS ether of amino acid 4 was selectively cleaved with $KFAl_2O_3$ [18], furnishing the glycosyl acceptor **31** in 60 % yield (Scheme 8).

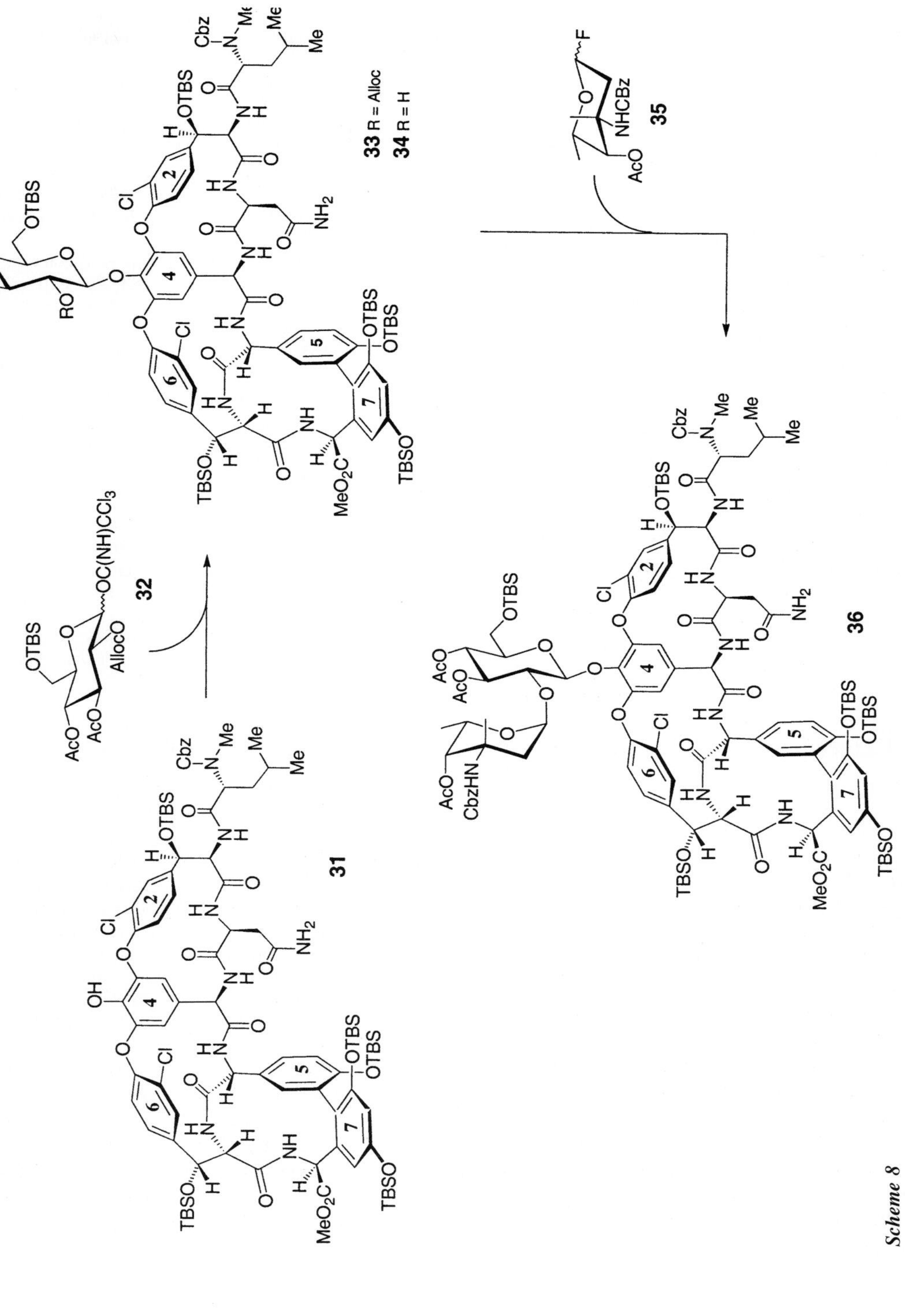

Scheme 8

Boron trifluoride etherate-promoted glycosylation with the 2-allyloxycarbonyl-protected trichloroacetimidate **32** as the donor yielded the β-glucoside **33** in 82 % yield. Selective deprotection of the 2-hydroxy group of the glucopyranoside was achieved with n-Bu_3SnH/Pd(0) after prior glycosylation with the protected vancosamine fluoride **35**, furnishing the α-glycosidically linked disaccharide **36** in 84 % yield (Scheme 8). The silyl ether protecting groups were then cleaved with HF/pyridine, and this was followed by removal of the acetyl protecting groups. The Cbz-groups were simultaneously hydrogenated with Raney nickel/H_2 in n-PrOH/H_2O, whereas the *C*-terminal methyl ester was cleaved by treatment with LiOH in THF/H_2O, yielding vancomycin **1** in 42 % yield (4 steps).

With the presented total syntheses of the vancomycin aglycon, for the first time comprehensive methods and strategies for the synthesis of vancomycin antibiotics and the construction of modified structures are available. Recently, bacterial strains have been discovered that displayed resistance against vancomycin [2]. Against this background, new biological studies with synthetically modified members of the vancomycin class of natural substances could be promising for the future.

References

[1] C. M. Harris, T. M. Harris, *J. Am. Chem. Soc.* **1982**, *104*, 4293 and literature cited therein.

[2] D. A. Evans, M. R. Wood, B. W. Trotter, T. I. Richardson, J. C. Barrow, J. L. Katz, *Angew. Chem.* **1998**, *110*, 2864; *Angew. Chem. Int. Ed.* **1998**, *37*, 2700 and literature cited therein.

[3] D. A. Evans, C. J. Dinsmore, P. S. Watson, M. R. Wood, T. I. Richardson, B. W. Trotter, J. L. Katz, *Angew. Chem.* **1998**, *110*, 2868; *Angew. Chem. Int. Ed.* **1998**, *37*, 2704 and literature cited therein.

[4] K. C. Nicolaou, S. Natarajan, H. Li, N. F. Jain, R. Hughes, M. E. Solomon, J. M. Ramanjulu, C. N. C. Boddy, M. Takayanagi, *Angew. Chem.* **1998**, *110*, 2872; *Angew. Chem. Int. Ed.* **1998**, *37*, 2708.

[5] K. C. Nicolaou, N. F. Jain, S. Natarajan, R. Hughes, M. E. Solomon, H. Li, J. M. Ramanjulu, M. Takayanagi, A. E. Koumbis, T. Bando, *Angew. Chem.* **1998**, *110*, 2879; *Angew. Chem. Int. Ed.* **1998**, *37*, 2714 and literature cited therein.

[6] K. C. Nicolaou, M. Takayanagi, N. F. Jain, S. Natarajan, A. E. Koumbis, T. Bando, J. M. Ramanjulu, *Angew. Chem.* **1998**, *110*, 2881; *Angew. Chem. Int. Ed.* **1998**, *37*, 2717.

[7] a) D. A. Evans, C. J. Dinsmore, D. A. Evrard, K. M. DeVries, *J. Am. Chem. Soc.* **1993**, *115*, 6426; b) D. A. Evans, C. J. Dinsmore, *Tetrahedron Lett.* **1993,** *34*, 6029.

[8] a) Y. Suzuki, S. Nishiyama, S. Yamamura, *Tetrahedron Lett.* **1989,** *30*, 6043; D. A. Evans, J. A. Ellman, K. M. DeVries, *J. Am. Chem. Soc.* **1989**, *111*, 8912.

[9] a) D. A. Evans, C. J. Dinsmore, A. M. Ratz, D. A. Evrard, J. C. Barrow, *J. Am. Chem. Soc.* **1997**, *119*, 3417; b) D. A. Evans, J. C. Barrow, P. S. Watson, A. M. Ratz, C. J. Dinsmore, D. A. Evrard, K. M. DeVries, J. A. Ellman, S. D. Rychnovsky, J. Lacour, *J. Am. Chem. Soc.* **1997**, *119*, 3419; c) D. A. Evans, C. J. Dinsmore, A. M. Ratz, *Tetrahedron Lett.* **1997,** *38*, 3189.

[10] Review: A. V. Rama Rao, M. K. Gurjar, K. L. Reddy, A. S. Rao; *Chem. Rev.* **1995**, *95*, 2135.

[11] A. V. Rama Rao, T. K. Chakraborty, K. L. Reddy, A. S. Rao; *Tetrahedron Lett.* **1992**, *33*, 4799.

[12] D. L. Boger, R. M. Borzilleri, S. Nukui, R. T. Beresis, *J. Org. Chem.* **1997**, *62*, 4721.

[13] a) R. Beugelmans, G. P. Singh, J. Zhu, *Tetrahedron Lett.* **1993**, *34*, 7741; b) review: J. Zhu, *Synlett* **1997**, 133.

[14] D. A. Evans, P. H. Carter, C. J. Dinsmore, J. C. Barrow, J. L. Katz, D. W. Kung, *Tetrahedron Lett.* **1997**, *38*, 4535.

[15] K. C. Nicolaou, J. M. Ramanjulu, S. Natarajan, S. Bräse, H. Li, C. N. C. Boddy, F. Rübsam, *Chem. Commun.* **1997**, 1899 and literature cited therein.

[16] K. C. Nicolaou, C. N. C. Boddy, S. Natarajan, T.-Y. Yue, H. Li, S. Bräse, J. M. Ramanjulu, *J. Am. Chem. Soc.* **1997**, *119*, 3421 and literature cited therein.

[17] K. C. Nicolaou, H. J. Mitchell, N. F. Jain, N. Winssinger, R. Hughes, T. Bando, *Angew. Chem.* **1999**, *111*, 253; *Angew. Chem. Int. Ed.* **1999**, *38*, 240.

[18] E. A. Schmittling, J. S. Sawyer, *Tetrahedron Lett.* **1991**, *32*, 7207.

B. Synthesis of Non-Natural Compounds and Materials

An Update on the New Inductees in the „Hall of Phane“ – No Phane, No Gain!

Graham J. Bodwell

Department of Chemistry, Memorial University of Newfoundland, Canada

Since their emergence as a distinct class of compounds in the 1950s, the general appeal of cyclophanes has shown no signs of waning. They possess unusual and aesthetically pleasing structures. They pose unique synthetic challenges. They undergo interesting conformational processes. They exhibit peculiar chemical reactivity, spectroscopic properties and physical properties. They can also be designed to be chiral. From the ever smaller and more strained cyclophanes to the ever larger and more complex cyclophanes, chemists continue to push back the frontiers of all aspects of this captivating field of chemistry.

The [n]cyclophanes are the archetypal small cyclophanes and, within this area, it is the [*n*]paracyclophanes (Fig. 1) that have received by far the most attention. A detailed computational study of [4]paracyclophane [1], which has only been prepared as a transient species via matrix isolation [2], was recently reported. An energy difference of 9 kcal mol^{-1} between it and its Dewar benzene isomer was predicted. This paper also provides a comprehensive summary of the literature of the [n]paracylophanes.

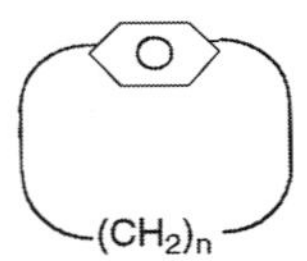

Figure 1. Structural diagram of the [n]paracyclophanes.

[5]Paracyclophane, which exists as the minor component of an equilibrium with its valence-isomeric Dewar benzene, is the smallest member of this series to exhibit sufficient stability to be studied directly [3]. In order to further stabilize the [5]paracyclophane unit, Bickelhaupt [4] recently synthesized a benzannulated version, namely [5](1,4)naphthalenophane (**2**, Scheme 1). Even though this compound also exists in equilibrium with the Dewar naphthalene **1** from which it was produced, the proportion of the cyclophane (35 % of the mixture) is higher than that in any previously reported case. A particularly interesting feature of this system is that there are two observable bridge conformers of **2** in solution in a ratio of 95 : 5. Unfortunately, despite considerable effort, assignment of the two conformers was not possible.

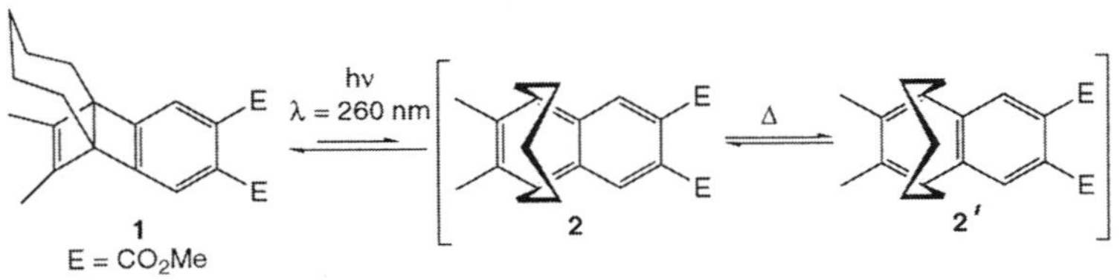

Scheme 1

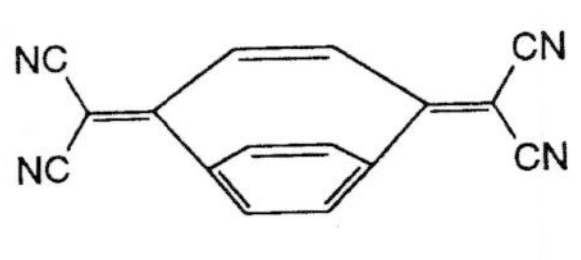

3 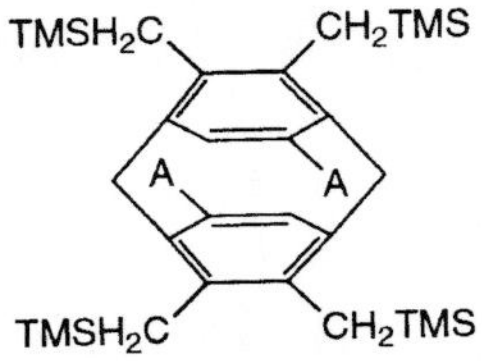   4 A = $CONMe_2$

Scheme 2

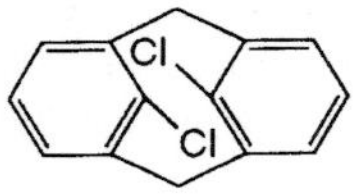

5 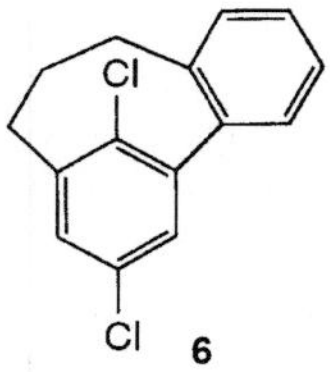   6 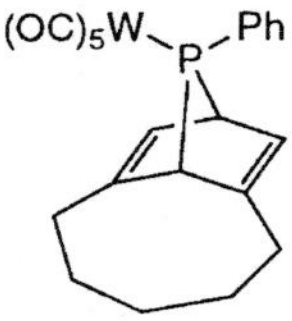   7

Scheme 3

Seemingly impossibly strained, but kinetically stabilized, cyclophanes **3** and **4** have been prepared by the group of Tsuji (Scheme 2). A small, but observable, proportion (6 : 94) of [4]paracyclophanediene **3** was formed upon low-temperature irradiation of the corresponding Dewar benzene [5]. Weak ^{1}H NMR signals at δ values of 5.85 and 7.97 were assigned to the bridge and ring protons, respectively. The bend angles α and β were calculated (B3LYP/6–31G*) to be 28.6° and 43.0°. The observation of [1.1] paracyclophane was first reported by Tsuji in 1993 [6] and the heavily substituted version **4** in 1998 [7], both by irradiation of the appropriate Dewar benzenes. The fairly air-sensitive **4** was stable enough to allow the determination of its structure in the crystal. The angles α (24.3° and 25.6°) and β (22.9° and 26.8°) are the largest ever measured for a paracyclophane, but slightly less than those calculated for [5]paracyclophane [8]. The ring protons of **4** were observed at δ 7.67. Other advances in paracyclophane chemistry have been covered recently by de Meijere and König [9].

In the area of metacyclophanes, the existence of 7,14–dichloro [1.1]metacyclophane (**5**) was postulated in 1997 by Bickelhaupt et al. (Scheme 3) [10]. Related work from this group led to the synthesis of the benzannulated [5]metacyclophane **6** [11] and the 7-phosphanorbornadiene **7** by an apparent [4 + 1] cycloaddition to [5]metacyclophane [12].

The aromatic portion of a cyclophane does not necessarily have to be benzenoid. Some very appealing non-benzenoid cyclophanes have been prepared by Gleiter et al. over the last decade (Scheme 4). Examples of these are the cyclopropenyliophanes **8–10** [13] and the cyclopropenonophanes **11–13** [14]. This and related work is discussed in a recent review [15].

Over the years, a number of [n]cyclophanes based on polycyclic aromatic hydrocarbons have been prepared [16]. However, the absolute limits to which any of these systems can be bent has not been systematically investigated. Hafner's [n](1, 6)- (**15**) and [n](2, 6)azulenophanes ($n = 11-13$, **16**) are a promising step in this direction [17]. They were produced by an intramolecular azulene-forming reaction of the 4–(ω-cyclopentadienylalkyl)-1-methylpyridinium salts **14** (Scheme 5). The crystal structure of

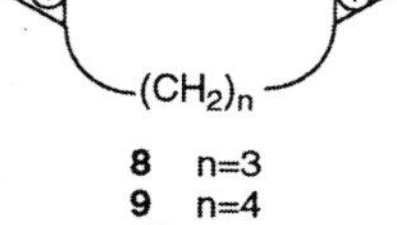

8 n=3
9 n=4
10 n=5

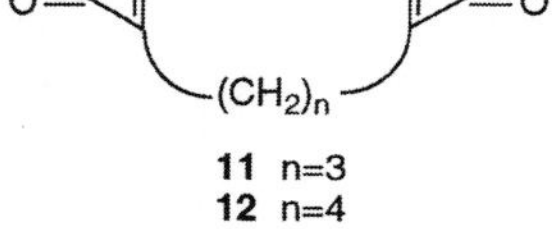

11 n=3
12 n=4
13 n=5

Scheme 4

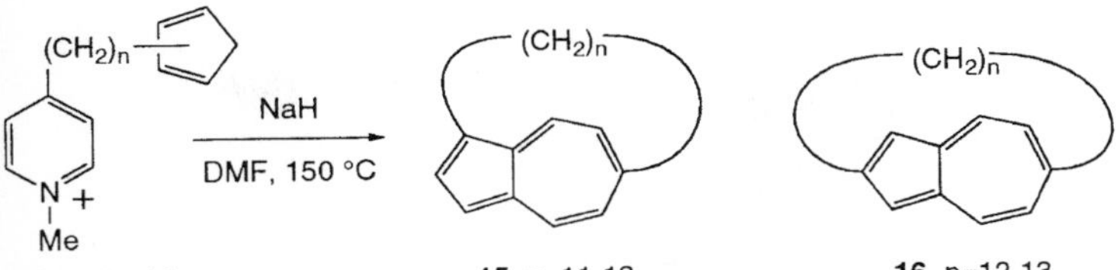

Scheme 5

Scheme 6

[13](2,6)azulenophane (**16**) (n = 13) revealed a 3.7° angle between the mean planes of the two rings. Hopefully, lower homologs of this series will soon be prepared.

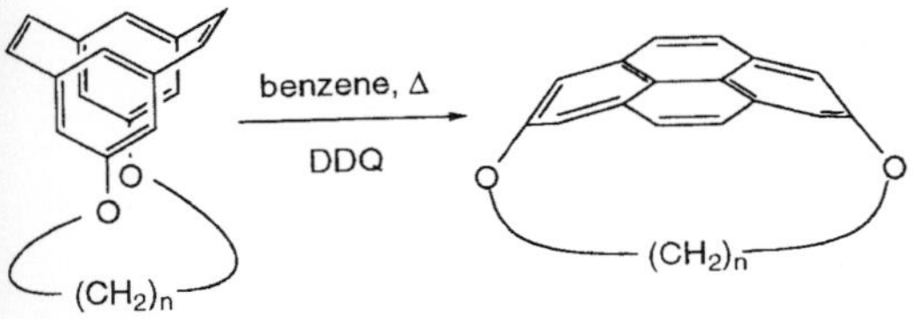

Our group was able to impart much greater deviations from planarity in the pyrene framework [18]. The remarkably distorted 1,8-dioxa[8](2,7)pyrenophane **19** was prepared by the valence isomerization and dehydrogenation of the tethered [2.2]metacyclophanediene **17** (Scheme 6). In its crystal structure an overall bend of nearly 90° was observed. The curvature of the aromatic surface approximates to that expected for the corresponding part of the equatorial belt of D_{6h} C_{84}. Bond angles of up to 119° were observed in the aliphatic bridge as well as a very high field proton resonance at δ−1.47 ppm. The next smaller member of this series **20** was initially elusive, but modification of the work-up allowed the isolation of the [7](2,7)pyrenophane, albeit it in somewhat reduced yield [19]. Nevertheless, a crystal structure was determined, and this revealed an enormous end-to-end bend of 109.1°, slightly more than that of the corresponding part of the equatorial belt of D_{5h} C_{70}. The highest field protons of the bridge appear at δ−2.10.

Related work in our group led to the synthesis of the cyclophanes **21**–**23** (Scheme 7) [20]. It was found that **21** adopted the *syn* conformation exclusively and **23** adopted the *anti* conformation exclusively. However, cyclophane **22** was observed to exist in a ca. 6 : 1 *anti*:*syn* ratio at equilibrium. The two conformers can be separated by flash chromatography and the return to the equilibrium ratio monitored by ^{1}H NMR. Noteworthy here is the direct observation of an *anti* to *syn* flip of a [2.2]metacyclophane. There have been only two other reports of such *anti* to *syn* flips [21]. Also noteworthy is the chemical shift of the internal proton of the inner ring of *anti*-**22**, which appears at δ 3.03.

It has long been known that appropriately substituted [2.2]paracyclophanes are chiral, chemically stable and do not racemize under normal reaction conditions. With these seemingly ideal prerequisites for use in chiral synthesis, it is perhaps surprising that only three examples have appeared in the literature, all of them in recent years (Scheme 8). Reich employed [2.2]paracyclophane-derived selenides such as **24** to administer chirality transfer in selenoxide [2, 3] sigmatropic rearrangements. Using this methodology, he was able to synthesize optically active linalool **25**

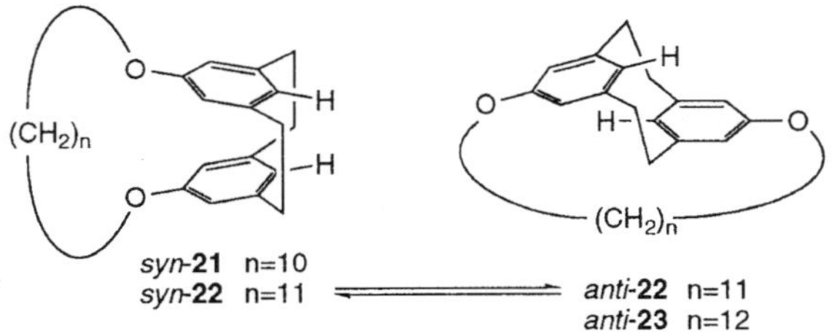

Scheme 7

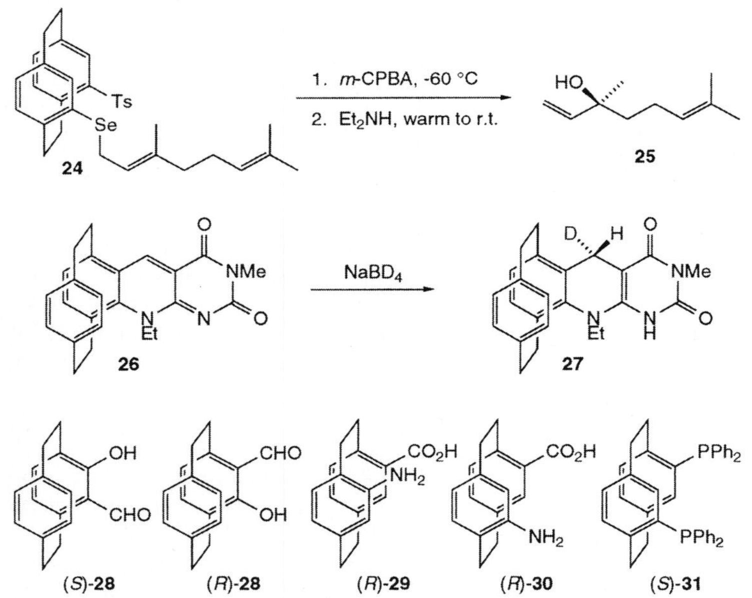
1. m-CPBA, -60 °C
2. Et2NH, warm to r.t.
Ts
Se
24
HO
25
NMe
N
Et
N
O
26
NaBD4
D
H
27
OH
CHO
(S)-28
(R)-28
CO2H
NH2
(R)-29
(R)-30
PPh2
(S)-31

Scheme 8

(S:R ratio = 5 : 1) [22]. Yanada and Yoneda constructed the deazaflavinophane **26**, which exhibits complete facial selectivity in its oxidation and reduction reactions, e.g. the reduction with $NaBD_4$ to afford **27** [23]. Belokon and Rozenberg used scalemic 4–formyl-5-hydroxy[2.2]paracyclophane (FHPC) **28** in the synthesis of α-amino acids (*ee* 45–98 %) [24]. An alternative approach to FHPC was more recently reported by Hopf [25]. Other interesting advances in the area of chiral cyclophanes include the homochiral [2.2]paracyclophane-derived amino acids **29** and **30** [26], as well as (S)-PHANEPHOS (**31**) [27], which has been shown to be an effective ligand for highly enantioselective Ru-catalyzed asymmetric hydrogenations of β-ketoesters and Rh-catalyzed asymmetric hydrogenations of dehydroamino acids.

Many cyclophanes are conformationally mobile, and the study of their behavior in solution has proved to be most intriguing. A particularly striking example of conformational analysis was recently reported by Fukazawa [28]. The [4.4]paracyclophane derivative **34** was synthesized by a photoinduced double $S_{RN}1$ reaction of the precursors **32** and **33** (Scheme 9). In the low-temperature ^1H NMR spectrum of **34**, signals due to three distinct conformers were observed. Assignment of the structure of the major conformer was then accomplished in a unique way. The crystal structure was determined, and this corresponded to the calculated lowest energy

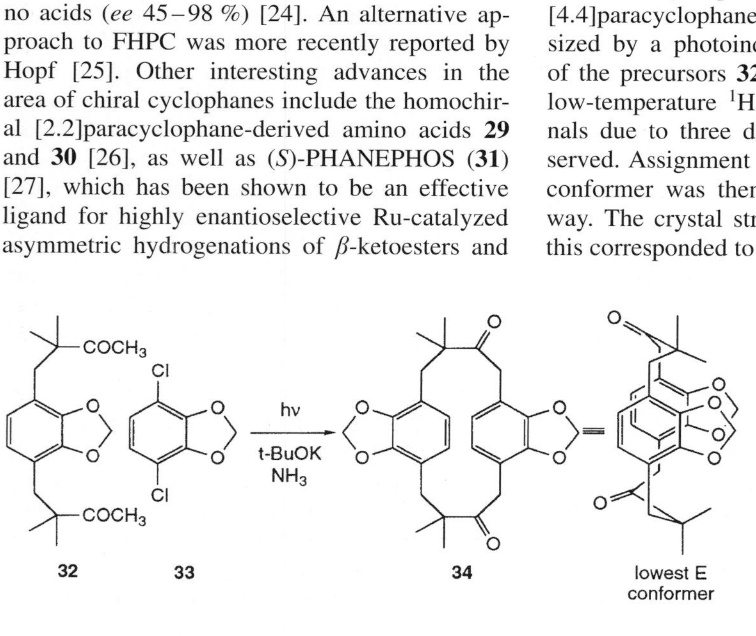
COCH3
Cl
hν
t-BuOK
NH3
32
33
34
lowest E
conformer

Scheme 9

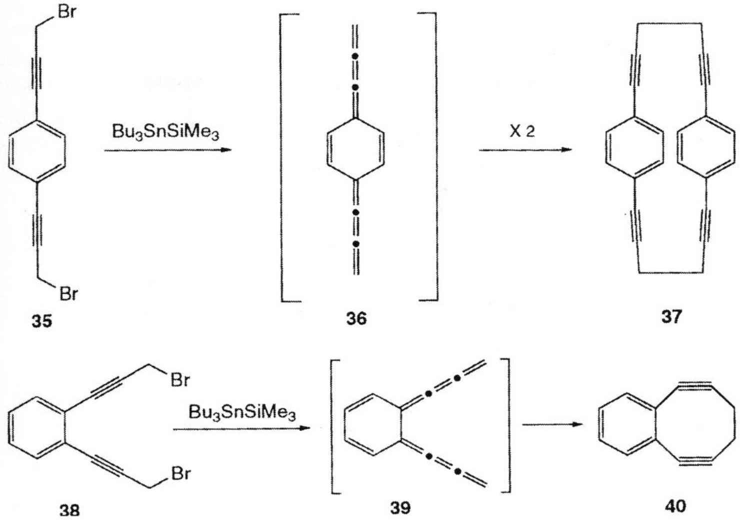

Br Br Bu3SnSiMe3 X 2 35 36 37 Br Br Bu3SnSiMe3 38 39 40

Scheme 10

conformer. When crystals of **34** were dissolved in precooled (–80 °C) solvent, the ^1H NMR spectrum of the resulting solution exhibited signals arising only from the major conformer of the previous mixture.

[6.6]Paracyclophane-1, 5, 12, 16–tetrayne (**37**) was synthesized by Hopf via the reaction of dibromide **35** with $Bu_3SnSiMe_3$ (Scheme 10) [29]. The intermediacy of the novel cumulated *para*-quinodimethane **36** was invoked. Reaction of the *ortho*-substituted isomer **38** under the same conditions presumably gave the intermediate **39**, which did not afford the corresponding orthocyclophane, but rather gave the highly strained 3,4-benzocyclooct-3-ene-1,5–diyne **40**.

In addition to Hopf's elegant work, a variety of other alkyne-containing cyclophanes have been prepared in the 1990s (Scheme 11). Examples of these are cyclophanes **41** (Bodwell) [30], **42** (Ensley) [31], **43** (Gleiter) [32], **44–46** (Oda) [33], **47** (Fox) [34], and **48–49** (Rubin) [35].

Seventeen years after Boekelheide's landmark synthesis of $[2_6](1, 2, 3, 4, 5, 6)$cyclophane (also known as superphane) [36], Shinmyozu very recently described the synthesis of the homologous superphane $[3_6](1, 2, 3, 4, 5, 6)$cyclophane (**53**, Scheme 12) [37]. Formation of the final bridge of this "molecular pinwheel" was accomplished with an intramolecular aldol condensation. While the double bond of enone **50** could be easily hydrogenated, the carbonyl of the resulting ketone **51** proved to be quite inert towards reduction. This was finally achieved by the action of SmI_2/1 M KOH. Treatment of the alcohol **52** with $LiAlH_4/AlCl_3$ finally afforded the long sought after cyclophane **53**. The well averaged ^1H NMR signals of the bridge protons were said to be consistent with a correlated inversion of all six trimethylene bridges in solution at room temperature, but a subsequent computational study concluded that a synchronous mechanism is ruled out [38]. No report has yet appeared to confirm or refute Osawa's prediction [39] that **53** will isomerize under irradiation to the [6]prismane derivative **54**, but a photochemical study of a lesser bridged cyclophane has recently been published [40].

Other superphanes of note are CpCo-capped cyclobutadiene and cyclopentadienone-containing superphanes prepared by Gleiter's group. Examples of these are **55–57** (Scheme 13) [41].

Tani recently described the coupling of the tetrabromide **58** with the tetraselenocyanate **59** to give the parallel and crossed tetraselenabiphenylophanes **60** and **61** in a ratio of 1 : 3 (Scheme 14) [42]. The synthesis of the thia-analog of **61** has been previously reported using a thiol-bromide coupling [43]. However, its formation was not accompanied by the isomer corresponding to **60**. While desulfurization of the thia-analog of

61 was not achieved, **60** and **61** were successfully deselenated in 40–50 % yield upon photolysis in the presence of $P(NMe_2)_3$ to give the structurally interesting bridge-contracted phanes **62** and **63**.

The *in*-cyclophanes **64-67** described by Pascal are excellent vehicles for the study of functional group interactions resulting from enforced proximity (Scheme 15). While original work was directed towards compounds containing alkane CH groups projected towards the center of an aromatic ring (viz. **64**, $\delta_H = -4.03$) [44], it has since been expanded to include second row elements. The silaphane **65** [45] exhibits a proton resonance at δ 1.04, some 5 ppm higher field

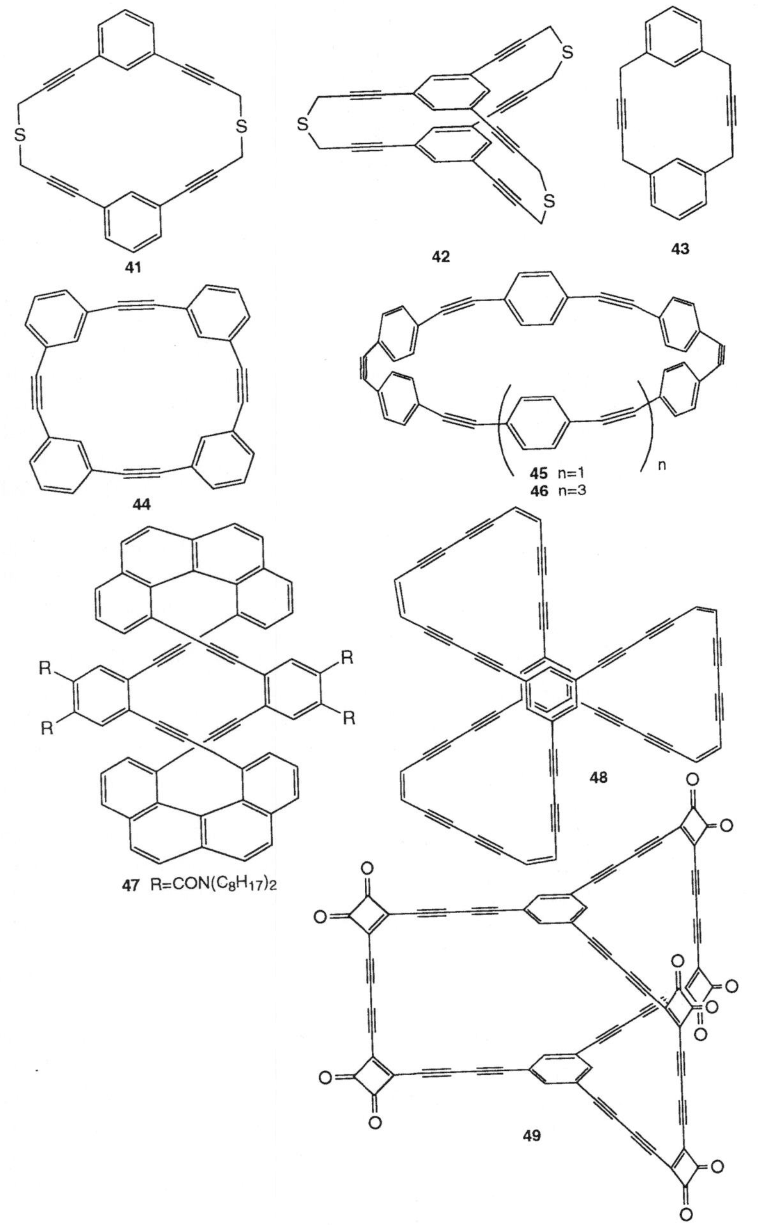

Scheme 11

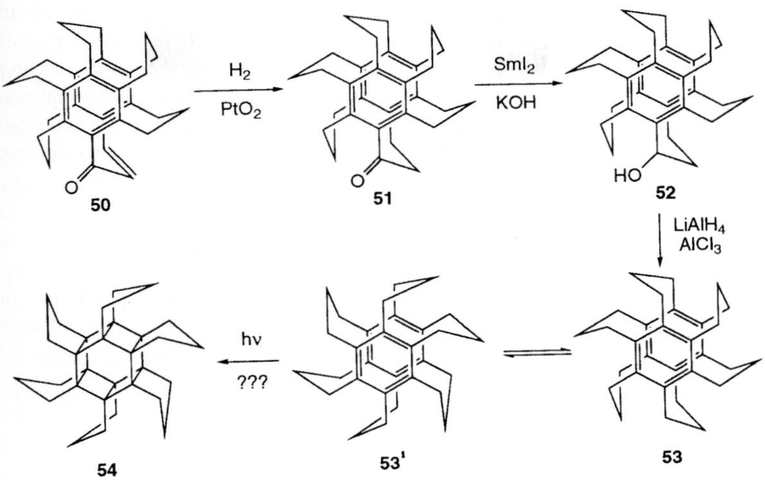

Scheme 12

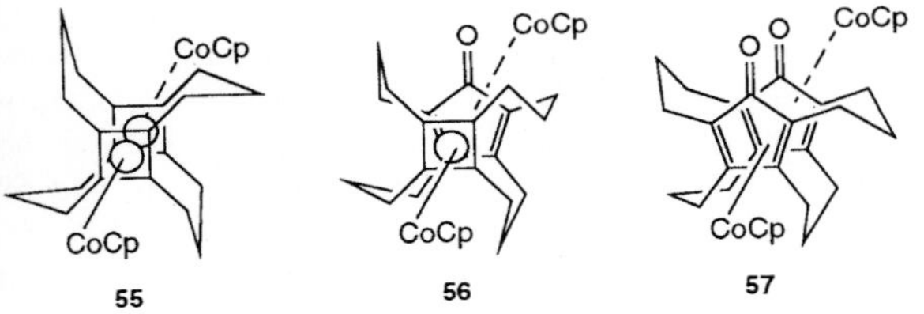

Scheme 13

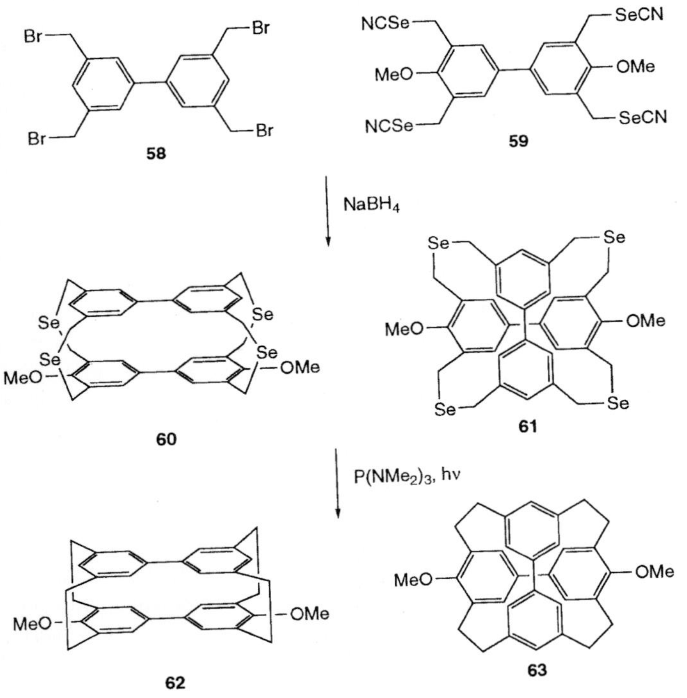

Scheme 14

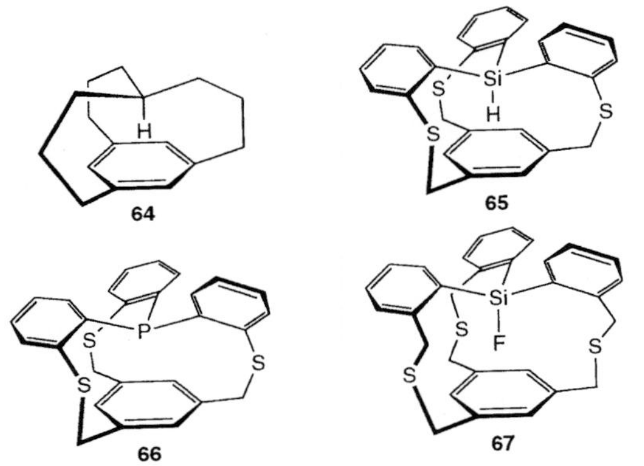

Scheme 15

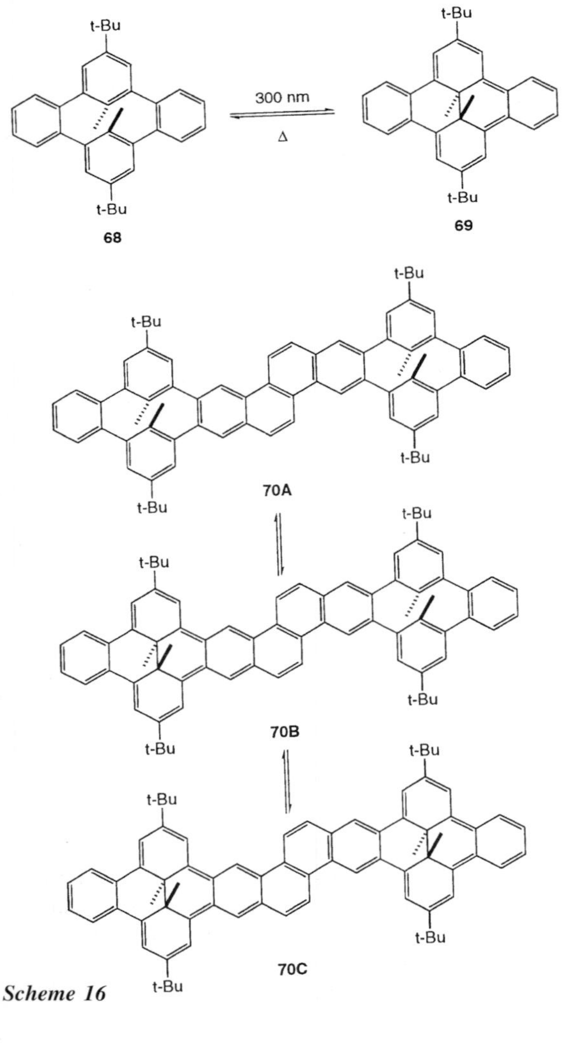

Scheme 16

than that of an appropriate model compound. In addition there is a 280 cm^{-1} hypsochromic shift of the infrared Si–H stretching frequency arising from steric compression. For the phosphaphane **66** [46], which is resistant to protonation with HBr, the ^{13}C and ^{31}P NMR spectra indicate that there is an electronic interaction between the phosphine and the basal ring. However, the presence of an attraction between them (i.e. a bond) could not be determined. Subsequent functionalization of the basal ring with a nitro or amino substituent allowed for HPLC resolution of the enantiomers which exhibit exceptionally high optical activity [46]. The most recent addition to this family is fluorosilaphane **67** [47]. The Si–F bond is particularly short (1.59 Å).

The elusive dibenzannulated dimethyldihydropyrene **69** was prepared by the group of Mitchell (Scheme 16) [48]. A Diels-Alder cycloaddition between an aryne and furan was employed to introduce the key benzene rings to the metacyclophanediene framework **68**. The colorless **68** switches to green **69** on UV irradiation, and **69** reverts thermally or photochemically. More recently, the same group reported the synthesis of the three-way molecular switch **70** [49].

Vögtle's group continues to produce a steady stream of fascinating cyclophanes. Among these are the nanometer-scale molecular ribbons **71**, which were synthesized using an iterative synthetic approach (Scheme 17) [50]. The longest one reported to date, **71** ($n = 7$), has nine layers. Layered cyclophanes based on the [3.3]orthocyclophane skeleton, e.g. **72**, have also been described [51]. The conversion of some of Vögtle's ribbons into molecular belts such as **73–75** has also been reported [52]. Cyclic side-products containing up to 40 (!) benzene rings were observed by plasma desorption mass spectrometry, although none of these higher oligomers has been isolated. The eventual conversion of these macrocycles into fully aromatic belts **76** should provide an entry into the direct study of nanotube fullerene fragments. The "spheriphanes" **77** and **78** were also prepared by the same group [53]. The former of these exhibits a high affinity for silver cations and, by virtue of its sixty carbon framework, may serve as a direct precursor to C_{60} [54]. The same can also be said for Rubin's cyclophanes **48** and **49**.

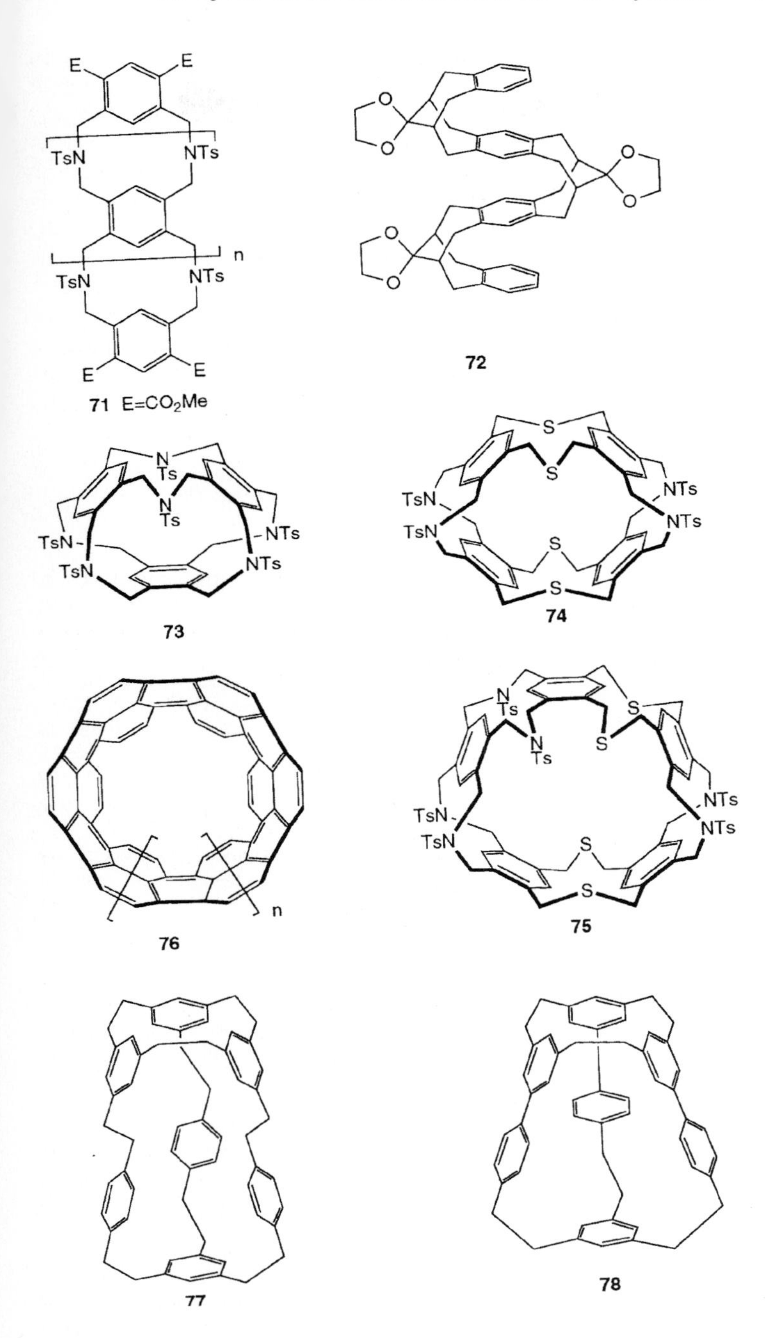

Scheme 17

One of the most spectacular cyclophanes to be prepared in recent years is the "Kuratowski cyclophane" **79** (Scheme 18) [55]. This nanodimensional multicyclophane is the first reported example of an achiral molecule possessing non-planar $K_{3,3}$ topology. The straightforward synthesis of this intricately entwined macrocyclophane should pave the way to a whole new family of molecules possessing novel architecture and vast dimensions. A further contribution from

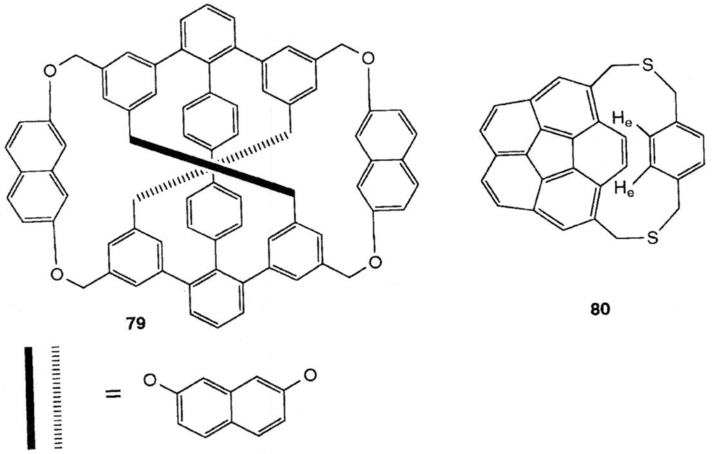

Scheme 18

Siegel's group is the synthesis of the first corannulene cyclophane **80** [56]. A striking feature of its ^{1}H NMR spectrum is that the *endo* aromatic protons (H_e) of the benzene ring, which are positioned almost directly over the center of the corannulene system, appear at remarkably high field ($\delta = 1.89$). Furthermore, with no peak broadening up to 148 °C, it was established that dynamic conformational processes in the thioether bridges and bowl-to-bowl inversion of the corannulene moiety both have lower limits of 18 ± 1 kcal mol^{-1}.

From its rudimentary beginnings over four decades ago, cyclophane chemistry has developed into an ever-evolving, multifaceted discipline which overlaps with many other areas of chemistry. A rough cross-section (by no means exhaustive) of some of the recent developments in the mainstream of this field, as highlighted here, certainly attests to that.

References

[1] B. Ma, H. M. Sulzbach, R. B. Remington, H. F. Schaefer III, *J. Am. Chem. Soc.* **1995**, *117*, 8392–8400.

[2] G. B. M. Kostermans, M. Bobeldijk, W. H. de Wolf, F. Bickelhaupt, *J. Am. Chem. Soc.* **1987**, *109*, 2471–2475; T. Tsuji, S. Nishida, *ibid.* **1988**, *110*, 2157–2164; F. Bickelhaupt, *Pure Appl. Chem.* **1990**, *62*, 373–382.

[3] L. W. Jenneskens, F. J. J. de Kanter, P. A. Kraakman, L. A. M. Turkenburg, W. E. Koolhaas, W. H. de Wolf, F. Bickelhaupt, Y. Tobe, K. Kakiuchi, Y. Odaira, *J. Am. Chem. Soc.* **1985**, *107*, 3716–3717.

[4] D. S. van Es, F. J. J. de Kanter, W. H. de Wolf, F. Bickelhaupt, *Angew. Chem.* **1995**, *107*, 2728–2730; *Angew. Chem. Int. Ed. Engl.* **1995**, *34*, 2553–2555.

[5] M. Okuyama, T. Tsuji, *Angew. Chem.* **1997**, *109*, 1157–1158; *Angew. Chem. Int. Ed. Engl.* **1997**, *36*, 1085–1086.

[6] T. Tsuji, M. Ohkita, T. Konno, S, Nishida, *J. Am. Chem. Soc.* **1997**, *119*, 8425–8431; T. Tsuji, M. Ohkita, S. Nishida, *J. Am. Chem. Soc.* **1993**, *115*, 5284–5285.

[7] H. Kawai, T. Suzuki, M. Ohkita, T. Tsuji, *Angew. Chem.* **1998**, *110*, 827–829; *Angew. Chem. Int. Ed. Engl.* **1998**, *37*, 817–819.

[8] S. Grimme, *J. Am. Chem. Soc.* **1992**, *114*, 10542; M. von Arnim, S. D. Peyerimhoff, *Theor. Chim. Acta* **1993**, *85*, 43.

[9] A. de Meijere, B. König, *Synlett* **1997**, 1221–1232.

[10] M. J. van Eis, F. J. J. de Kanter, W. H. de Wolf, F. Bickelhaupt, *J. Org. Chem.* **1997**, *62*, 7090–7091.

[11] M. J. van Eis, F. J. J. de Kanter, W. H. de Wolf, F. Bickelhaupt, *J. Am. Chem. Soc.* **1998**, *120*, 3371–3375.

[12] M. J. van Eis, C. M. D. Komen, F. J. J. de Kanter, W. H. de Wolf, K. Lammertsma, F. Bickelhaupt, M. Lutz, A. L. Spek, *Angew. Chem.* **1998**, *110*, 1656–1658; *Angew. Chem. Int. Ed. Engl.* **1998**, *37*, 1547–1550.

[13] R. Gleiter, M. Merger, *Tetrahedron Lett.* **1992**, *33*, 3473–3476; R. Gleiter, M. Merger, T. Oeser, H. Irngartiner, *Tetrahedron Lett.* **1995**, *36*, 6425–6428.

[14] R. Gleiter, M. Merger, A. Altreuther, H. Irngartiner, *J. Org. Chem.* **1996**, *61*, 1946–1953.
[15] R. Gleiter, M. Merger, *Angew. Chem.* **1997**, *109*, 2532–2546; *Angew. Chem. Int. Ed. Engl.* **1997**, *36*, 2426–2439.
[16] For some examples, see J. A. Reiss in *Cyclophanes, Vol II* (Eds.: P. M. Keehn, S. M. Rosenfeld), Academic Press, New York, **1983**, p. 443–484.
[17] P. Schuchmann, K. Hafner, *Tetrahedron Lett.* **1995**, *36*, 2603–2606.
[18] G. J. Bodwell, J. N. Bridson, T. J. Houghton, J. W. J. Kennedy, M. R. Mannion, *Angew. Chem.* **1996**, *108*, 1418–1420; *Angew. Chem. Int. Ed. Engl.* **1996**, *35*, 1320–1321.
[19] G. J. Bodwell, J. N. Bridson, T. J. Houghton, J. W. J. Kennedy, M. R. Mannion, *Chem. Eur. J.* **1999**, *5*, 1823–1827.
[20] G. J. Bodwell, J. N. Bridson, T. J. Houghton, J. W. J. Kennedy, M. R. Mannion, *Angew. Chem.* **1996**, *108*, 2280–2281; *Angew. Chem. Int. Ed. Engl.* **1996**, *35*, 2121–2123.
[21] Y.-H. Lai, Z.-L. Zhou, *J. Org. Chem.* **1994**, *59*, 8275–8278; T. Yamato, H. Kamimura, T. Furukawa, *J. Org. Chem.* **1997**, *62*, 7560–7564.
[22] H. J. Reich, Y. E. Yelm, *J. Org. Chem.* **1991**, *56*, 5672–5679.
[23] R. Yanada, H. Higashikawa, Y. Mura, T. Taga, F. Yoneda, *Tetrahedron Asymmetry* **1992**, *3*, 1387–1390.
[24] D. Y. Antonov, Y. N. Belokon, N. S. Ikonnikov, S. A. Orlova, A. P. Pisarevsky, N. I. Raevski, V. I. Rozenberg, E. V. Sergeeva, Y. T. Struchkov, V. I. Tararov, E. V. Vorontsov, *J. Chem. Soc., Perkin Trans. 1* **1995**, 1873–1879.
[25] H. Hopf, D. G. Barrett, *Liebigs Ann.* **1995**, 449–451.
[26] A. Pelter, R. A. N. C. Crump, H. Kidwell, *Tetrahedron Lett.* **1996**, *37*, 1273–1276.
[27] P. J. Pye, K. Rossen, R. A. Reamer, N. N. Tsou, R. P. Volante, P. J. Reider, *J. Am. Chem. Soc.* **1997**, *119*, 6207; P. J. Pye, K. Rossen, R. A. Reamer, R. P. Volante, P. J. Reider, *Tetrahedron Lett.* **1998**, *39*, 4441–4444; P. W. Dyer, P. J. Dyson, S. L. James, C. M. Martin, P. Suman, *Organometallics* **1998**, *17*, 4344–4346.
[28] Y. Fukazawa, H. Kitayama, K. Yasuhara, K. Yoshimura, S. Usui, *J. Org. Chem.* **1995**, *60*, 1696–1703.
[29] H. Hopf, P. G. Jones, P. Bubenitschek, C. Werner, *Angew. Chem.* **1995**, *107*, 2592–2594; *Angew. Chem. Int. Ed. Engl.* **1995**, *34*, 2367–2368.
[30] G. J. Bodwell, T. J. Houghton, D. Miller, *Tetrahedron Lett.* **1998**, *39*, 2231–2234.
[31] H. E. Ensley, S. Mahadevan, J. Mague, *Tetrahedron Lett.* **1996**, *37*, 6255–6258.
[32] M. Ramming, R. Gleiter, *J. Org. Chem.* **1997**, *62*, 5821–5829.
[33] T. Kawase, N. Ueda, H. R. Darabi, M. Oda, *Angew. Chem.* **1996**, *108*, 1658–1660; *Angew. Chem. Int. Ed. Engl.* **1996**, *35*, 1556–1558; T. Kawase, H. R. Darabi, M. Oda, *Angew. Chem.* **1995**, *108*, 2803–2805; *Angew. Chem. Int. Ed. Engl.* **1996**, *35*, 2664–2666.
[34] J. M. Fox, D. Lin, Y. Itagaki, T. Fujita, *J. Org. Chem.* **1998**, *63*, 2031–2038.
[35] Y. Rubin, T. C. Parker, S. I. Khan, C. L. Holliman, S. W. McElvany, *J. Am. Chem. Soc.* **1996**, *118*, 5308–5309; Y. Rubin, T. C. Parker, S. J. Pastor, S. Jalisatgi, C. Boulle, C. L. Wilkins, *Angew. Chem.* **1998**, *110*, 1353–1356; *Angew. Chem. Int. Ed. Engl.* **1998**, *37*, 1226–1229.
[36] Y. Sekine, M. Brown, V. Boekelheide, *J. Am. Chem. Soc.* **1979**, *101*, 3126–3127; Y. Sekine, V. Boekelheide, *ibid.* **1981**, *103*, 1777–1785.
[37] Y. Sakamoto, N. Miyoshi, T. Shinmyozu, *Angew. Chem.* **1996**, *108*, 585–586; *Angew. Chem. Int. Ed. Engl.* **1996**, *35*, 549–550; see also T. Shinmyozu, S. Kusumoto, S. Nomura, H. Kawase, T. Inazu, *Chem. Ber.* **1993**, *126*, 1815–1818.
[38] H. F. Bettinger, P. v. R. Scheleyer, H. F. Schaeffer III, *J. Am. Chem. Soc.* **1998**, *120*, 1074–1075.
[39] O. J. Cha, E. Osawa, S. Park, *J. Mol. Struct.* **1993**, *300*, 73–81.
[40] Y. Sakamoto, T. Kumagai, K. Matohara, C. Lim, T. Shinmyozu, *Tetrahedron Lett.* **1999**, *40*, 919–922.
[41] R. Roers, F. Rominger, C. Braunweiler, R. Gleiter, *Tetrahedron Lett.* **1998**, *39*, 7831–7834. See also R. Roers, F. Rominger, R. Gleiter, *Tetrahedron Lett.* **1999**, *40*, 3141–3144.
[42] K. Tani, H. Seo, M. Maeda, K. Imagawa, N. Nishiwaki, M. Ariga, Y. Tohda, H. Higuchi, H. Kuma, *Tetrahedron Lett.* **1995**, *36*, 1883–1886.
[43] F. Vögtle, G. Hohner, E. Weber, *J. Chem. Soc., Chem. Commun.* **1973**, 366–367; K. Matsumoto, M. Kugimiya, *Z. Kristallogr.* **1975**, *141*, 260–274.
[44] R. A. Pascal Jr., R. B. Grossman, D. Van Engen, *J. Am. Chem. Soc.* **1987**, *109*, 6878–6880. See also R. A. Pascal Jr., C. G. Winans, D. Van Engen, *J. Am. Chem. Soc.* **1989**, *111*, 3007–3111; A. Pascal Jr., R. B. Grossman, *J. Org. Chem.* **1987**, *52*, 4616–4617.
[45] R. P. L'Esperance, A. P. West Jr., D. Van Engen, R. A. Pascal Jr., *J. Am. Chem. Soc.* **1991**, *113*, 2672–2676.
[46] A. P. West Jr., N. Smyth, C. M. Craml, D. M. Ho, R. A. Pascal Jr., *J. Org. Chem.* **1993**, *58*, 3502–3506.
[47] S. Dell, N. J. Vogelaar, D. M. Ho, R. A. Pascal, Jr., *J. Am. Chem. Soc.* **1998**, *120*, 6421–6422.
[48] R. H. Mitchell, Y. Chen, *Tetrahedron Lett.* **1996**, *37*, 5239–5242.
[49] R. H. Mitchell, T. R. Ward, Y. Wang, P. W. Dibble, *J. Am. Chem. Soc.* **1999**, *121*, 2601–2602.

[50] S. Breidenbach, S. Ohren, F. Vögtle, *Chem. Eur. J.* **1996**, *2*, 832–837; S. Breidenbach, S. Ohren, M. Nieger, F. Vögtle, *J. Chem. Soc., Chem. Commun.* **1995**, 1237–1238.
[51] S. Mataka, K. Shigaki, T. Sawada, Y. Mitoma, M. Taniguchi, T. Thiemann, K. Ohga, N. Egashira, *Angew. Chem.* **1998**, *110*, 2626–2628; *Angew. Chem. Int. Ed. Engl.* **1998**, *37*, 2532–2534.
[52] W. Josten, D. Karbach, M. Nieger, F. Vögtle, K. Hägele, M. Svoboda, M. Przybylski, *Chem. Ber.* **1994**, *127*, 767–777; W. Josten, S. Neumann, F. Vögtle, M. Nieger, K. Hägele, M. Przybylski, F. Beer, K. Müllen, *ibid.* **1994**, *127*, 2089–2096; A. Schröder, D. Karbach, R. Güther, F. Vögtle, *ibid.* **1992**, *125*, 1881–1887.
[53] J. Gross, G. Harder, F. Vögtle, H. Stephan, K. Gloe, *Angew. Chem.* **1995**, *107*, 523–526; *Angew. Chem. Int. Ed. Engl.* **1995**, *34*, 481–484.
[54] R. Taylor, G. J. Langley, H. W. Kroto, D. R. M. Walton, *Nature* **1993**, *366*, 728–731; F. T. Edelmann, *Angew. Chem.* **1995**, *107*, 1071–1075; *Angew. Chem. Int. Ed. Engl.* **1995**, *34*, 981–985.
[55] C.-T. Chen, P. Gantzel, J. S. Siegel, K. K. Baldridge, R. B. English, D. M. Ho, *Angew. Chem.* **1995**, *107*, 2870–2873; *Angew. Chem. Int. Ed. Engl.* **1995**, *34*, 2657–2660.
[56] T. J. Seiders, K. K. Baldridge, J. S. Siegel, *J. Am. Chem. Soc.* **1996**, *118*, 2754–2755.

Well-Rounded Research: Nanotubes through Self-Assembly

Burkhard König

Institut für Organische Chemie, Universität Regensburg, Germany

Tube-shaped structures with nanometer dimensions are currently being reported from completely different fields of chemistry: on one hand, carbon chemists keep surprising us with new shapes [1] and properties [2] of carbon nanotubes; on the other, synthetic chemists are introducing highly functionalized nanotubes, which are of some interest as models for biological ion channels. This account is concerned with topical developments in the latter field.

The construction of synthetic nanotubes is achieved through self-assembly of suitable subunits. Among the various possibilities for such self-alignment, stacking of macromolecules is a particularly promising starting point. As early as 1974, DeSantis and co-workers predicted a roughly planar structure for cyclopeptides made up of alternating D and L amino acids [3]. In this conformation the amide chain would be orientated at a right angle to the plane of the macrocycle; thus the carbonyl groups would be set in ideal positions for the formation of intermolecular hydrogen bonds between the stacked rings. The side chains would protrude outwards in the D/L motif to give an open channel whose diameter is determined only by the number of amino acid residues. Based on this model, Lorenzi et al. constructed cyclopeptides consisting of D- and L-valine units in 1987; however, their low solubility prevented structure investigations [4]. A breakthrough was then achieved in 1993 by Ghadiri's group with the octapeptide cyclo[-(D-Ala-Glu-D-Ala-Gln)$_2$-]

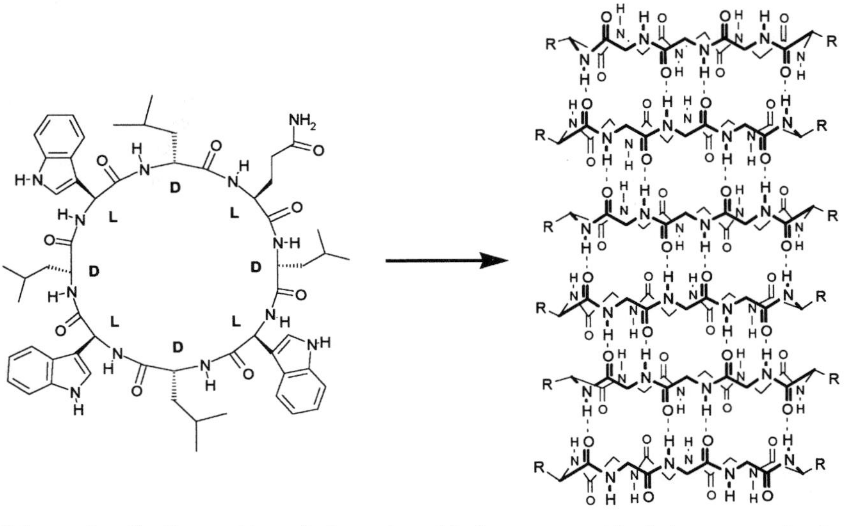

Scheme 1. Cyclic peptides of alternating chirality spontaneously form nanotubes. For reasons of clarity most side chains were omitted.

[5]. The glutamic acid residue renders this compound soluble in alkaline media; stacking to form nanotubes occurs after acidification (Scheme 1). The more hydrophobic side chains in cyclo[-(Trp-D-Leu)$_3$-Gln-D-Leu-] even enabled the construction of a transmembrane ion channel with a proton transport activity similar to that of gramicidin A or amphotericine B [6]. Measurements of single-channel conductivity showed fast transport of sodium and potassium ions; the channels pore diameter of 7.5 Å led to weak potassium selectiv-ity [7].

Larger cyclopeptides yield nanotubes with a larger inner diameter [8]. Thus, macrocycles with 10–12 amino acid residues result in tubes in whose cavities molecules can be transported. The pore size of the channel leads to exclusion selectivity: a nanotube build from cyclic decapeptides with 10 Å diameter is permeable by glucose, whereas a tube of cyclic octapeptides is not [9]. Partial amide *N*-alkylation has been shown to block peptide aggregation and limit self-assembly to cylindrical *β*-sheet peptide dimer formation. The dimerization process tolerates a number of *N*-alkyl substituents, whereas disubstitution of the peptide *α*-position prevents association [10]. Crystalline cyclic peptide nanotubes have been detected at the air-water interface by X-ray diffraction. Films of cyclic octapeptides were transferred onto a solid support and visualized by scanning force microscopy [11]. In a detailed biophysical investigation the angle of peptide nanotube orientation relative to the lipid bilayer was investigated using IR spectroscopy. In accordance with the structure–function model hypothesis for a transport-competent channel the central axis of nanotubes composed of cyclo[(L-Trp-D-Leu)$_3$-L-Gln-D-Leu] is aligned parallel to the lipid bilayer hydrocarbon chains, at approximately 7° from the axis normal to the bilayer plane [12]. The biological activity of self-assembling transmembrane channels was determined by in vitro assays. These channels are antibacterially active against gram-positive strains and cytotoxic in a test with human kidney cells [7].

The very efficient non-covalent approach to nanotubes by molecular self-assembly of cyclic peptides is limited by the kinetic instability of the resulting constructs. An elegant solution to this difficulty is the covalent capture of assembled cyclopeptides that bear alkene groups using ring-closing metathesis. The ring-closing reaction was mediated by the Grubbs catalyst, a ruthenium carbene complex (Scheme 2). Alternatively, self-assembled cyclic peptides with thiol groups were covalently captured upon oxidative formation of disulfide bridges [13].

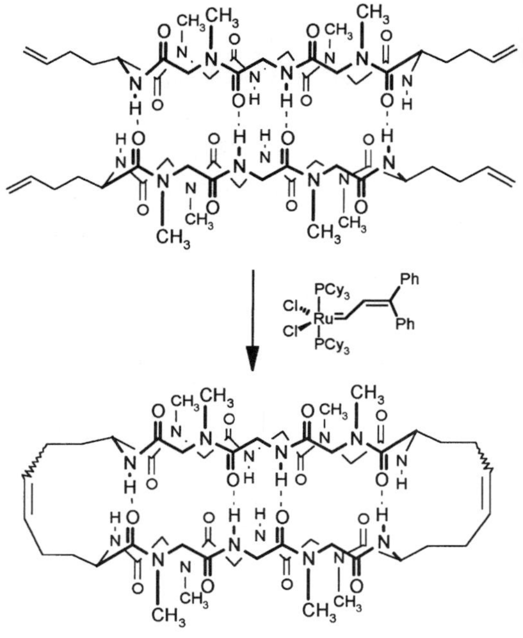

Scheme 2. Covalent capture of a self-assembled cyclic peptide dimer using the Grubbs ruthenium catalyst.

Molecular modeling suggested that, like their cyclic D,L-*α*-peptide counterparts, cyclic peptide subunits composed of homochiral *β*^{3}-amino acids could adopt flat, disklike conformations with amino acid side chains occupying equatorial positions on the exterior of the peptide ring, while axial and interior positions remain unobstructed. Cyclic *β*^{3}-octapeptides were synthesized and examined in liposome-based proton transport assays and single-channel conductance experiments [14]. The investigations revealed for some compounds a greater potassium conductivity than that of gramicidin under similar conditions, which indicates an effective nanotube and ion channel formation [15].

Not only cyclopeptides are suitable for the construction of nanotubes. [16] Stacked cyclo-

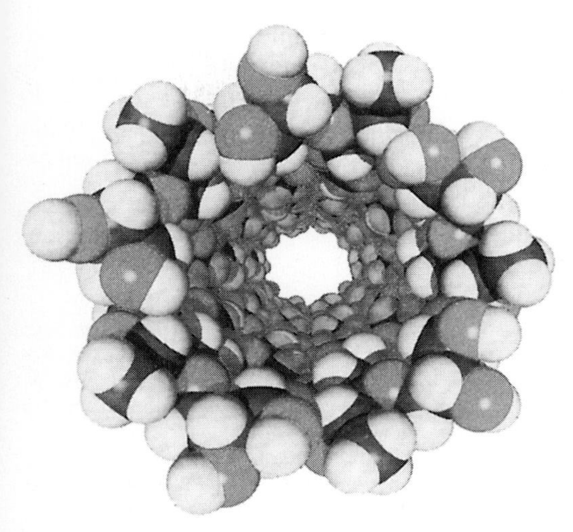

Figure 1. View into the cavity of an isolated nanotube consisting of stacked cyclodextrins **3–RR** in the solid.

dextrins [17] also form tube-shaped structures with an internal diameter of up to 1.3 nm, as has been proven by Stoddart, Williams, and co-workers through X-ray structure analysis (Fig. 1) [18]. The similarity of the design principles used here is striking: the cyclodextrin macrocycles consist of alternating D and L sugar units! A whole series of new cyclodextrin derivatives from alternating D and L rhamnopyranose and D and L mannopyranose units has been presented by this research group (Scheme 3). Among these are also the first achiral cyclic oligosaccharides (**RR** and **MM** derivatives) as well as the largest cyclooligosaccharide known to date (5–**RR**), with 14 sugar units. The synthesis of the cyclodextrin analog **1-RM** and **2-RM** succeeded efficiently by polycondensation-cycloglycosilation of the disaccharide **1**, in which the cyanoethylidene group serves as glycosyl donor and the trityloxy function as glycosyl acceptor. The key intermediate **1** is available in 15 steps from L-rhamnose and D-mannose.

The results obtained by Fujita, Lichtenthaler, and co-workers show that cyclooligosaccharides with sugar units other than glucose do not necessarily have to assemble as tubes. Via the 2,3 anhydro compound as intermediate, α- and β-cycloaltrin were obtained from the corresponding cyclodextrins (Scheme 4) [19]. The X-ray structure analysis of the α-cycloaltrin revealed a highly unusual structure with alternating ${}^{1}C_{4}/{}^{4}C_{1}$ chair conformations and a cavity open on only one side. Instead of the now impossible formation of tubes, the layers lie on top of each other in a staggered way with water molecules in the gaps.

Other than with ion channels, it is not the inner void, but the groups pointing outwards that are crucial for the function of a tube-shaped self-replicating system [20] built up on the base of α-helical peptides (Scheme 5). For this purpose a peptide consisting of 32 amino acids whose sequence resembles the GCN4 leucine zipper region was used. Ghadiri and co-workers showed that this compound catalyzes its own formation by accelerating amide bond formation between appropriate peptide strands. The most important interaction responsible for the specific interhelical recognition between template and reactant is the hydrophobic interaction of the leucine and valine residues, which increases because of electrostatic forces. If this interaction is interrupted, for example, through the addition of guanidinium hydro-

	n	R^1	R^2
1-MM	3	CH_2OH	CH_2OH
1-RM	3	Me	CH_2OH
2-RM	4	Me	CH_2OH
1-RR	3	Me	Me
2-RR	4	Me	Me
3-RR	5	Me	Me
4-RR	6	Me	Me
5-RR	7	Me	Me

Scheme 3. D/L-Cyclooligosaccharides synthesized so far. **M** = mannopyranose, **R** = rhamnopyranose.

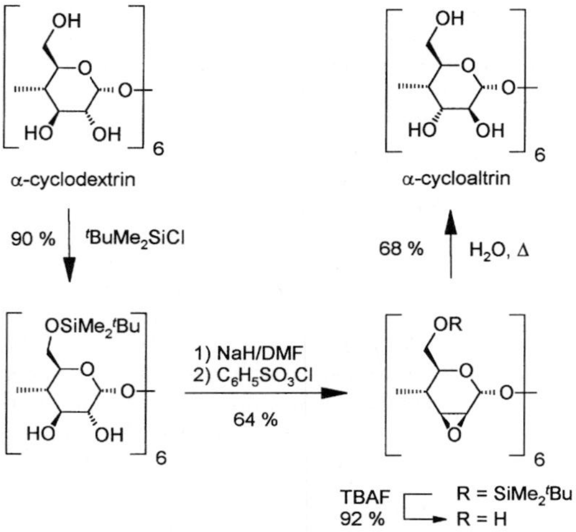

Scheme 4. Synthetic sequence leading from α-cyclodextrin to α-cycloaltrin. TBAF = tetrabutylammonium fluoride.

chloride or through exchange of a valine or leucine residue for alanine, the autocatalysis comes to a standstill. If several, slightly different self-replicating peptides are present in solution autocatalytic peptide networks were observed, which exhibit simplest forms of symbiosis, sequence-selective reproduction and dynamic correction of replication errors [21].

The above-mentioned examples show impressively how fast the development of functional nanostructures on the basis of molecular recognition is progressing [22]. Even if the complexity falls far short of that of biological self-assembly processes such as the RNA-induced formation of the coat of the tobacco mosaic virus from 2130 identical proteins [23], the next step might already lead to applications in nano- and microtechnology.

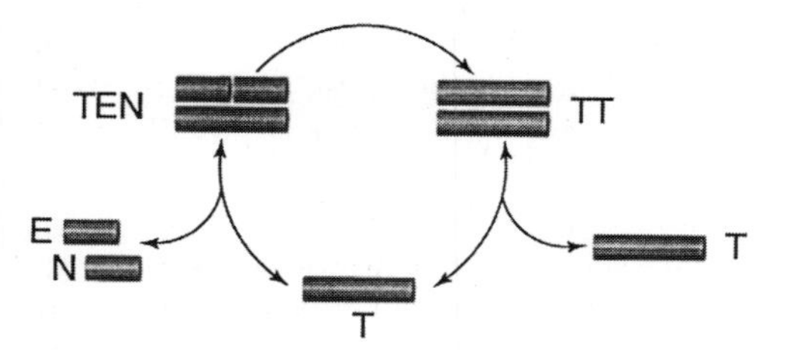

Scheme 5. Schematic representation of the minimal autocatalytic reaction cycle of a self-replication α-helical peptide. The peptide fragments E and N are preorientated at the template through hydrophobic, interhelical interactions. The amide bond formation leads to an identical copy of the template, which further accelerates the peptide ligation in the autocatalytic cycle.

References

[1] a) H. Terrones, M. Terrones, W. K. Hsu, *Chem. Soc. Rev.* **1995**, *24*, 341 – 350; b) C. N. R. Rao, R. Seshardri, A. Govindaraj, R. Sen, *Mater. Sci. Eng., R.* **1995**, *R15*, 209 – 262; c) T. W. Ebbesen, NATO ASI Ser., Ser. C **1994**, *443*, 11 – 25; d) S. Iijima, T. Ishihashi, *Nature* **1993**, *363*, 603 – 605; e) A. Thess, *Science* **1996**, *273*, 483 – 487; f) R. Tenne, *Adv. Mater.* **1995**, *7*, 965 – 972 und 989 – 995; g) for the synthesis and structure of a hydrocarbon picotube, see S. Kammermeier, P. G. Jones, R. Herges, *Angew. Chem.* **1996**, *108*, 2834 – 2838; *Angew. Chem. Int. Ed. Engl.* **1996**, *35*, 2669 – 2671.

[2] S. J. Tans, M. H. Devoret, H. Dai, A. Thess, R. E. Smalley, L. J. Geerligs, C. Dekker, *Nature* **1997**, *386*, 474 – 477.

[3] P. DeSantis, S. Morosetti, R. Rizzo, *Macromolecules* **1974**, *7*, 52 – 58.

[4] a) L. Tomasic, G. P. Lorenzi, *Helv. Chim. Acta* **1987**, *70*, 1012 – 1016; b) X. Sun, G. P. Lorenzi bzw. **1994**, *77*, 1520 – 1526.

[5] a) J. D. Hartgernik, J. R. Granja, R. A. Milligan, M. R. Ghadiri, *J. Am. Chem. Soc.* **1996**, *118*, 43 – 50; b) M. R. Ghadiri, J. R. Granja, R. A. Milligan, D. E. McRee, N. Khazanovich, *Nature* **1993**, *366*, 324 – 327; c) M. R. Ghadiri, K. Kobayashi, J. R. Granja, R. K. Chadha, D. E. McRee, *Angew. Chem.* **1995**, *107*, 76 – 78; *Angew. Chem. Int. Ed. Engl.* **1995**, *34*, 93 – 95; d) K. Kobayashi, J. R. Granja, M. R. Ghadiri, *ibid.* **1995**, *107*, 79 – 81 and **1995**, *34*, 95 – 97; e) The structure of the nanotube was confirmed by transmission electron microscopy, electron scattering, and infrared spectroscopy.

[6] a) M. R. Ghadiri, J. R. Granja, L. K. Buehler, *Nature* **1994**, *369*, 301 – 304; b) R. M. Ghadiri, *Adv. Mater.* **1995**, *7*, 675 – 677; c) Apart from the self-assembly process of several subunits described here, working artificial ion channels have been obtained by insertion of a single complex molecule. The synthetic requirements are less with self-assembly; however, with covalent structures the transmembrane channel stability is greater. For recent reviews of chemical models of transmembrane channels, see: U. Koert, *Chem. unserer Zeit* **1997**, *31*, 20 – 26; d) G. W. Gokel, O. Murillo, *Acc. Chem. Res.* **1996**, *29*, 425 – 432; e) N. Kimizuka, T. Kawasaki, K. Hirata, T. Kunitake, *J. Am. Chem. Soc.* **1995**, *117*, 6360 – 6361; f) a modified peptide as a

transmembrane channel: J.-C. Meillon, N. Voyer, *Angew. Chem.* **1997**, *109*, 1004 – 1006; *Angew. Chem. Int. Ed. Engl.* **1997**, *36*, 967 – 969 ; g) an ion-binding, tube-shaped calix[4]arene dimer: P. Schmitt, P. D. Beer, M. G. B. Drew, P. D. Sheen, *ibid.* **1997**, *109*, 1926 – 1928; **1997**, *36*, 1840 – 1842.

[7] Recent results show that ion conductivity changes discontinuously with the number of stacked units: M. R. Ghadiri, personal communication (1997).

[8] N. Khazanovich, J. R. Granja, D. E. McRee, R. A. Milligan, M. R. Ghadiri, *J. Am. Chem. Soc.* **1994**, *116*, 6011 – 6012.

[9] J. R. Granja, M. R. Ghadiri, *J. Am. Chem. Soc.* **1994**, *116*, 10785 – 10786.

[10] T. D. Clark, J. M. Buriak, K. Kobayashi, M. P. Isler, D. E. McRee, M. R. Ghadiri, *J. Am. Chem. Soc.* **1998**, *120,* 8949 – 8962.

[11] H. S. K. H. Rapaport, K. Kjaer, P. B. Howes, S. Cohen, J. Als-Nilsen, M. R. Ghadiri, L. Leiserowitz, M. Lahav, *J. Am. Chem. Soc.* **1999**, *121,* 1186 – 1191.

[12] H. S. Kim, J. D. Hartgerink, M. R. Ghadiri, *J. Am. Chem. Soc.* **1998**, *120,* 4417 – 4424.

[13] T. D. Clark, K. Kobayashi, M. R. Ghadiri, *Chem. Eur. J.* **1999**, *5*, 782 – 792.

[14] T. D. Clark, L. K. Buehler, M. R. Ghadiri, *J. Am. Chem. Soc.* **1998**, *120,* 651 – 656.

[15] The actual mechanism of channel-mediated ion transport is, so far, not established unequivocally. Alternative possibilities exist, such as formation of tubular bundles with holes large enough to serve as a conduit for water and ions.

[16] For the formation of tubular ion channels from oligonucleotide analogs, see: L. Chen, N. Sakai, S. T. Moshiri, S. Matile, *Tetrahedron Lett.* **1998**, *39*, 3627 – 3630.

[17] a) Three general types of structures have been observed in solid cyclodextrins (CD): cage structures, which are found in all CD hydrates, channel structures and layer structures for host-guest complexes: K. Harata in *Comprehensive Supramolecular Chemistry*, Vol 3, (Eds.: J. Szejtli, T. Osa), Elsevier, Oxford, **1996**, 279 – 304; b) For a recent review of cyclodextrin chemistry, see: V. T. D Souza, K. B. Lipkowitz (Eds.), *Chem. Rev.* **1998**, *98,* 1741 – 2076.

[18] a) P. R. Ashton, C. L. Brown, S. Menzer, S. A. Nepogodiev, J. F. Stoddart, D. J. Williams, *Chem. Eur. J.* **1996**, *2*, 580 – 591; b) S. A. Nepogodiev, G. Gattuso, J. F. Stoddart, *Proceedings of the 8th International Cyclodextrin Symposium* (Eds.: J. Szejtli, L. Szente), Kluwer Academic Press, Dordrecht, **1996**, pp. 89 – 94; c) P. R. Ashton, S. J. Cantrill, G. Gattuso, S. Menzer, S. A. Nepogodiev, A. N. Shipway, J. F. Stoddart, D. J. Williams, *Chem. Eur. J.* **1997**, *3*, 1299 – 1314; d) G. Gattuso, S. Menzer, S. A. Nepogodiev, J. F. Stoddart, D. J. Williams, *Angew. Chem.* **1997**, *109*, 1615 – 1617; *Angew. Chem., Int. Ed. Engl.* **1997**, *36*, 1451 – 1454.

[19] a) K. Fujita, H. Shimada, K. Ohta, Y. Nogami, K. Nasu, T. Koga, *Angew. Chem.* **1995**, *107*, 1783 – 1784; *Angew. Chem. Int. Ed. Engl.* **1995**, *34*, 1621 – 1622; b) Y. Nogami, K. Nasu, T. Koga, K. Ohta, K. Fujita, S. Immel, H. J. Lindner, G. E. Schmitt, F. W. Lichtenthaler, *ibid.* **1997**, *109*, 1987 – 1991; and **1997**, *36*, 1899 – 1902.

[20] a) D. H. Lee, J. R. Granja, J. A. Martinez, K. Severin, M. R. Ghadiri, *Nature* **1996**, *382*, 525 – 528; b) K. Severin, D. H. Lee, J. A. Martinez, M. R. Ghadiri, *Chem. Eur. J.* **1997**, *3*, 1017 – 1024; c) The kinetic analysis of the self-replication progress revealed parabolic growth; for other self-replicating systems, see: d) D. N. Reinhoudt, D. M. Rudkevich, F. de Jong *J. Am. Chem. Soc.* **1996**, *118*, 6880 – 6889 and references therein.; e) G. von Kiedrowski, *Nature*, **1994**, *369*, 221 – 224; f) A. Terfort, G. von Kiedrowski, *Angew. Chem.* **1992**, *104*, 626 – 628; *Angew. Chem. Int. Ed. Engl.* **1992**, *31*, 654 – 656.

[21] a) D. H. Lee, K. Severin, Y. Yokobayashi, M. R. Ghadiri, *Nature* **1997**, *390*, 591 – 594; correction: *ibid.*, **1998**, *394*, 101; b) K. Severin, D. H. Lee, J. A. Martinez, M. Vieth, M. R. Ghadiri, *Angew. Chem.* **1998**, *110*, 133 – 135; *Angew. Chem. Int. Ed. Engl.* **1998**, *37*, 126 – 128; c) K. Severin, D. H. Lee, A. J. Kennan, M. R. Ghadiri, *Nature* **1997**, *389*, 706 – 709; d) S. Yao, I. Ghosh, R. Zutshi, J. Chmielewski, *Nature*, **1998**, *396*, 447 – 450; e) for a new surface-promoted replication process (SPREAD), see: A. Luther, R. Brandsch, G. von Kiedrowski, *Nature*, **1998**, *396*, 245 – 248; f) for a coverage of recent achievements in "self-replicating systems", see: *Chem. Eng. News*, **1998**, December 7, 40.

[22] a) Recent research by Whitesides et al. shows that the strategy of spontaneous association of molecules is not restricted to nanometer dimensions, but is also suitable for the construction of aggregates of millimeter size: A. Terfort, N. Bowden, G. M. Whitesides, *Nature* **1997**, *386*, 162 – 164; b) N. Bowden, A. Terfort, J. Carbeck, G. M. Whitesides, *Science* **1997**, *276*, 233 – 235.

[23] a) K. Namba, G. Strubbs, *Science* **1986**, *231*, 1401 – 1406. b) The *in* vitro reconstitution of the intact virus from the isolated components shows impressively that the information about the supramolecular structure is stored in the subunits and that the process of self-assembly is highly cooperative: A. Klug, *Angew. Chem.* **1983**, *95*, 579 – 596; *Angew. Chem., Int. Ed. Engl.* **1983**, *22*, 565 – 582; c) H. Conrat-Fraenkel, R. C. Williams, *Proc. Natl. Acad. Sci. USA* **1955**, *41*, 690 – 698.

From Random Coil to Extended Nanocylinder: Dendrimer Fragments Shape Polymer Chains

Holger Frey

Institut für Makromolekulare Chemie und Freiburger Materialforschungszentrum (FMF), Albert-Ludwigs-Universität Freiburg, Germany

Dendrimers usually exhibit spherical (isotropic) shape. However, wedge-like dendrimer fragments ("dendrons") that have been attached to linear polymers as side groups can be used to create anisotropic "nanocylinders", leading to uncoiling and extension of the polymer chains. Synthetic macromolecules of this type can be visualized directly on surfaces and their contour length determined from the images. Unexpected acceleration effects in the "self-encapsulated" polymerization of dendron monomers used to prepare such polymers as well as the structural consequences of dendritic "pieces of cake" on linear polymer chains are discussed.

Dendrimers, perfectly branched, highly symmetrical tree-like macromolecules have evolved from a curiosity to an important trend in current chemistry, attracting rapidly increasing attention from an unusually broad community of scientists [1]. Various dendrimer construction strategies have been developed on the basis of classical organic chemistry, heteroatom chemistry [2], and transition metal complexation [3,4]. Based on these now well-established synthetic "algorithms", increasing research efforts are currently directed at peculiar supramolecular structures and self-assembly processes of dendrimers as well as the elucidation of structure–property relationships, as documented by several reviews [5]. For instance, flexible dendrimers with mesogenic branching points [6] or end groups [7] have been constructed that are able to induce anisotropic liquid crystalline order despite the isotropic dendrimer topology. Furthermore, self-assembly of dendrimers on surfaces to ultrathin films with a thickness of a few nanometers has been investigated by a number of groups [8], aiming at application in catalysis, sensors or chromatographic separation [9].

For polymer chemists it is interesting to know how well-known linear polymers can be linked with dendritic architectures and what the supramolecular consequences of this approach might be. Combination of dendrimers with linear polymers in hybrid linear–dendritic block copolymers has been employed to achieve particular self-assembly effects. Block copolymers with a linear polyethylene oxide block and dendritic polybenzylether block form large micellar structures in solution that depend on the size (i.e., the generation) of the dendritic block [10]. Amphiphilic block copolymers have been prepared by the combination of a linear, apolar polystyrene chain with a polar, hydrophilic poly(propylene imine) dendrimer [11] as well as PEO with Boc-substituted poly-α,ε-L-lysine dendrimers, respectively [12]. Such block copolymers form large spherical and cylindrical micelles in solution and have been described as "superamphiphiles" and „hydra-amphiphiles“, respectively.

In contrast to these successful efforts to link dendrimers and linear polymer chains in the manner of block copolymers, only very recently has a breakthrough been achieved in the synthesis of linear polymers with dendritic side groups, and this will be highlighted in this contribution (Scheme 1). Considerable synthetic effort is required in preparing suitable dendritic monomers. The amazing kinetic and structural features observed for such monomers and the respective polymers are nevertheless highly rewarding.

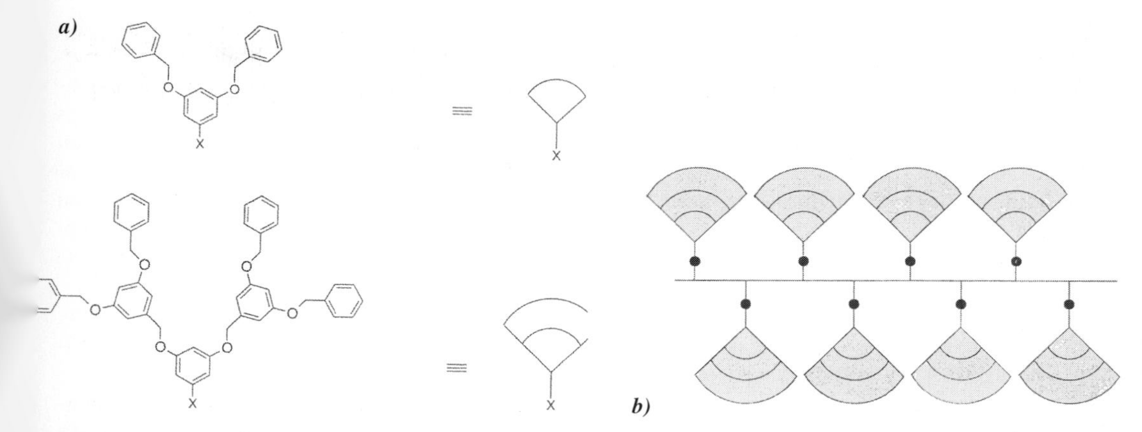

Scheme 1. a) Fréchet-type [18] dendritic wedges used as conical or fan-like building units; b) schematic image of a polymer chain that bears dendritic side groups [27].

Synthesis and Self-Organization of Dendrons

The term "dendron" designates a dendrimer segment possessing an AB_w-type structure, which is not attached to a central B_n core unit, as in the case of a perfect dendrimer (A and B represent the functionalities that are linked to build up the dendrimer structure). Thus, a dendron is structurally characterized by exactly one focal unit A and $w = m^g$ end groups B (m is the branching multiplicity, usually 2 or 3, g represents the generation number). Obviously, the size and number of end groups of a dendron can be varied with generation, which gives access to a large variety of nanosize building blocks for supramolecular chemistry. The conical dendron structure may be compared to a molecular "piece of cake" (Scheme 1). In most examples of dendrons used as supramolecular building blocks, dendritic polybenzylethers based on construction reactions developed by Hawker and Fréchet [13a] have been employed (Scheme 1a). Some examples of polymerizable dendrons are shown in Figure 1. Recently, a number of papers have detailed the synthesis of dendrons with various functional groups [13b-d].

Supramolecular organization of dendrons can lead to columnar or spherical superstructures, depending on directional and attractive interactions between such wedge-shaped molecules as well as size and shape. For instance, hydrogen-bonding interactions between di-isophthalic acid moieties attached to the focal point were used by Zimmerman [14] et al. to obtain discrete self-assembled hexamers of polybenzylether dendrimers. Remarkably, the self-assembled structures were SEC-stable and the self-assembly process was solvent dependent. In the case of small dendrons, less stable chain-like aggregates were obtained, illustrating the correlation between specific recognition processes and shape of the dendrimer wedge.

Self-organization of tapered and dendritic, i.e., fan- and cone-shaped dendron molecules in the solid was reported by Percec et al. in a series of papers [15]. Detailed diffraction experiments evidenced that spherical or cylindrical, (i.e. co-

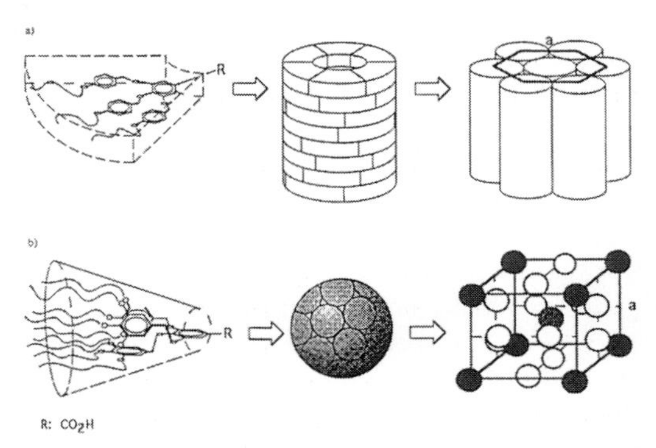

Scheme 2 .Shape-directed self-assembly of cone- and wedge-like dendrons to hexagonal-columnar and cubic superstructures according to Percec et al. [15a].

lumnar) superstructures are formed, which are further packed in cubic or hexagonal columnar phases, as illustrated in Scheme 2. It was demonstrated that first- and second-generation dendrons possessed shapes that are fragments of a disklike molecule, i.e., a quarter and a half of a disk, respectively. The third-generation dendron, in contrast, represented one-sixth of a sphere. From these results, Percec et al. concluded that in general at a certain critical generation, depending on the dendron architecture and functional groups in its core and on its periphery, a change from cylindrical to spherical superstructures occurs [15c].

2. Preparation of polymers with dendron side-chains

Obviously, there are three approaches that can be used to prepare polymers with dendron side groups: (i) divergent construction in analogy to divergent dendrimer synthesis, using a multifunctional linear polymer as B_n-type core instead of a small "point-like" core; (ii) attachment of prefabricated dendron building blocks to a reactive polymer chain via polymer-analogous reaction, similar to the final coupling step to the core in convergent dendrimer synthesis [13], and (iii) polymerization of dendron monomers (Fig 1).

The idea to string dendron "pieces of cake" to a polymer chain as a core, was mentioned by Tomalia and Kirchhoff [16a] in a patent as early as 1987 ("comb-burst" polymers). In a recent paper, the results of this work based on strategy (i) have been published in more detail. A polyethylenimine prepared by living cationic polymerization of 2–ethyl-2-oxazoline and subsequent deprotection was used as core for the synthesis of rod-shaped poly(amidoamine) structures [16b]. Unfortunately, little information is given by the authors concerning control of molecular weights and polydispersities of the dendronized polyethylenimine that would permit this concept (i) to be compared with strategies (ii) and (iii) discussed below.

The second strategy (ii) was investigated by Schlüter et al., using Fréchet-type polybenzylether wedges and hydroxy-functional poly(*p*-phenylene) (PPP) [17a]. The key problem encountered in this case lies in the limited conversion achievable in polymer-analogous reactions, particularly if higher generation dendritic wedges (i.e., larger than G2) are employed [17b,c], leading to incompletely covered polymer chains, e.g. for a G3 dendron only 90 % of the reactive groups at the PPP backbone are linked with dendrimer fragments. This drawback of the polymer-analogous approach has motivated intense efforts along the lines of concept (iii), based on polymerizable dendrons.

If a polymerizable function is attached to a dendron at the focal point, "dendron monomers" are obtained (Fig 1). Dependent on the dendrimer generation employed, large polymerizable building blocks of variable size with conical topology may be prepared. Can such dendronmonomers actually be polymerized, and if so, what are the properties of the resulting polymer chains with unusually large substituents? A styrene-based copolymer with approximately 40 w% (2 mol%) of repeat units carrying dendrons was reported by

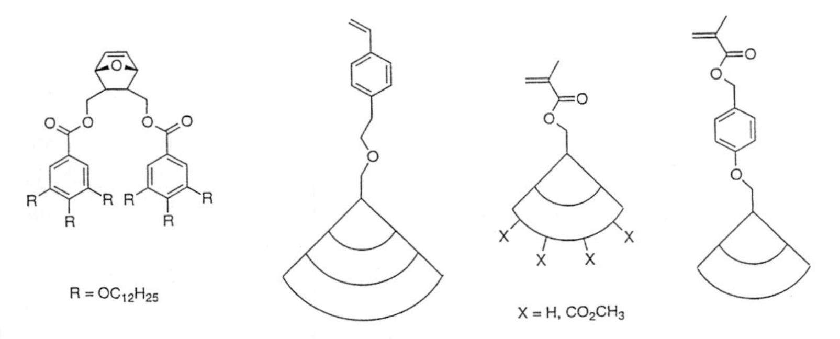

Figure 1. Examples of dendron monomers described by Percec et al., Schlüter et al. and Xi et al.

Hawker and Fréchet in 1992 [18]. However, only in the last 4 years has the maturing of dendrimer synthesis and related characterization methods made it possible to study dendron (macro)monomers and to exploit their potential for the synthesis of dendron-bearing polymers.

One would intuitively expect polymerization of such monomers to be troublesome, because of the steric requirements of the large dendrimer fragments as well as the shielding effect of the polymerizable moiety, which incrases with increasing generation number. These problematic aspects are well known in macromonomer chemistry [19] and are expected to be even more severe for dendron monomers. This assumption has been confirmed by a number of authors [20] who studied the polymerization of methacrylates with various dendritic fragments attached. Indeed, polymerization of monodendrons to macromolecules with a high degree of polymerization was only possible when a spacer was inserted between the polymerizable unit and the bulky dendron group, using very long reaction times.

In contrast, recent kinetic investigation of the polymerization of spacerless G2 dendron-substituted styrene and methylmethacrylate, respectively, in solution lead to the unexpected conclusion that above a certain critical monomer concentration a strong increase in the rate of the free radical polymerization is observed [21]. The results can be explained by self-organization of the growing polymer chain to a spherical or columnar superstructure in solution, depending on the degree of polymerization (DP, Fig. 2). The rate constants and low initiator efficiency lead one to conclude that the self-assembled structure acts like a supramolecular nanoreactor that leads to a strongly enhanced local concentration of polymerizable groups. Thus, the kinetics of the polymerization is determined by self-assembly and can be viewed as "self-encapsulated" and "self-accelerated".

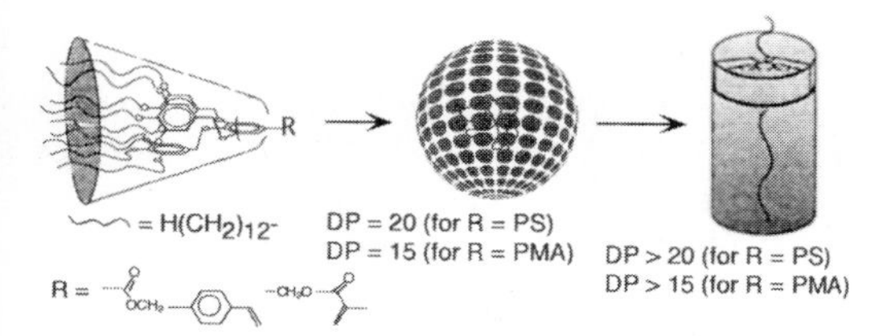

Figure 2. Dependence of self-assembly of dendronized polymer chains on the degree of polymerization; monomers and short polymer chains assemble in spheres; longer chains assemble in cylinders. This effect leads to rapid polymerization of such dendron monomers due to "self-encapsulation" [21].

Polycondensation of suitable monomers represents an alternative route to high-molecular-weight dendronized polymers e.g., PPP with attached dendron segments has been prepared using Suzuki cross-coupling of dendron-substituted dibromobenzenes with alkyl chain-substituted diboronic acids in a polycondensation type of reaction [17c]. In this case, high-molecular-weight dendronized PPP with DP exceeding 100 was obtained. Similarly, polyaddition reactions of dendronized diols with semirigid and flexible diisocyanates have been employed by Jahromi et al. to prepare polyurethanes with pending dendrons [17d].

Properties and Visualization of Dendronized Polymers

If dendrimer fragments are attached to polymer chains, the conformation of the polymer chain is strongly affected by the large dendrimer wedges attached. "Dendronized" polymers can be considered as a subclass of comb polymers, i.e., linear polymer chains densely substituted with polymeric side chains, which are known to be extremely rigid in solution, exhibiting Kuhn lengths l_k of 10–20 nm (in some cases as much as 120 nm! [22d]), in comparison to 1–2 nm for polystyrene or polymethylmethacrylate [22]. For that reason, such comb polymers are sometimes designated "cylindrical brushes". In recent work, Schmidt et al. have given a detailed summary of peculiarities encountered in the characterization of this class of macromolecules [22d].

Consequently, cylindrical shape is also expected for polymers with large dendron side chains. Of course, biomolecules such as DNA or RNA with rigid cylindrical or wormlike shapes are well known, although in these cases the supermolecular structure (i.e., the secondary and tertiary structure) is the result of well-controlled secondary bonds.

For flexible synthetic polymers, the chain conformation is commonly controlled by the degree of polymerization (DP), with low-DP polymers having a rather extended chain conformation and high-DP polymers adopting random coil conformations in solution. Recently, Percec, Möller et al. reported extensive studies concerning the chain conformation and supramolecular structure of dendron-substituted polystyrene and polymethacrylate [23]. Remarkably, at low DP (i.e., short chains) the conical monodendrons assemble to produce a spherical superstructure with random-coil backbone conformation (Fig. 3a). On increasing the DP, the self-assembly pattern of the dendrons changes, leading to cylindrical polymers with rather extended backbone (Fig. 3b). It is remarkable that this correlation between polymer conformation and the DP is opposite to that usually seen in most synthetic and natural macromolecules. A detailed understanding of this effect was obtained by Percec et al. on the basis of "libraries" of dendron monomers of different shape [23b].

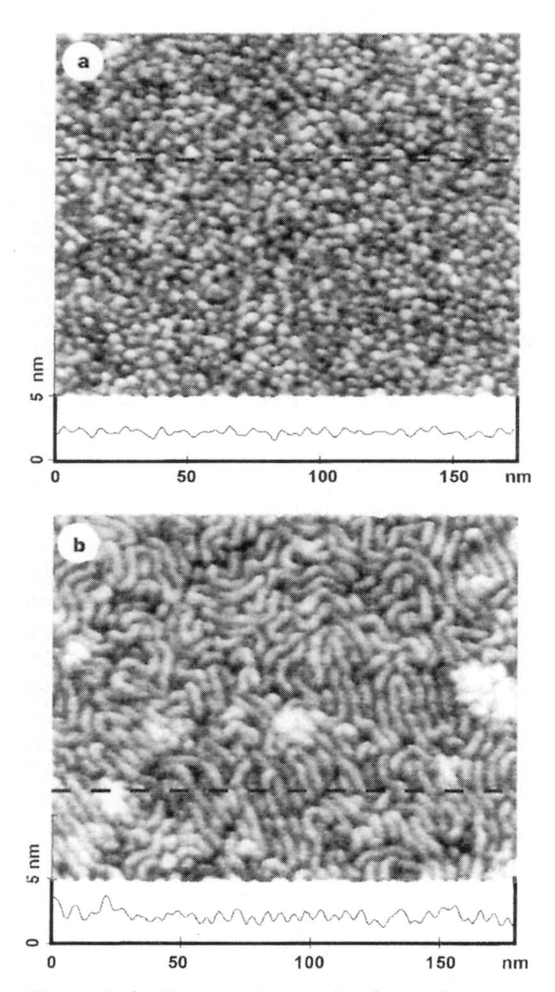

Figure 3a, b. Topographic scanning force microscopy (SFM) images (tapping mode) of monomolecular films of dendron-bearing polystyrene; a) M_n = 10 800; b): M_n = 186 500. Spherical and cylindrical features are observed; cylindrical features are due to single dendronized macromolecules [23].

Usually, synthetic macromolecules cannot be visualized directly, because of the very small size, particularly the diameter (several angstroms) of the backbone of such polymer molecules. Therefore, the present knowledge of the shape, size and conformation of polymers was mostly obtained from the interpretation of scattering results (X-rays, electrons, neutrons). However, in the case of dendronized synthetic polymers it is a fascinating consequence of the stiffness and densely covered surface of such cylindrical macromolecules that single polymer chains can actually be visualized. Möller et al. demonstrated impressively that the imaging of dendron-substituted polystyrene and polymethacrylate can be employed for an analysis of the molecular size distribution and conformation, using dendron-substituted polystyrene and polymethacrylate [24a]. The image shown in Fig. 4a was obtained by scanning force microscopy (SFM) and shows single chains of dendronized polystyrene deposited on a pyrolytic graphite substrate. Clearly to be seen, the molecules exhibit short, straight segments with bends of a characteristic angle of 60° and 120° corresponding to the threefold symmetry of the graphite. In addition, within near region, the molecules tend to align parallel, forming hairpin bends. The height of the chains is in the region of 1.6 nm, and the lateral diameter of the chains is 5.3 nm (SFM, tapping mode), which illustrates that the macromolecules are collapsed on the surface. The worm-like contour of the macromolecules can be approximated by segments of 10–20 nm, and a distribution of the contour lengths may be calculated from the images, which is a remarkable feature of these novel comb poly-

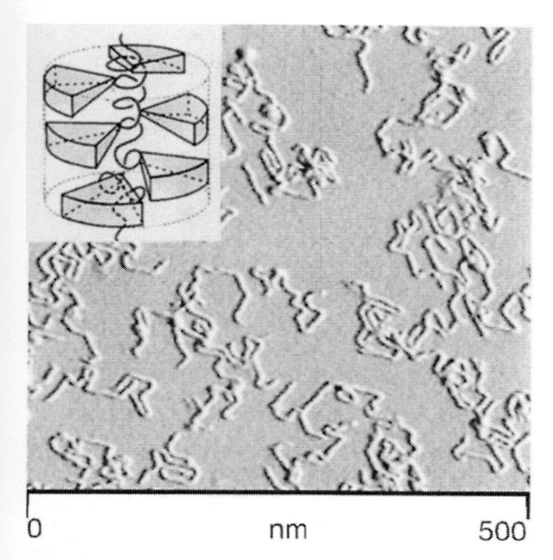

Figure 4. Single "worm-like" polymer chains of dendronized polystyrene deposited on pyrolytic graphite, imaged by topographic scanning force microscopy (SFM) [24]; the inset illustrates the structure of the dendron-substituted chains

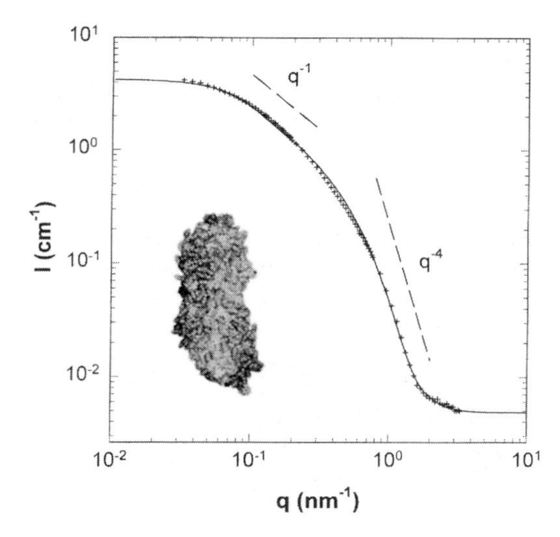

Figure 5. SANS scattering curve $I(q)$ vs. q of a G3–dendronized polystyrene, obtained by Förster, Schlüter et al. [25]. The scattering curve at larger q follows the Porod q^{-4} law, arising from the scattering at the surface of a cylinder, evidencing the stiff, worm-like character of the macromolecules.

mers. It is an intriguing question whether the molecular weight distribution can be determined in the solid state. Comparison of the apparent length of the macromolecules determined from the images shows that for large dendron side groups this is in good agreement with expectation, whereas for lower generation dendrons attached, the contour length determined from the images is considerably lower than the calculated length of the extended chains. This demonstrates that the degree of extension (i.e., uncoiling) of the chains is determined by the size of the dendrons attached. Besides these data, detailed information on packing defects, such as hairpin structures and intersections, could be obtained from the images [24b].

As interaction with the substrate may affect the images, it is an interesting question, whether the polymers actually possess cylindrical shape in solution. Small angle neutron scattering (SANS) experiments have been employed to answer this question. Figure 4b shows the scattering curve of a dendron-substituted polystyrene (G3–dendrons) [25a]. It is possible to describe the scattering curve by assuming that the dendron-substituted polystyrene exhibits a rod-like structure, as shown in the inset of the figure. Thus, the persistence length of the molecule is on the order of the contour length, supporting the presence of stiff, cylindrical macromolecules in solution. Detailed neutron-scattering studies by Förster et al. have shown that the chain diameter of the dendronized macromolecules indeed increases with increasing dendron generation, showing that the diameters of the dendronized polymer chains are comparable to those of DNA ($d = 24$ Å). In this study, also the first example of a charged dendronized polymer chain was analyzed.

Conclusion

In summary, it is obvious that the attachment of dendron building blocks to common monomers leads to dramatic kinetic and structural consequences. Once more, it should be kept in mind that it is only the shape of the side groups attached that governs the conformation and structure of the resulting polymer chains, not hydrogen-bonding interactions, which are ubiquitous in self-assembly in biological processes.

Based on the breakthrough in the synthesis and visualization of "dendronized" polymers achieved, is a safe bet that this novel class of extremely stiff macromolecules will stimulate further interdisciplinary efforts to understand their physical behavior in bulk and in solution as well as assess their usefulness for a future nanotechnology based on "molecular objects". On the other hand, these developments can lead to libraries of shapes that can be combined to create any nanometer-size cylindrical object with programmed length and diameter [26]. Block copolymers with one "dendronized" cylindrical block and one coiled block can also be envisaged [27]. Dendron side groups may also function as protective layers for "insulated molecular wires", as suggested by Diederich et al. [28]. In addition, dendrons bearing protected functionalities developed by Schlüter et al. offer further possibilities for the functionalization of such nanocylinders [20c]. We now look forward to seeing the shapes of other exciting molecular objects based on dendrimer wedges to come!

References

[1] a) D. A. Tomalia, A. M. Naylor, W. A. Goddard III, *Angew. Chem.* **1990**, *102*, 119–157; *Angew. Chem Int. Ed. Engl.* **1990**, 102, 138–175; b) D. A. Tomalia, H. D. Durst, *Top. Curr. Chem.* **1993**, 165, 193–313; c) G. R. Newkome, C. N. Moorefield, F. Vögtle, *Dendritic Macromolecules: Concepts, Syntheses, Perspectives*; 1. Aufl., VCH, Weinheim, Germany, **1996**; d) J. M. J. Fréchet, C. J. Hawker, *React. Funct. Polym.* **1995**, *26*, 127–150; e) B. I. Voit, *Acta Polymer.* **1995**, *46*, 87–99; f) H. Frey, K. Lorenz, L. Lach, *Chemie unserer Zeit* **1996**, 75–85; g) O. A. Matthias, A. N. Shipway, J. F. Fraser-Stoddart, *Prog. Polym. Sci.* **1998**, *23*, 1–56; h) M. Fischer, F. Vögtle, *Angew. Chem. Int. Ed. Engl.* **1999**, *38*, 885–905; i) A. W. Bosman, H. M. Janssen, E. W. Meijer, *Chem. Rev.* **1999**, *99*, 1665–1688

[2] a) H. Frey, C. Lach, K. Lorenz, *Adv. Mater.* **1998**, *10*, 279–293; b) C. Schlenk, H. Frey, *Monatsh. Chem.* **1999**, *130*, 3–14.

[3] a) C. Gorman, *Adv. Mater.* **1998**, *10*, 295–309; b) G. R. Newkome, E. He, C. N. Moorefield, *Chem. Rev.* **1999**, *99*, 1689–1746.

[4] W. T. S. Huck, F. C. J. M. van Veggel, D. N. Reinhoudt, *Angew. Chem. Int. Ed. Engl.* **1996**, *35*, 1304–1306; *Angew. Chem. Int. Ed. Engl.* **1996**, *35*, 1213–1215.

[5] a) F. Zeng, S. C. Zimmerman, *Chem. Rev.* **1997**, *97*, 1681–1712; b) S. C. Zimmerman, *Curr. Op, Coll. Interf. Sci.* **1997**, *2*, 89–99; c) J. S. Moore, *Acc. Chem. Res.* **1997**, *30*, 402–413.

[6] a) V. Percec, P. W. Chu, G. Ungar, J. P. Zhou, *J. Am. Chem. Soc.* **1995**, *117*, 11441–11454; b) J. F. Li, K. A. Crandall, P. W. Chu, V. Percec, R. G. Petschek, C. Rosenblatt, *Macromolecules* **1996**, *29*, 7813–7819.

[7] a) K. Lorenz, D. Hölter, B. Stühn, R. Mülhaupt, H. Frey, *Adv. Mater.* **1996**, *8*, 414–416; b) S. A. Ponomarenko, E. A. Rebrov, A. Y. Bobrovsky, N. I. Boiko, A. M. Muzafarov, *Liq. Cryst.* **1996**, *21*, 1–12; c) K. Lorenz, H. Frey, R. Mülhaupt, *Macromolecules*, **1997**, *30*, 6860–6868; d) M. W. P. L. Baars, S. H. M. Söntjens, H. M. Fischer, H. W. I. Peerlings, E. W. Meijer, *Chem. Eur. J.* **1998**, *4*, 2456–2466; e) J. Barberá, M. Marcos, J. L. Serrano, *Chem. Eur. J.* **1999**, *5*, 1834–1840.

[8] a) S. Watanabe, S. L. Regen, *J. Am. Chem. Soc.* **1994**, *116*, 8855–8856; b) M. Wells, R. M. Crooks, *ibid.* **1996**, *118*, 3988–3989; c) S. S. Sheiko, G. Eckert, G. Ignat'eva, A. M. Muzafarov, J. Spickermann, H. J. Räder, M. Möller, *Macromol. Rapid Commun.* **1996**, *17*, 283–297; d) M. Collaud Coen, K. Lorenz, J. Kressler, H. Frey, R. Mülhaupt, *Macromolecules* **1996**, *29*, 8069–8076.

[9] a) S. A. Kuzdzal, C. A. Monnig, G. R. Newkome, C. N. Moorefield, C. N. *J. Chem. Soc. Chem. Commun.* **1994**, *18*, 2139–2140; b) P. G. H. M. Muijselaar, H. A. Claessens, C. A. Cramers, J. F. G. A. Jansen, E. W. Meijer, E. M. de Brabender-Van den Berg, S. Vanderwal, *HRC J. High Res. Chromat.* **1995**, *18*, 121–123.

[10] a) I. Gitsov, K. L. Wooley, C. J. Hawker, P. T. Ivanova, J. M. J. Fréchet, *Macromolecules* **1993**, *26*, 5621–5627; b) I. Gitsov, J. M. J. Fréchet, *ibid.* **1993**, *26*, 6536–6546; c) J. M. J. Fréchet, I. Gitsov, *Macromol. Symp.* **1995**, *98*, 441–465.

[11] a) J. C. M. van Hest, D. A. P. Delnoye, M. W. P. L. Baars, M. H. P. van Genderen, E. W. Meijer, *Science* **1995**, *268*, 1592–1595; b) J. C. M. van Hest, M. W. P. L. Baars, R. C. Elissenroman, M. H. P. van Genderen, E. W. Meijer, *Macromolecules* **1995**, 28, 6689–6691.

[12] T. M. Chapman, G. L. Hillyer, E. J. Mahan, K. A. Shaffer, *J. Am. Chem. Soc.* **1994**, *116*, 11195–11196.

[13] a) C. J. Hawker, J. M. J. Fréchet, *J. Am. Chem. Soc.* **1990**, *112*, 7638–7647; b) R. Klopsch, S. Koch, A. D. Schlüter, *Eur. J. Org. Chem.* **1998**, *1*, 1275–1283; c) A. Ingerl, I. Neubert, R. Klopsch, A. D. Schlüter, *Eur. J. Org. Chem.* **1998**, 2551–2556; d) S. M. Grayson, M. Jayaraman, J. M. J. Fréchet, *Chem. Commun.* **1999**, 1329–1330.

[14] S. C. Zimmerman, F. W. Zeng, D. E. C. Reichert, S. V. Kolotuchin, *Science* **1996**, *271*, 1095–1098.

[15] a) V. S. K. Balagurusamy, G. Ungar, V. Percec, G. Johansson, *J. Am. Chem. Soc.* **1997**, *119*, 1539–1555; b) S. D. Hudson, H.-T. Jung, V. Percec, W. D. Cho, G. Johansson, G. Ungar, V. S. K. Balagurusamy, *Science* **1997**, *278*, 449–452; c) V. Percec, W.-D. Cho, P. E. Mosier, G. Ungar, D. J. P. Yeardley, *J. Am. Chem. Soc.* **1998**, *120*, 11061–11070.

[16] a) D. A. Tomalia, P. M. Kirchhoff, US patent 4,694,064 (1987); b) R. Yin, Y. Zhu, D. A. Tomalia, H. Ibuki, *J. Am. Chem. Soc.* **1998**, *120*, 2678–2679.

[17] a) R. Freudenberger; W. Claussen, A. D. Schlüter, H. Wallmeier, *Polymer* **1994**, *35*, 4496–4501; b) B. Karakaya, W. Claussen, A. Schäfer, A. Lehmann, A. D. Schlüter, *Acta Polym.* **1996**, *47*, 79–84; c) B. Karakaya, W. Claussen, K. Gessler, W. Saenger, A. D. Schlüter, *J. Am. Chem. Soc.* **1997**, *119*, 3296–3301; d) S. Jahromi, B. Coussens, N. Meijerink, A. W. M. Braam, *J. Am. Chem. Soc.* **1998**, *120*, 9753–9762.

[18] C. J. Hawker, J. M. J. Fréchet, J. M. J. *Polymer* **1992**, *33*, 1507–1513.

[19] *Chemistry and Industry of Macromonomers* (Hrsg.: Y. Yamashita), Hüthig & Wepf, Basel, Heidelberg, New York **1993**.

[20] a) G. Draheim, H. Ritter, *Macromol. Chem. Phys.* **1995**, 196, 2211–2222; b) Y. M. Chen, C.-F. Chen, Y.-F. Liu, Y. F. Li, F. Xi, *Macromol. Rapid Commun.* **1996**, 17, 401–407; c) I. Neubert, R. Klopsch, W. Claussen, A. D. Schlueter, *Acta Polym.* **1996**, 47, 455–459.

[21] V. Percec, C.-H. Ahn, B. Barboiu, *J. Am. Chem. Soc.* **1997**, *119*, 12978–12979.

[22] a) Y. Tsukahara, Y. Tsutsumi, Y. Yamashita, S. Shimada, *Macromolecules* **1990**, *23*, 5201–5208; b) M. Wintermantel, M. Gerle, K. Fischer, M. Schmidt; I. Wataoka, H. Urakawa, K. Kajiwara, Y. Tsukahara, *ibid.* **1996**, *29*, 978–983; c) S. S. Sheiko, M. Gerle, K. Fischer, M. Schmidt, M. Möller, *Langmuir* **1997**, *13*, 5368–5372; d) M. Gerle, K. Fischer, S. Roos, A. H. E. Müller, M. Schmidt, S. S. Sheiko, S. Prokhorova, M. Möller, *Macromolecules* **1999**, *32*, 2629–2637.

[23] a) V. Percec, C.-H. Ahn, G. Ungar, D. J. P. Yeardley, M. Möller, S. S. Sheiko, *Nature* **1998**, *391*, 161–164; b) V. Percec, C.-H. Ahn, W.-D. Cho, A. M. Jamieson, J. Kim, T. Leman, M. Schmidt, M. Gerle, M. Möller, S. A. Prokhorova, S. S. Sheiko, S. Z. D. Cheng, A. Zhang, G. Ungar, D. J. P. Yeardley, *J. Am. Chem. Soc.* **1998**, *120*, 8619–8631.

[24] a) S. A. Prokhorava, S. S. Sheiko, M. Möller, C.-H. Ahn, V. Percec, *Macromol. Rapid. Commun.* **1998**, *19*, 359–366; b) S. A. Prokhorava, S. S. Sheiko, C.-H. Ahn, V. Percec, M. Möller, *Macromolecules* **1999**, *32*, 2653–2660.

[25] a) W. Stocker, B. L. Schürmann, J. P. Rabe, S. Förster, P. Lindner, I. Neubert, A. D. Schlüter, *Adv. Mater.* **1998**, 10, 793–797; b) W. Stocker, B. Karakaya, B. L. Schürmann, J. P. Rabe, A. D. Schlüter, *J. Am. Chem. Soc.* **1998**, *120*, 7691–7695; c) S. Förster, I. Neubert, A. D. Schlüter, P. Lindner, *Macromolecules* **1999**, *32*, 4043–4049.

[26] V. Percec, Makromolekulares Kolloquium Freiburg, Feb. 1998.

[27] A. D. Schlüter in Dendrimers; *Top. Curr. Chem.*, F. Vögtle, Ed., **1998**, pp. 165–191.

[28] A. P H. J. Schenning, R. E. Martin, M. Ito, F. Diederich, C. Boudon, J. P. Gisselbrecht, M. Gross, *Chem. Comm.* **1998**, 1013–1014.

C. Solid Phase Synthesis and Combinatorial Chemistry

Combinatorial Methods – Prospects for Catalysis?

Reinhard Räcker and Oliver Reiser

Institut für Organische Chemie, Universität Regensburg, Germany

In no time the concept of combinatorial chemistry has become a valuable tool in the process of drug discovery. The popularity of this approach is based on the possible synthesis and screening of libraries containing millions of compounds. In the field of asymmetric catalysis, combinatorial methods could help to discover new efficient catalysts as the number of possible metal-ligand combinations is immense. In an ideal case a substrate would be screened against a library of catalysts or a library of substrates against one catalyst to find out the most efficient conditions for the reaction in question. Up to now this has been essentially a utopian idea, but the first examples of the efficient detection of new catalysts and their reactions using combinatorial methods were recently described in the literature [1].

Principles of Combinatorial Chemistry

Several reviews have been published lately which provide detailed information on general aspects of combinatorial chemistry [2]. For this reason, only a short survey of the basic strategies is given here (Fig. 1): The first step is the synthesis of a library of compounds, i. e. the aim is to generate a large number of compounds at the same time.

Despite this seemingly unselective approach, the rules applied are the same as for the preparation of single compounds. The reactions used for the generation of the compound mixture have to proceed with high yields and without side reactions to assure the desired composition of the library. For the synthesis of a large number of

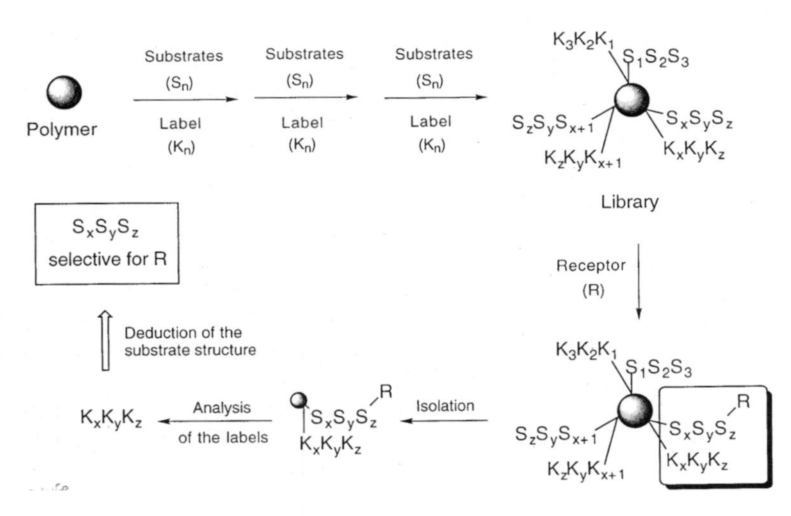

Figure 1. Principles of combinatorial chemistry.

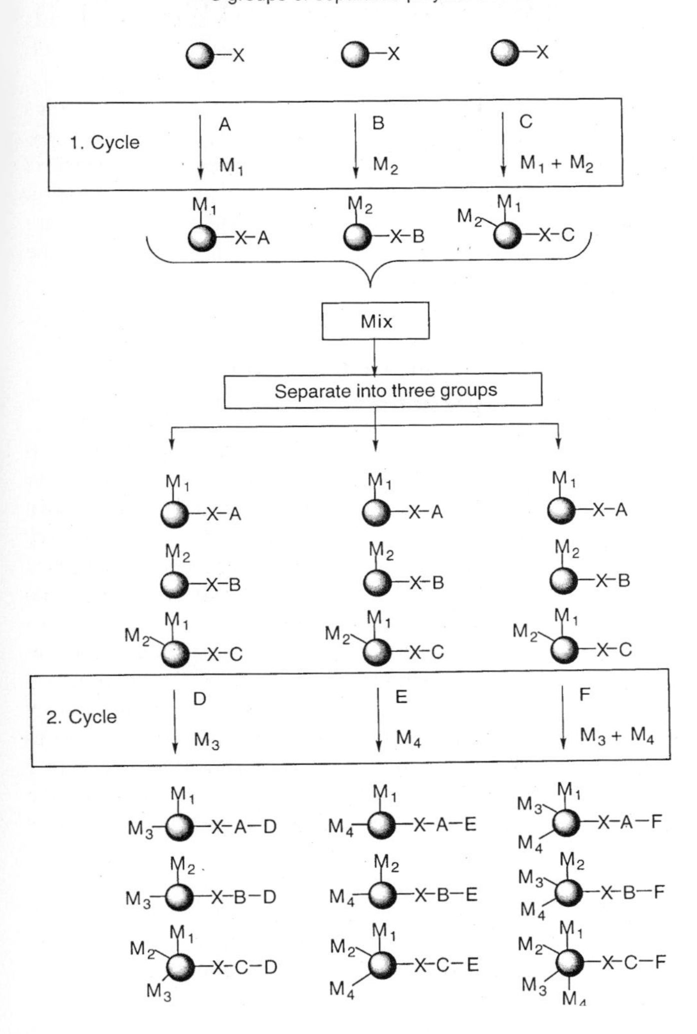

Figure 2. Encoded libraries prepared through split synthesis.

compounds, the single reaction steps have to be iterative, i. e. in each step a new functionality has to be created which allows another transformation in the subsequent step. The products must be separated from the reagents in an efficient manner, which can be achieved most easily by filtration. As a consequence, the products have to be solids, and this can be realized most efficiently if the reactions are carried out with polymer-bound substrates (solid-phase synthesis).

All of these demands were fulfilled for the synthesis of peptides long before combinatorial chemistry gained the importance that it has today, and therefore it is not surprising that peptides and their analogs are the most widely used compounds in combinatorial libraries. Nowadays, a large number of different reactions have been optimized for solid phase synthesis [3].

Furthermore, it has to be possible to determine the exact structure of a single compound within a library. As each compound in a combinatorial library is synthesized on a picomole scale, standard analytical methods are usually not applicable. These days, small amounts of peptides can be directly analyzed and identified using sequencing methods, but usually the compounds of a library have to be labeled in order to allow their identification.

A library obtained in this way can then be examined with regard to its interaction with a receptor. It is challenging to find an easy way to identify substances which are capable of interacting with the receptor. If a binding interaction takes place, the identification of the responsible structure can be carried out as follows: the receptor is labeled with a dye molecule so that an interaction can be detected optically by a change in color caused by the substrate–receptor complex. Finally, the structure of an active molecule can be deduced from the label to which it is correlated.

Combinatorial Libraries via "Split Synthesis"

One of the most powerful and systematic methods for the synthesis of an encoded combinatorial library is the so called "split synthesis" procedure (Fig. 2) [4].

Polymer beads serve as the solid phase, and synthesis is set up in such a way that eventually every bead carries just one compound (about 100 pmol $\approx 10^{13}$ molecules). As a consequence the obtained polymer-bound products are spatially separated. In each cycle, one reagent (A–F, e. g. an amino acid) and an encoding tag molecule (M_1–M_4) are attached to separate samples of resin beads. After this, the beads are thoroughly mixed and split up into equal quantities, and this is followed by the next reaction cycle. Diazoketones (**1**), which can be selectively incorporated into the excess polymer backbone using the corresponding acylcarbenes, being activated by rhodium catalysis, proved to be good tag molecules.

As part of the subsequent analysis, the tag molecule **2** is cleaved from the resin, and its identity is determined by an especially sensitive capillary gas chromatography method (ECGC). By this means the tag molecules may be used in much smaller amounts (1 mol%) than the substrate molecules. To keep the number of tag molecules low, they are employed as mixtures. However, it is crucial that every single compound in a library is correlated to a distinctively different combination of tags.The libraries described in the following paragraph were prepared according to the method depicted above.

New Metal Complexes through Combinatorial Chemistry

For the discovery of new ligands for transition-metal complexes, two different approaches were applied. Still and co-workers modified the known ligand cyclen in a combinatorial way by attaching peptide chains, and thus obtained **3** [5]. The screening of this ligand library against Cu(II) or Co(I) ions revealed different binding affinities depending on the peptide which was used for modification.

Jacobsen and co-workers synthesized a library of compounds with no predefined functionalities or subunits (Fig. 3): Two amino acids (positions 1 and 2), a turn-inducing fragment and a terminating group were used as variables [6].

This library was exposed to a solution of Ni(II) acetate or Fe(III) chloride in order to find the most efficient ligands for these ions among the synthesized compounds. The detection of the most stable complexes could easily be accomplished using classic color reactions (dimethylglyoxime for nickel and potassium rhodanide for iron). The colored resin beads were then selected under the microscope. Actually, certain structures proved to have a very high affinity for each of these ions. In the case of sufficiently dilute Ni(II)-solutions the metal

OMe N2 Cl m n HO Cl m n

n = 2–11, m = 2–5

1 **2** *Scheme 1*

CH_2CH_2NHR NH O $C H_2$ N N CH_2CH_2NHR N CH_2CH_2NHR

3 R = peptide *Scheme 2*

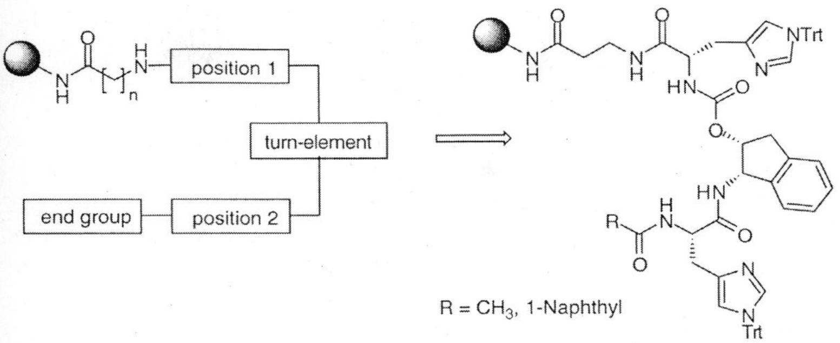

Figure 3. Composition of a ligand library and two out of four ligands with the highest affinity to Ni(II).

could only be detected on 6 out of 24 000 polymer beads.

However, the properties of the new complexes derived from these projects have not been investigated. Furthermore it is questionable whether this approach allows the discovery of new *catalysts* because the screening procedure led to the most stable complexes. However stability is a property that does not favor catalytic activity. Nevertheless, a recent study by the same authors revealed that libraries made by this approach can indeed point at least towards new lead structures for catalysts [7]. For the identification of efficient catalysts from a large library, strategies of parallel testing and deconvolution were employed. However, to distinguish between highly reactive and less reactive species remains problematic because of the large numbers of compounds and small quantities of each. A new strategy makes use of thermographic methods, which detect the heat which is set free in the exothermic reactions at the metal centers [8]. Also, resonance-enhanced multiphoton ionization (REMPI) techniques have been used to selectively and quantitatively determine the product formation in catalyzed reactions [9]. Yet another efficient assay for the discovery of active catalysts was shown by Hartwig et al. Palladium/ligand combinations

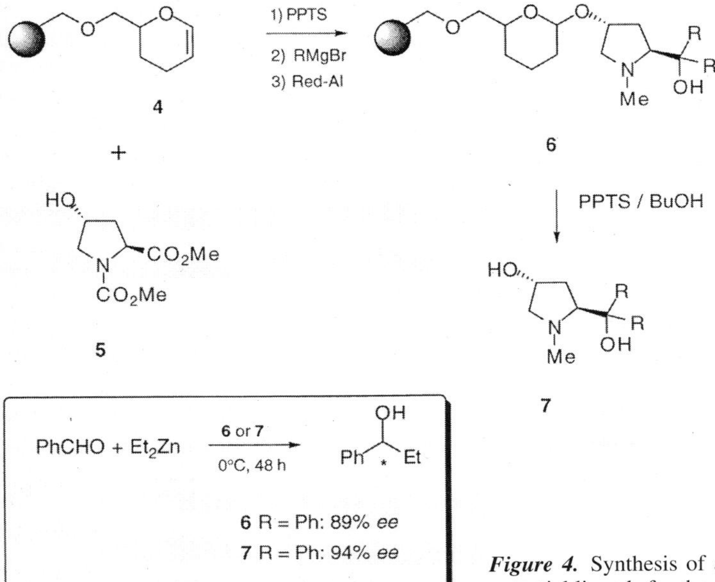

Figure 4. Synthesis of amino alcohols on a solid phase as potential ligands for the addition of diethylzinc to aldehydes.

were tested for Heck reactions in a way that a tethered fluorophore was coupled to an aryl halide attached to a solid support [10]. A successful coupling is signaled by fluorescence of the solid support.

Attempts at a More Efficient Screening of Catalysts

The synthesis of 2-pyrrolidinemethanol ligands on a polymer support was recently demonstrated by Ellman and coworkers and used for the preparation of single compounds, but not for combinatorial synthesis (Fig. 4) [11].

The resulting amino alcohols were tested as chiral ligands for the addition of diethylzinc to aldehydes while still bound to the polymer support or after cleavage from it. In both cases high enantioselectivities were observed depending on the ligand structure. However, this paper showed that there is no correlation between the selectivity of the polymer-bound and the free catalyst. Mikami et al. also chose a combinatorial approach for the optimization of ligands for the addition of diethylzinc to aldehydes [12]. The authors postulate a zinc complex containing a diol ligand and a diimine ligand as the actual catalyst. Various diol-diimine combinations were prepared in a parallel synthesis in solution, and the enantiomeric excesses obtained with the corresponding complexes in the diethylzinc addition to aldehydes were determined using HPLC-circular dichroism techniques (HPLC-CD).

A library of 63 peptides, each containing one of the two phosphine-substituted amino acids **8** or **9**, was prepared by Gilbertson and coworkers in a parallel synthesis on polymer beads [13]. After complexation of *all* peptides with rhodium, the obtained polymer-bound complexes were employed in the asymmetric hydrogenation of **10** (Fig. 5).

For that purpose the set-up was reduced in size to an extent that allowed the simultaneous hydrogenation with all catalysts in spatially separated vessels. The enantioselectivities which were determined for each catalyst were only moderate (< 20 % *ee*) in all cases, but nevertheless this approach is interesting because of the possibility of carrying out a high-throughput screening of ligands.

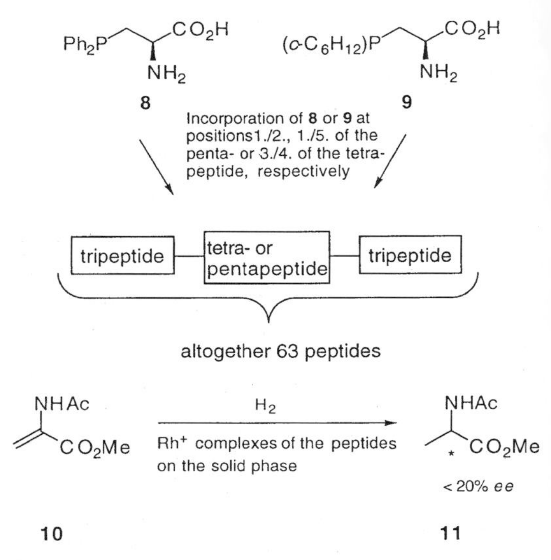

Figure 5. Synthesis of a peptide library with phosphine-containing amino acids as potential ligands for the asymmetric hydrogenation.

In a similar approach, Burgess et al. investigated a C-H-insertion reaction with different metal-ligand combinations on microtiter plates [14]. By these means it was possible to test 96 catalyst systems at a time, indeed in single reactions, but still faster than by conventional methods. The uncertainty of the results due to the small scale of the experiments is a drawback of this screening procedure, as was (laudably) revealed by the authors themselves. Other approaches for screening catalysts in parallel have also been successful [15].

A remarkable strategy for the optimization of catalysts which also utilizes the parallel synthesis of ligands was presented by Snapper, Hoveyda et al. (Fig. 6) [16].

Based on the fundamental perception that the titanium-catalyzed addition of trimethylsilyl cyanide to *meso*-epoxides can be carried out in an asymmetric fashion in the presence of a dipeptide Schiff base, the authors were able to raise the enantioselectivity of the cyclohexene oxide ring opening systematically through a successive variation of the building blocks AA1→AA2→ aldehyde in **12**. In this way *tert*-leucine was recognized in the first screening to be optimal for AA1 (although a closer look at the original publication reveals that *β*-trityl-L-asparagine is

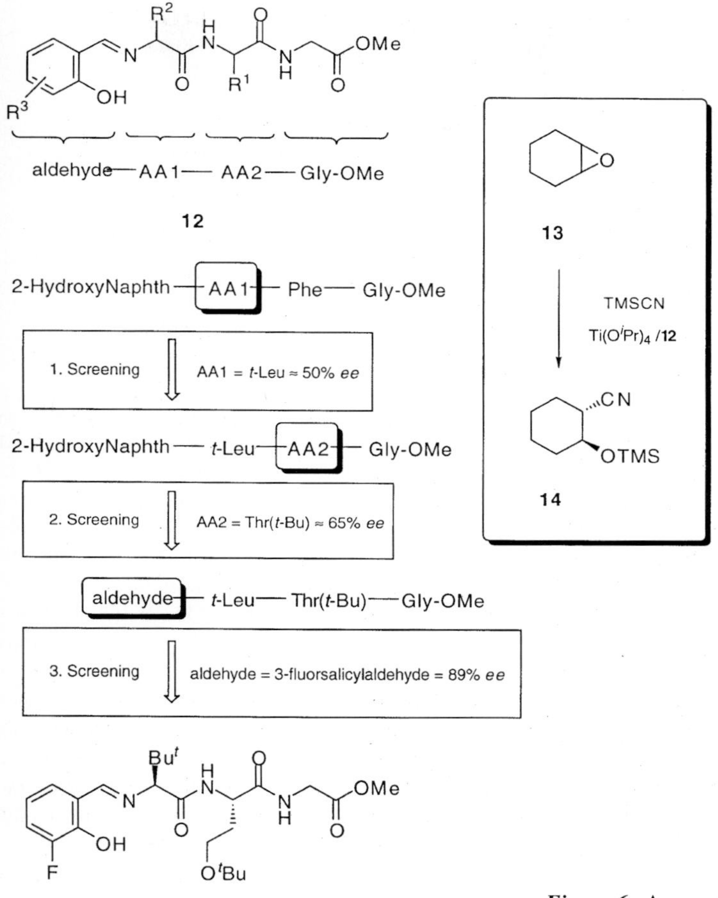

Figure 6. A new screening procedure for catalysts via variation of single subunits.

at least equally suitable). In the second screening step *O-tert*-butyl- L-threonine was found to be the best choice for AA2 while keeping *tert*-Leu as AA1, and in the third step *tert*-Leu and Thr(*t*-Bu) were kept and 3–fluorosalicylaldehyde was determined to be the most suitable aldehyde. This led to ligand **15**, which allowed the titanium-catalyzed addition of TMSCN to **13** with remarkable 89 % *ee*.

As not all combinations are tested when following this procedure, the question remains whether the best ligands were really found. This would certainly be the case if the effects of the single building blocks were additive but, in general this should not be true. For the ligand system just described, an independent influence of the ligand subunits seems indeed to exist, since the authors also find **15** as the best ligand when they reverse the screening strategy to aldehyde →AA2→AA1. The authors also showed that the catalyst optimization leads to equal results no matter whether the catalyst is polymer-bound or not [17]. Similarly, catalysts for the asymmetric Strecker reaction have been discovered [18].

The high-throughput screening of asymmetric catalysts requires efficient techniques for the determination of enantiomeric excesses. Siuzdak and Finn recently developed a method for that purpose which makes use of kinetic resolution and mass spectrometry [19]. Various chiral secondary alcohols and amines were esterified on

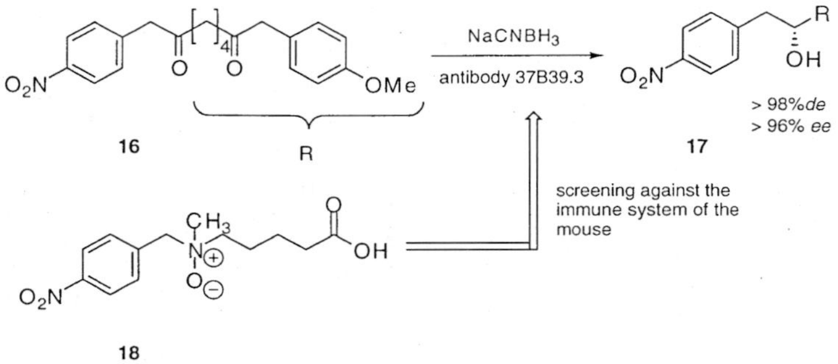

Figure 7. Regio- and enantioselective reduction of a dicarbonyl compound using a catalytic antibody.

a nanomole scale with a 1 : 1 – mixture of two chiral carboxylic acids that differ by their molecular weight, i. e. by one methyl group. The authors showed that even if the ratio of the reaction rates for the formation of diasteriomeric esters or amides is as small as 1.2, the enantiomeric excess of the sample can be determined with an accuracy of $\pm 10\,\%$. The relative amounts of the derivatized compounds were determined by electrospray ionization mass spectrometry (ESI-MS). Reetz et al. developed a method for the determination of the enantioselectivity in asymmetric catalytic reactions of chiral substances and of prochiral compounds containing enantiotopic groups [20]. Their approach is based on the use of isotope-labeled *pseudo*-enantiomers or *pseudo*-prochiral compounds. Enzymatic kinetic resolution processes and the desymmetrization of *meso*-compounds were investigated, and the conversion as well as the enatiomeric excess were determined by ESI-MS.

Catalytic Antibodies

The discovery of catalysts using combinatorial methods was successfully demonstrated for the generation of catalytic antibodies [21]. This method exploits the great number of indeed highly complex compounds of the immune system. The strategy for the detection of a catalyst for a certain reaction is as follows: the compound should be the most suitable catalyst which best stabilizes the *transition state* of the reaction, i. e. lowers its energy most. As a transition state is an energy maximum and can thus not be isolated, a stable analogon is prepared which is as similar to the transition state as possible, and the substance with the highest affinity to it is searched for from the immune system. For example, to detect an appropriate catalyst for the reduction of **16** to **17** Schultz et al. used the *N*-oxide **18,** which should simulate the tetrahedral geometry as well as the charge distribution of the hydride attack on a carbonyl group (Fig. 7) [22].

Actually an antibody was found which not only allowed the preparation of **17** with an excellent regioselectivity (> 75 : 1) in favor of the carbonyl group adjacent to the nitrobenzoyl unit but also effected an outstanding enantioselectivity (> 96 % *ee*) in favor of the (*S*)-enantiomer because of the chirality of antibodies. With conventional chemical methods such a transformation would have been difficult to carry out.

This procedure made the identification of a multitude of effective catalysts for organic reactions possible. The general availability of antibodies and their high substrate specifity pose a problem, however. For this reason the detection of small compounds as catalysts for organic reactions will remain an important goal for synthesis.

A real combinatorial testing of libraries of small compound catalysts is hampered by the problem that the information about which catalyst reacted with the substrate to which product needs to be transferred. In the case of the highly inventive encoding techniques for libraries (e. g. via radio frequencies or bar codes), this problem may be solved in the future [23].

Although there is general euphoria about combinatorial chemistry, the question whether the chosen approach is really important for the target envisaged should be critically considered. Furthermore, it should be more severely questioned whether selectivity is still an outstanding goal in research on catalysis. Almost universal ligands like the binaphthyls [24] or the bisoxazolines [25] were developed through rational design, and because of their non-modular structure they would probably not have been detected using combinatorial methods. While high selectivities can be achived for almost all known catalytic reactions after optimization, most catalytic reactions suffer from low turnover rates and turnover numbers of the catalysts. Increased investigations and new approaches to raise reactivity are necessary, and, for this, combinatorial chemistry can contribute even more to the discovery of new and highly reactive catalytic systems. *Give me rate, I probably can give you selectivity later* is a rule from many lectures by B. Sharpless which one should bear in mind.

Acknowledgement: The authors acknowledge support from the Fonds der Chemischen Industrie.

References

[1] Earlier reviews: a) C. Gennari, H. P. Nestler, U. Piarulli, B. Salom, *Liebigs Ann./Recl.* **1997**, 637–647; b) H. B. Kagan, *J. Organomet. Chem.* **1998**, *567*, 3–6; c) W. F. Maier, *Angew. Chem.* **1999**, *111*, 1294–1296; d) T. Bein, *Angew. Chem.* **1999**, *111*, 335–338, e) R. Schlögl, *Angew. Chem. Int. Ed. Engl.* **1998**, *37*, 2333–2336; f) P. Cong, R. D. Doolen, Q. Fan, D. M. Giaquinta, S. Guan, E. W. McFarland, D. M. Poojary, K. Self, H. W. Turner, W. H. Weinberg, *Angew. Chem.* **1999**, *111*, 508–512; g) O. Reiser, *Nachr. Chem. Tech. Lab.* **1996**, *44*, 1182–1188; h) S. Borman, *Chem. Eng. News* **1996**, *74*, 37.

[2] a) F. Balkenhohl, C. von dem Bussche-Hünnefeld, A. Lansky, C. Zechel, *Angew. Chem.* **1996**, *108*, 2437; b) A. W. Czarnik et al., *Combinatorial Chemistry (Special Issue), Acc. Chem. Res.* **1996**, *29*, 112; c) L. A. Thompson, J. A. Ellmann, *Chem. Rev.* **1996**, *96*, 955.

[3] J. S. Früchtel, G. Jung, *Angew. Chem.* **1996**, *108*, 19–46.

[4] *Review:* W. C. Still, *Acc. Chem. Res.* **1996**, *29*, 155.

[5] M. T. Burger, W. C. Still, *J. Org. Chem.* **1995**, *60*, 7382.

[6] M. B. Francis, N. S. Finney, E. N. Jacobsen, *J. Am. Chem. Soc.* **1996**, *118*, 8983.

[7] M. B. Francis, E. N. Jacobsen, *Angew. Chem.* **1999**, *111*, 987–991.

[8] a) S. J. Taylor, J. P. Morken, *Science* **1998**, *280*, 267–270; b) D. E. Bergbreiter, *Chemtracts* **1997**, *10*, 683–686; c) A. Holzwarth, H.-W. Schmidt, W. F. Maier, *Angew. Chem., Int. Ed. Engl.* **1998**, *37*, 2644–2647; (d) M. T. Reetz, M. H. Becker, K. M. Kühling, A. Holzwarth, *Angew. Chem.* **1998**, *110*, 2792–2795.

[9] S. M. Senkan, *Nature* **1998**, *394*, 350–353; b) S. M. Senkan, S. Ozturk, *Angew. Chem.* **1999**, *111*, 867–871.

[10] K. H. Shaughnessy, P. Kim, J. F. Hartwig, *J. Am. Chem. Soc.* **1999**, *121*, 2123–2132.

[11] G. Liu, J. A. Ellman, *J. Org. Chem.* **1995**, *60*, 7712.

[12] K. Ding, A. Ishii, K. Mikami, *Angew. Chem.* **1999**, *111*, 519–523.

[13] S. R. Gilbertson, X. Wang, *Tetrahedron Lett.* **1996**, *37*, 6475.

[14] K. Burgess, H.-J. Lim, A. M. Porte, G. A. Sulikowski, *Angew. Chem.* **1996**, *108*, 192.

[15] a) T. Berg, A. M. Vandersteen, K. D. Janda, *Bioorg. Med. Chem. Lett.* **1998**, *8*, 1221–1224; b) C. Gennari, S. Ceccarelli, U. Piarulli, C. A. G. N. Montalbetti, R. F. W. Jackson, *J. Org. Chem.* **1998**, *63*, 5312–5313.

[16] B. M. Cole, K. D. Shimzu, C. A. Krueger, J. P. A. Harrity, M. L. Snapper, A. H. Hoveyda, *Angew. Chem.* **1996**, *108*, 1776.

[17] K. D. Shimizu, B. M. Cole, C. A. Krueger, K. W. Kuntz, M. L. Snapper, A. H. Hoveyda, *Angew. Chem.* **1997**, *109*, 1782–1785.

[18] M. S. Sigman, E. N. Jacobsen, *J. Am. Chem. Soc.* **1998**, *120*, 4901–4902.

[19] J. Guo, J. Wu, G. Siuzdak, M. G. Finn, *Angew. Chem.* **1999**, *111*, 1868–1871.

[20] M. T. Reetz, M. H. Becker, H.-W. Klein, D. Stöckigt, *Angew. Chem.* **1999**, *111*, 1872–1875.

[21] a) L. C. Hsieh-Wilson, X.-D. Xiang, P. G. Schultz, *Acc. Chem. Res.* **1996**, *29*, 164; b) C. Gao, B. J. Lavey, C.-H. L. Lo, A. Datta, P. Wentworth, K. D. Janda, *J. Am. Chem. Soc.* **1998**, *120*, 2211–2217; c) K. D. Janda, L.-C. Lo, C.-H. L. Lo, M.-M. Sim, R. Wang, C.-H. Wong, R. A. Lerner, *Science* **1997**, *275*, 945–948.

[22] L. C. Hsieh, S. Yonkovich, L. Kochersperger, P. G. Schultz, *Science* **1993**, *260*, 337.

[23] For a recent review on combinatorial catalysis, see: B. Jandeleit et al. *Angew. Chem.* **1999**, *111*, 2648–2689, *Angew. Chem.* Inl. Ed. Engl. **1999**, *38*, 2494–2532.

[24] cf. T. Wabnitz, O. Reiser in *Organic Synthesis Highlights IV* (Ed. H. G. Schmalz), Wiley-VCH, Weinheim, **2000**, pp. 155.

[25] cf. M. Glos, O. Reiser in *Organic Synthesis Highlights IV* (Ed. H. G. Schmalz), Wiley-VCH, Weinheim, **2000**, pp. 17.

The Renaissance of Soluble Polymers

M. Reggelin

Institut für Organische Chemie, Universität Mainz, Germany

In 1963 R. B. Merrifield published his seminal paper on the possibility of peptide synthesis on a solid, polymeric support [1]. On the basis of this work, which helped its author to get the Nobel Prize in 1984, the field of chemical synthesis of biopolymers was developed and has been an area of great interest since then. It is therefore not surprising that the annual number of citations of this publication (▼ in Fig. 1, up to Autumn 1997 : 3511 in total!) increased rapidly between 1975 and 1990 and has now stayed at a very high level with only a slight tendency to fall.

Solid-phase synthesis seemed to be the key to a fascinating possibility: the chemical synthesis of proteins able to function. Maybe this "El Dorado" feeling at the end of the 1960s and beginning of the 1970s is responsible for the lack of work on alternatives; it simply seemed unnecessary. On the other hand, as time passed it became clear that even the brilliant solid-phase method had its problems:

1) Loss of reactivity due to heterogeneous reaction conditions.
2) Incompatibility between the swelling behavior of the solid support and the solvents needed in the reaction.
3) Non-linear reaction kinetics due to locally different reactivities of the polymeric matrix.
4) The solid support hampering reaction monitoring.

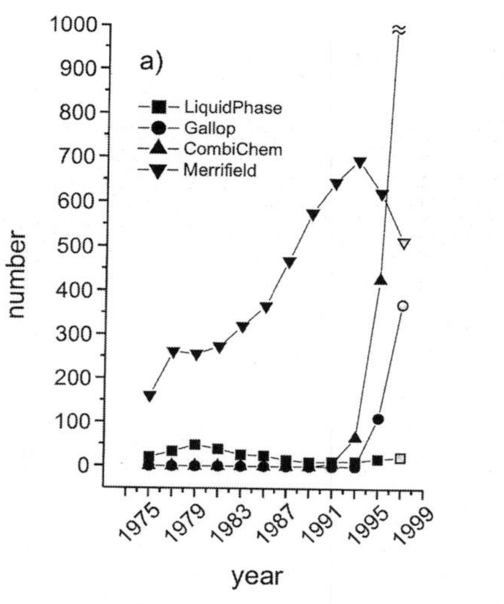

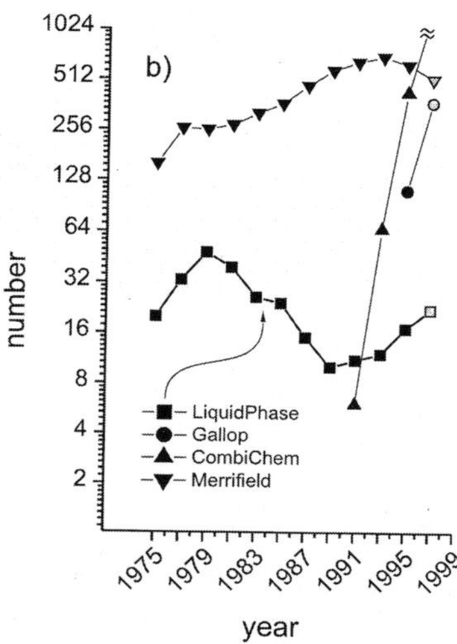

Figure 1. (a) Literature statistics illustrating the revival of interest in soluble polymers, time interval: 2 years, linear representation. ▼: number of citations of Ref. [1]; ■: sum of the citations of Refs. 3a, 3b, 6, and 13a; ▲: citations in the keywords "Combinatorial Chem? or Combinatorial Librar?"; •: Number of citations of Ref. [14]. (b) logarithmic representation (basis 2).

Being aware of these difficulties in 1965, a russian group proposed (in a paper now virtually unheard of) the use of *soluble* polystyrene as polymeric support [2]. Although a tetrapeptide was synthesized in 65 % yield, the method did not gain acceptance, since the polystyrene used became crosslinked during the synthesis and the by-products tended to precipitate with the polymer. It was not until the years 1971 and 1972 that E. Bayer and M. Mutter proposed a practicable concept of peptide synthesis on a linear soluble polymer [3]. It is proven by the number of citations of this work, which steadily increased in the following years (■ in Fig. 1), that this protocol has been accepted as a interesting alternative. Maybe a certain disillusionment concerning the potential of solid-phase synthesis of proteins arising at that time was beneficial for this acceptance [4]. So let us take a closer look at the first "Golden Age" of this interesting methodology.

The Best of Both Worlds

The recipe for success was to select the "right" polymer. If polyethylene glycol [PEG; or more correctly poly(oxyethylene)] was used, the problems associated with solid phase variation, namely those concerning reactivity and diffusion, were overcome, and conditions typical of chemistry in solution could be restored at the same time. Additionally, the operative advan-

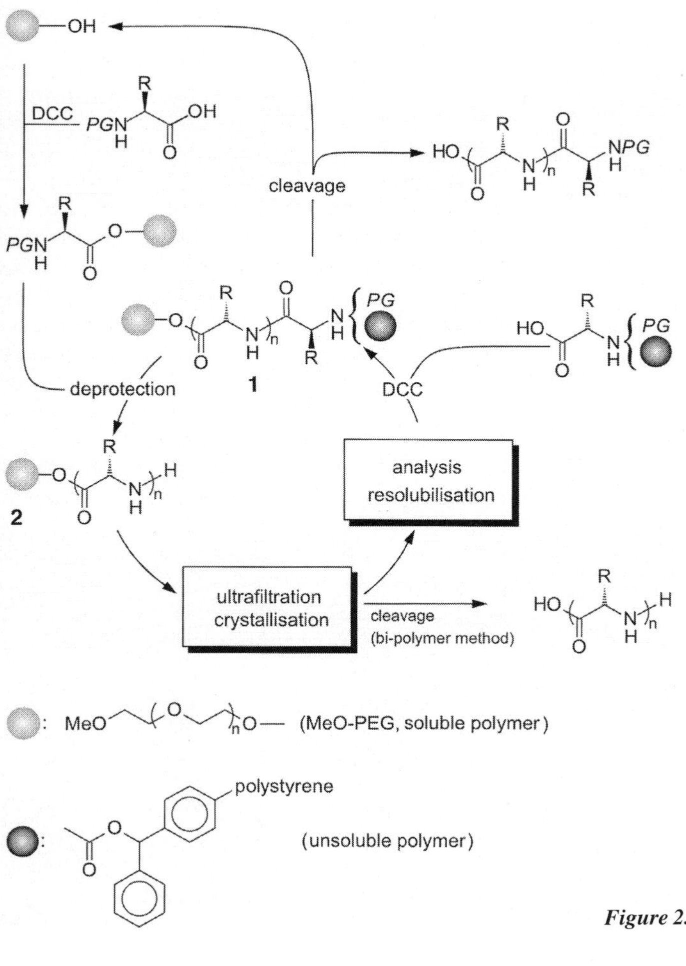

Figure 2. Peptide synthesis on soluble polymers.

tages of the solid-phase protocol are preserved, because the polymer can be precipitated after completion of the reaction in a homogenous phase. Indeed, kinetic studies on the aminolysis of BOC-protected *p*-nitrophenyl esters of glycine with glycine esters showed that the presence of the polymer in the alcohol component (MW: 2000 – 20 000) didn't have any effect on the reaction rate compared to the presence of an ethylester of lower molecular weight [5]. Figure 2 depicts the protocol of the polypeptide synthesis using soluble MeO-PEG *X* (MeO-PEG *X* stands for mono-methylated PEG with an average molar mass of *X*; typically $X = 5000 - 15\ 000$).

Protected at the *N*-terminus (PG polymer), the amino acid is condensed to the free OH group of the polymer and subsequently deprotected. Thus, an amino acid (**2**) immobilized at the *C*-terminus [or the conjugate of a growing peptide chain and a soluble polymer, (**1**), PG polymer] is obtained which can be separated in (at least) [6] two ways from the by-products and excessive reagents. The initial suggestion by Bayer and Mutter was ultra-filtration through a semi-permeable membrane [3], but the precipitation technique introduced in 1974 is more covenient, faster and more flexible with regard to the solvents used [7]. Due to their helical structure, PEG and MeO-PEG with molecular weights between 2000 and 20 000 exhibit a strong tendency to crystallize [8]. Since this is true for polypeptide conjugates as well [9], precipitation by addition of ether or *t*-butyl methyl ether seems to be the method of choice [10]. If the degree of purity obtained by precipitation and washing is not sufficient, the polymer can be further purified, almost without loss, by reprecipitation from ethanol, in which it shows a strong temperature gradient of solubility [9]. Another advantage of the liquid-phase method is the opportunity to apply all spectroscopic methods known for reactions in solution for evaluation of purity and effectiveness of coupling. After the reaction has been monitored, the polymer is redissolved and coupled to the next *N*-terminal-protected amino acid. Now the cycle of deprotection and coupling can either be run again or the product is cleaved from the polymer (e.g. with NH_3, hydrazine or NaOH). It is obvious that the application of MeO-PEG as a soluble support has the advantages of both worlds: quick, diverse chemistry in solution and convenient separation of the excess reagents using the properties of the polymer. Of course, it cannot escape notice that *polymer-bound* by-products or erratic sequences generated by failed coupling reactions cannot be avoided. In order to cope with this problem, Frank et al. developed an interesting procedure (Fig. 2) [11]. The idea behind this is to grow the peptide on the soluble polymer, as described, but to attach *solid-supported* monomers for elongation. In this way, only successfully extended polymers become insoluble bipolymers (**1**, PG = polymer). Excess non-coupling PEG-peptides can be separated by washing, and thus, even with incomplete coupling, erratic sequences will never be obtained [12].

Limitations

Iterative peptide synthesis on a soluble support is feasible as long as the properties of the polymer-bound peptides are determined by the polymer. In the case of *random coil* sequences or helical peptides, this is guaranteed for up to about 20 amino acids. But with peptides which tend to form β-sheets, this limit is hit much earlier. Also, problems arose in the automation, giving a competitive edge to the solid-phase variant. Certainly these circumstances contributed to the change from maximum interest in the 1970s (Fig. 1, ■) to a low level of interest [13]. In the early 1990s. However, interest was resurrected and seems to have been growing disproportionately since 1993 (Fig. 1b, ■). The temporary coincidence with combinatorial chemistry, which has been in an explosive phase of expansion since about 1991 (Fig. 1, ▲ and • [14]), strongly suggests a relationship between the renaissance of soluble polymers and this development.

Liquid-Phase Combinatorial Synthesis (LPCS)

The early days of combinatorial synthesis have been dominated by technical aspects such as the *development of methods* for the generation and analysis of molecular diversity [15]. After these problems had been solved, e.g. for biopolymer libraries, people started thinking about the *content* of the libraries which had to be synthesized. As a logical consequence of this change in interest, today's library synthesis increasingly aims at small molecules with favorable pharmacological features [15, 16]. In order to adapt the most effective strategy for the synthesis of equimolarly composed combinatorial libraries (the "split-mix" method) [17], to this class of compounds, reactions optimized in solution had to be transferred to the solid phase. However, this implicates research and development activities, and this obviously discourages the non-chemist managers of the chemical industry. Chemical innovation and industrial success no longer follow the same time scale. One consequence of this development is the preferential use of parallel synthesis in chemical industry. Here, the opportunities possibly inherent in combinatorial synthesis cannot be exploited. The tentative transfer of reactions from solution to the solid phase (which of course was also encouraged by the tentative acceptance by chemists at universities) and the associated problems led to an increased consideration of alternatives. In this state it was obviously recollected that both are possible: the diversity of chemistry in solution *and* the polymer-mediated advantages in the reaction process. This and the perception that additionally there is the possibility of library synthesis following the "split-mix" protocol mark the beginning of the renaissance of soluble polymers. The following para-

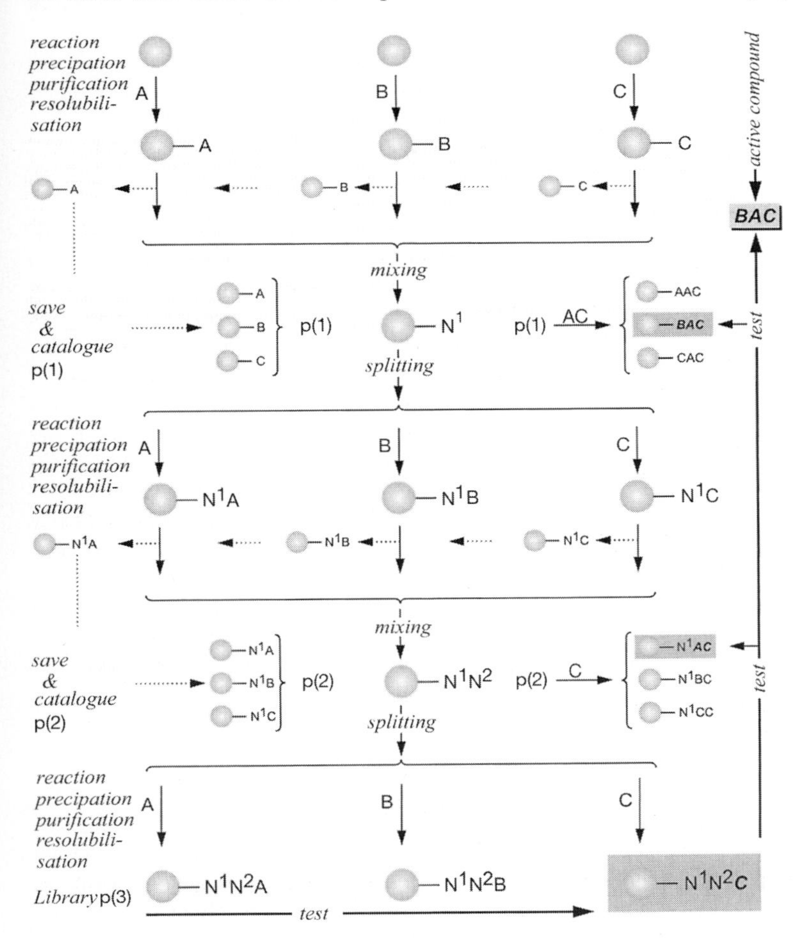

Figure 3. Liquid phase combinatorial synthesis (LPCS) and recursive deconvolution.

graph describes Janda's approach to combinatorial chemistry using soluble polymeric supports (Fig. 3) [18].

The three building blocks **A**, **B** and **C** are coupled to the soluble polymer MeO-PEG 5000, whereby three reaction channels were opened up. In each reaction channel, precipitation is achieved by the addition of ether, separation and purification by filtration and/or recrystallization from ethanol. In every channel, the precipitate is redissolved in (e. g.) dichloromethane, and a portion of the solution is set aside and labeled as partial library p(1), consisting of three members (MeOPEG-N_1) in three different vials. The remaining polymer solutions are then combined and again divided among the three reaction vessels. After another coupling cycle, portions of the purified PEG-bound products were again sampled to generate a second sub-library [p(2), 3 containers, each with 3 substances = 9 compounds]. This procedure is repeated until all coupling reactions necessary have been carried out [(Pn); in this case $n = 3$, 3 containers with 9 substances = 27 compounds]. This approach corresponds to Furkas "split-mix" protocol [17] with the modification of generating sub-libraries, which are very important for the analysis of the library. They facilitate the search for active members in a combinatorial library, which is an advancement of Houghten's iterative deconvolution method [19] and was named "recursive deconvolution" by Janda. The way it works is depicted on the righthand side of Fig. 3. The pronounced solubilizing power of MeO-PEG [9, 20] opens up the opportunity to test the library in a multiplicity of solvents in a homogeneous phase for (biological) activity, which is another important advantage compared to the solid phase version! The deconvolution starts with the final library p(3). In the example, the highest activity is found in the group of 9 compounds denoted with **C**. The next step uses this information. The sub-library p(2) is coupled to **C**, and the new library containing 3 × 3 compounds, in which positions 2 and 3 are known, is again tested for activity. In this example the highest activity is displayed by the group denoted by **AC**. For that reason in the last step p(1) is coupled to **AC,** and **BAC** is identified to be the most active compound. Janda et al. validated this concept by the synthesis of a sulfonamide library and a pentapeptide library (1024 members) [18]. By application of an affinity assay the library was searched for binders for a monoclonal antibody which itself exhibits a strong affinity to the β-endorphin sequence (Tyr-Gly-Gly-Phe-Leu). Besides the native epitope a set of other peptides with high affinity towards the antibody was found. It is important to note that the MeO-PEG peptide conjugates bind comparably to their polymer-free counterparts [20c]! The advantages of this method are enormous:

a) No problematic "on-bead-assays" are necessary.
b) The soluble conjugate of polymer and substrate allows for the direct application of known biological tests.
c) The "split-mix" technique is applicable for the generation of large, equimolarly composed libraries.
d) Sub-libraries render repeated "split-mix" syntheses during the deconvolution superfluous.
e) In each deconvolution cycle the number of simultaneously tested compounds decreases by a factor of $X^n/X^{n-1} = X$ (X: number of building blocks, n: number of cycles). This increases the signal/noise ratio, which reduces the incidence of "false-positives".
f) If the final library contains several active compounds, they all can be identified, since the corresponding deconvolution pathways can be run independently (even in parallel!).
g) Recursive deconvolution not only indicates which compound is the active one, but also yields the active compound itself. The amount of substance synthesized can be controlled by the amount of the (very cheap!) polymer used.

Conclusions

Chemical synthesis assisted by soluble polymers originally developed as an alternative to solid phase synthesis is now undergoing a second period of expansion. Just like in the early days, the desire to combine the advantages of solution chemistry with those of reactions on a solid phase has triggered this development. Soluble polymers can provide this and have therefore become important for one of the most dynamic fields of contemporary chemistry: the combinatorial synthesis of non-biopolymer libraries. The results obtained by Janda et al. could be considered a milestone in the search for libraries exhibiting a maximum of constitutional and configurational diversity. The whole arsenal, even stereoselective reactions, could be made available for combinatorial chemistry. Thereby the "dreariness of heteroatom acylation" would finally come to an end!

References

[1] R. B. Merrifield, *J. Am. Chem. Soc.* **1963**, *85*, 2149.

[2] M. M. Shemyakin, Yu. A. Ovchinnikov, A. A. Kinyushkin, *Tetrahedron Lett.* **1965**, 2323.

[3] (a) M. Mutter, H. Hagenmaier, E. Bayer, *Angew. Chem.* **1971**, *83*, 883. (b) E. Bayer, M. Mutter, *Nature* **1972**, *237*, 512.

[4] The small proteins (55–129 amino acids, e. g. the ribonucleases A and T_1 as well as lysozyme) synthesized in the years 1968–1973 did not satisfy chemical purity demands. E. Bayer, *Angew. Chem.* **1991**, *103*, 117.

[5] E. Bayer, M. Mutter, R. Uhmann, J. Polster, H. Mauser, *J. Am. Chem. Soc.* **1974**, *96*, 7333.

[6] Alternatives are centrifugation and gel permeation chromatography: K. E. Geckeler, *Adv. Polym. Sci.* **1995**, *121*, 31.

[7] M. Mutter, E. Bayer, *Angew. Chem.* **1974**, *86*, 101.

[8] (a) J. M. Harris in *Poly(Ethylene Glycol) Chemistry: Biotechnical and Biomedical Applications* (Ed.: J. M. Harris), Plenum Press, New York, **1992**, p. 2. (b) In the crystal, PEG is present as a 7_2-helix: H. Tadokoro, Y. Chatani, T. Yoshihara, S. Tahara, S. Murahashi, *Macromol. Chem.* **1964**, *73*, 109.

[9] V. N. R. Pillai, M. Mutter, *Acc. Chem. Res.* **1981**, *14*, 122.

[10] The solubilities of MeO-PEG 5000 (in wt %) at room temperature in various solvents are: water: 55; dichloromethane: 53; chloroform: 47; DMF: 40; pyridine: 40; methanol: 20; benzene: 10; ethanol (60 %): 50; ethanol (100 %): 0.1; ethanol (100 %, 34 °C): 20; ether: 0.01. E. Bayer, M. Mutter, J. Polster, R. Uhmann, *Pept., Proc. Eur. Pept. Symp. 13th* **1974**, 129.

[11] H. Frank, H. Hagenmaier, *Experientia* **1975**, *31*, 131.

[12] In a recent publication (*Angew. Chem.* **1997**, *109*, 1835) H. Han and K. D. Janda picked up this approach for the transfer of the asymmetric Sharpless dihydroxylation into a "multi-polymeric environment".

[13] This does not mean that interest disappeared completely. But it is a fact however that the procedure described so far as well as the ideas developed by Bayer (1975 and 1986) concerning catalysis with soluble polymers [(a) E. Bayer, V. Schurig, *Angew. Chem.* **1975**, *87*, 484; (b) E. Bayer, W. Schumann, *J. Chem. Soc., Chem. Commun.* **1986**, 949] have only been taken up in the 1990s. The application of soluble polymers in the synthesis of oligosaccharides [(c) S. P. Douglas, D. M. Whitfield, J. J. Krepinsky, *J. Am. Chem. Soc.* **1991**, *113*, 5095] and oligonucleotides [(d) G. M. Bonora, G. Biancotto, M. Maffini, C. M. Scremin, *Nucleic Acids Res.* **1993**, *21*, 1213] took place in the same time period.

[14] In the early days of combinatorial chemistry two important reviews were published: (a) M. Gallop, R. W. Barrett, W. J. Dower, S. P. A. Fodor, E. M. Gordon, *J. Med. Chem.* **1994**, *37*, 1233. (b) E. M. Gordon, R. W. Barrett, W. J. Dower, S. P. A. Fodor, M. A. Gallop, *J. Med. Chem.* **1994**, *37*, 1385. The number of citations on these two articles correlates to the number of matches on a keyword survey on the topic of combinatorial chemistry, as expected (Fig. 1).

[15] For an excellent review, see: F. Balkenhohl, C. von dem Bussche-Hünnefeld, A. Lansky, C. Zechel, *Angew. Chem.* **1996**, *108*, 2436.

[16] (a) M. Reggelin, V. Brenig, *Tetrahedron Lett.* **1996**, 6851. (b) J. B. Backes, J. A. Ellman, *J. Am. Chem. Soc.* **1994**, *116*, 11171. (c) B. A. Bunin, J. A. Ellman, *J. Am. Chem. Soc.* **1992**, *114*, 10997.

[17] (a) A. Furka, F. Sebestyen, M. Asgedom, G. Dibo, *Int. J. Pept. Prot. Res.* **1991**, *37*, 487. (b) F. Sebestyen, G. Dibo, A. Kovacs, A. Furka, *Bioorg. Med. Chem. Lett.* **1993**, *3*, 413.

[18] (a) Review: D. J. Gravert, K. D. Janda, *Chem. Rev.* **1997**, *97*, 489. (b) H. Han, M. M. Brenner, K. D. Janda, *Proc. Natl. Acad. Sci. U.S.A.* **1995**, *92*, 6419. (c) H. Han, K. D. Janda, *J. Am. Chem. Soc.* **1996**, *118*, 2539. (d) Recursive deconvolution: E. Erb, K. D. Janda, S. Brenner, *Proc. Natl. Acad. Sci. U.S.A.* **1994**, *91*, 11422.

[19] R. A. Houghten, C. Pinilla, S. E. Blondelle, J. R. Appel, C. T. Dooley, J. H. Cuervo, *Nature* **1991**, *354*, 84.

[20] This feature was often used to study the folding of a growing peptide chain. (a) R. C. de L. Milton, S. C. F. Milton, P. A. Adams, *J. Am. Chem. Soc.* **1990**, *112*, 6039. (b) M. Mutter, H. Mutter, R. Uhmann, E. Bayer, Biopolymers, **1976**, *15*, 917. (c) P. Koziej, M. Mutter, H.-U. Gremlich, G. Hölzemann, *Z. Naturforsch.* **1985**, *40b*, 1570. In the articles (b) and (c) the conformational preferences of PEG-bound substance P is described, a peptide which underwent a renaissance as well: K. J. Watling, J. E. Krause „The rising sun shines on substance P and related peptides", *TiPS* **1993**, *14*, 81.

Polymeric Catalysts

M. Reggelin

Institut für Organische Chemie, Universität Mainz, Germany

Highly efficient selection of substrates and their stereocontrolled transformations are characteristic features of many biological systems. Normally, to obtain these effects, nature uses large, catalytically active polymers. The mass transport as well as the topicity of the substrate/reagent interaction is controlled by the secondary or tertiary structure of the macromolecule. At the same time, decoupling of the residence time of the catalyst and the reactants results in high turnover numbers [1], separation of the micromolecular products is straightforward and construction of continuously working systems is possible. Obviously preparative working chemists soon tried to copy this very attractive scenario [2]. An early example is the addition of methanol to methyl phenyl ketene in the presence of an immobilized cinchona-alkaloid (Fig. 1) [3].

Under the influence of polycinchonine acrylate (**1**, 1 mol%) the *S*-configured ester (**3**) was formed in 35 % *ee* at – 78 °C in toluene. Interestingly, if instead of the polymer the monomeric derivative cinchonine propionate (**2**) is used under the same reaction conditions the product with inverted absolute configuration [(*R*-**3**), 4.4 % *ee*] is formed! This result, which was reported in the original article without any comment, can be interpreted as follows: the polymer exhibits a chiral (probably helical) superstructure, which dominates the chiral induction in the reaction [4].

Definition and Classification

Apparently chiral polymeric catalysts offer a *new quality* of asymmetric catalysis. If one succeded

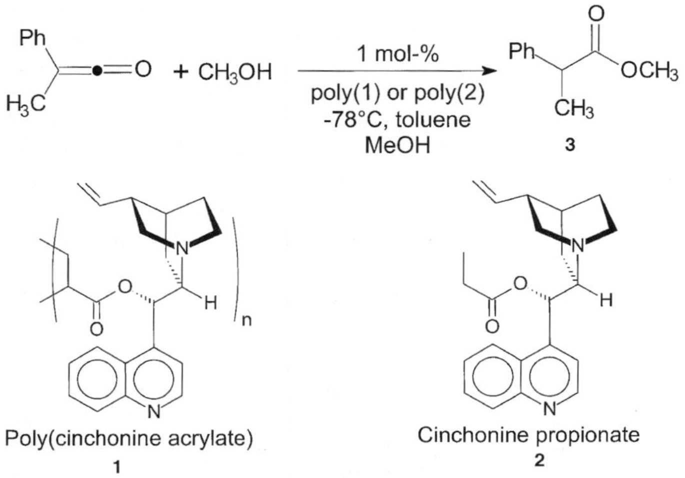

Ph, H_3C, O + CH_3OH → 1 mol-%, poly(1) or poly(2), -78°C, toluene, MeOH → Ph, O, OCH_3, CH_3, **3**

Poly(cinchonine acrylate) **1** — Cinchonine propionate **2**

Figure 1. Nucleophile catalysis by polymeric cinchona-alkaloids.

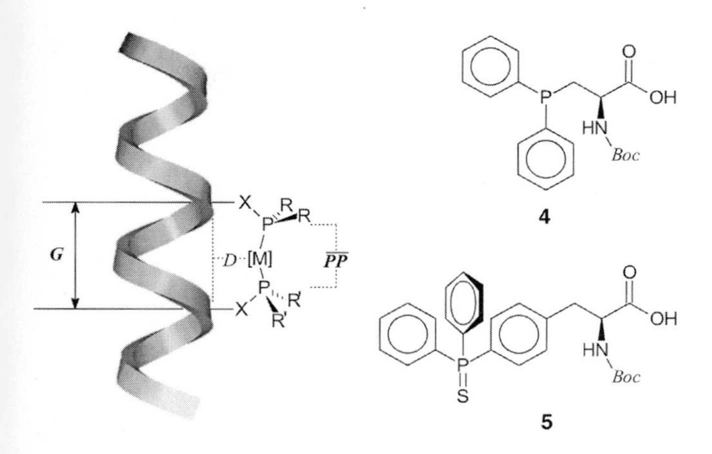

Figure 2. Type C catalysts from stereoregular polymers with spatially variable complexation locations.

in the preparation of polymers or oligomers with a stable chiral conformation (e.g. a helix with uniform sense of chirality), this chiral hyperstructure could play the dominant role in the stereo asymmetric transformation. These considerations lead to a plausible classification of polymeric catalysts, particularly those with chirally ligated metals:

Type A: a chiral ligand with spacially invariant complexation positions bound to a non-stereoregular polymer.

Type B: a chiral ligand with spacially invariant complexation positions bound to a stereoregular polymer or part of the polymer backbone.

Type C: a stereoregular polymer (oligomer) displaying spacially variable complexation positions.

Type A includes the classical approach to chiral polymeric catalysts: a monomeric ligand is synthesized and attached to a "random coil" polymer. If a binding position for the metal is incorporated in every constitutional repeating unit, the catalytically active centres are randomly oriented, which is quite a delicate situation. Their micro-environment is dependent on the position on the polymer, and it can be expected that this is true for their asymmetric induction as well. Such a catalyst can hardly be optimized rationally.

Type B requires the synthesis of a configuratively and/or conformationally homogeneous polymer to which the atoms or groups of atoms constituting the complexation sites are attached. This can be achieved in at least two ways. Firstly it is possible to form the polymer with a uniform secondary structure (for example a helical polypeptide) first and to attach the coordination sites subsequently. On the other hand, a chiral, non- racemic monomer or even a prochiral substrate can be polymerized into a conformationally stable hyper-structure ("helix sense selective polymerisation") [5].

Last but not least, type C is a system whose synthetic potential is almost completely unexplored. The modification of α-amino acid dodecamers with phosphine ligands in variable positions reported by Gilbertson et al. is the only example so far (Fig. 2) [6].

The constitutional range of monomers (amino acids and phosphine-modified species such as the serine derivative **4** and the tyrosine derivative **5** and the option to vary the distance between the complexating atoms (here P) by the spacer (*X*) allow for a versatile and transparent manipulation of the metal's ligand sphere. Moreover, this type of polymeric catalyst is easily transferrable into a combinatorial environment, which can at least partially compensate for the high synthetic effort. Although the results in the asymmetric hydrogenation of dehydro-amino acids obtained by these helical oligomers is not yet convincing (only up to about 12 % *ee*), this concept seems to be promising.

Type A Catalysts

Most polymeric catalysts belong to this category and therefore the selection of examples reviewed had to be very strict [7]. Chiral metal 1,3-di-

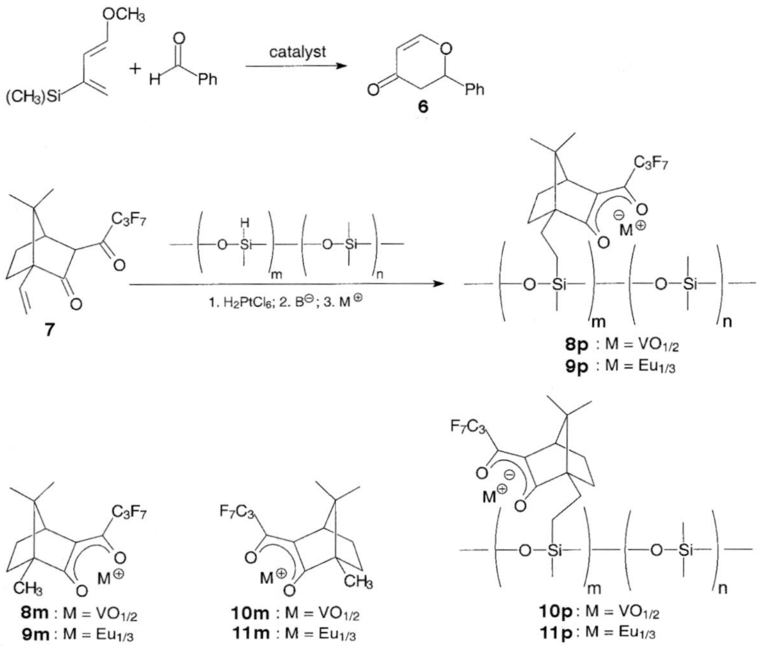

Figure 3. Chirasil metals in the Danishefsky reaction.

ketonates which are immobilized on polysiloxanes (so called chirasil metals) are materials of manifold applicability [8]. Here, we focus on their application in hetero-Diels-Alder reactions (Fig. 3) [9].

In the reaction between an activated diene and a prochiral aldehyde, which is catalyzed by chiral Lewis acids, 5,6-dihydropyranones such as **6** are formed enantiomerically enriched. By attachment of a chiral auxiliary (3-heptafluorobutanoyl camphor derivatives such as **7** to a soluble polysiloxane by hydrosilylation the catalyst should be recyled easily by precipitation or ultrafiltration while the cycloaddition reaction can be performed in homogeneous solution [10].

Starting with both enantiomers of camphor, Schurig and his co-workers synthesized the monomeric complexes **8m**–**11m** as well as the corresponding polymers **8p**–**11p**. In the course of studies on their catalytic activities a surprising discovery was made (Fig. 4).

While the monomeric vanadium complex **8m** furnishes the expected levorotatory enantiomer of **6** in high yield [11] (*ee* = 59 %), an excess of (+)-**6** (Fig. 4 a, *ee* = 40.5 %) is found in the reaction using the polymeric catalyst **8p** with the same sense of chirality in the auxiliary! In contrast to this oxovanadium system, the polymer-induced inversion of the sense of asymmetric induction does not occur with the europium complexes **9m** or **9p** (Fig. 4 b). In both cases, formation of the levotatory enone (−)-(**6**) is favored, although the yield of the polymer mediated catalysis was rather poor (only 6.4 %). In accordance with these observations the symmetry-related results are observed with the compounds **10m** and **10p** (Fig. 4 c) derived from the camphor enantiomer. A satisfying explanation for this rather strange behavior, which is obviously related to the polymer-induced change in the coordination sphere of the metal, is not given. "It is open to speculation why the change of enantioselectivity between monomeric and polymeric catalysts occurs only in bis-chelated oxo-vanadium(IV)- but not in tris-chelated europium(III) complexes" [9]. In a footnote, the possibility of a polymer-induced epimerization at the vanadium atom is discussed. This would entail an

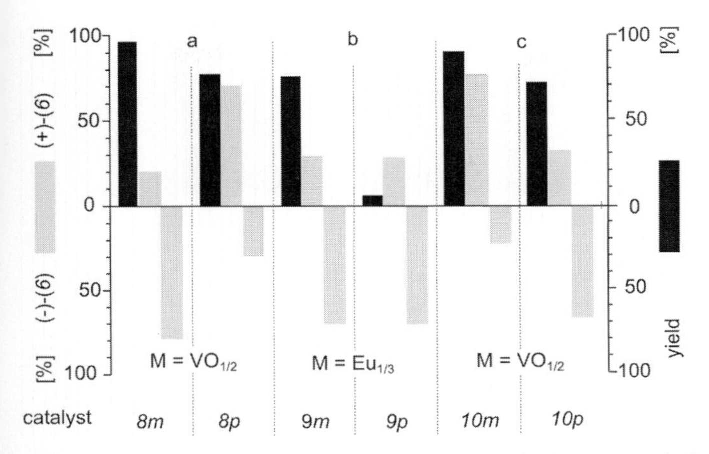

Figure 4. Sense and magnitude of the chiral induction in the asymmetric Danishefsky reaction with monomeric (**m**) and polymeric (**p**) catalysts.

inversion of the absolute topicity of attack onto the aldehyde. In addition to this speciality it is noteworthy that the enantioselectivities of the polymer-catalyzed reactions are generally lower than those found in the reactions with the monomeric catalysts. Both of these findings refer, ultimately, to the position-dependence of the micro-environment which is typical of type A catalysts. Generally speaking, there are two ways to avoid these complications. Either a stereoregular polymer with a uniform micro-environment should be synthesized (changeover from type A to type B) or the catalytically active unit is bound only once (e. g. terminally) to a stereoirregular polymer. The second approach was very successfully put into practice by K. Janda and C. Bolm, who published their results on a polymer-based variant of the Sharpless asymmetric dihydroxylation reaction. Since its discovery in 1988 [12]. there have been many attempts to simplify this reaction by immobilization of the catalyst and to develop ecologically and economically improved procedures. Unfortunately, prolongation of reaction times and decrease of enantioselectivity and yield have been the rule [12 a, 8]. Janda et al. [13] emphasized the incompatibility of a ligand-accelerated catalysis (LAC) [14] and a polymer-bound ligand as being decisive for the lack of

Table 1: Polymer-assisted asymmetric dihydroxylations (n.a. = not available).

	Olefin	Ligand[a)]	*t* [h]	Yield [%]	Reoxidant	*ee* [%]	Ref.
1		12	5	89	NMO[b)]	88	13a
2		13	n. a.	95	$K_3Fe(CN)_6$	99	13b
3		14	≤ 5	91	$K_3Fe(CN)_6$	99	15
4		16	15	93	$K_3Fe(CN)_6$	99	17
5		12	5	80	NMO[a)]	60	13a
6		13	n. a.	88	$K_3Fe(CN)_6$	98	13b
7		14	≤ 5	92	$K_3Fe(CN)_6$	98	15
8		16	20	83	$K_3Fe(CN)_6$	99	17

[a)]see Figure 5. [b)]*N*-methyl morpholin-*N*-oxide.

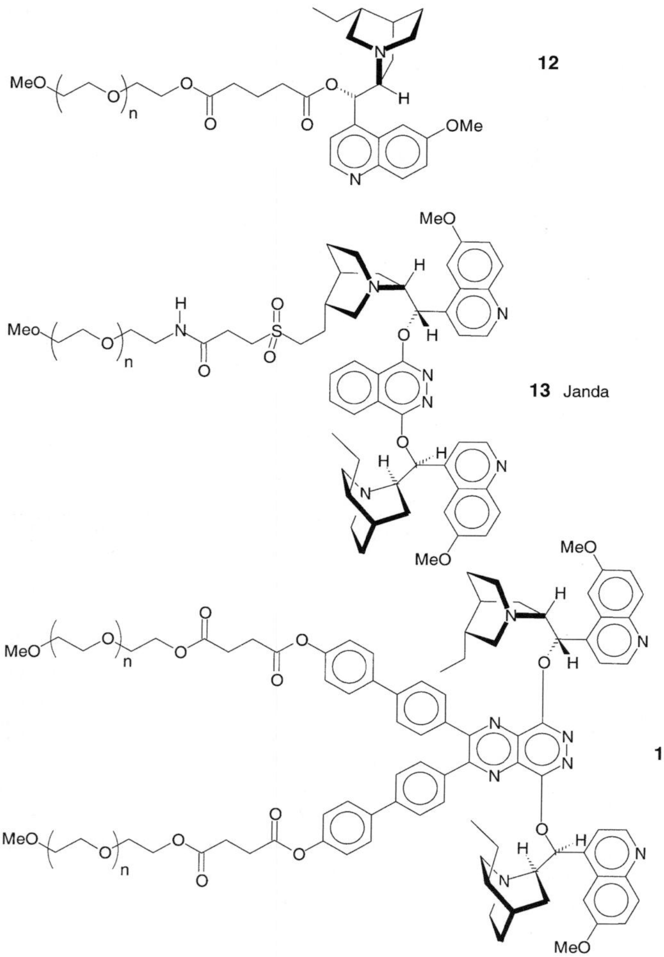

12

13 Janda

14 Bolm

Figure 5. Polymeric dihydroquinine (DHQD) catalysts.

success. In order to facilitate the hampered access of the ligand to all reaction compartments which are within reach of the metal, the reoxidant and the olefin, MeO-PEG (MeO-polyethylene glycol) [10] was chosen as being a soluble polymer (Table 1 and Fig. 5).

Janda's phthalazine derivative (**13**) [13 b] and the pyridazine derived ligand (**14**) developed by Bolm [15] almost reach the yield and enantiomeric excess achieved with the original catalytic system in a homogenous reaction (*tert*-butanol/water or acetone/water) [12 b]. After addition of diethyl ether or *tert*-butyl methyl ether, the polymer-bound ligand is precipitated quantitatively. Thereby, the product as well as the ligand can be obtained in a pure state. Neither the yield nor the *ee* are diminished on re-use of the recovered polymer.

Type B Catalysts

The most self-evident method for formation of a stereoregular polymer is the polymerization of enantiomerically pure monomers with structural properties supporting the assembly of a conformationally uniform and stable secondary structure. Alternatives such as asymmetric polymerization generating stereogenic centers ("asymmetric synthesis polymerization") or *P*- or *M*-he-

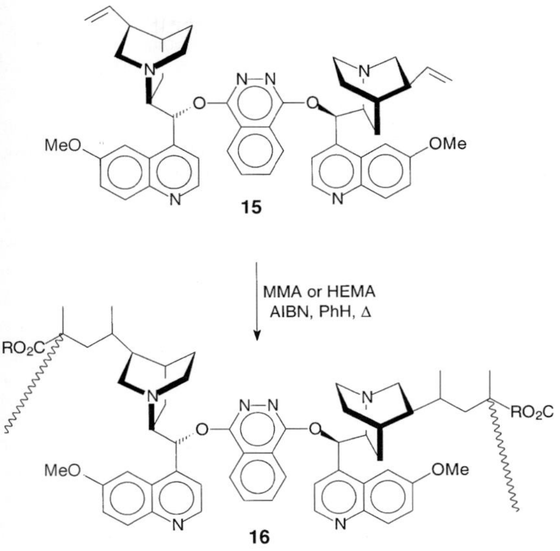

Figure 6. A bis(quininyl)-modified phthalazine as chiral cross-linker.

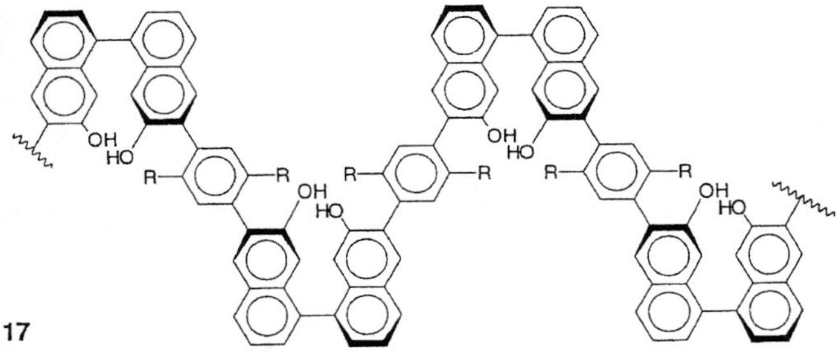

Figure 7. Polybinaphthols as stereoregular polymers with a well-defined micro-environment of the catalytically active centers.

lix-selective polymerization ("helix sense-selective polymerization") or finally the enantiomer-differentiating polymerization have not yet been applied in the synthesis of chiral polymeric catalysts [16]. An interesting hybrid is presented by Song et al. (Fig. 6, Table 1, items 4 and 8) [17]. Sterically non-demanding methacrylates are used as monomers, but under the influence of an enantiomerically pure cross-linking unit (probably) a polymer with a uniform micro-environment around the catalytically active centers is produced.

The strategy is impressively simple: the phthalazine derivative **15** can readily be prepared from quinine in one step. Being a divinyl derivative, it can be submitted as a cross-linking unit in the radical polymerization of methyl methacrylate (MMA) or 2-hydroxy methacrylate (HEMA). Thereby, an immobilized (DHQ-PHAL) derivative **16** is obtained, which is suited for the asymmetric dihydroxylation of *trans*-stilbene (>99 % *ee*) and (*E*)-cinnamic acid methyl ester (>99 % *ee*, Table 1). The insoluble catalyst can be recovered by simple filtration, and its repeated usage is possible. Despite this encouraging result, the constitutional spectrum of olefins suitable for this reaction remains to be seen. A purebred type B catalyst is exemplified by the polybinaphthol derivative **17** (Fig. 7) [18].

These polymers ($M = 6700-24\,300$) synthesized by Pu et al. in a polymerizing Suzuki coupling of substituted *p*-diboronic acids with 3,3'-diiodine-substituted binaphthol derivatives are characterized as stereoregular macromolecules with a well-defined micro-environment around the complexing positions (so called "minor groove binaphthol polymers") [18]. The hexyloxy groups in the *p*-phenyl bridges guarantee solubility in organic solvents and behave as ligands for the metals. In the addition reaction of diethylzinc to various aldehydes catalyzed by this polymer (5 mol%), the corresponding alcohols (**18**) were obtained in excellent yield and high enantiomeric excess (74–93 % *ee*). This applies as well to aliphatic aldehydes, which are known to be problematic since they produce mostly bad results in the catalysis with polymeric aminoalcohols (type A) [19]. The polymer can be separated from the products by precipitation from methanol.

Responsive Polymeric Catalysts

In the introduction of this article it was mentioned that biomacromolecular catalysts not only control the topicity of the reagent/substrate interaction but also interfere regulatively in the mass transport. Substrates are recognized and transported to the location of reaction and transformed in a stereocontrolled fashion, and the products are released. In addition, regulative mechanisms or features of the material control these processes and, for example, are responsible for the pH- and temperature-dependence of an enzyme-catalyzed reaction. Now there are synthetic polymers which respond to external events with a reversible change in physical or even chemical properties. In this context, materials scientists are interested, for example, in the regulation of the specific resistance, the tensile strength or any other mechanical property. Moreover, when this concept is extended to variation of chemical behavior, the so-called "smart ligands" [20] have to be mentioned. These ligands control catalytical activity as a function of temperature [20 a], pH value [20 b] or solvent. A *thermoresponsive*

Ph2P PPh2 PPh2 PPh2 Rh_2^{2+}

n-1 m n-1

19 n = 11, m = 34, LCST = 25°C

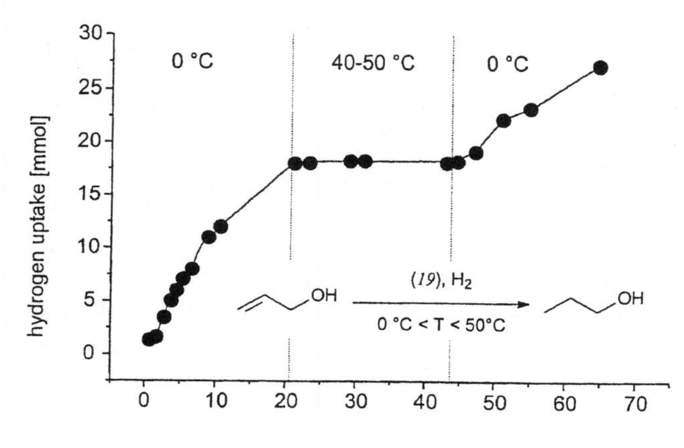

Figure 8. A phosphane modified PEO-PPO-PEO block copolymer as thermoresponsive catalyst in the hydrogenation of allylic alcohols.

catalyst can be put into practice when ligands are connected to a polymer which exhibits an inverse temperature dependence of solubility. These polymers are subject to a temperature-dependent phase transition, i.e. they are soluble below a critical temperature (LCST: "lower critical solution temperature") and insoluble above this temperature (Fig. 8).

One example of such a "smart ligand" is the rhodium complexating bis(2–diphenyl phosphinoethyl)amide derivative (**19**) of a PEO-PPO-PEO block co-polymer [PEO: poly(ethylene oxide); PPO: poly(propylene oxide)]. In water its LCST is 25 °C and it can be used for hydrogenation of allylic alcohols. At 0 °C the rate of hydrogen uptake is about 2 mmol H_2 / mmol Rh · h. If the reaction mixture is heated above the LCST, the hydrogenation reaction slows down and nearly comes to an end at 40 – 50 °C. If the reaction mixture is cooled down again, the polymer becomes hydrated and resolubilized (following the most simple explanation), and the reaction proceeds at maximum velocity again.

In a very recent publication, Bergbreiter, and co-workers studied the extension of this concept using the pH value as a regulatory variable [20 b]. A phosphine-modified polyacid (a copolymer from methyl vinyl ether and maleic acid anhydride, Gantrez®) was taken as a ligand for rhodium. The rate of hydrogenation of various unsaturated substrates was reversibly modified by tuning the pH value.

Conclusions

Polymeric catalysts provide advantages in reaction processing. They can easily be separated from low-molecular-weight products, they offer the possibility of performing reactions continuously, and expensive chiral ligands can be unproblematically recovered and recycled. But furthermore, if they are used in asymmetric catalysis, new qualities which are connected with the macromolecular state and are able to control every aspect of the products structure may arise. In the context of polymeric catalysts, fascinating ideas obtrude themselves. Exothermic reactions can be controlled by their coupling to the "Anti-Arrhenius"-behavior of "smart ligands". Temperature- or pH-dependent changes in selectivity can influence asymmetric syntheses and so on.

The complete potential of polymeric catalysts will probably be revealed where configurative features of the polymer backbone and its specific interactions with the environment have a decisive effect on the conversion of matter. Here, the generation of chiral micro-environments which are uniform throughout the whole polymer and the control of mass transport and reactivity by the properties of the polymeric material will be most important.

References

[1] The "turnover number" describes the amount of product formed in relation to the amount of catalyst used.

[2] An early review: G. Manecke, W. Stark, *Angew. Chem.* **1978**, *90*, 691.

[3] (a) T. Yamashita, H. Yasueda, N. Nakamura, *Chem. Lett.* **1974**, 585. (b) T. Yamashita, H. Yasueda, Y. Miyauchi, N. Nakamura, *Bull. Chem. Soc. Jpn.* **1977**, *50*, 1532.

[4] Similar effects are found in the enantiomer-differentiating polymerization of oligomers ($n \approx 30$) of chiral, sterically crowded methacrylic acid esters: E. Yashima, Y. Okamoto, K. Hatada, *Macromolecules* **1988**, *21*, 854.

[5] (a) Y. Okamoto, T. Nakano, *Chem. Rev.* **1994**, *94*, 349. (b) Y. Okamoto, E. Yashima, *Prog. Polym. Sci.* **1990**, *15*, 263.

[6] (a) S. R. Gilbertson, G. Chen, M. McLaughlin, *J. Am. Chem.Soc.* **1994**, *116*, 4481. (b) S. R. Gilbertson, X. Wang, *J. Org. Chem.* **1996**, *61*, 434. (c) S. R. Gilbertson, G. W. Starkey, *J. Org. Chem.* **1996**, *61*, 2922.

[7] One of the first (the first?) representatives of this class of chiral polymeric catalysts is worth mentioning: W. Dumont, J.-C. Poulin, T.-P. Dang, H. B. Kagan, *J. Am. Chem. Soc.* **1973**, *95*, 8295.

[8] (a) As chiral stationary phases in gas chromatography: M. Schleimer, M. Fluck, V. Schurig, *Anal. Chem.* **1994**, *66*, 2893. (b) As chiral, precipitable shift reagents in NMR spectroscopy: H. Weinmann, diploma thesis, University of Tübingen, **1993**.

[9] F. Keller, H. Weinmann, V. Schurig, *Chem. Ber. Recueil* **1997**, *130*, 879.

[10] M. Reggelin, *Nachr. Chem. Tech. Lab.* **1997**, *45*, 1002.

[11] A. Togni, *Organometallics* **1990**, *9*, 3106.

[12] (a) E. N. Jacobsen, J. Marko, W. S. Mungall, G. Schroder, K. B. Sharpless, *J. Am. Chem. Soc.* **1988**, *110*, 1968. (b) H. Becker, K. B. Sharpless, *Angew. Chem.* **1996**, *108*, 447.
[13] (a) H. Han, K. D. Janda, *J. Am. Chem. Soc.* **1996**, *118*, 7632. (b) H. Han, K. D. Janda, *Tetrahedron Lett.* **1997**, 1527.
[14] D. J. Berrisford, C. Bolm, K. B. Sharpless, *Angew. Chem.* **1995**, *107*, 1159.
[15] C. Bolm, A. Gerlach, *Angew. Chem.* **1997**, *109*, 773.
[16] T. Nakano, Y. Okamoto in *Catalysis in Precision Polymerization* (Ed.: S. Kobayashi), Wiley, **1997**, p. 293.
[17] C. E. Song, J. W. Yang, H. J. Ha, S.-gi Lee, *Tetrahedron: Asymmetry* **1996**, *7*, 645.
[18] (a) W.-S. Huang, Q.-S. Hu, X.-F. Zheng, J. Anderson, L. Pu, *J. Am. Chem. Soc.* **1997**, *119*, 4313. (b) Application as a chiral Lewis acid in the Mukayama aldol reaction: Q.-S. Hu, D. Vitharana, X.-F. Zheng, C. Wu, C. M. S. Kwan, L. Pu, *J. Org. Chem.* **1996**, *61*, 8370. (c) Q.-S. Hu, X.-F. Zheng, L. Pu, *J. Org. Chem.* **1996**, *61*, 5200.
[19] In contrast the polymer-bound Ti-TADDOLates are very effective: D. Seebach, R. E. Marti, T. Hinterinann, *Helv. Chim. Acta.* **1996**, *79*, 1710.
[20] (a) D. E. Bergbreiter, L. Zhang, V. M. Mariagnanam, *J. Am. Chem. Soc.* **1993**, *115*, 9295. (b) D. E. Bergbreiter, Y.-S. Lin, *Tetrahedron Lett.* **1997**, 3703.

Combinatorial Chemistry for the Synthesis of Carbohydrate/Carbohydrate Mimics Libraries

Prabhat Arya, Robert N. Ben and Kristina M. K. Kutterer

Steacie Institute for Molecular Sciences, National Research Council of Canada, Ottawa, Canada

The use of combinatorial libraries in the identification and elucidation of structure-activity relationships has become a powerful tool in the pharmaceutical sector [1]. Traditionally, novel lead compounds have been obtained as natural products from a number of sources including extracts from plants, animals, insects or microorganisms. When an extract shows a desired biological activity, the active compound is identified, isolated and then subjected to further biological testing. Optimization of the chemical structure in order to enhance biological activity is a labor intensive, time consuming process, which dictates that each new structure be independently synthesized. This overall approach has made the development of new therapeutics a very lengthy and expensive process. In contrast, combinatorial chemistry has provided an attractive alternative to these traditional synthetic approaches since it allows for the synthesis of a large number of structurally diverse compounds within a short period of time. The approach utilizes a large array of building blocks which are systematically assembled in such a way that all possible combinations are represented. Typically, a solution or solid phase approach may be used in conjunction with either a "split" or "parallel" synthetic strategy. While the technology required to assemble a small molecule library is not new, combinatorial chemistry was not fully exploited until recently, since efficient methods for screening such libraries were virtually nonexistent. Many of these screening strategies, as well as technical aspects of combinatorial chemistry, have been summarized in several well written review articles [2].

Unlike protein-protein and nucleotide-protein interactions, progress in understanding the role of cell surface carbohydrates in biological and pathological processes has been slow [3]. While comparatively little is known on this subject, it is these weak, non-covalent interactions between cell surface carbohydrate ligands and various protein receptors that form the basis of recognition events which are fundamental to a vastly diverse range of biological and pathological processes. For instance, interactions of this nature have been implicated in cell to cell communication, bacterial and viral infections, chronic inflammation, cancer/metastasis and rheumatoid arthritis [3]. Given their chemical nature, oligosaccharides are very complex and diverse, which makes their synthesis both labor intensive and expensive. As a result, the discovery of new biologically active oligosaccharide ligands is a complex problem. This aside, even when a promising compound has been identified, optimization to enhance activity is a difficult and time consuming process. The synthesis of oligosaccharide libraries using a combinatorial approach offers a feasible solution to these problems.

In contrast to peptide and nucleotide libraries, preparation of an oligosaccharide library is not a facile process. The synthesis of such a library is complicated by the issues of stereochemistry at the anomeric position and the fact that multiple hydroxyl groups are present. Traditionally, these groups would be dealt with using a less than elegant orthogonal protection/deprotection strategy. As an alternative to this, Hindsgaul et al. [4a] have demonstrated that a random glycosylation approach could be used to form small di- and tri-saccharide libraries. The random glycosylation strategy utilizes a glycosyl donor which is protected with only one type of protecting group, and a glycosyl acceptor in which all hydroxyl groups are unprotected (Scheme 1). Using this approach, Hindsgaul coupled a benzylated glycosyl donor to a disaccharide acceptor which possessed six free hydroxyl groups. After 3 h at room temperature, a complex mixture was obtained in which about 30 % of the starting disaccharide acceptors were fucosylated. Separation of the mixture using reverse phase chromatography furnished individual trisaccharides which were analyzed by NMR. Analysis confirmed that all six expected products were present in 8–23 % yield. Ideally, a statistical mixture would contain 17 % of each product.

In an alternate solution phase approach, a latent-active glycosylation method was developed by Boons and co-workers (Scheme 2) [4b]. This strategy uses a glycosyl donor and acceptor, derived from a common building block. Coupling of glycosyl donor and acceptor would produce a disaccharide which could be deacetylated and reacted with other glycosyl donors in a combinatorial manner. Boons demonstrated the feasibility of such a strategy by synthesizing a small trisaccharide library containing anomeric mixtures which was purified by gel filtration column chromatography.

Another important discovery in the area of combinatorial synthesis with di- and tri-saccharides was made by Kahne and co-workers at Princeton [4c]. Kahne's approach utilized solid phase technology to synthesize a saccharide library (Scheme 3), which was especially challenging since, in order for a solid phase approach to be successful, bonds between monomers must be formed in high yields. This is not trivial since most high-yielding coupling reactions in carbohydrate chemistry are not general in nature [5].

Kahne employed a novel coupling procedure which utilized anomeric sulfoxides as glycosyl donors. Such compounds were attractive intermediates, since it has been previously shown that they could be activated at low temperatures, independent of other protecting groups, and gave nearly quantitative yields (~90 %) in solid phase synthesis [5b]. Using this approach, a sizable library which contained approximately 1300 di- and tri-saccharides, possessing a diverse array of linkages, was synthesized in three steps. The approach utilized a split and mix synthesis and first involved the separate coupling of six dif-

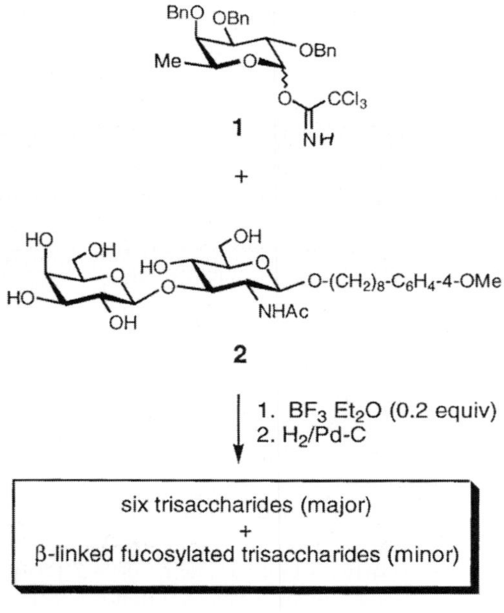

1. BF_3 Et_2O (0.2 equiv)
2. H_2/Pd-C

six trisaccharides (major)
+
β-linked fucosylated trisaccharides (minor)

Scheme 1. Hindsgaul's random glycosylation.

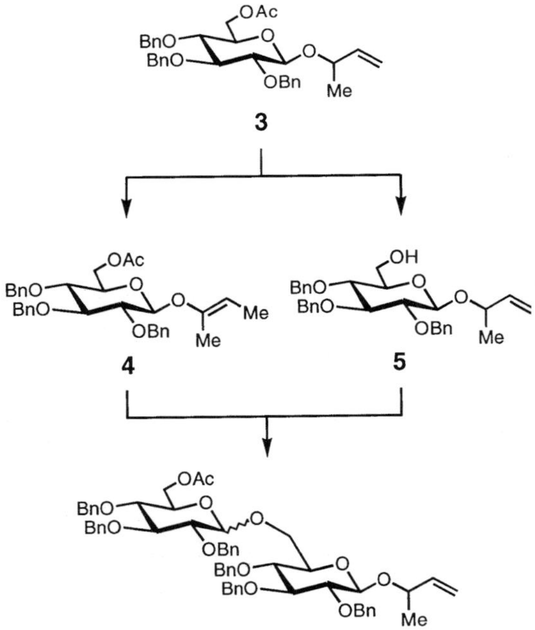

Scheme 2. Boons's latent-active glycosylation approach.

ferent monomers onto TentaGel resin beads. Next, twelve glycosyl sulfoxide donors were coupled separately to mixtures of beads containing the six monomers. The beads were then combined, and a reductive procedure for the conversion of the azido group present on the glycosyl acceptor to an amine was performed. At this point, the beads were divided again into eighteen groups and reacted with various acylating reagents, after which all the protecting groups were removed. In order to screen the library, a colorimetric assay was performed using a lectin. The library contained approximately six copies of the 1300–compound library, which was exposed to biotinylated lectin, then streptavidin-linked alkaline phosphatase, and was finally stained with nitro blue tetrazolium. The beads which stained exhibited the greatest degree of binding and were then removed with the aid of a simple light microscope. Remarkably, only 0.3 % (25 beads) stained and out of these, thirteen were shown to have the same disaccharide core acylated with various hydrophobic groups. The remaining twelve "hits" were discarded since analysis revealed that none of these compounds exhibited any degree of commonality. Surprisingly, the natural ligand for the lectin was not identified as a "hit" even though it was present in the library. Through separate experiments it was proven that all of the thirteen "hits" were, in fact, better ligands than the natural substrate.

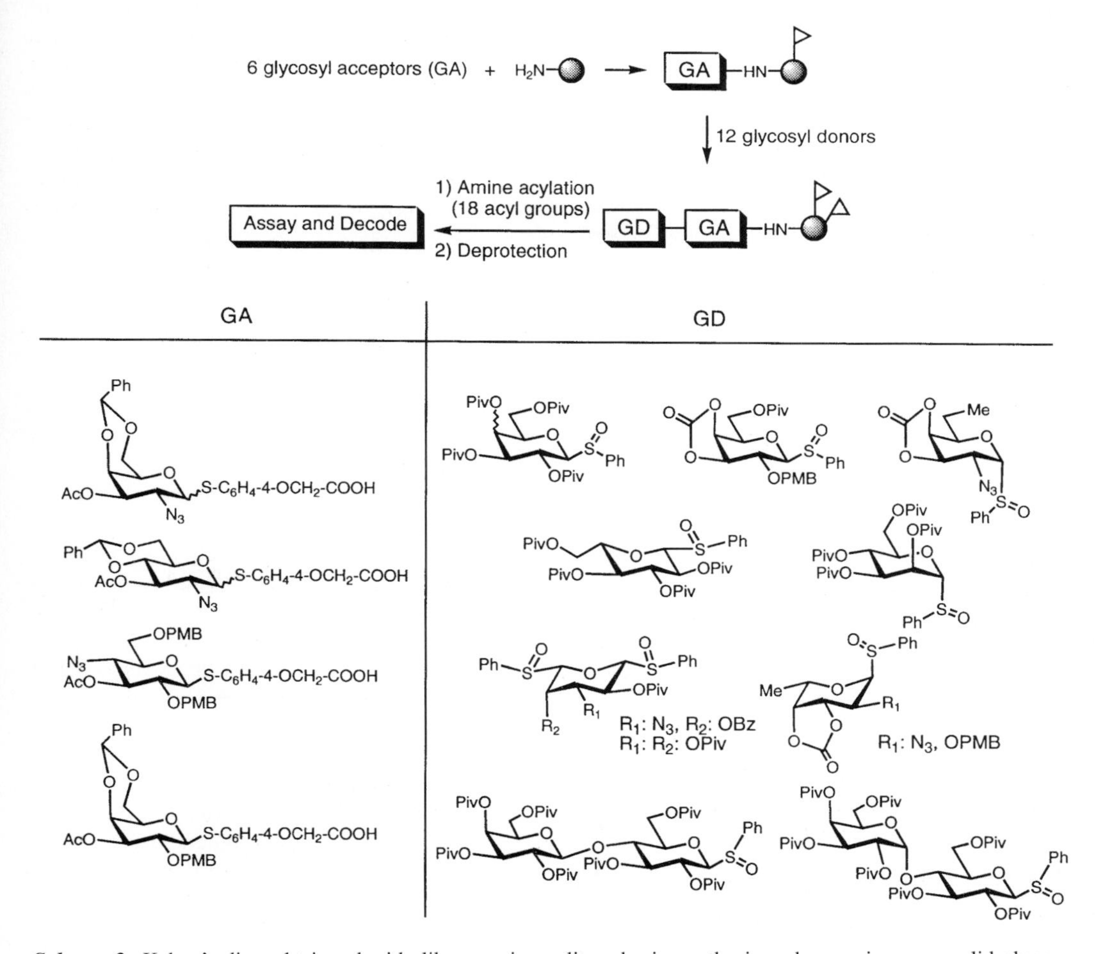

Scheme 3. Kahne's di- and trisaccharide library using split and mix synthesis and screening on a solid phase.

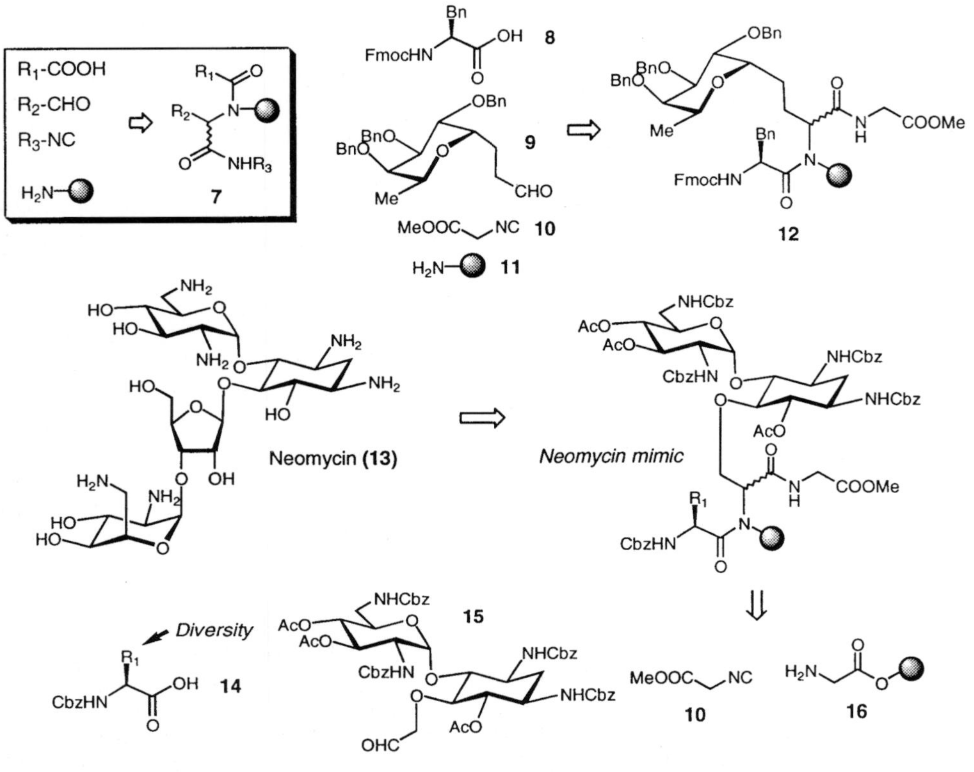

Scheme 4. Glycomimetics by Ugi four-component condensation.

As pointed out by Kahne, it is very interesting that the lectin discriminates so well in its binding of certain di- and trisaccharides. One of the paradoxes of carbohydrate binding is that carbohydrate-binding proteins may bind different substrates in solution but may function with remarkable specificity in cell-to-cell recognition. This work also emphasized that the presentation of the saccharide on the surface of the bead is also critical to lectin binding since solution affinity experiments demonstrated that both the natural ligand and all thirteen compounds identified as "hits" bound the lectin in solution phase. This is indeed surprising since presentation effects complicate most on-bead screening techniques.

The techniques of Hindsgaul's random glycosylation, Boons's latent-active glycosylation and Kahne's solid phase methodology now make the synthesis of di- and trisaccharide combinatorial libraries a feasible process. Also, a colorimetric assay can make the screening of such libraries a facile, efficient process since the same approach could be applied to other lectins. There can be little doubt that these recent advances in di- and trisaccharide combinatorial chemistry will have a great impact in the understanding of cell surface interactions and aid in the design of polyvalent compounds which inhibit these interactions.

While these advances make the synthesis of di- and trisaccharide combinatorial libraries a relatively facile process, the synthesis of oligosaccharide libraries (such as tetrasaccharides and higher derivatives) is still quite challenging. This is largely due to the fact that the present glycosylation methods are not quantitative.

Because of the complex nature of carbohydrates, interest in combinatorial approaches to obtain carbohydrate mimics has grown in recent years. By applying combinatorial strategies, the

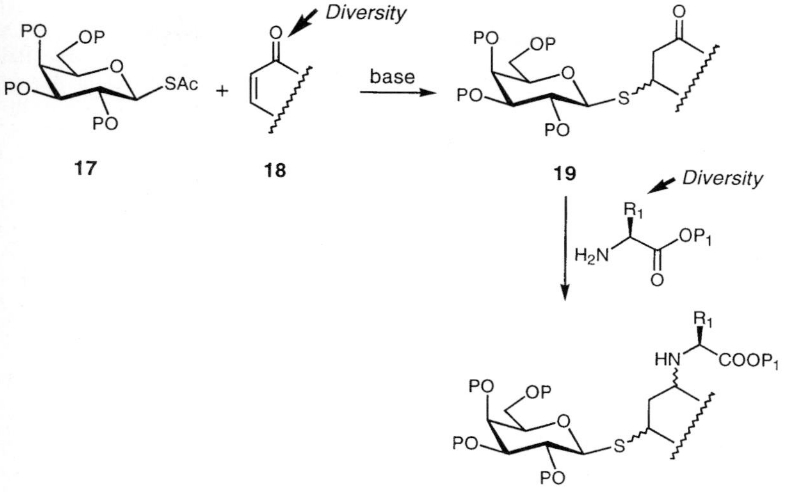

Scheme 5. Glycohybrids by Hindsgaul et al.

ultimate challenge is to obtain compounds that are simple in nature and possess similar or even better binding potential with the target receptors. A variety of approaches to carbohydrate diversity have been reported recently, mostly incorporating commercially available sugars as known essential recognition elements. Armstrong et al. have utilized the Ugi four-component condensation reaction in a combinatorial manner to obtain diverse libraries of carbon-linked glycoside peptides [7a,b]. The Ugi condensation reaction is a powerful strategy in which an aldehyde, an amine, an isocyanide and a carboxylic acid are reacted in one pot (Scheme 4). A library of neomycin, an aminoglycoside antibiotic, mimetics was obtained by the Ugi condensation on a soluble PEG polymer [7c].

In another approach, Hindsgaul et al. reported a combinatorial strategy to obtain "glycohybrids" (Scheme 5) [8]. Glycohybrids are derived from monosaccharides via a Michael reaction, followed by the derivatization of the carbonyl group with several amino acids. This chemistry was further extended to the solution phase parallel synthesis to obtain a library of several compounds.

We have developed an automated, multi-step, solid phase strategy for the synthesis of libraries

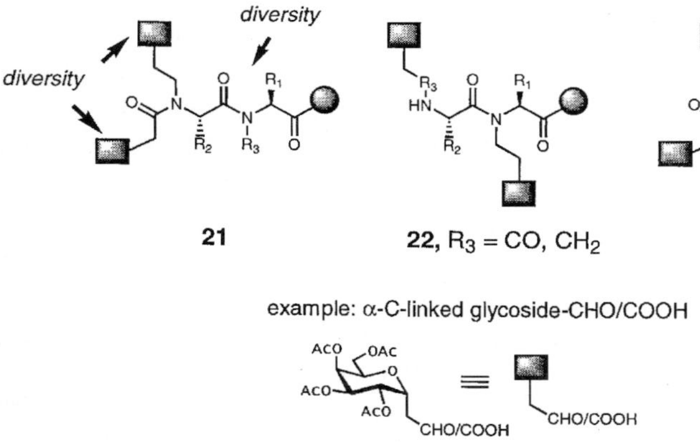

Scheme 6. Programmed approach to neoglycopeptides.

of neoglycopeptides (Scheme 6) [9]. In our approach, different α- or β-carbon-linked carbohydrate based aldehyde and carboxylic acid derivatives can be incorporated either at the N-terminal moiety or at the internal amide N of short peptides/pseudopeptides in a highly flexible and control-oriented manner. Using neoglycopeptide derivatives, the contribution of the secondary groups (i.e., peptide/pseudopeptide backbone) to overall binding, through additional sub-site oriented interactions with protein receptors or by mimicking portions of the complex carbohydrate, could also be explored. Compounds such as **21** (Scheme 6) are obtained by reacting a library of short resin-bound peptides (e.g. dipeptide) with a glycoside aldehyde via reductive amination, followed by coupling with a glycoside carboxylic acid. In another approach, reductive amination using a glycoside aldehyde is followed by peptide coupling leading to an internal N-glycosyl derivative. After deprotection, the terminal amine moiety is then glycosylated as previously described (see compounds **22** and **23**). The strength of this method for the generation of diverse neoglycopeptides is apparent in that it is possible to obtain 400 derivatives of compound **21** from 2 glycoside aldehydes, 2 glycoside carboxylic acids and 10 amino acids.

To summarize, several groups are currently involved in developing combinatorial approaches for carbohydrates and carbohydrate mimics. This is a relatively new area of research and its impact towards understanding the biological function of carbohydrates will be seen in coming years.

References

[1] a) *Molecular Diversity and Combinatorial Chemistry-Libraries and Drug Discovery,* (Eds.: K. D. Janda, I. M. Chaiken), ACS Series, Washington, **1996**. b) *Combinatorial Peptide and Nonpeptide Libraries-A Handbook*, (Eds. G. Jung), VCH, New york, **1996**. c) *Combinatorial Chemistry-Synthesis and Application*, (Eds.: S. R. Wilson, A. W. Czarnik), John Wiley & Sons, Inc. New York, **1997**.

[2] Recent review articles on combinatorial chemistry: a) M. A. Gallop, R. W. Barret, W. J. Dower, S. P. A. Fodor, E. M. Gordon, *J. Med. Chem.* **1994**, *37*, 1233–1251; b) E. M. Gordon, R. W. Barret, W. J. Dower, S. P. A. Fodor, M. A. Gallop, *ibid.* **1994**, *37*, 1385–1401; c) L. A. Thompson, J. A. Ellman, *Chem. Rev.* 1996, *96*, 555–600; d) F. Balkenhohl, C. von dem Bussche-Hünnefeld, A. Lansky, C. Zechel, *Angew. Chem.* **1996**, *108*, 2436–2487; *Angew. Chem. Int. Ed. Engl.* **1996**, *35*, 2288–2337; e) M. J. Sofia, *Drugs Discovery Today*, **1996**, *1*, 27–34.

[3] a) N. Sharon, H. Lis, *Sci. Am.* **1993,** *268*(1), 82–89; b) P. Sears, C.-H. Wong, *Proc. Natl. Acad. Sci. USA* **1996**, *93*, 12086–12093. c) R. A. Dwek, *Chem. Rev.* **1996**, *96*, 683. c) M. Mammen, S.-K. Choi, G. M. Whitesides, *Angew. Chem. Int. Ed. Engl.* **1998**, *37*, 2754–2794. d) J. C. McAuliffe, O. Hindsgaul, *Chem. Ind.*, **1998**, 170–174.

[4] a) O. Kanie, F. Barresi, Y. Ding, J. Labbe, A. Otter, L. S. Forsberg, B. Ernst, O. Hindsgaul, *Angew. Chem.* **1995**, *107*, 2912–2915; *Angew. Chem. Int. Ed. Engl.* **1995**, *34*, 2720–2722; Y. Ding, J. Labbe, O. Kanie, O. Hindsgaul, Bioorg. Med. Chem., **1996**, *4*, 683–692. b) G.-J. Boons, B. Heskamp, F. Hout, *Angew. Chem.* **1996**, *106*, 3053–3056; *Angew. Chem. Int. Ed. Engl.* **1996**, *35*, 2845–2847; c) R. Liang, L. Yan, J. Loebach, M. Ge, Y. Uozumi, K. Sekanina, N. Horan, J. Gildersleeve, C. Thompson, A. Smith, K. Biswas, W. C. Still, D. Kahne, *Science* **1996**, *274*, 1520–1522; R. Liang, J. Loebach, N. Horan, M. Ge, C. Thompson, L. Yan, D. Kahne, *Proc. Natl. Acad. Sci. USA* **1997**, *94*, 10554–10559.

[5] a) M. Schuster, P. Wang, J. C. Paulson, C.-H. Wong, *J. Am. Chem. Soc.*, **1994**, *116*, 1135–1136; b) L. Yan, C. M. Taylor, R. Goodnow, Jr., D. Kahne, *J. Am. Chem. Soc.*, **1994**, *116*, 6953–6954; c) S. P. Douglas, D. M. Whitfield, J. J. Krepinsky, *J. Am. Chem. Soc.*, **1995**, *117*, 2116–2117; d) J. Y. Roberge, X. Beebe, S. J. Danishefsky, *Science* **1995**, *269*, 202–204; e) J. A. Hunt, W. R. Roush, *J. Am. Chem. Soc.*, **1996**, *118*, 9998–9999; e) K. C. Nicolaou, N. Winssinger, J. Pastor, F. DeRoose, *J. Am. Chem. Soc.*, **1997**, *119*, 449–450.

[6] a) E. E. Simanek, G. J. McGarvey, J. A. Jablonowski, C.-H. Wong, *Chem. Rev.* **1998**, *98*, 833–862. b) Z.-G. Wang, O. Hindsgaul, *Glycoimmunology 2* **1998**, 219–236. c) M. J. Sofia, *Mol. Diversity* **1998**, *3*, 75–94.

[7] a) R. W. Armstrong, A. P. Combs, P. A. Tempest, S. D. Brown, K. A. Keating, *Acc. Chem. Soc.* **1996**, *29*, 123–131. b) D. P. Sutherlin, T. M. Stark, R. Hughes, R. W. Armstrong, *J. Org. Chem.* **1996**, *61*, 8350–8354. (c) W. K. C. Park, M. Auer, H. Jaksche, C.-H. Wong, *J. Am. Chem. Soc.* **1996**, *118*, 10150–10155.

[8] U. J. Nilsson, E. J.-L. Fournier, O. Hindsgaul, *Bioorg Med. Chem.* **1998**, 1563–1575.

[9] P. Arya, K. M. K. Kutterer, *J. Comb. Chem.* submitted for publication, **1999**.

Combinatorial Biosynthesis of Polyketides

Kai Donsbach and Karola Rück-Braun

Institut für Organische Chemie, Universität Mainz, Germany

Macrolides and polyethers such as erythromycin A (**4**), FK 506, rapamycin or avermectin A^{1a} (**5**, Scheme 1) are products of modular type I polyketide-synthases. These compounds are distinguished by extraordinary structural diversity and complexity [1,2]. Because of their biological potency, members of this structural class as well as the aromatic polycyclic products of type II polyketide-synthases, tetracyclines and antharacyclines, e.g. adriamycin (**6**), became useful as pharmaceuticals (antibiotics, cytostatics, immunosuppressives) [1,2].

A multifunctional biosynthetic machinery mediates the synthesis of these complex natural products from acetyl- and propionyl-coenzyme A [3]. In the case of type I polyketide-synthases, the β-oxo-esters made by polycondensation steps are modified for example by reduction or dehydration after the chain elongation. Additional specific enzymatic transformations, e.g. oxidations and glycosylations, usually take place after the decoupling at the completed macrocyclic ring framework [1,3].

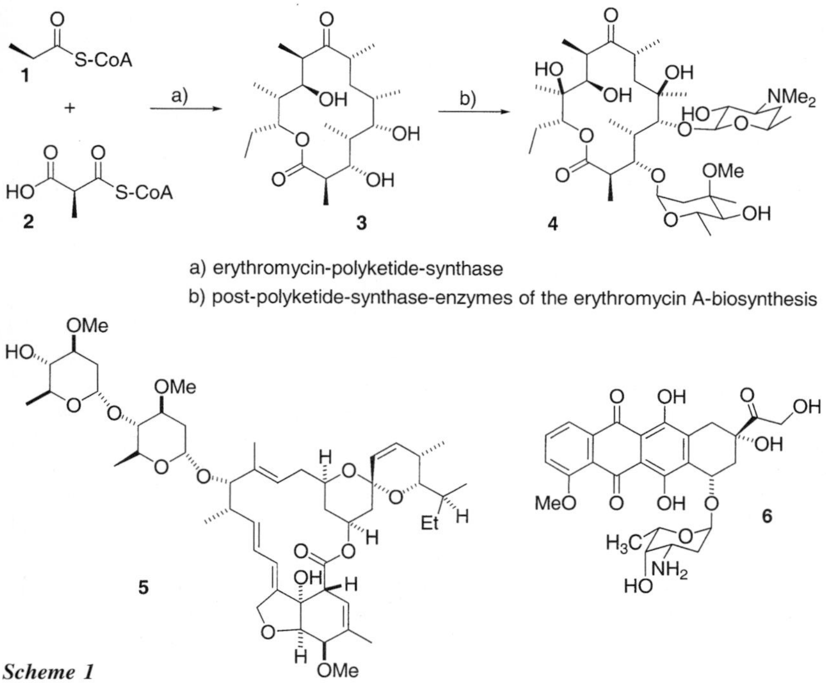

a) erythromycin-polyketide-synthase

b) post-polyketide-synthase-enzymes of the erythromycin A-biosynthesis

Scheme 1

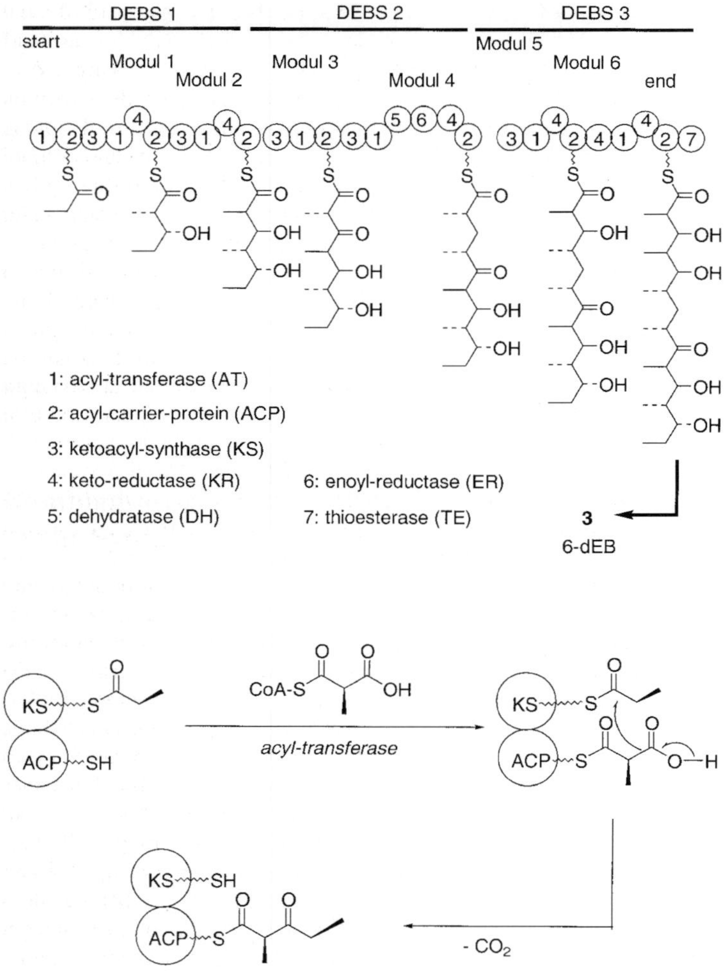

1: acyl-transferase (AT)
2: acyl-carrier-protein (ACP)
3: ketoacyl-synthase (KS)
4: keto-reductase (KR)
5: dehydratase (DH)
6: enoyl-reductase (ER)
7: thioesterase (TE)

Scheme 2

Polyketides are only available in traces from micro-organisms, fungi and plants. Therefore, these secondary metabolites are a challenge for analysts and geneticists as well as synthetic chemists. In recent years, a milestone in polyketide research was the discovery of the total sequence of the erythromycin polyketide-synthase (PKS) gene by L. Katz and co-workers [4c]. The knowledge of the genetic code of the polyketide-synthases opened up a new approach to specifically altering nature's strategy of biosynthesis, and thus provides a predictable access to structurally new polyketides by fermentation. As a consequence of the initial work of Katz and co-workers [4] concerning gene-technological modifications of the erythromycin PKS genes in the year 1991 and the pioneering work on the function of polyketide-synthases in the last few years, combinatorial biosynthesis of modified and unnatural polyketides has come close to realization. For example, in the case of erythromycin polyketide-synthase it became possible to integrate polyketide biosynthesis genes in foreign organisms [5]. Today, artificial polyketides, which carry structural elements of different natural polyketides are already available by fermentation. Moreover, combinations of chemical and biological synthetic methods give way to new synthetic pathways to erythromycins.

Biosynthesis of 6-Desoxy-erythronolide B

The molecular backbone of the antibiotic erythromycin A [6-desoxy-erythronolide B (**3**)] is built up repetitively from one propionyl-coenzyme A (**1**) and six methyl-malonyl-coenzyme A (**2**) constituents by the action of polyketide-synthase, which itself consists of three proteins (DEBS 1–3) (Schemes 1 and 2). Each protein contains two modules with several separate, catalytically active domains. In the first section, DEBS 1 carries an additional loading zone, and DEBS 3 contains a thioesterase in the final segment, catalyzing the decoupling of the product by building the lactone ring [6].

The β-ketoacyl-synthases/acyltransferases (KS/AT) in each module effect the chain elongation by methyl-malonyl-coenzyme A units catalyzing a Claisen ester condensation followed by decarboxylation (Scheme 2). Subsequent domains are module-specific ketoreductases (KR), dehydratases (DH) or enoyl-reductases (ER), which regulate the functionalization of the newly prepared β-oxoesters. The stepwise growing chain is picked up by an acyl-carrier protein (ACP).

The modular organization of this and other polyketide-synthases and the composition in defined domains of enzymatic activity allows the inactivation, the deletion (loss of function), the enlargement (gain of function) and the exchange of separate enzymatic units through genetic manipulation of polyketide-synthase genes.

Such modifications of single modules open up the possibility of generating structural diversity using designed unnatural organisms to synthesize the desired polyketides. The synthesis of new structures, however, depends decisively on the substrate tolerance of the enzymes that follow upon the modified segments in the polyketide-synthases.

Defined Di- and Triketides: Potential and Possibilities of Genetic Manipulations

To examine the function and substrate specificity of the polyketide-synthases, Khosla et al. developed a genetically manipulated organism, *Streptomyces coelicolor* CH999, lacking the natural actinorhodin polyketide gene cluster. Instead, this cell line was variably equipped with the desired polyketide biosynthetic genes by plasmid-technology [7]. Thus, it was possible to create an organism (CH999/pCK12) based on *streptomyces coelicolor* CH999 carrying a bimodular system consisting of DEBS 1 and the erythromycin thioesterase from DEBS 3 for the hydrolytic fission of the acyl-enzyme intermediate. This mutant catalyzes the conversion of propionyl-coenzyme A with two molecules of methyl-malonyl-coenzyme A to the triketide lactone **7** (Scheme 3). The synthetic pathway contains one reduction step per module.

Especially the NADPH-dependent β-ketoacyl-reductase domains KR1 and KR2, which synthesize 5-(*R*)- and 3-(*S*)-configured β-hydroxy-acylthioester building blocks, respectively, were intensively used to study the impact of genetic manipulations [8–11].

For instance, the β-ketoreductase in module 1 of DEBS 1 of erythromycin polyketide-synthase was successfully replaced by the β-ketoreductase/dehydratase domains of rapamycin polyketide-synthase [8,9]. Thereby, a mutant with a shortened erythromycin polyketide-synthase consisting of the first three modules and the thioesterase from DEBS 3 was generated from the cell line *Streptomyces coelicolor* CH999. This organism (CH999/pCK13) produces the tetraketide-lactone **8** and the decarboxylated tetrahemiketal **9** [20 mg/L and 5 mg/L] by fermentation (Scheme 3). However, the cell lines derived from this mutant and the rapamycin polyketide-synthase (RAPS) domains KR4/DH4 from the RAPS module 4 furnish the polyketides **10** and **11** with a *trans*-configured double bond from propionyl- and acetyl-coenzyme A starting units after the final decarboxylation [ca. 5–10 mg/L]. After implementation of the DH1/ER1/KR1 domains from RAPS module 1, the artificial mutant (CH999/pKA0410) produces the tetraketide lactone **12** as a new metabolite [20 mg/L] (Scheme 3) [9]. In addition, mutants were generated from cell line CH999/pCK13, which carried a ketoreductase from module 2 or module 4 of the rapamycin polyketide-synthase instead of the natural KR2 domain. These domains effect the formation of

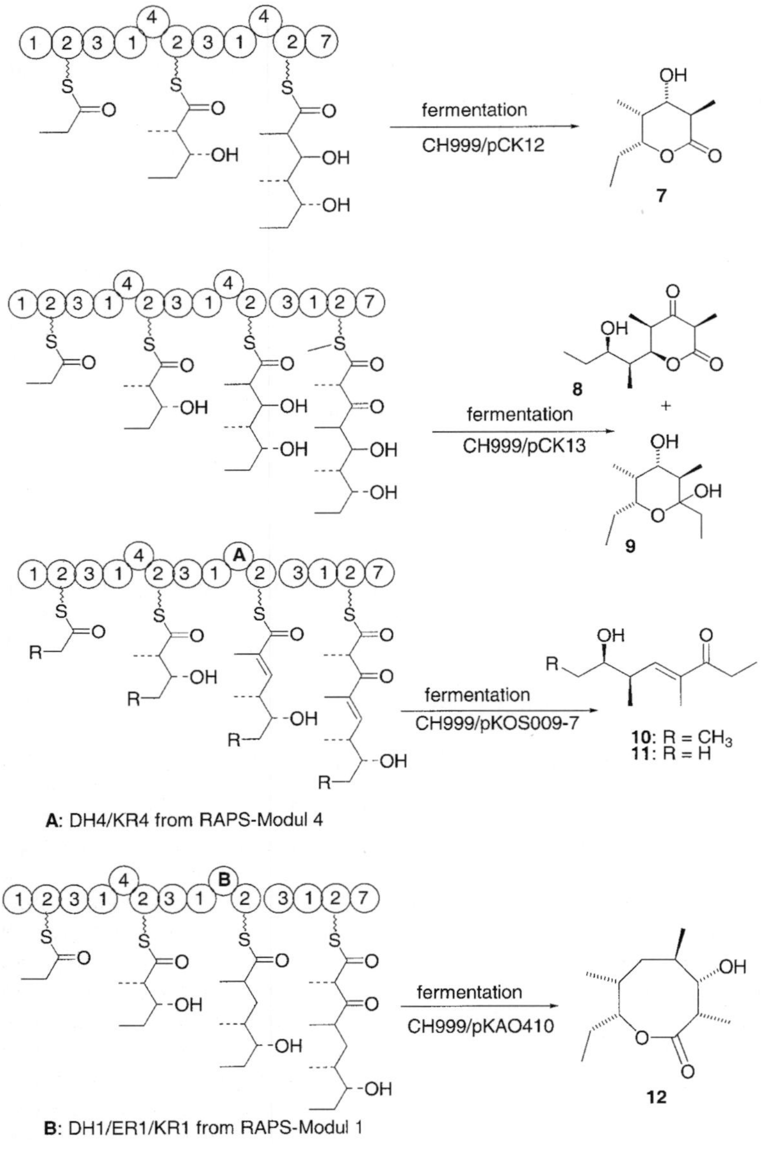

A: DH4/KR4 from RAPS-Modul 4

B: DH1/ER1/KR1 from RAPS-Modul 1

Scheme 3

(*R*)-configured building blocks in their natural environment.

The product of the fermentation process of both cell lines is the triketide lactone **13** in Scheme 4 with 3-(*R*)-configuration [ca. 20 mg/L]. The structure of **13** was proved by the synthesis of an authentic sample by applying chemical methods. The synthesis of such triketides is well known by the methods of Evans et al. (Scheme 4) [12].

Probably the formation of **13** is effected by a defined specificity of the KS3 domains from DEBS 2 (module 3) for (*S*)-configured intermediates. For the first time, this example and comparable studies proved that the individual properties of the ketoreductase domains deter-

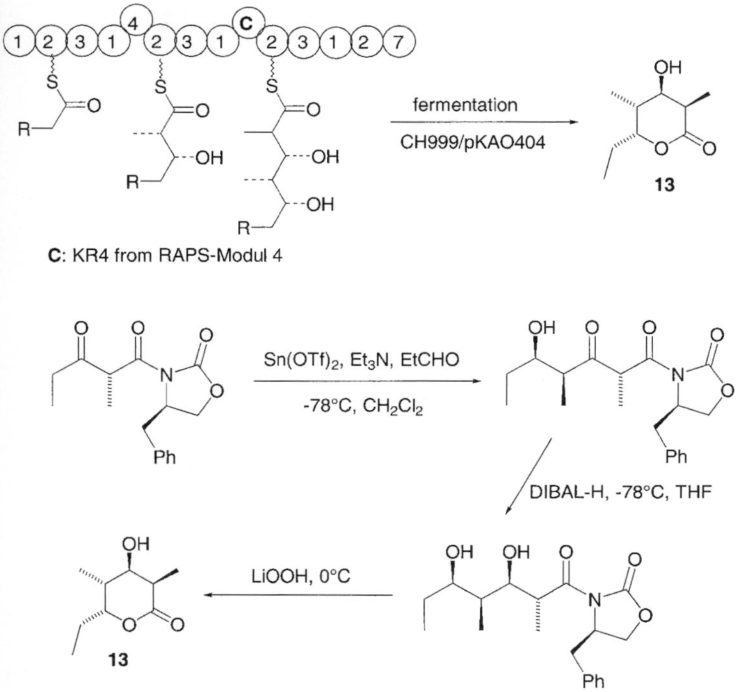

Scheme 4

mine the stereochemistry at the newly formed β-hydroxy carbon. In addition, the stereochemistry and the substitution pattern of the β-(ketoacyl)-ACP substrates have no influence on the reduction steps [10, 11].

In the course of mechanistic investigations covering these enzymatic reductions, labeling experiments were carried out with biologically produced, selectively deuterated NADPH-molecules {4-(*R*)-[4-^{2}H]NADPH and 4-(*S*)-[4-^{2}H] NADPH} [11]. The formation of hydroxy derivatives of opposite stereochemistry is caused by the ketoreductase domains KR1 and KR2 from the protein DEBS 1 of the erythromycin polyketide-synthase. However, both domains have a preference for the 4-pro-(*S*)-hydride of the NADPH molecule. Probably the binding of the cofactor in KR domains takes place in an identical manner, whereas the individual β-ketoacylthioester building blocks in the domains KR 1 and KR 2 of DEBS 1 capture a different orientation relative to the cofactor [11].

Erythromycin A-Analogs with Avermectin Starting Units

Recently, Leadlay and co-workers reported the synthesis of biologically active erythromycin A analogs [13]. They were able to exchange the loading zone of DEBS 1 in the erythromycin polyketide-synthase, which exclusively accepts in vitro and in vivo propionyl- and acetyl-coenzyme A as starting units, with the appropriate subunit of the avermectin polyketide-synthase from *Streptomyces avermitias* [13]. This domain is distinguished by an extremely broad substrate specificity and, except for isobutyryl- and 2-methylbutyryl-coenzyme A, tolerates more than 40 branched carboxylic acid derivatives. The gene of the enzyme hybrid was expressed in *Saccharopolyspora erythrea*. The artificial cell line produces new, antibiotic erythromycin A analogs bearing an isopropyl or a *sec*-butyl group at carbon C13 of erythromycin instead of the native ethyl group (Scheme 1).

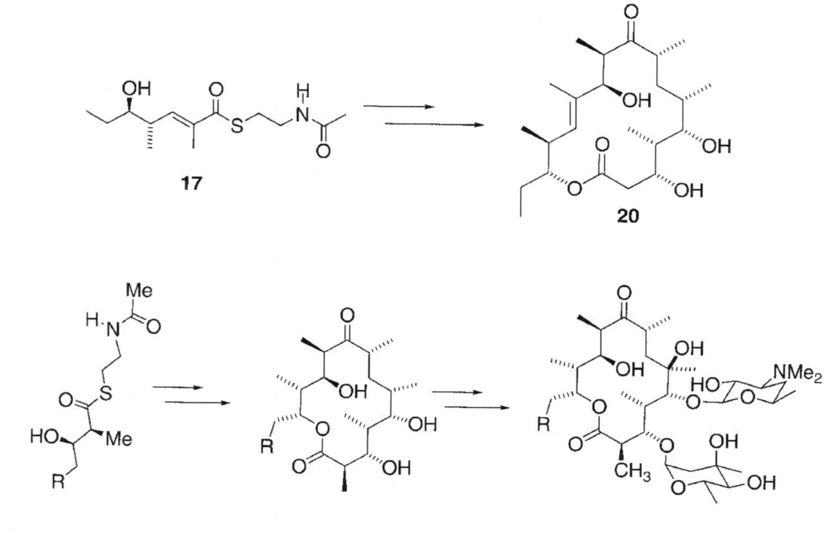

14: R = Me
15: R= *n*-Propyl
16: R= Phenyl

4: R = Me
18: R = *n*-Propyl
19: R = Phenyl

21: R = *n*-Propyl
22: R = Phenyl

Scheme 5

Erythromycin D Analogs from Synthetic Diketides by Fermentation

In 1997, Khosla and co-workers succeeded in realizing a new chemo-enzymatic access to artificial erythromycin D analogs on a multimilligram scale [14]. Therefore, a mutant (CH999/pJRJ2) of *Streptomyces coelicolor* CH999 was equipped with erythromycin polyketide-synthase genes in which the function of the ketosynthase KS1 in the gene segment of DEBS 1 was inactivated in order to inhibit the usual biosynthetic pathway of 6-desoxy-erythronolide B (6-dEB). The resulting cell line produced 6–desoxy-erythronolide B (6-dEB, **4**) by fermentation conditions only upon addition of *β*-hydroxy-*α*-methyl-valerianic acid (**14**), a cell-permeable *N*-acetyl cysteamine thioester, mimicking the genuine thioester derivative. Starting from 100 mg substrate, they were able to isolate 30 mg of product. In a similar fashion, the synthetic diketides **15**, **16** and **17** (Scheme 5) of different substitution pattern were accepted by the domains of the erythromycin polyketide-synthase, and new, artificial 6-erythronolide B analogs were obtained [**18**: 55 mg/L, **19**: 22 mg/L]. Substrate **16** bears an aromatic residue instead of the terminal methyl group of **14**. Even compound **17,** showing the structure of a biosynthetic tylosin intermediate, is tolerated and cyclized to the 16-membered macrolactone **20** [25 mg/L] (Scheme 5). Moreover, Khosla and co-workers were able to produce a mutant of *Saccharopolyspora erythrea* (A 34), which itself cannot synthesize 6–dEB, but carries the post-polyketide-synthase enzymes of the erythromycin pathways. This organism produced new antibacterial erythromycin D analogs after addition of **18** or **19** to the growing culture by hydroxylation of carbon atom C6 of the 6-desoxy-erythronolide B building blocks and two subsequent glycosylations. Apparently, the post-polyketide-synthase enzymes of the erythromycin pathway also display a considerable substrate tolerance.

In analogy to the synthetic route shown for the triketide **13** (Scheme 4), diketide substrates of variable substitution pattern are available by chemical methods from enolates by aldol reactions and alkylations [12].

Recently, the first solid-phase synthesis of such building blocks was reported [12c]. Thus, in connection with the combinatorial biosynthetic route according to Khosla, a broad access to a large variety of erythromycin analogs is opened. It remains to be seen whether genetically modified organisms will be able to produce polyketides within the bounds of their primary metabolism in the near future.

By the methods presented for the combinatorial biosynthesis of polyketides, a multitude of modified and artificial polyketide substances should be available. New antibiotics, potentially with fewer side effects and consequently broader applicability, are desperately needed in the light of increasing resistance of bacteria towards established medications.

References

[1] Review: J. Staunton, B. Wilkinson in *Topics in Current Chemistry 195* (Ed. F.J. Leeper, J.C. Vederas) Springer Verlag, Berlin **1998**, 49 and references cited therein.
[2] Review: J. Rohr, *Angew. Chem.* **1995**, *107*, 963.
[3] Review: C.W. Carreras, R. Pieper, C. Khosla in *Topics in Current Chemistry 188* (Ed. J. Rohr) Springer Verlag, Berlin **1997**, 85 and references cited therein.
[4] Review: L. Katz, *Chem. Rev.* **1997**, *97*, 2557.
[5] J. Cortes, K.E.H. Wiesmann, G.A. Roberts, M.J.B. Brown, J. Staunton, P.F. Leadlay, *Science* **1995**, *268*, 1487.
[6] K. J. Weissman, C.J. Smith, U. Hanefeld, R. Aggarwal, M. Bycroft, J. Staunton, P.F. Leadlay, *Angew. Chem.* **1998**, *110*, 1503.
[7] C.W. Carreras, R. Pieper, C. Khosla, *J. Am. Chem. Soc.* **1996**, *118*, 5158.
[8] R. McDaniel, C.M. Kao, H. Fu, P. Hevezi, C. Gustafsson, M. Betlach, G. Ashley, D.E. Cane, C. Khosla, *J. Am. Chem. Soc.* **1997**, *119*, 4309 and references cited therein.
[9] C.M. Kao, M. McPherson, R.N. McDaniel, H. Fu, D.E. Cane, C. Khosla, *J. Am. Chem. Soc.* **1997**, *119*, 11339.
[10] C.M. Kao, M. McPherson, R.N. McDaniel, H. Fu, D.E. Cane, C. Khosla, *J. Am. Chem. Soc.* **1998**, *120*, 2478 and references cited therein.
[11] M. McPherson, C. Khosla, D.E. Cane, *J. Am. Chem. Soc.* **1998**, *120*, 3267.
[12] a) D.A. Evans, J.S. Clark, R. Metternich, V.J. Novack, G.S. Sheppard, *J. Am. Chem. Soc.* **1990**, *112*, 866; b) D.A. Evans, H.P. Ng, J.S. Clark, D.L. Rieger, *Tetrahedron* **1992**, *48*, 2127; c) M. Reggelin, V. Brenig, *Tetrahedron Lett.* **1996**, *37*, 6851.
[13] A.F.A. Marsden, B. Wilkinson, J. Cortés, N.J. Dunster, J. Staunton, P.F. Leadlay, *Science*, **1998**, *279*, 199.
[14] J.R. Jacobsen, R. Hutchinson, D.E. Cane, C. Khosla, *Science* **1997**, *277*, 367.

Index

B

C

D

E

F

G

H

I

J

K

L

M

N

O

P

Q

R

S

U

V

W

X

Y

Z